Lecture Notes in Physics

The Editorial Policy for Proceedings

The series Lecture Notes in Physics reports new developments in physical research and teaching – quickly, informally, and at a high level. The proceedings to be considered for publication in this series should be limited to only a few areas of research, and these should be closely related to each other. The contributions should be of a high standard and should avoid lengthy redraftings of papers already published or about to be published elsewhere. As a whole, the proceedings should aim for a balanced presentation of the theme of the conference including a description of the techniques used and enough motivation for a broad readership. It should not be assumed that the published proceedings must reflect the conference in its entirety. (A listing or abstracts of papers presented at the meeting but not included in the proceedings could be added as an appendix.)
When applying for publication in the series Lecture Notes in Physics the volume's editor(s) should submit sufficient material to enable the series editors and their referees to make a fairly accurate evaluation (e.g. a complete list of speakers and titles of papers to be presented and abstracts). If, based on this information, the proceedings are (tentatively) accepted, the volume's editor(s), whose name(s) will appear on the title pages, should select the papers suitable for publication and have them refereed (as for a journal) when appropriate. As a rule discussions will not be accepted. The series editors and Springer-Verlag will normally not interfere with the detailed editing except in fairly obvious cases or on technical matters.
Final acceptance is expressed by the series editor in charge, in consultation with Springer-Verlag only after receiving the complete manuscript. It might help to send a copy of the authors' manuscripts in advance to the editor in charge to discuss possible revisions with him. As a general rule, the series editor will confirm his tentative acceptance if the final manuscript corresponds to the original concept discussed, if the quality of the contribution meets the requirements of the series, and if the final size of the manuscript does not greatly exceed the number of pages originally agreed upon.
The manuscript should be forwarded to Springer-Verlag shortly after the meeting. In cases of extreme delay (more than six months after the conference) the series editors will check once more the timeliness of the papers. Therefore, the volume's editor(s) should establish strict deadlines, or collect the articles during the conference and have them revised on the spot. If a delay is unavoidable, one should encourage the authors to update their contributions if appropriate. The editors of proceedings are strongly advised to inform contributors about these points at an early stage.
The final manuscript should contain a table of contents and an informative introduction accessible also to readers not particularly familiar with the topic of the conference. The contributions should be in English. The volume's editor(s) should check the contributions for the correct use of language. At Springer-Verlag only the prefaces will be checked by a copy-editor for language and style. Grave linguistic or technical shortcomings may lead to the rejection of contributions by the series editors.
A conference report should not exceed a total of 500 pages. Keeping the size within this bound should be achieved by a stricter selection of articles and not by imposing an upper limit to the length of the individual papers.
Editors receive jointly 30 complimentary copies of their book. They are entitled to purchase further copies of their book at a reduced rate. As a rule no reprints of individual contributions can be supplied. No royalty is paid on Lecture Notes in Physics volumes. Commitment to publish is made by letter of interest rather than by signing a formal contract. Springer-Verlag secures the copyright for each volume.

The Production Process

The books are hardbound, and the publisher will select quality paper appropriate to the needs of the author(s). Publication time is about ten weeks. More than twenty years of experience guarantee authors the best possible service. To reach the goal of rapid publication at a low price the technique of photographic reproduction from a camera-ready manuscript was chosen. This process shifts the main responsibility for the technical quality considerably from the publisher to the authors. We therefore urge all authors and editors of proceedings to observe very carefully the essentials for the preparation of camera-ready manuscripts, which we will supply on request. This applies especially to the quality of figures and halftones submitted for publication. In addition, it might be useful to look at some of the volumes already published. As a special service, we offer free of charge LATEX and TEX macro packages to format the text according to Springer-Verlag's quality requirements. We strongly recommend that you make use of this offer, since the result will be a book of considerably improved technical quality. To avoid mistakes and time-consuming correspondence during the production period the conference editors should request special instructions from the publisher well before the beginning of the conference. Manuscripts not meeting the technical standard of the series will have to be returned for improvement.

For further information please contact Springer-Verlag, Physics Editorial Department V, Tiergartenstrasse 17, W-6900 Heidelberg, FRG

W. Hollik R. Rückl J. Wess (Eds.)

Phenomenological Aspects of Supersymmetry

Proceedings of a Series of Seminars
Held at the Max-Planck-Institut für Physik
Munich, FRG, May to November 1991

Springer-Verlag
Berlin Heidelberg GmbH

Editors

Wolfgang Hollik
Max-Planck-Institut für Physik
Föhringer Ring 6, W-8000 München 40
Fed. Rep. of Germany

Reinhold Rückl
Julius Wess
Sektion Physik, Ludwig-Maximilians-Universität
Theresienstraße 37, W-8000 München 2
Fed. Rep. of Germany
and
Max-Planck-Institut für Physik
Föhringer Ring 6, W-8000 München 40
Fed. Rep. of Germany

ISBN 978-3-662-13910-3 ISBN 978-3-540-47281-0 (eBook)
DOI 10.1007/978-3-540-47281-0

Originally published by Springer-Verlag Berlin Heidelberg New York in 1992
Softcover reprint of the hardcover 1st edition 1992

Typesetting: Camera ready by author/editor
2158/3140-543210 - Printed on acid-free paper

Preface

The standard model of strong and electroweak interactions describes particle physics phenomena up to the scale of the vector boson masses. In spite of its success, it is generally believed that the standard model is not the ultimate theory. The large number of free parameters is still unsatisfactory, and gravity is not included at all.

As there is no experimental sign of new physics beyond the standard model, every generalization of the model is motivated by theoretical arguments only. Supersymmetry (SUSY) is a particularly attractive extension of the minimal model. It could solve the fine-tuning problem related to the hierarchy of scales, and it could also provide a link to gravity. Moreover, SUSY could play a significant role in building a consistent grand unified theory (GUT) of the strong and electroweak interactions. If bosons and fermions are to be part of such a scheme, supersymmetry is essential.

The most impressive motivation for SUSY comes from extrapolation of the LEP high precision data. Including SUSY matter in the running couplings, and masses of the SUSY-partners in the range 100 GeV - 10 TeV, the data display a spectacular coupling constant unification at the scale of 10^{16} GeV. Although this is very suggestive it is far from being a proof that nature knows about SUSY.

The direct searches for SUSY particles at LEP, the Tevatron, and at the future hadron colliders LHC and SSC are the primary tests of the low-energy consequences of SUSY. However, the investigation of the scalar sector is of similar importance since the minimal model predicts at least one relatively light Higgs boson. For a detailed experimental search it is mandatory to incorporate the radiative corrections in the Higgs sector since they influence the theoretical predictions considerably. Complementary to the direct searches, the indirect searches for SUSY through virtual effects in rare FCNC processes and in high precision observables are also important tests of SUSY.

For a critical assessment of the present status of SUSY phenomenology a series of seminars was organized from May to November 1991 at the Max-Planck-Institut für Physik in Munich. The intention was to present the most recent results and to initiate discussions between field theorists and phenomenologists. The subjects covered include an introduction to the basic features of SUSY models, in particular the minimal supersymmetric standard model. This is followed by discussions of SUSY's implications for direct searches at future colliders, effects in rare processes, quantum effects in the Higgs sector, and cosmology. A special section is devoted to dynamical electroweak symmetry breaking where SUSY can be used to solve the fine-tuning problem in a natural way. This predicts a mass of the top quark consistent with the bounds obtained from the precision LEP data and W mass measurements.

Munich, April 1992

W. Hollik
R. Rückl
J. Wess

Contents

THE SUPERSYMMETRIC STANDARD MODEL

Hans Peter Nilles

Physik Department, Technische Universität München,
D-8046 Garching
and
Max Planck Institut für Physik, Werner Heisenberg Institut,
P.O. Box 40 12 12, D-8000 München

ABSTRACT

We give an introduction to $N = 1$ supersymmetry and supergravity and review the attempts to construct models in which the breakdown scale of the weak interactions is related to supersymmetry breaking. Special emphasis is put on the discussion of the minimal supersymmetric extension of the standard model of strong and electroweak forces. Grand unified models are analyzed in their supersymmetric and nonsupersymmetric versions.

1. THE STANDARD MODEL

Although the standard model [1] of strong and electroweak interactions describes particle physics phenomena up to the 100 GeV scale very well, it is in general believed that it is not the ultimate theory of everything. Gravity is not included and many parameters are used to fit the data. But there is not the slightest experimental sign of new physics beyond the standard model. Every generalization of the model is thus motivated only by theoretical argumentations. To appreciate these constructions we need to have a closer look at the standard model itself. We shall skip details and concentrate on those points that make us believe that there must be new physics beyond the TeV-scale.

The standard model is based on the gauge interactions of the strong and electroweak forces with gauge group $SU(3) \times SU(2) \times U(1)$. It thus contains 12 spin 1 gauge bosons: eight gluons of $SU(3)$, three $SU(2)$ weak gauge bosons and the hypercharge gauge boson of $U(1)$. The photon will be a particular combination of the neutral $SU(2)$ gauge boson and the hypercharge boson. The fermions of the theory consist of three generations of quarks and leptons, where we assume the existence of the top quark for which direct experimental evidence ist still lacking. The spin-1/2 fermions of a family have the following transformation properties with respect to $SU(3) \times SU(2) \times U(1)$:

$$\begin{aligned} U^a = \begin{pmatrix} u \\ d \end{pmatrix} &= (3, 2, 1/6) \\ \bar{u} &= (\bar{3}, 1, -2/3) \\ \bar{d} &= (\bar{3}, 1, 1/3) \\ L^a = \begin{pmatrix} \nu_e \\ e \end{pmatrix} &= (1, 2, -1/2) \\ \bar{e} &= (1, 1, 1) \end{aligned} \tag{1.1}$$

where $a = 1,2$ is an $SU(2)$ index and the first two entries in the brackets denote the dimensions of the $SU(3) \times SU(2)$ representations while the last entry denotes $U(1)$ hypercharge. Electric charge is given by $Q = T_3 + Y$. Thus the up-quark, for example, has $Q(u) = 1/2 + 1/6 = 2/3$ whereas for the down quark we obtain $Q(d) = -1/3$.

The socalled Higgs sector contains a scalar $SU(2)$-doublet

$$h = \begin{pmatrix} h^0 \\ h^- \end{pmatrix} = (1, 2, -1/2) \tag{1.2}$$

with potential $V = \mu^2(h^\dagger h) + \lambda(h^\dagger h)^2$ and one also introduces Yukawa couplings for the interactions of the scalars with the fermions

$$L_Y = g_d U h \bar{d} + g_e L h \bar{e} + g_u U h^\dagger \bar{u} \tag{1.3}$$

in all combinations that are allowed by $SU(3) \times SU(2) \times U(1)$ gauge symmetry. A spontaneous breakdown of $SU(2) \times U(1)$ occurs for negative μ^2 and the neutral component of h receives a vacuum expectation value (vev)

$$< h >= \frac{1}{\sqrt{2}} \begin{pmatrix} v \\ 0 \end{pmatrix} \tag{1.4}$$

where $v = (-\mu^2/\lambda)^{1/2}$. $SU(2) \times U(1)_Y$ is broken to $U(1)_Q$ and three gauge bosons become massive

$$\begin{aligned} M_{W^\pm} &= \frac{1}{2} g_2 v \\ M_Z &= \frac{1}{2} v \sqrt{g_1^2 + g_2^2} \end{aligned} \tag{1.5}$$

where g_1 and g_2 are the coupling constant of $SU(2)$ and $U(1)$, respectively. The $U(1)$ gauge coupling constant is given by

$$e = g_2 \sin\theta_W = g_1 \cos\theta_W \tag{1.6}$$

where θ_W denotes the weak mixing angle. The mass of the physical Higgs-scalar is given by $\sqrt{-2\mu^2}$. Yukawa couplings then allow, in presence of the spontaneous breakdown of $SU(2) \times U(1)$, mass terms for the fermions. The term $g_d h U \bar{d}$, e.g. leads to $g_d v d \bar{d} = m_d d \bar{d}$. The masses and mixings for the three families of quarks and leptons are parametrized by the 3×3 Kobayashi-Maskawa [2] matrix.

Let us now count the parameters of the model. We have three gauge couplings g_1, g_2 and g_3 usually parametrized by $\alpha_{\text{e.m.}}$, α_{strong} and $\sin\theta_W$. In the gauge sector we have in addition a Θ -parameter multiplying $F^{\mu\nu} F^{\rho\sigma} \epsilon_{\mu\nu\rho\sigma}$ in the action. Its actual value seems to by very close to zero as can be deduced from the absence of the electric dipole moment of the neutron. Nonetheless we have to treat Θ as an arbitrary parameter and it still has to be understood why its value is so small.

In the Higgs sector we have introduced two parameters μ^2 and λ of which one combination defines the scale of $SU(2) \times U(1)$ breakdown while the other determines the Higgs mass. The 9 fermion masses (not including the possibility for neutrino-Majorana masses) are parametrized by the Yukawa couplings. The same applies to quark mixing consisting of 3 angles and one phase in the Kobayashi-Maskawa matrix, the latter giving

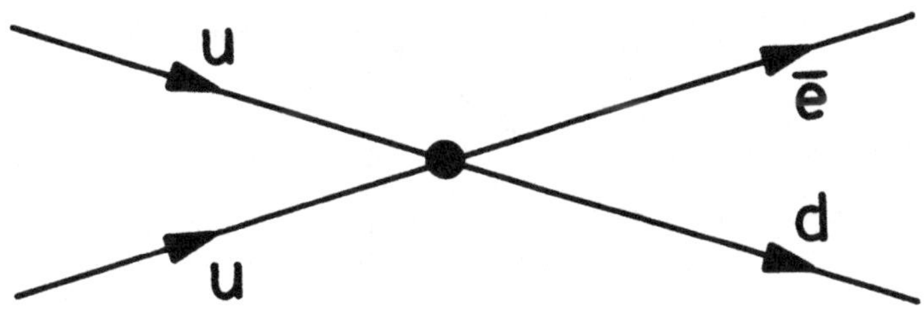

Fig. 1.1: A dimension 6 operator that could lead to proton decay.

rise to CP-violation. We do not know yet whether there is a corresponding mixing in the lepton sector. In any case we can conclude that the above mentioned quantities are completely free parameters in the standard model. Any attempt to understand their specific values will require a generalization of the model. Apart from these questions we have eventually also to address the more fundamental puzzles out of which I shall mention some in the following. Why is the gauge group $SU(3) \times SU(2) \times U(1)$, why is $SU(2)$ broken and why at a scale of 100 GeV and not at the Planck mass? Why is the mass of the proton 1 GeV and is this scale related to other physical scales? Why do we have this repetition of families, why 3 families and why does a family not contain exotic representations of $SU(3) \times SU(2) \times U(1)$ (like e.g. a 3 of $SU(2)$)? Why are neutrinos massless (are they?) and why is the electron mass so small compared to the W-mass? These and many more related questions are the subject of discussions of the physics beyond the standard model.

One important property of the standard model is the chirality of the fermion spectrum. Fermion masses are protected by $SU(2) \times U(1)$, i.e. they can be nonzero only after $SU(2) \times U(1)$ breakdown. Thus all fermion masses are proportional to the vev of the Higgs-field (1.4) and this explains why fermion masses cannot be very large compared to M_W. It does, of course, not explain why the mass of the electron is so small compared to M_W and also the smallness of neutrino masses remains a mystery. Only the top quark seems to be as heavy as allowed by $SU(2) \times U(1)$. We will regard this chirality of fermions as a very important property of the standard model and will therefore in the course of these lectures only discuss extensions that share these remarkable properties.

Another important symmetry of the standard model is baryon (B)- and lepton (L)-number conservation. From the requirement of gauge invariance and renormalizability (i.e absence of nonrenormalizable terms in the action) the model has automatic B and L conservation. Among other things this implies the stability of the proton. Possible violations could come from higher dimensional (nonrenormalizable) terms as e.g. the one displayed in Fig. 1.1. This operator has dimension 6 and therefore the coefficient $1/M_x^2$ has the dimension of inverse $(\text{mass})^2$. M_x denotes the scale of the new physics that is responsible for proton decay. From the long lifetime of the proton we conclude that M_x must be larger than 10^{15} GeV, a very large scale. For other processes, like lepton number violation, the corresponding scale could still be in the TeV region. It is a central question in all discussions of the physics beyond the standard model to isolate these new processes and discuss the corresponding scales.

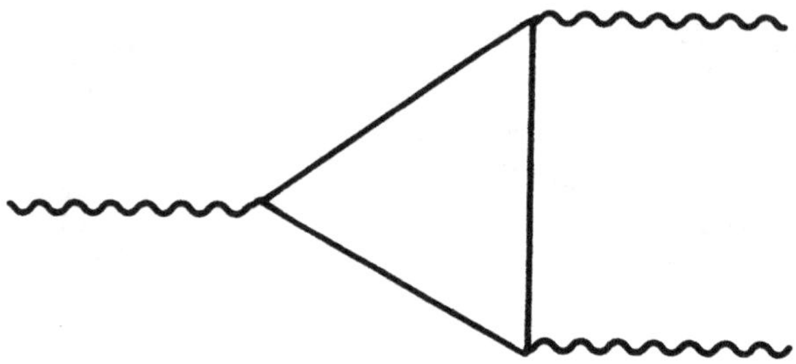

Fig. 1.2: A triangle graph responsable for an anomaly. Wavy lines denote gauge bosons while the solid lines denote fermions.

In models with chiral fermions there is the potential danger of gauge anomalies [3]. Such anomalies can be determined by the calculation of triangle graphs as displayed in Fig. 1.2. These give several constraints in the cases where all three gauge bosons come from the same gauge group, but in addition we also have mixed anomalies e.g. photon-gluon-gluon, photon-graviton-graviton as well as a global $SU(2)$-anomaly [4] (implying an even number of doublets). In the standard model we find a cancellation of anomalies within a given family, i.e. the spectrum and the quantum numbers of the fermions are arranged in such a way that the sum of the graphs of Fig. 1.2 cancels exactly. Does this tell us something about the quantum numbers with respect to $U(1)_Y$ [5]? Let us for a moment consider one family as given in (1.1) but with arbitrary hypercharges: $(3,2,p)$, $(3,1,q)$, $(3,1,r)$, $(1,2,s)$ and $(1,1,t)$. After normalization $t=1$ the absence of anomalies gives a unique solution $p=1/6$, $q=-2/3$, $r=1/3$ and $s=-1/2$, thus hypercharge is quantized in the standard model. A word of caution should be added here. In our argumentation we have assumed the $SU(3)\times SU(2)$ representations of the particle content of the standard model and also demanded anomaly cancellation within one family. The authors of ref. [6] have generalized the analysis. They tried to find a minimal set of particles in $SU(3)\times SU(2)\times U(1)$ which is chiral and anomaly free (and participating in all 3 interactions). If one just considers triangular anomalies such a set is given by $(3,1,Q)$, $(3,1,-Q)$, $(\bar{3},1,q)$, $(\bar{3},1,-q)$ and $(1,2,0)$ but it suffers from a global $SU(2)$ anomaly. Including this constraint the minimal set is found to be exactly one family of quarks and leptons. We do not know at the moment what this fact can teach us, but at least we have to conclude that hypercharge quantization does not necessarily imply grand unification. Let us remark in closing this section that the behaviour above is strictly true only in the standard model. If we add e.g. a right-handed neutrino things change [7]. Add $(1,1,u)$ to the set described above and repeat the analysis. You will find that hypercharge is no longer quantized and instead obtain a one parameter family of solutions (the usual choice $u=0$, of course, remains a solution). The discussion above shows, that it still pays off to have a closer look at the standard model.

In the following we shall be concerned with a discussion of physics beyond the standard model. Since there is no phenomenological indication that contradicts the standard model, our argumentation has to remain purely theoretical. Efforts in such a direction include:

- grand unification
- L-R symmetric gauge groups
- extra $U(1)$'s
- extra Higgs-bosons
- technicoulor
- composite quarks and leptons
- supersymmetry
- Kaluza-Klein theories
- supergravity
- strings.

In these lectures, of course, I shall concentrate on a presentation of those extensions of the standard model that are based on the incorporation of supersymmetry and supergravity. Grand unified models, whether supersymmetric or not, are also included in our discussion.

2. GRAND UNIFICATION

The aim is to find a group that contains $SU(3) \times SU(2) \times U(1)$ and thus unifies the gauge coupling constants g_1, g_2 and g_3 of the standard model.

One very popular model [8] makes use of the astonishing fact that one family of quarks and leptons fits into the $\bar{5}$ and 10 representation of $SU(5)$. We have the following identifications

$$\psi_\alpha = \bar{5}_\alpha = (\bar{d}_1, \bar{d}_2, \bar{d}_3; e, \nu) \tag{2.1}$$

for the complex conjugate of the fundamental representation and

$$\chi^{\alpha\beta} = 10^{\alpha\beta} = \begin{pmatrix} 0 & \bar{u} & \bar{u} & u & d \\ & 0 & \bar{u} & u & d \\ & & 0 & u & d \\ & & & 0 & \bar{e} \\ & & & & 0 \end{pmatrix} \tag{2.2}$$

for the two-index antisymmetric representation. The hypercharge generator is defined as

$$Y = \frac{1}{3} \begin{pmatrix} -1 & 0 & 0 & 0 & 0 \\ 0 & -1 & 0 & 0 & 0 \\ 0 & 0 & -1 & 0 & 0 \\ 0 & 0 & 0 & 3/2 & 0 \\ 0 & 0 & 0 & 0 & 3/2 \end{pmatrix} \tag{2.3}$$

while the $SU(3) \times SU(2)$ embedding is obvious.

So far the classification of the fermions. The questions remains whether $SU(5)$ is just a global symmetry to classify representations or whether it is a gauge symmetry. In the latter case we would explain the three gauge coupling constants $g_{1,2,3}$ by a single constant g_5. Present observation tells us, however, that g_1, g_2 and g_3 are vastly different,

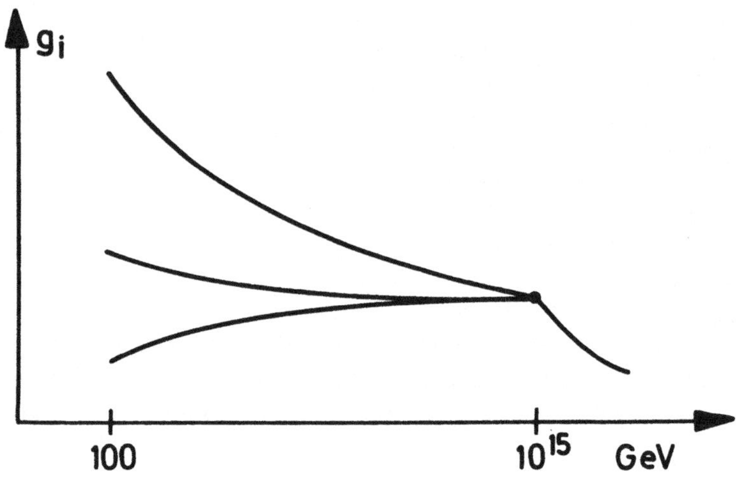

Fig 2.1: The desired evolution of coupling constants in grand unified models.

an apparent paradox. The resolution lies in the fact that coupling constants are scale dependent [9]. At the one-loop level they evolve according to

$$\frac{1}{\alpha_i(\mu)} = \frac{1}{\alpha_i(M_x)} + \frac{b_i}{2\pi} \log\left(\frac{M_x}{\mu}\right) \tag{2.4}$$

with

$$b_i = -\frac{11}{3}C + \frac{4}{3}N_f l_f(R) + \frac{1}{3}N_s l_s(R) \tag{2.5}$$

where C is the quadratic Casimir operator of the adjoint representation, N_f (N_s) the number of Dirac fermions (complex scalar bosons) in representation R of which $l_{f,s}(R)$ is the quadratic Casimir (normalized to 1/2 for the fundamental representation). The question therefore is: at which scale do the couplings $g_{1,2,3}$ coincide with a single coupling constant g_5? The evolution of the coupling constants depends crucially on the fermion spectrum. For minimal SU(5) we obtain

$$\begin{aligned} b_3 &= -11 + \frac{4}{3}N_G \\ b_2 &= -\frac{22}{3} + \frac{4}{3}N_G + \frac{1}{6}H \\ b_1 &= \frac{4}{3}N_G + \frac{1}{10}H \end{aligned} \tag{2.6}$$

where N_G is the number of families and H the number of light Higgs-doublets. For $N_G = 3$ and $H = 1$ we obtain a behaviour as shoun in Fig. 2.1. At a scale of approximately $M_x \approx 10^{15}$ GeV the three coupling constants aproach each other and we obtain $\alpha(M_x) =$

$g_5^2/4\pi \approx 1/45$ and $g_5 \approx g_3 \approx g_2 \approx \sqrt{5/3}g_1$. The weak mixing angle at this large scale M_x is given by $\sin^2\theta_W = 3/8$. The behaviour of Fig. 2.1 can now be interpreted in the following way. Above M_x $SU(5)$ is a good symmetry while at M_x $SU(5)$ is broken to $SU(3) \times SU(2) \times U(1)$ and below that scale the coupling constants evolve differently. $SU(5)$ leads to 24 gauge bosons twelve of which receive a mass of order M_x

$$(24)_\alpha^\beta = \begin{pmatrix} \text{gluons}(8,1) & (3,2) \\ (\bar{3},\bar{2}) & (W,Z,\gamma) \end{pmatrix} \tag{2.7}$$

and they transform as $(3,2)+(\bar{3},\bar{2})$ of $SU(3) \times SU(2)$. It is important to realize that quarks and leptons now sit in the same $SU(5)$ representations and that these new gauge bosons denoted by X and Y can cause transitions between them. This leads to B and L violation and proton decay.

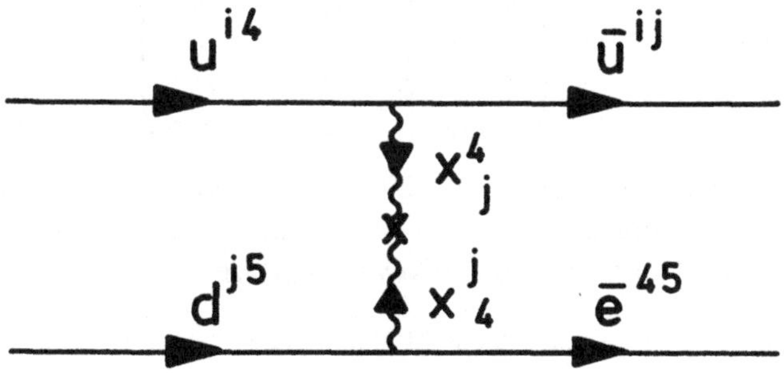

Fig. 2.2: B and L violating process mediated by $SU(5)$ gauge bosons

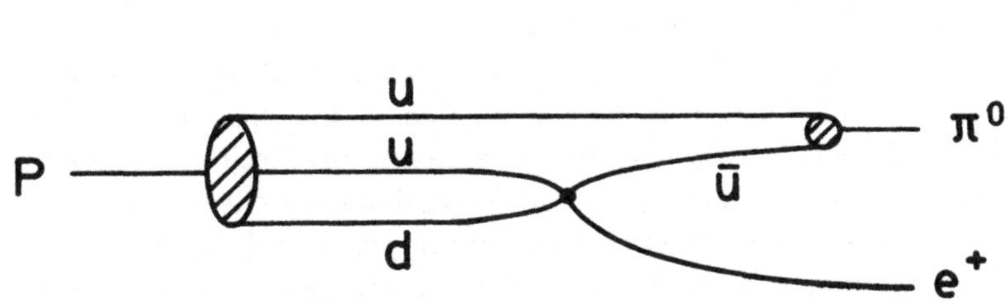

Fig. 2.3: $p \to \pi^0 e^+$ in minimal $SU(5)$.

Consider the process in Fig. 2.2 which causes the transition $ud \to \bar{u}e^+$. It leads to proton decay via a dimension six operator to $p \to e^+\pi^0$ as shown in Fig.2.3. The rate for proton decay is suppressed by a factor M_x^{-4} leading to a proton lifetime of less than 10^{30} years in minimal SU(5). This is in contrast with the experimental observations and minimal SU(5) is ruled out. We shall later discuss grand unified models which do not suffer from this desease. For the moment we shall however still discuss the simplest model for paedagocical reasons. $B - L$ is still conserved in the $SU(5)$ model. This comes from the fact that one family of quarks and leptons consists of two irreducible representations of $SU(5)$. Thus we have a global nonanomalous U(1) with charge $S(\bar{5}) = -3$ and $S(10) = 1$. This leads to

$$B - L = \frac{1}{5}(S + 4Y) \tag{2.8}$$

as a conserved quantum number (Y is hypercherge as given in (2.3)). In the present model this implies that there is always an antilepton in the final state of proton decay.

The breakdown of grand unified groups is achieved by vev's of scalar fields in connection with a Higgs effect. In $SU(5)$ the breakdown to $SU(3) \times SU(2) \times U(1)$ can be achieved with a scalar ϕ in the adjoint representation, provided the vev has the following pattern:

$$\phi_\alpha^\beta = v \begin{pmatrix} 1 & 0 & 0 & 0 & 0 \\ 0 & 1 & 0 & 0 & 0 \\ 0 & 0 & 1 & 0 & 0 \\ 0 & 0 & 0 & -3/2 & 0 \\ 0 & 0 & 0 & 0 & -3/2 \end{pmatrix} \tag{2.9}$$

A second step of breakdown to $SU(3) \times U(1)$ in the standard model requires a 2 of $SU(2)$ and in the grand unified model this generalizes to a doublet component in the 5 of $SU(5)$. We then have the following Yukawa-couplings

$$g_{ab} H_5 \times (10)_a \times (10)_b \tag{2.10}$$

giving mass to up-type quarks, whereas

$$g'_{ab} H_5^* \times (\bar{5})_a \times (10)_b \tag{2.11}$$

gives masses to down-type quarks and leptons. Thus masses for down quarks and leptons come from the same coupling and one obtains relations between them e.g. $m_b \approx 3m_\tau$ [10] at low energies. Similar relations, however, should also hold for the masses of the first two families and it remains to be seen whether grand unified predictions can help in the understanding of quark and lepton masses. Such mass relations exist in many grand unified models and mostly lead to embarrasing predictions. Therefore one adopts the viewpoint that only the heaviest family should satisfy these relations.

In all grand unified there exists a fine-tuning problem [11] concerning the relation of the two scales of gauge symmetry breakdown (usually called the gauge hierarchy problem). Let us discuss it in the $SU(5)$ model. In the Higgs potential we have among others the following terms

$$m^2 H_5 H_5^* + m' H_5 \phi_{24} H_5^* \tag{2.12}$$

and ϕ is supposed to receive a vev as displayed in (2.9) with $v \sim 10^{15}$ GeV. With respect to $SU(3) \times SU(2)$ H splits into $(3,1) + (1,2)$ with $(\text{masses})^2$ $m^2 + m'v$ and $m^2 - 3/2m'v$ respectively. The $SU(3)$ triplet of H should become heavy ($M \geq 10^{11}$ GeV) because of the potential danger of fast proton decay through the process displayed in Fig. 2.4. The moderate limit of 10^{11} GeV comes the fact that here the vertices are of the strength of Yukawa couplings which are smaller than the gauge couplings that appear in Fig. 2.2. There is no problem to achieve such large masses of the triplet since already v is supposed to be in the 10^{15} GeV range (which actually is the natural range of Higgs masses in the model). The problem lies in the fact that the Higgs doublet (as a part of the standard model) should become light; not exceeding the TeV scale, i.e. some 13 orders of magnitude below the scales m, m' and v. This can be achieved by a fine tuning of the parameters such that $m^2 - (3/2)m'v$ becomes small. There is in principle nothing wrong with such a finetuning but one feels uneasy about it since one does not understand the relation between the two breakdown scales and also because the fine-tuning is not stable in perturbation theory. We shall come back to a discussion of this and related problems in the next chapter.

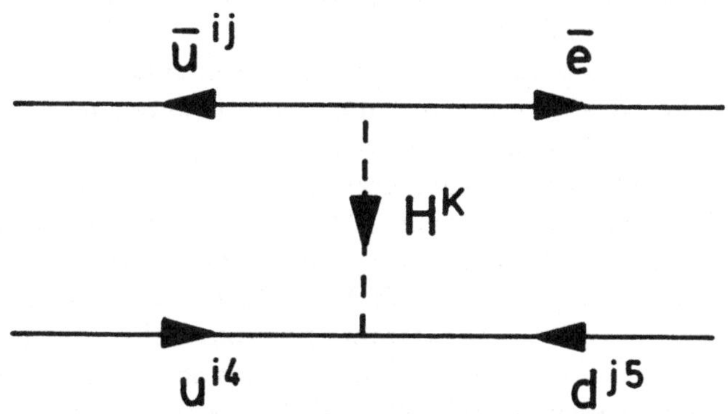

Fig. 2.4: Proton decay via Higgs exchange.

The simplest grand unified model based on $SU(5)$ is ruled out experimentally. The first indication in this direction came from the absence of proton decay but also more recently precision electroweak data are inconsistent with simple $SU(5)$. We shall discuss the results explicitely in chapter 8 where we also consider supersymmetric grand unified models. In the nonsupersymmetric case a way out is to consider larger groups and try to move the grand unification scale to higher values. There is then a complication in group theoretical technicalities and we shall not have the time here to discuss the issues in detail. A complete treatment can be found in ref. [12].

To embed the standard model spectrum in a larger group we have to keep two things in mind. First we need to consider complex representations because of the chirality of the fermion spectrum. Secondly one has then to ensure the cancellation of anomalies. Prime candidates for grand unified models are then $SU(2N+1)$ with an anomaly free set of antisymmetric tensor representations, $O(4N+2)$ with a spinor representation and E_6 with the fundamental 27-dimensional representation. The next simplest candidate to $SU(5)$ is $O(10)$. The adjoint representation is 45-dimensional and decomposes as $24+10+10+1$ with respect to $SU(5)$. In fact $SU(5)\times U(1)$ is a subgroup of $O(10)$ and the U(1) corresponding to $B-L$ is now gauged. One chain of breakdown pattern to the standard model is then via $SU(5)\times U(1)$ but there are also others like $SU(4)\times SU(2)\times SU(2)$. A remarkable property of $O(10)$ is the fact that one family of quarks and leptons fits into one irreducible representation: the 16-dimensional spinor representation with

$SU(5)$ decomposition $\bar{5}+10+1$. The additional $SU(5)$ singlet has all the properties of a righthanded neutrino. The standard model Higgs can be embedded in the fundamental 10-dimensional representation of $O(10)$ and there is a single Yukawa coupling $10 \times 16 \times 16$ and one would expect e.g. a relation between m_{top} and m_{bottom}. The Higgs field 10, however, contains two $SU(2)$ Higgs-doublets: one coupling to up-type and the other to down-type quarks. Since the two could have different vev's we cannot expect $m_{\text{top}}/m_{\text{bottom}}$ to be fixed. Including the assumption that the masses of the light quarks are given by the same mechanism one would expect mass relations as $m_{\text{top}}/m_{\text{bottom}} \approx m_{\text{charm}}/m_{\text{strange}}$ and as before we conclude that grand unified models do not make a decisive contribution to our understanding of quark and lepton masses. To work out the spontaneous breakdown patters of $O(10)$ is already quite complicated. One needs representations of dimension 45, 16 and/or 54 and 126 and we shall not discuss these questions in detail [12].

The next step to more complicated groups is E_6 which has recently again attracted interest because of the consideration of superstring inspired models. Here we have $E_6 \supset SO(10) \times U(1)$ and 78 gauge bosons decomposing as $45 + 16 + \bar{16} + 1$ of $O(10)$. The fundamental representation is 27-dimensional and contains $16+10+1$ of $O(10)$: one family but also a lot of additional fermions. The Yukawa coupling term $(27)^3$ is unique and the breakdown requires at least a 351-dimensional representation. Higher rank groups like $SU(2N+1)$ and $O(4N+2)$ with $N \geq 3$ have also been considered for model building, among other motivations as attempts to explain the number of families. One aim would be, that three families of quarks and leptons are contained in one irreducible representation of a large gauge group. While such representations can be found, they usually predict many more exotic fermions and it seems that this approach is not the correct one to explain the number of families.

Let us summarize some of the universal properties of grand unified models. One of them is charge quantization but remember from our discussion in the last chapter that this might already be a property of the standard model. Unification of gauge coupling constants was the prime motivation and it can be achieved. In the most promising models one obtains $\sin^2\theta_W = 3/8$ at the grand unification scale M_x. Quarks and leptons sit in the same multiplet and therefore we necessarily obtain baryon number violation. This implies that M_x must have a rather large value and the simplest model is ruled out. The model leads to "naive" quark-lepton mass relations which are usually not very sucessful (maybe with the exception of $m_b/m_\tau \approx 3$). Such questions now also concern possible neutrino masses. In all models except for simple $SU(5)$ we have a righthanded neutrino and the possibility of a Dirac-mass term. In $O(10)$ we would obtain $m_\nu \sim m_{\text{up}}$ according to the naive mass relation. How can one avoid a problem with such a neutrino mass? It can be achieved by assuming a large Majorana mass M for the righthanded neutrino and a mass matrix [13]

$$\begin{pmatrix} 0 & m \\ m & M \end{pmatrix} \tag{2.13}$$

with $M \gg m$, leading (approximately) to the eigenvalues M and m^2/M. M can be as large as the scale of $O(10)$-breakdown and exceed even 10^{15}GeV. Assuming natural values for m as e.g. $m \sim 1$GeV for the Dirac mass term we would have a mass of 10^{-6}eV for the light neutrino.

It still remains an open question whether grand unified models are realized in nature. The long lifetime of the proton requires a rather large scale of M_x and the simplest model is ruled out. In many models M_x approaches the Planck-mass and one has to worry about the fact that the influence of gravitational interactions becomes inportant. We shall later come back to these questions again.

3.THE PROBLEM OF THE WEAK SCALE

In the grand unified models discussed above we had to perform a fine tuning to obtain the breakdown scale M_W of $SU(3) \times SU(2) \times U(1)$ small compared to the grand unified scale M_x . More precisely we had to require

$$m^2 - \frac{3}{2}m'v \sim (100\text{GeV})^2 \tag{3.1}$$

while m, m' and v were of order of 10^{15} GeV or larger. (3.1) was required to keep the Higgs doublet mass in the range of a hundred GeV. There is nothing wrong with such a choice since m, m', and v are free input parameters and not constrained from other theoretical requirements. Nonetheless one might feel uneasy to perform such a fine tuning to a precision of 13 orders of magnitude. Also such a finetuning is not stable under radiative corrections, and has to be retuned order by order in perturbation theory. Again this can be done without leading to inconsistencies. The main question, however, remains why M_W is so small compared to M_x . We think that we should find a simple reason to explain why there is this hierarchy of scales. Such a reason could be a symmetry as we encountered in the discussion of fermion masses. Here chiral symmetry protected the masses. The fact that we do not understand the smallness of M_W and have to rely on a finetuning of parameters is called the gauge hierarchy problem.

The problem exists already within the standard model. Recall the Higgs potential

$$V(h) = \mu^2|h|^2 + \lambda|h|^4. \tag{3.2}$$

The Higgs mass is $m = \sqrt{-2\mu^2}$ and $M_W = g_2 < h > \approx 80$ GeV . Experimental bounds [14] on m come from LEP $m \geq 40$GeV while an upper bound of 1 TeV can be argued from unitarity constraints. Observe that the mass scale of the standard model M_W is solely set by the parameters μ^2 and λ in the Higgs sector.

Theoretically the model is very appealing; it is not just based on an effective Lagrangian, like e.g. the Fermi theory of weak interactions, but it is a renormalizable field theory. This has drastic consequences for the possible range of validity of the model; would it be nonrenormaliziable it necessarily would only be defined with a cutoff Λ (of dimension of a mass) and its region of validity would be bounded from above by Λ. Above Λ one expects new things to happen which are not described by the model. Since the standard model is renormalizable it could, however, be valid in a much larger energy range. Strangely enough this very nice property of the model constitutes one of its problems. The mass scale of 100 GeV is put in by hand and there is no idea about its origin: it is a completely free imput parameter. In a more complete theory one would like to understand the origin of M_W in terms of more fundamental parameters like e.g. the Planck scale $M_P \sim 10^{19}$ GeV, but such a complete theory would need more structure than present in the standard model.

A reconfirmation of the statement that M_W is a completely free parameter is found in the discussion of perturbation theory. The parameter μ^2 in (3.2) receives a contribution due to the graph of Fig. 3.1 which is quadratically divergent. There is nothing wrong with quadratic divergencies as they do not spoil the consistency of the theory; we regularize them and define the theory in terms of the renormalized parameters. The actual correction to μ^2 depends on the regularization scheme and the renormalized quantity is an arbitrary parameter even if we would have understood its value at the tree level. This is true for all quadratically divergent quantities. These divergences introduce a new mass scale in the theory which has nothing to do with the scales already present; it is an arbitrary parameter which we can choose at our will. To understand the origin of these masses the quadratic divergencies have to be absent; i.e. they have to be cut off at a larger scale by a new physical structure. With such a *physical* cutoff Λ we would have

$$\delta\mu^2 \sim \lambda\Lambda^2 \tag{3.3}$$

and to understand the order of magnitude of μ^2 it would not be appropriate to have Λ of the order of the Planck mass M_P but rather in the TeV region. An understanding of the order of magnitude of M_W would therefore require new physics in the TeV-region.

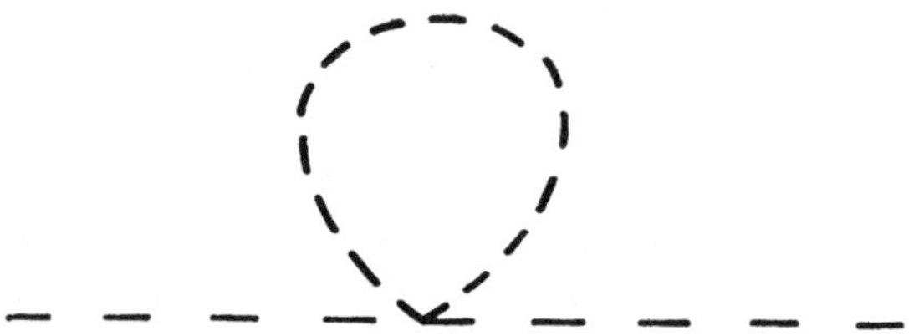

Fig. 3.1: Quadratically divergent contribution from the scalar self interaction.

Having agreed that the standard model might have this subtle theoretical problem one has to look for ways out. The presence of quadratic divergencies is originated by the existence of fundamental scalar particles. One way out is to remove these scalars from the theory. Since we have to break $SU(2) \times U(1)$ spontaneously (and want to maintain Lorentz invariance) some scalar objects have to exist; they could be composite as postulated in the technicolour approach [15]. A new gauge interaction becomes strong in the region of a few hundred GeV; leading to the formation of condensates and many composite bound states. This is the new physics in the TeV-region.

But this is not the only possible solution and we could try to insist to live with fundamental scalar particles. Remember for this purpose the situation with spin 1 particles. Models containing spin 1 particles have usually serious theoretical problems unless there is a gauge symmetry that make these fundamental spin 1 particles acceptable. Observe that this gauge symmetry also stabilizes the mass of these particles; in the symmetric limit they have to vanish. Could we also have such a situation for scalar

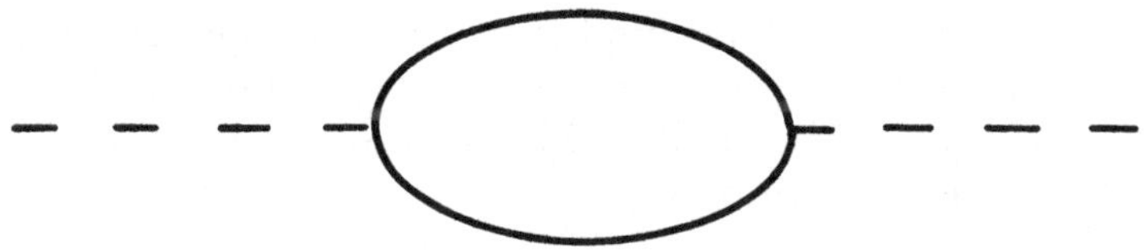

Fig. 3.2: A contribution to μ^2 from supersymmetric partners

masses? In the standard model, of course, such a situation is not present. We can take the limit $\mu^2 \to 0$ and this does not enhance the symmetry of the action.

The only known way to protect scalar masses is supersymmetry. This symmetry relates bosons and fermions and therefore makes bosons as well behaved as fermions which implies the absence of quadratic divergencies. Supersymmetry provides us with the physical cuttoff discussed earlier. In addition to the contribution to μ^2 given in Fig. 3.1 we have now a contribution of Fig. 3.2 with the supersymmetric partner of the Higgs boson in the loop. In the supersymmetric limit these two contributions cancel exactly. If supersymmetry is broken the masses of the boson-fermion multiplet are split. We get a contribution

$$\delta\mu^2 \approx \lambda(m_B^2 - m_F^2) \tag{3.4}$$

and we would require the quantity on the right-hand side to be in the TeV range. If we would remove the partner with mass m_F from the theory we would again recover the quadratic divergence of the standard model. Thus to solve the Higgs problem we have to consider new structure in the TeV-region.

4. INTRODUCTION TO SUPERSYMMETRY

Supersymmetry is a symmetry that transforms bosons into fermions and vice versa [16]. The generator of these transformations, the supercharge Q_α ($\alpha = 1, 2$), transforms as a left handed Weyl spinor, a $(\frac{1}{2}, 0)$ representation under Lorentz transformations. Its hermitean adjoint is denoted by $\bar{Q}_{\dot{\beta}}$; a right handed Weyl spinor. The algebra is defined by

$$[Q_\alpha, P_\mu] = 0 \tag{4.1}$$

$$\{Q_\alpha, \bar{Q}_{\dot{\beta}}\} = 2\sigma^\mu_{\alpha\dot{\beta}} P_\mu \tag{4.2}$$

where σ^μ denotes the Pauli matrices and P_μ is the energy momentum operator. From this algebra we can immediately read off two important consequences. From (4.1) for $\mu = 0$ we see that Q commutes with the Hamiltonian and states of nonzero energy are paired by the action of Q. Since Q is fermionic we have an equal number of fermionic and bosonic states degenerate in energy. Equation (4.2) relates the supercharge to the Hamiltonian and we have

$$H = P_0 \sim QQ^\dagger. \tag{4.3}$$

Thus H is semipositive definite. This implies that in supersymmetric theories the vacuum energy is well defined. Let $|0>$ denote the groundstate and let us suppose that supersymmetry is unbroken which implies $Q|0>=0$ i.e. the vacuum state is symmetric. We then conclude for the vacuum energy

$$E_{\text{vac}} = <0|H|0> = 0. \tag{4.4}$$

If, on the other hand, supersymmetry is spontaneously broken we have $Q_\alpha|0> = |\psi_\alpha> \neq 0$. This then implies that the supercurrent $J_{\mu\alpha}$ can create a fermion (goldstone fermion) out of the vacuum

$$<\psi_{\dot\beta}|J^\mu_\alpha|0> = f\sigma^\mu_{\dot\beta\alpha} \tag{4.5}$$

and consequently the vacuum energy is positive and given by

$$E_{\text{vac}} = f^2. \tag{4.6}$$

E_{vac} can therefore serve as an order parameter for supersymmetry. Supersymmetry is broken if and only if $E_{\text{vac}} \neq 0$. Together with (4.3) we can also conclude that whenever a supersymmetric ground state exists it is always the true vacuum of the theory. If supersymmetry is spontaneously broken there is a massless fermion in the theory (in analogy to ordinary symmetries where one has a Goldstone boson).

Representations of the supersymmetry algebra are most conveniently given in superspace. Superspace consists of ordinary space x^μ together with anticommuting Grassmann variables θ_α and $\bar\theta_{\dot\beta}$. The algebra can then be rewritten as

$$\begin{aligned}
[\theta Q, \bar Q\bar\theta] &= 2\theta\sigma_\mu\bar\theta P^\mu \\
[\theta Q, \theta Q] &= [\bar Q\bar\theta, \bar Q\bar\theta] = 0
\end{aligned} \tag{4.7}$$

Let us insert here the rules for manipulating with the Grassmann variables. Indices can be raised or lowered with the constant antisymmetric tensor $\epsilon_{\alpha\beta} = \epsilon^{\alpha\beta}$ with $\epsilon_{12} = +1$; and we have $\theta_\alpha = \epsilon_{\alpha\beta}\theta^\beta$. The antispinor is defined as $\bar\theta^{\dot\alpha} = (\theta^\alpha)^\dagger$ and $\bar\theta_{\dot\alpha} = \epsilon_{\dot\alpha\dot\beta}\bar\theta^{\dot\beta}$. The square is then defined as $\theta^2 = \theta^\alpha\theta_\alpha = -\theta_\alpha\theta^\alpha$ whereas $\bar\theta^2 = \bar\theta_{\dot\alpha}\bar\theta^{\dot\alpha}$. Differentiation is defined as

$$\partial_\alpha\theta^\beta = \frac{\partial\theta^\beta}{\partial\theta^\alpha} = \delta^\beta_\alpha \tag{4.8}$$

The reader should prove as an exercise that $\bar\partial_{\dot\beta}\bar\partial^{\dot\beta}(\bar\theta_{\dot\alpha}\bar\theta^{\dot\alpha}) = -4$. Intergration for a single θ is definied as $\int \mathrm{d}\theta = 0$ and $\int \theta \mathrm{d}\theta = 1$.

A supersymmetry transformation can now be defined as follows

$$S(x, \theta, \bar\theta) = \exp(\theta Q + \bar Q\bar\theta - x_\mu P^\mu) \tag{4.9}$$

A superfield $\phi(x, \theta, \bar\theta)$ is defined as a function of x, θ and $\bar\theta$ such that it transforms as follows under a supersymmetry transformation

$$S(y_\mu, \alpha, \bar\alpha)[\phi(x_\mu, \theta, \bar\theta)] = \phi(x_\mu + y_\mu - i\alpha\sigma_\mu\bar\theta + i\theta\sigma_\mu\bar\alpha, \theta + \alpha, \bar\theta + \bar\alpha). \tag{4.10}$$

The generators of the symmetry can be represented as differential operators in superspace

$$P_\mu = i\frac{\partial}{\partial x^\mu} \equiv i\partial_\mu \tag{4.11}$$

$$Q_\alpha = \frac{\partial}{\partial\theta^\alpha} - i\sigma^\mu_{\alpha\dot\beta}\bar\theta^{\dot\beta}\partial_\mu \tag{4.12}$$

$$\bar Q_{\dot\alpha} = -\frac{\partial}{\partial\bar\theta^{\dot\alpha}} + i\theta^\beta\sigma^\mu_{\beta\dot\alpha}\partial_\mu . \tag{4.13}$$

Slightly different representations (so-called right- and left-handed) can be defined by

$$\phi(x_\mu, \theta, \bar\theta) = \phi_L(x_\mu + i\theta\sigma_\mu\bar\theta, \theta, \bar\theta) = \phi_R(x_\mu - i\theta\sigma_\mu\bar\theta, \theta, \bar\theta). \tag{4.14}$$

In the L-representation the charges are given by

$$\begin{aligned} Q_L &= \partial_\theta \equiv \frac{\partial}{\partial\theta} \\ \bar Q_L &= -\partial_{\bar\theta} + 2i\theta\sigma_\mu\partial^\mu \end{aligned} \tag{4.15}$$

and we have covariant derivatives

$$\begin{aligned} D_L &= \partial_\theta + 2i\sigma_\mu\bar\theta\partial^\mu \\ \bar D_L &= -\partial_{\bar\theta}. \end{aligned} \tag{4.16}$$

If $\phi(x, \theta, \bar\theta)$ is a general complex superfield a chiral superfield is defined through

$$\bar D\phi = 0. \tag{4.17}$$

Going to the L-representation we see that superfields obeying constraint (4.17) are independent of $\bar\theta$. We can now expand $\phi(x, \theta)$. Since θ_α is anticommuting the expansion in θ will terminate after three terms ($\alpha = 1, 2$)

$$\phi(x, \theta) = \varphi(x) + \theta^\alpha\psi_\alpha + \theta^\alpha\theta_\alpha F(x). \tag{4.18}$$

The field $\varphi(x)$ is a complex scalar field and $\psi(x)$ is a Weyl fermion. $F(x)$ is a complex scalar field which, as we will see later, is auxiliary and does not constitute dynamical degrees of freedom. A chiral superfield describes therefore a (spin0, spin1) multiplet. Supersymmetry transformations transform the different components into each other

$$\delta\varphi = \alpha^\beta\psi_\beta \tag{4.19}$$

$$\delta\psi_\beta = 2\alpha_\beta F(x) + 2i\sigma^\mu_{\beta\dot\alpha}\bar\alpha^{\dot\alpha}(\partial_\mu\varphi) \tag{4.20}$$

$$\delta F = -i(\partial_\mu\psi^\beta)\sigma^\mu_{\beta\dot\alpha}\bar\alpha^{\dot\alpha}. \tag{4.21}$$

They transform lower components to higher ones $\varphi \rightarrow \psi \rightarrow F$ and in addition we have terms with derivatives of lower component fields. Observe that the variation of F is a total derivative; this will be important when we try to construct Lagrangians. Before, however, we have also to introduce multiplets that contain spin 1 fields. They are contained in a general superfield $\phi(x, \theta, \bar\theta)$ that fulfills the reality condition $\phi^* = \phi$. I will not write this superfield in full generality but only a subset of its components.

They are sufficient for our purposes because all spin 1 bosons we will introduce in the examples under consideration are gauge bosons. For the general vector superfield we can then describe the physical states in a special gauge (the Wess Zumino gauge)

$$V(x,\theta,\bar{\theta}) = -\theta\sigma_\mu\bar{\theta}V^\mu + i\theta\theta\bar{\theta}\bar{\lambda} - i\bar{\theta}\bar{\theta}\theta\lambda + \frac{1}{2}\theta\theta\bar{\theta}\bar{\theta}D \qquad (4.22)$$

where V^μ is a spin 1 field and λ is a two component spinor: the gauge fermion. D again will turn out to be an auxiliary field. This then supplies us with a (spin 1, spin 1/2) supermultiplet. The supersymmetry transformations are given by

$$\delta V_\mu = i\alpha\sigma_\mu\bar{\lambda} + i\bar{\alpha}\sigma_\mu\lambda \qquad (4.23)$$

$$\delta\lambda = \alpha\sigma^{\mu\nu}(\partial_\mu V_\nu - \partial_\nu V_\mu) + \bar{\alpha}D \qquad (4.24)$$

$$\delta D = -\alpha\sigma^\mu\partial_\mu\bar{\lambda} + \bar{\alpha}\sigma^\mu\partial_\mu\lambda \qquad (4.25)$$

Again, the variation of the highest component is a total derivative. To construct manifestly invariant actions we can make use of this fact. If we take the $F(D)$ component of a (composite) chiral (vector) superfield as a density L the action

$$S = \int \mathrm{d}^4x L \qquad (4.26)$$

will be invariant under supersymmetry transformations. Let me give some examples. If we consider a chiral superfield ϕ any power n of it will also be a chiral superfield. Take $n = 2$

$$\phi^2 = \varphi^2 + 2\varphi\theta\psi + \theta\theta[2\varphi F - \frac{1}{2}\psi\psi] \qquad (4.27)$$

and the F component is thus given by

$$(\phi^2)_F = 2\varphi F - \frac{1}{2}\psi\psi, \qquad (4.28)$$

thus $\int \mathrm{d}^4x(\phi^2)_F$ will be invariant under supersymmetry transformations. Next multiply ϕ by its conjugate. $\phi^*\phi$ is a general vector superfield. Its highest component is given by

$$(\phi^*\phi)_D = FF^* - \varphi\partial^\mu\partial_\mu\varphi^* + \frac{i}{2}\psi\sigma_\mu\partial^\mu\bar{\psi} \qquad (4.29)$$

and $\int \mathrm{d}^4x(\psi^*\psi)_D$ is invariant. The expression in (4.29) is well suited to describe kinetic terms for a complex scalar φ and a Weyl fermion ψ. In general we can therefore write the Lagrangian density as $L = L_F + L_D$ a sum of F- and D-terms. The F-terms are usually called superpotential. In a renormalizable theory the superpotential dos not contain powers of degree higher than three in the fields. Let me discuss a simple model in detail

$$L = (\phi^*\phi)_D + m(\phi^2 + \phi^{*2})_F + \lambda(\phi^3 + \phi^{*3})_F \qquad (4.30)$$

This reads in components

$$L = (\partial_\mu\varphi)(\partial\varphi^*) + \frac{i}{2}\psi\sigma_\mu\partial^\mu\bar{\psi} + FF^* +$$

$$+m(2\varphi F - \frac{1}{2}\psi\psi + \text{h.c.}) + \lambda(3\varphi^2 F - \frac{3}{2}(\psi\psi)\varphi + \text{h.c.}). \tag{4.31}$$

We see that there are no kinetic terms for F. F is an auxiliary field. It can be eliminated via the equations of motion

$$F^* + 2m\varphi + 3\lambda\varphi^2 = 0 \tag{4.32}$$

Observe that in general the auxiliary field is given by the derivative of the superpotential with respect to the scalar field

$$F^* = -\frac{\partial g(\varphi)}{\partial\varphi} \tag{4.33}$$

where we have denoted the superpotential by g. Using (4.31) and (4.32) we can write the scalar potential of the model as

$$V = FF^* = \left|\frac{\partial g}{\partial\varphi}\right|^2 = |2m\varphi + 3\lambda\varphi^2|^2 \tag{4.34}$$

and we see that the potential is semipositive definite as required by the algebra. We also know from there that supersymmetry requires vanishing vacuum energy. From $V = FF^*$ we then conclude that supersymmetry is spontaneously broken if and only if an auxiliary field receives a vacuum expectation value. We will come back to this point in a moment.

Let me first give some formulas for supersymmetric gauge theories. The relevant superfield to consider is $V \sim (V_\mu, \lambda, D)$. One defines

$$W_\alpha = \bar{D}\bar{D}[\exp(-gV)D_\alpha \exp(gV)] \tag{4.35}$$

a spinorial chiral superfield (g now is the gauge coupling). The gauge kinetic terms can then be given by

$$L = \frac{1}{2g^2}(W^\alpha W_\alpha)_F = -\frac{1}{4}G^{\mu\nu}G_{\mu\nu} + \frac{1}{2}D^2 -$$

$$-\frac{i}{2}[\lambda\sigma_\mu(\partial^\mu\bar{\lambda} + ig[V^\mu, \bar{\lambda}]) - (\partial^\mu\bar{\lambda} + ig[V^\mu, \bar{\lambda}])\sigma_\mu\lambda] \tag{4.36}$$

where the field strength is given as usual by

$$G_{\mu\nu} = \partial_\mu V_\nu - \partial_\nu V_\mu + ig[V_\mu, V_\nu]. \tag{4.37}$$

Observe that there are no kinetic terms for D, the auxiliary field. The minimal coupling of matter to the gauge system is given by

$$(\phi^* \exp(2gV)\phi)_D = |D_\mu\varphi|^2 - \frac{i}{2}\bar{\psi}\sigma_\mu D^\mu\psi +$$

$$+g\varphi^* D\varphi + ig[\varphi^*(\lambda\psi) - (\bar{\lambda}\bar{\psi})\varphi] + FF^* \tag{4.38}$$

where $D_\mu = \partial_\mu + igV_\mu$ is the gauge covariant derivative. Let us now discuss supersymmetry breakdown. Remember that the order parameter is the vacuum energy. The general form of the scalar potential is given by the auxiliary fields

$$V = F_i F_i^* + \frac{1}{2}D^2. \tag{4.39}$$

Since D is real we always have $V \geq 0$ and supersymmetry is broken if and only if auxiliary fields receive a vev. The fermionic partner of this auxiliary field is the goldstone fermion as we can see from the supersymmetry transformations (4.19-21) and (4.23-25) e.g.

$$\delta\psi = \{Q, \psi\} = F + \ldots. \tag{4.40}$$

Let us now discuss a special example in detail

$$L = (XX^*)_D + (YY^*)_D + (ZZ^*)_D + ([\lambda X(Z^2 - M^2) + gYZ]_F + \text{h.c.}). \tag{4.41}$$

We first derive the equations of motion for the auxiliary field $F_i^* = -\partial g/\partial\varphi_i$ and obtain

$$\begin{aligned} F_X^* &= -\lambda(z^2 - M^2) \\ F_Y^* &= -gz \\ F_Z^* &= -gy - 2\lambda xz \end{aligned} \tag{4.42}$$

Supersymmetry would be unbroken if the equations $F_X = F_Y = F_Z = 0$ have a solution. This is not the case here. $F_X = 0$ requires $< z >= M$ and this implies $F_Y \neq 0$. Thus supersymmetry is broken and the vacuum energy is strictly positive. To find the groundstate we have to minimize the potential

$$V = \lambda^2|z^2 - M^2|^2 + g^2|z|^2 + |gy + 2\lambda xz|^2. \tag{4.43}$$

Let me suppose $M^2 < g^2/2\lambda^2$. The minimum of the potential is then found at $z = y = 0$, $F_X = \lambda M^2$ and $E_{\text{vac}} = \lambda^2 M^4$. The vev of x is undetermined. The fermion ψ_X is massless; it is the goldstone fermion of spontaneously broken supersymmetry; the partner of that auxiliary field F_X that has received a vev. The fermions ψ_Y and ψ_Z pair up to give a Dirac fermion of mass g. For the bosons x is massless and y has mass g. Only the complex scalar z shows a splitting. The two real scalar have (mass)^2

$$\begin{aligned} &g^2 + 2\lambda^2 M^2 \\ &g^2 - 2\lambda^2 M^2 \end{aligned} \tag{4.44}$$

This is the only place where we see in the spectrum that supersymmetry is broken. In general the splitting of the multiplet is determined by its coupling to the goldstone fermion and in the model at hand only the Z-multiplet couples to the goldstino multiplet at the tree graph level through the term λXZ^2 in the superpotential. In particular this always implies that the fermions do not feel the supersymmetry breakdown at the tree level. In the following we will denote by M_S the supersymmetry breakdown scale as the vev of the auxiliary field; here $M_S^2 = F_X = \lambda M^2$. The vacuum energy will then be given by M_S and the splittings of the multiplets are given by

$$\Delta m^2 \sim \lambda M_S^2 \tag{4.45}$$

where λ denotes the coupling of the corresponding multiplet to the goldstino multiplet.

Another striking property of spontaneous supersymmetry breakdown can be read off from formula (4.44). The sum of the masses of the two real scalars is the same as in the case of unbroken supersymmetry and this is in fact a very general property of tree level mass splittings in theories with spontaneously broken supersymmetry. It is most conveniently expressed in terms of the supertrace of the (mass)^2-matrix

$$\mathrm{STr}m^2 = \sum_J (-1)^{2J}(2J+1)m_J^2 \tag{4.46}$$

where J denotes the spin and m_J is the mass matrix of the states with spin J. In the model discussed before, even after the breakdown of supersymmetry [17] we have

$$\mathrm{STr}m^2 = 0. \tag{4.47}$$

As we have seen, fermion masses are not affected by the breakdown and the splitting in the scalar sector is given by

$$\begin{aligned} m_1^2 &= m_F^2 + \Delta \\ m_2^2 &= m_F^2 - \Delta \end{aligned} \tag{4.48}$$

The result in (4.47) is a very general result of spontaneously broken global supersymmetry, there is only one exception in the presence of a U(1) gauge group with nonvanishing trace of the charge matrix. Let us discuss this as our next example. Consider

$$L = \frac{1}{2g^2}[W^\alpha W_\alpha]_F + [\phi^* \exp(2gV)\phi]_D + [2\xi V]_D \tag{4.49}$$

After elimination of auxiliary fields one obtains

$$\begin{aligned} F &= 0 \\ D &= -\xi - g\varphi^*\varphi \end{aligned} \tag{4.50}$$

and the potential is given by

$$V = \frac{1}{2}D^2 = \frac{1}{2}|\xi + g\varphi^*\varphi|^2. \tag{4.51}$$

We can now distinguish two cases. Consider first $g\xi < 0$: we obtain $< D >= 0$ and supersymmetry in unbroken; but we have

$$< \varphi^*\varphi >= -\frac{\xi}{g} \neq 0 \tag{4.52}$$

and $U(1)$ is spontaneously broken. The second choice is $g\xi > 0$ and we obtain $< D > \neq 0$ and supersymmetry is spontaneously broken [18]. In the first case a supersymmetric version of the Higgs-effect takes place. The massless gauge boson and the complex scalar combine to give a massive gauge boson and a real scalar of the same mass. In addition through the term

$$g(\varphi^*(\lambda\psi) - \varphi(\bar\lambda\bar\psi)) \tag{4.53}$$

also λ and ψ combine to a Dirac fermion of mass $g < \varphi >$, degenerate with the mass of the bosons.

In the second case, supersymmetry is broken and $< D >= \xi$. λ is the goldstino and the gauge boson remains massless since $U(1)$ is unbroken. Only φ couples to λ and

D and feels the breakdown of supersymmetry. We obtain $m_\varphi^2 = g\xi$ and $m_\psi = 0$. The supertrace of the mass matrix is therefore in general nonvanishing

$$\mathrm{STr}m^2 = 2\mathrm{Tr}Q < D >, \tag{4.54}$$

a counterexample to (4.46).

Supersymmetric field theories have better ultraviolet behaviour than usual field theories. We have seen that already in our discussion of quadratic divergencies in chapter 3. But not only the quadratic divergencies are absent, we found $\delta m^2 = 0$; not even a finite contribution in the limit of unbroken supersymmetry. The improved ultraviolet behaviour manifests itself in socalled nonrenormalization theorems [19]. They imply that masses as bilinear terms in the superpotential are not renormalized in perturbation theory. Previously it was believed that all terms in the superpotential would be stable under radiative corrections, since it can be shown that all contributions from loops involve an integration over full superspace $\int \mathrm{d}^2\theta \mathrm{d}^2\bar{\theta}$. Recently it has been observed, however, that in the presence of massless field the trilinear terms could receive finite corrections [20]. Theories with N-extended supersymmetries ($N > 1$ supercharges) have less and less divergencies, $N = 4$ theories are finite.

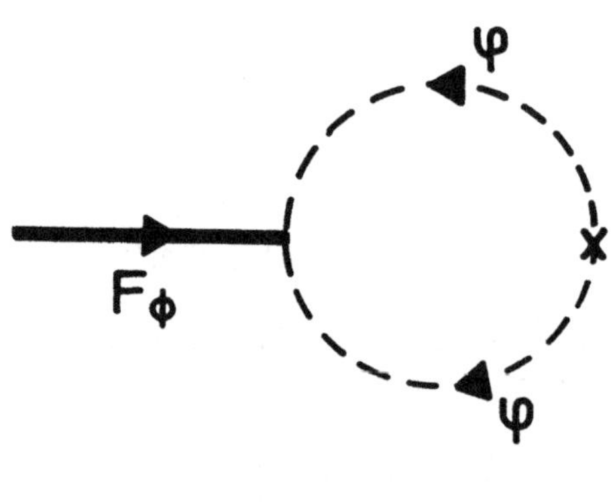

Fig. 4.1: A potential contribution to the F_ϕ-tadpole

These nonrenormalization theorems are crucial for the survival of supersymmetry in perturbation theory*. To illustrate this fact consider our example with superpotential

$$\lambda X Z^2 + gYZ - \lambda M^2 X. \tag{4.55}$$

If we choose $M^2 = 0$ we have unbroken supersymmetry. If a coefficient M^2 could be generated in perturbation theory, this would lead to a breakdown of supersymmetry. Such a term, however, is not generated. Consider e.g. the Wess-Zumino model $g =$

* Apparently [21] a finite correction of the trilinear terms in the superpotential does not invalidate this argumentation.

$m\phi^2 + \lambda\phi^3$ and compute the coefficient of $\int d^2\theta\phi = F_\phi$ from the graph given in Fig. 4.1. The coupling comes from the term

$$3\lambda(F\varphi^2 + F^*\varphi^{*2}). \tag{4.56}$$

Let us write new $F_\phi = f + ig$ and $\varphi = a + ib$. (4.56) then leads to $6\lambda[f(a^2 - b^2) - 2gab]$. In addition we have to consider the mass insertion $4m^2\varphi^*\varphi = 4m^2(a^2 + b^2)$. This implies that the graph in 4.1 does not give a contribution to g since the mass insertion does not contain an ab-term. For f, however, we have two contributions, from the coupling to a^2 and to b^2. They cancel because of the opposite sign of the fa^2 and fb^2 couplings and therefore an F_ϕ counterterm is not produced in perturbation theory.

If, however, supersymmetry is spontaneously broken, these nonrenormalization theorems do no longer hold and there are in general finite corrections proportional to the mass scale of supersymmetry breakdown. In the above example, this would happen if the masses of a a and b were not degenerate. Our prime motivation to consider supersymmetry in connection with the standard model is the absence of quadratic divergencies. They are absent even in the case of spontaneously broken supersymmetry. Actually we have to be a bit more careful. The coefficient of a Fayet-Iliopoulos term $\int d^4\theta V$ is in general quadratically divergent and we have to demand $\sum_i Q_i = 0$ where Q_i are the charges of the $U(1)$ gauge theory. The question remains whether quadratic divergencies are absent in an even more general framework than spontaneously broken supersymmetry. It turns out that certain so-called soft breaking terms can be included in a supersymmetric theory without reintroducing quadratic divergencies. They consist of $\varphi^*\varphi$, $\varphi^2 + \varphi^{*2}$, $\varphi^3 + \varphi^{*3}$ terms as well as a gaugino mass term $\lambda\lambda$. For a more detailed discussion see ref. [22].

5. THE PARTICLE CONTENT OF A SUPERSYMMETRIC STANDARD MODEL

Let us recall the particle content of the standard model. Apart from the gauge bosons G^a_μ, W^i_μ, B_μ in the adjoint representation we have quarks and leptons in three families with quantum numbers

$$\begin{aligned} Q &= \begin{pmatrix} u \\ d \end{pmatrix} = (3, 2, 1/6) \\ \bar{u} &= (\bar{3}, 1, -2/3) \\ \bar{d} &= (\bar{3}, 1, 1/3) \\ L &= \begin{pmatrix} \nu_e \\ e \end{pmatrix} = (1, 2, -1/2) \\ \bar{e} &= (1, 1, 1) \end{aligned} \tag{5.1}$$

together with a Higgs doublet

$$h = \begin{pmatrix} h^0 \\ h^- \end{pmatrix} = (1, 2, -1/2) \tag{5.2}$$

The spectrum of this model is not supersymmetric and we have to add new degrees of freedom. There are no fermions in the adjoint representation of $SU(3) \times SU(2) \times U(1)$

and we thus have to add gauge fermions (gauginos), which together with the gauge bosons form a massless vector superfield $V = (V_\mu, \lambda, D)$. Quarks and leptons require spin 0 partners in chiral superfields e.g. $\bar{E} = (\varphi_{\bar{e}}, \bar{e}, F_{\bar{e}})$ where $\varphi_{\bar{e}}$ is a complex scalar with $\bar{e}$ quantum numbers. Next observe that the lepton doublet has the same quantum numbers as the Higgs: could it be that $\varphi_e = h^-$? Unfortunately it does not work. One reason is the absence of lepton number violation and other reasons will become clear in a moment. We thus have to add scalar partners to all quarks and leptons. To the Higgs scalar we have to join the partner spin 1/2 fermions. With these fermions $SU(2) \times U(1)$ is longer anomaly free and we have to add a second Higgs chiral superfield $\bar{H} = (1, 2, +1/2)$. In short, every particle in the standard model requires a new supersymmetric partner and one has to add a second Higgs superfield.

To construct the Lagrangian we first write the kinetic terms and the gauge couplings in the usual supersymmetric way as discussed in the last chapter. We still have to discuss the superpotential which contains mass terms and the supersymmetric generalization of the Yukawa couplings. If we write the most general superpotential consistent with the symmetries and renormalizability it will contain two sets of terms

$$g = g_{\mathbf{wanted}} + g_{\mathbf{unwanted}}. \tag{5.3}$$

Let me first discuss the term

$$g_{\mathbf{wanted}} = \mu H \bar{H} + g_E^{ij} L_i^a H^b \epsilon_{ab} \bar{E}_j + g_D^{ij} Q_i^a H^b \epsilon_{ab} \bar{D}_j + g_U^{ij} Q_i^a \bar{H}_a \bar{U}_j \tag{5.4}$$

where $i, j = 1, \ldots 3$ is a family index and a, b are $SU(2)$ indices (colour indices are suppressed). It is not really clear whether we want μ from a theoretical point of view but we need it to break certain global symmetries that might be problematic. I will come back to this point later. Observe that we really need two Higgs superfields to give masses to all quarks and leptons. We can here no longer couple the up-type quarks to h^* as we did in the nonsupersymmetric case. It is then also clear that in the breakdown of $SU(2) \times U(1)$ both Higges have to aquire a vev to provide masses to all quarks and leptons.

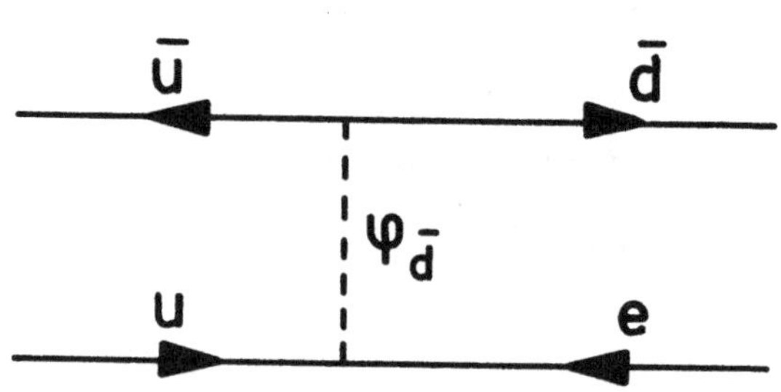

Fig. 5.1: Proton decay through exchange of the scalar partner of $\bar{d}$.

Unlike in the standard model where the requirement of gauge symmetry and renormalizability automatically led to baryon and lepton number conservation we are here not in such a nice situation. This comes from the fact that the Higgs and the lepton doublet superfields have the same $SU(3) \times SU2) \times U(1)$ quantum numbers. Consequently we have additional terms in (5.3) that we can write as (forgetting family indices)

$$g_{\text{unwanted}} = Q^a L^b \epsilon_{ab} \bar{D} + L^a \bar{E} L^b \epsilon_{ab} + \bar{U}\bar{D}\bar{D}. \tag{5.5}$$

These terms violate baryon and lepton number explicitly and lead to proton decay at unacceptable rates (as long as we assume the partner of the d-quark to be lighter than the grand unification scale). The terms in (5.5) have to be forbidden and we want to achieve this with help of a symmetry. We can turn the question the other way around. Suppose we drop (5.5) from the superpotential; does the symmetry increase? In fact it does. The new symmetry is a global symmetry that, however, does not commute with supersymmetry (called R-symmetry) [23]. Different components in the same supermultiplet have different charges. The concept of R-symmetry can best be explained in superspace. Suppose we have a symmetry that transforms θ to $e^{i\alpha}\theta$; so θ has charge $R = 1$. Suppose we have a chiral superfield ϕ transforming also with $R = 1$. Then it is obvious that the scalar component transforms as

$$\varphi \to e^{i\alpha}\varphi \tag{5.6}$$

with $R = 1$. But what happens to the fermion? Since $R(\phi) = 1$ we have

$$(\theta\psi) \to e^{i\alpha}(\theta\psi) \tag{5.7}$$

but the phase comes already from the θ transformation and obviously $R(\psi) = 0$. The F-component of the superfield has $R(F) = -1$. Invariance of the Lagrangian requires $\int \mathrm{d}^2\theta g$ to have $R = 0$ whereas $\mathrm{d}^2\theta$ transforms with $R = -2$. In the given example only the term ϕ^2 is allowed in the superpotential. So far our discussion of the implication of R-symmetry on chiral superfields. The vector superfield is real and consequently $R = 0$. From this we conclude

$$\begin{aligned} R(V_\mu) &= 0 \\ R(\lambda) &= 1 \end{aligned} \tag{5.8}$$

and this is a general and important statement. Gauginos transform nontrivially under any R-symmetry. The R-symmetry, in particular, forbids Majorana masses for the gauge fermions.

Let us now go back to the superpotential (5.4) and (5.5). There is an R-symmetry with e.g. $R(\theta) = 1$ and

$$\begin{aligned} R(H, \bar{H}) &= 1 \\ R(Q, L, \bar{U}, \bar{D}, \bar{E}) &= 1/2 \end{aligned} \tag{5.9}$$

which leaves g_{wanted} in (5.4) as the most general superpotential. In other words this means that if we drop the terms in (5.5) a continuous global R-symmetry appears. To forbid these terms in principle a smaller symmetry like R-parity

$$R_p = (-1)^{3B+L+2S} \tag{5.10}$$

(where B, L are baryon, lepton number and S is the spin) would be sufficient, but in general a continuous R-symmetry occurs. This continuous R-symmetry is somewhat problematic since it forbids gaugino Majorana masses and at least for the case of the gluino we have experimental evidence that its mass cannot vanish. Thus the R-symmetry has to be broken. Since only a spontaneous breakdown of this symmetry is acceptable, this then would lead to an embarrassing Goldstone boson. Actually in our case it will be an axion since the R-symmetry is anomalous [24]. This then tells us that this spontaneous breakdown cannot happen at an energy scale like 100GeV. (The only exeption are supercolor models, where the axion can be made heavy through a supercolor anomaly but I will not discuss these models here.) The breakdown scale of the R-symmetry has to be larger to make the axion invisible [25], i.e. a breakdown scale of something like 10^{10} to 10^{11}GeV. In a simple way this, however, can only be realized if also the supersymmetry breakdown scale M_S is large. Now remember that the multiplet splitting is given by $\Delta m^2 \sim gM_S^2$ where g is the coupling to the goldstino. We thus need small couplings to have the supersymmetric partners of quarks and leptons in the TeV-range to provide us with a physical cutoff that stabilizes M_W. These couplings have to be really small, compare them e.g. with the gravitational coupling constant κ. We have

$$\delta m \sim \kappa M_S^2 = M_S^2/M_P \tag{5.11}$$

which is in the TeV-region for $M_S = 10^{11}$GeV. Actually if we assume that all particles couple universally to gravity our requirement of the mass splittings implies M_S to be approximately 10^{11}GeV. It is thus natural to assume that the small coupling required from our discussion about R-symmetry is actually the gravitational coupling constant.

We consider this as a hint to include gravity in our framework. This will lead us to the local version of supersymmetry which includes gravity automatically. It will turn out that such considerations help us in lifting the restriction of the supertrace formula in (4.47) and its desastrous consequences for model building. We shall not discuss this here in detail and refer the reader to ref. [26] for a review.

Local supersymmetry [27] will also resolve the paradox with the cosmological constant in models of spontaneously broken supersymmetry. We shall see that one can have $E_{\text{vac}} = 0$ in models of spontaneously broken local supersymmetry.

6. LOCAL SUPERSYMMETRY (SUPERGRAVITY)

In local supersymmetry the transformation parameter is no longer constant but depends on space-time [27]. We have already acquired some experience in the framework of gauge symmetries: the local form of ordinary global symmetries; and for supersymmetry we proceed in the same way. In usual symmetries we had a scalar transformation parameter Λ. The requirement of local invariance then leads to the introduction of a gauge field A_μ with transformation property $\delta A_\mu = \partial_\mu \Lambda$. In supersymmetry we have a spinorial parameter ϵ_α. Local supersymmetry then requires the introduction of a gauge particle $\Psi_{\mu\alpha}$ (the gravitino) with transformation property $\delta\Psi_{\mu\alpha} = \partial_\mu\epsilon_\alpha(x)$. Thus the gauge particle of local supersymmetry is a spin 3/2 particle and for reasons that will become clear in a moment it is called the gravitino. These statements can also be made plausible when we discuss the Higgs effect. In ordinary global symmetries a spontaneous breakdown implied the existence of Goldstone bosons. In the local version these bosons then supply the gauge bosons with the missing degress of freedom to make them massive. In supersymmetry the goldstone particle is a spin 1/2 fermion. This then can provide the two degress of freedom in the transition of a massless to massive spin 3/2 particle: the super-Higgs effect.

The next point to discuss shows a conceptual difference between ordinary symmetries and supersymmetry. While in ordinary theories it was sufficient for the local symmetry to introduce a spin 1 gauge boson in supersymmetry this is not the case. The gauge particle is a spin 3/2 fermion and supersymmetry requires a bosonic partner. The construction of local supersymmetry has shown that this partner is a spin 2 boson that has to have all the properties of the graviton. This then implies that local supersymmetry necessarily includes gravity. We could have guessed that already from the algebra

$$[\epsilon(x)Q, \bar{Q}\bar{\epsilon}(x)] = 2\epsilon(x)\sigma_\mu\bar{\epsilon}(x)P^\mu. \tag{6.1}$$

On the right hand side we have a space-time translation that differs from point to point: a general coordinate transformation.

We have now to discuss explicit Lagrangians containing chiral matter and gauge fields coupled to the $(2,\frac{3}{2})$-supergravity multiplet. In general this requires a lot of tedious calculations which I shall not repeat here. Also the general form of the Lagrangian is quite lengthy and I refer to the literature where the complete Lagrangian can be found [29]. I will instead concentrate on a analysis of the scalar potential of these theories which we need for our further discussion.

Remember that in the global case the most general Lagrangian was defined by three functions of the superfields: the gauge kinetic terms W^2, the matter field kinetic terms $S(\phi^* \exp(gV)\phi)$ and the superpotential $g(\phi)$. In the local case the most general action can be defined by $f_{\alpha\beta}(\phi)W^\alpha W^\beta$ (with indices α, β labeling the adjoint representation of the gauge group) and the Kähler potential

$$G = 3\log\left(-\frac{S}{3}\right) - \log(|g|^2). \tag{6.2}$$

The kinetic terms of the scalar particles z_i are then given by

$$G^i_j D_\mu z_i D^\mu z^{j*} = \frac{\partial^2 G}{\partial z_i \partial z^{j*}} D_\mu z_i D^\mu z^{j*} \tag{6.3}$$

where z_i is the lowest component of a chiral superfield ϕ_i. The scalar potential reads

$$V = -\exp(-G)[3 + G_k (G^{-1})^k_l G^l] + \frac{1}{2} f^{-1}_{\alpha\beta} D^\alpha D^\beta. \tag{6.4}$$

In these lectures I will use what is called minimal kinetic terms

$$G^i_j = -\delta^i_j. \tag{6.5}$$

This simplifies our formulas considerably and allows us nonetheless to see all the essential properties of the potential. The Kähler potential can therefore be written as

$$G = -\frac{z_i z^{i*}}{M^2} - \log \frac{|g|^2}{M^6} \tag{6.6}$$

where we have explicitly written out the mass scale M related to the gravitational coupling constant κ:

$$M = \frac{1}{\kappa} = \frac{M_{\text{Planck}}}{\sqrt{8\pi}} \approx 2.4 \times 10^{18}\text{GeV}. \tag{6.7}$$

The first derivative of the Kähler potential is then given by

$$G^i = -\frac{z^{i*}}{M^2} - \frac{g^i(z_i)}{g(z)} \tag{6.8}$$

and we can rewrite the potential in terms of the superpotential $g(z)$ as

$$V = \exp\left(\frac{z_i z^{i*}}{M^2}\right)\left[\left|g^i + \frac{z^{i*}}{M^2} g\right|^2 - \frac{3}{M^2}|g|^2\right]. \tag{6.9}$$

Contrary to the case of global supersymmetry the potential is no longer semipositive definite. I still have to tell you under which conditions supersymmetry is spontaneously broken. As in the global case this breakdown is signaled by a vacuum expectation value of an auxiliary field. There we had (compare (4.33)) the auxiliary field F given as the derivative of the superpotential; here we have an additional term

$$F^i = g^i + \frac{z^{i*}}{M^2} g \tag{6.10}$$

where in the limit $M \to \infty$ we recover the global result. Supergravity is now spontaneously broken if and only if an auxiliary field receives a vev. The supergravity breakdown scale is found to be

$$M_S^2 = < F > \exp\left(\frac{z_i z^{i*}}{M^2}\right). \tag{6.11}$$

Observe that the vacuum energy is no longer an order parameter. We can have unbroken supergravity with $E_{\text{vac}} < 0$ (Anti de Sitter) or $E_{\text{vac}} = 0$ (Poincare supersymmetry) and $E_{\text{vac}} > 0$ always implies broken supergravity. The most important observation, however, is that we can have broken supergravity with vanishing vacuum energy (cosmological constant), a situation that could not occur in the framework of global supersymmetry. Here we need

$$\sum_i F^i F_i^* = \frac{3}{M^2}|g|^2 \tag{6.12}$$

and we will assume this to be fulfilled. In all cases I know of, this is an ad hoc adjustment of the cosmological constant to zero. If (6.12) is fulfilled and if $M_S \neq 0$ the gravitino becomes massive through the super-Higgs effect

$$m_{3/2} = M\exp(-G/2) = \frac{g}{M^2}\exp\left(\frac{z_i z^{i*}}{M^2}\right) \tag{6.13}$$

and we therefore have the relation

$$m_{3/2} = \frac{M_S^2}{\sqrt{3}M} \tag{6.14}$$

valid in the case of vanishing cosmological constant.

Let us now discuss some simple specific models with spontaneous supersymmetry breakdown. As a warm-up example consider one field z and a constant superpotential $g = m^3$. The potential is then given by

$$V = m^6 \exp\left(\frac{zz^*}{M^2}\right)\left[\frac{|z|^2}{M^4} - \frac{3}{M^2}\right] \tag{6.15}$$

which has stationary points at $z = 0$ and $|z| = \sqrt{2}M$. At $z = 0$ supersymmetry is unbroken but this is a local maximum of the potential. The minima with broken supersymmetry and $E_{\text{vac}} < 0$ are at $z = \pm\sqrt{2}M$.

Next we want to give an example with broken supersymmetry and $E_{\text{vac}} = 0$. We consider a superpotential

$$g(z) = m^2(z+\beta) \tag{6.16}$$

A nonvanishing vev of

$$F = \frac{\partial g}{\partial z} + \frac{z^*}{M^2}g = m^2\left(1 + \frac{z^*(z+\beta)}{M^2}\right) \tag{6.17}$$

would signal a spontaneous breakdown of supergravity. The equation

$$M^2 + zz^* + z^*\beta = 0 \tag{6.18}$$

has the solutions

$$z = -\frac{\beta}{2} \pm \frac{1}{2}\sqrt{\beta^2 - 4M^2}. \tag{6.19}$$

Since (6.18) only allows real solutions (we assume β to be real) (6.19) implies that supersymmetry is broken as long as $\beta < 2M$. Thus we can arrange for a supersymmetry breakdown but we still have the annoying task to fine tune the vacuum energy. Let us therefore first consider the case $\beta = 0$ in which the potential is proportional to

$$(M^2 + |z|^2)^2 - 3M^2|z|^2 \tag{6.20}$$

which is positive definite with minimum at $z = 0$. Increasing β implies decreasing the vacuum energy and also z aquires a nonvanishing vev. We can now increase β until the potential just touches zero. This is found to happen at $\beta = (2-\sqrt{3})M$ with a vev of $(\sqrt{3}-1)M$ for the z-field. The potential now is semipositive definite with $E_{\text{vac}} = 0$ and since $|\beta| < 2M$ supersymmetry is broken and we have found the desired example. The super-Higgs effect occurs. The gravitino swallows the fermion in the chiral superfield and has a mass

$$m_{3/2} = \frac{m^2}{M} \exp\left(\frac{(\sqrt{3}-1)^2}{2}\right) \tag{6.21}$$

and the two remaining scalars have masses

$$\begin{aligned} m_1^2 &= 2\sqrt{3} m_{3/2}^2 \\ m_2^2 &= 2(2-\sqrt{3}) m_{3/2}^2 . \end{aligned} \tag{6.22}$$

Supersymmetry is broken and E_{vac} remains zero. Observe that such a situation is not possible in the framework of global supersymmetry. Observe also, that in the present example we had to perform an explicit fine-tuning to obtain $E_{\text{vac}} = 0$.

The simple example given above should serve only as existence proof for the mechanism of supergravity breakdown in the presence of a vanishing cosmological constant; it does not give a convincing argument why the scale of supersymmetry breakdown has to be in the 10^{11} GeV region. The scale is set by putting the parameter m^2 in (6.16) at this value. Much effort has been devoted to the investigation of the question about the origin of supersymmetry breakdown. Up to now the only satisfactory suggestion in this direction consists in the consideration of a nonpertubative mechanism: *supersymmetry breakdown via gaugino condensation* [30]. The history of this discussion is quite interesting [31]. Preliminary investigations with the help of an effective Lagrangian approach indicated that the formation of gaugino condensates does not break global supersymmetry [32]. Using similar methods it was first observed in [30] that such a no-go theorem is not valid in the framework of local supersymmetry and an explicit example of the spontaneous breakdown of supergravity was given. This was actually the first example of a model where supersymmetry is broken in a hidden sector that coupled only gravitationaly to the observable sector and where this breakdown of weak interactions was induced by radiative corrections. Nonetheless later most efforts were devoted to investigations of toy models like those given in (6.16).

The fact that local supersymmetry is broken while global supersymmetry remains intact has some interesting consequences. To explain this consider the simplest model: pure supersymmetric gauge theory, just a gauge theory with fermions (the gauginos) in the adjoint representations of the gauge group. Such a theory is asymptotically free, the gauge coupling becomes strong at small energies and we assume, in analogy to QCD, that this leads to confinement and that gaugino bilinears condense. The magnitude of the condensate is given by the renormalization group invariant scale

$$< \lambda\lambda > \approx \Lambda^3 \approx \mu^3 \exp\left(-\frac{3}{b_0 g^2}\right) \tag{6.23}$$

where g denotes the gauge coupling constant and b_0 is the coefficient of the one-loop β-function. If this would break global supersymmetry one would expect the scale of this breakdown to be set by Λ; but in the global limit supersymmetry remains unbroken!

Instead local supersymmetry is broken. We thus observe a suppression of the scale since the global limit is obtained as $M \rightarrow \infty$. This leads to

$$M_S^2 \sim \frac{<\lambda\lambda>}{M} \tag{6.24}$$

and a gaugino mass of order $<\lambda\lambda>/M^2$. Thus a value of Λ even exceeding 10^{11} GeV leads to observable sector mass splittings in the TeV region.

While these first arguments were all based on effective Lagrangians, the results could be confirmed once the complete $N=1$ supersymmetric Lagrangian [29] had been constructed. To see whether supersymmetry is broken or not we have to consider the auxiliary fields, now also including fermion bilinears:

$$F_i = \exp(-G/2)(G^{-1})_i^j G_j + \frac{1}{4} f_{\alpha\beta k}(G^{-1})_i^k(\lambda^\alpha\lambda^\beta) + \ldots \tag{6.25}$$

where λ^α are the gauginos, $f_{\alpha\beta}$ the socalled gauge-kinetic function that multiplies $W^\alpha W^\beta$ and $f_{\alpha\beta k} = \partial f_{\alpha\beta}/\partial z^k$. A nontrivial vev $<\lambda\lambda> \neq 0$ thus breaks supersymmetry provided that the gauge kinetic function is nontrivial [33]. The appearance of this derivative of f leads to the same relation between the scale of supersymmetry breakdown and the gaugino condensate as given in (6.24), confirming the previous results.

But there is more to (6.25) than just that. The gauge coupling constant depends on this gauge kinetic function $g^2 \sim 1/f$. A nontrival f thus implies that the value of the gauge coupling constant is not a priori an input parameter but is dynamically determined [33]. This is true in any theory where supersymmetry is broken through gaugino condensates.

With the investigation of four-dimensional superstring theories the consideration of supersymmetry breakdown via gaugino condensation became popular again [34]. Partially this is due to the fact that in such theories no other smooth and spontaneous breakdown of supersymmetry is known. But again there is more to it. First of all, string models do usually contain several sectors that are only coupled gravitationally as can be clearly seen in the heterotic $E_8 \times E_8$ theory. Secondly the hidden sector candidates contain additional gauge groups that could give rise to gaugino condensates. But this is not yet enough. Through the existence of the dilaton field and its coupling to $W^\alpha W_\alpha$ we obtain a nontrivial gauge function f in a model independent way. Thus all the necessary requirements are met to render gaugino condensation to be *the* candidate for supersymmetry breakdown in hidden sector supergravity models. More details can be found in [31].

Before we close this section let us mention another interesting development in the construction of supergravity models. Up to now we have for the sake of simplicity discussed minimal kinetic terms for the scalar fields. Models with nonmininal kinetic terms can have interesting structure. Consider e.g. (compare (6.2))

$$G = 3\log(\phi + \phi^*) - \log|g|^2 \tag{6.25}$$

and take a constant superpotential. If you compute the potential as given in (6.4) you will find that it vanishes identically. Nontheless the quantity

$$e^{-G} = \frac{|g|^2}{(\phi + \phi^*)^3} \tag{6.26}$$

does not vanish and supersymmetry is broken. Such so-called no-scale models [35] might also have applications in the low energy limit of string theories.

7. LOW ENERGY SUPERGRAVITY MODELS

As we discussed in chapter 5 we should consider models that consist of two sectors: a hidden sector and an observable sector which are only coupled weakly through gravitational interactions. The observable sector consists of the fields discussed in chapter 5 which we will collectevely denote by y_a. The hidden sector is responsible for the breakdown of supersymmetry at a scale $M_S \sim 10^{11}$GeV and leads to a gravitino mass in the TeV region. Its fields will be denoted by z_i and we chose a superpotential

$$\tilde{g}(z_i, y_a) = h(z_i) + g(y_a). \tag{7.1}$$

Let us parametrize a general hidden sector by assuming that at the minimum

$$\begin{aligned} &< z_i > = b_i M \\ &< h > = mM^2 \\ &< h_i > = < \partial h / \partial z_i > = a_i^* mM \end{aligned} \tag{7.2}$$

while all abserable sector fields y_a should have vanishing vev's. In the simple example of last chapter we had $b = \sqrt{3} - 1$. The potential is given by

$$V = \exp\left(\frac{|z_i|^2 + |y_a|^2}{M^2}\right) \left[\left| h_i + \frac{z_i^* \tilde{g}}{M^2} \right|^2 + \left| g_a + \frac{y_a^* \tilde{g}}{M^2} \right|^2 - \frac{3}{M^2} |\tilde{g}|^2 \right]. \tag{7.3}$$

The vacuum energy vanishes provided that

$$\sum_i |a_i + b_i|^2 = 3 \tag{7.4}$$

and the gravitino mass is given by

$$m_{3/2} = \exp\left(\frac{1}{2}|b_i|^2\right) m, \tag{7.5}$$

thus m sets the scale of the gravitino mass. We furthermore define [36]

$$A = b_i^*(a_i + b_i) \tag{7.6}$$

which will turn out to be an important parameter besides the gravitino mass. In the previous example we had $A = 3 - \sqrt{3}$. The potential given in (7.3) is complicated but we have $m \ll M$ and we can simplify the expressions enormously by neglecting subleading terms. Formally this means that we take the limit $M \to \infty$ keeping, however, $m_{3/2}$ fixed. We then replace the hidden sector fields by there vev's and obtain the following potential for the observable sector fields

$$v = \left|\frac{\partial g}{\partial y_a}\right|^2 + m_{3/2}^2 \left|y_a\right|^2 + m_{3/2}\left[y_a \frac{\partial g}{\partial y_a} + (A-3)g + \text{h.c.}\right] \tag{7.7}$$

Thus the spontaneous breakdown of supergravity in the hidden sector manifests itself as explicit breakdown of global supersymmetry in the low energy limit of the observable sector. The first term in (7.7) is the usual potential of a globally supersymmetric theory while the other terms are soft breaking terms.

The second term gives universal scalar masses to all the partners of quarks and leptons. The supertrace formula is here given in general by [29]

$$\text{STr}M^2 = 2(N-1)m_{3/2}^2 \tag{7.8}$$

where N is the number of chiral superfields. This avoids the mass relations obtained in the globally supersymmetric models and its desastrous consequences for model building. The universality property of the mass terms is needed to ensure the absence of flavour changing neutral currents. It appears here because of the choice of minimal kinetic terms for the scalar fields.

The term $(A-3)g$ is of equal importance since it breaks all R-symmetries of the model. This implies that there are no problems with potential axions and that also gaugino Majorana masses are allowed (recall our discussion in chapter 5). This breakdown of R-symmetry is a direct consequence of the coupling to gravity.

One more technical remark. In general we will deal with a superpotential $g = g_3 + g_2$ where g_3 denotes the trilinear and g_2 the bilinear terms. The last term in (7.7) then reads $Am_{3/2}g_3 + (A-1)m_{3/2}g_2$. Apart grom the gaugino mass m_0 we find that $m_{3/2}$ and A are the important parameters parametrizing the effects of supersymmetry breakdown in this class of models.

Let us now specify the superpotential

$$g = \mu H\bar{H} + g_E HL\bar{E} + g_D HQ\bar{D} + g_U \bar{H}Q\bar{U} \tag{7.9}$$

The parameter μ has to be different from zero since otherwise we would have problems with a light higgsino (the supersymmetric partner of the Higgs-scalar) or axions. The value of μ is not directly related to the supersymmetry breakdown scale but one can construct models where μ is related to $m_{3/2}$ and we shall assume that also μ is in the TeV range [37].

Let us now address the question of $SU(2) \times U(1)$ breakdown. We have two Higgs multiplets and members of both have to receive nonvanishing vev's to give masses to all quarks and leptons, according to (7.9). The Higgs potential reads [38] (omitting the off-diagonal $SU(2)$ D-terms which are irrelevant for the discussion of the minimum)

$$V = m_1^2|h|^2 + m_2^2|\bar{h}|^2 + m_3^2(h\bar{h} + h^*\bar{h}^*) + \frac{g_1^2 + g_2^2}{8}(|h|^2 - |\bar{h}|^2)^2 \tag{7.10}$$

where the last term corresponds to the $SU(2) \times U(1)$ D-term and g_2 and g_1 denote the respective coupling constants. From (7.7) and (7.9) we obtain

$$\begin{aligned} m_1^2 &= m_2^2 = m_{3/2}^2 + \mu^2 \\ m_3^2 &= -B\mu m_{3/2} \\ B &= A - 1 \end{aligned} \tag{7.11}$$

The potential consists of quadratic and quartic terms. The quartic terms have a positive coefficient such that the potential at infinity is well behaved, with the exception, however, of certain flat directions for $|h| = |\bar{h}|$. To have the potential bounded from below we therefore have to impose a constraint on the coefficients of the quadratic terms

$$m_1^2 + m_2^2 \geq 2|m_3^2|. \tag{7.12}$$

Next we have to discuss the requirement of $SU(2) \times U(1)$ breakdown. Since there are no trilinear terms in (7.10) a stationary point at $h = \bar{h} = 0$ has to be unstable, i.e. the mass matrix at this point has to have a negative eigenvalue. The requirement for a nontrivial $SU(2) \times U(1)$ breaking absolute minimum is therefore

$$|m_3|^4 \geq m_1^2 m_2^2. \tag{7.13}$$

With the input parameters (7.11) we can observe now that the constraints (7.12) and (7.13) can only be fulfilled in the limiting case

$$m_{3/2}^2 + \mu^2 = B\mu m_{3/2} \tag{7.14}$$

i.e. at most we can arrive at a flat direction where $SU(2) \times U(1)$ breaking and non-breaking minima are degenerate. We would then have to look for radiative corrections to see whether $SU(2) \times U(1)$ breaking minima can be reached at all within this approach. This is actually a nice feature of the model. It tells you again that we have not put in $SU(2) \times U(1)$ breaking by hand. This breakdown will be intrinsically related to the supersymmetry breakdown and the dynamics of the model. But we still have to see whether it works. In addition we have to observe that all our input parameters are defined at a very large scale M. The value of the parameters in the 100 GeV region has still to be computed using renormalization group improved perturbation theory in the same way as we have to compute the evolution of the gauge coupling constants in a grand unified model. This we would have also to do if our input parameters would already allow a $SU(2) \times U(1)$ breakdown at the tree level. In the evolution from M to 100 GeV the parameters will change substantially and it would not be clear at all whether $SU(2) \times U(1)$ could not be restored. Before we do this calculation, however, let me give you a simple argument how an $SU(2) \times U(1)$ breakdown can be induced by radiative corrections [39]. This argument is not complete and has later to be backed up by the real calculation but it exhibits the essential points of the mechanism quite nicely.

Let us therefore look at the radiative corrections to the masses of the scalar particles. One way to see whether there is a chance to have $SU(2) \times U(1)$ breakdown is to see whether a m^2 of a Higgs scalar can become negative and at the same time the m^2 of all others scalars in the theory should remain positive. The contributions to the masses of the scalar particles can be classified as in Fig. 7.1. The sum of the two gauge

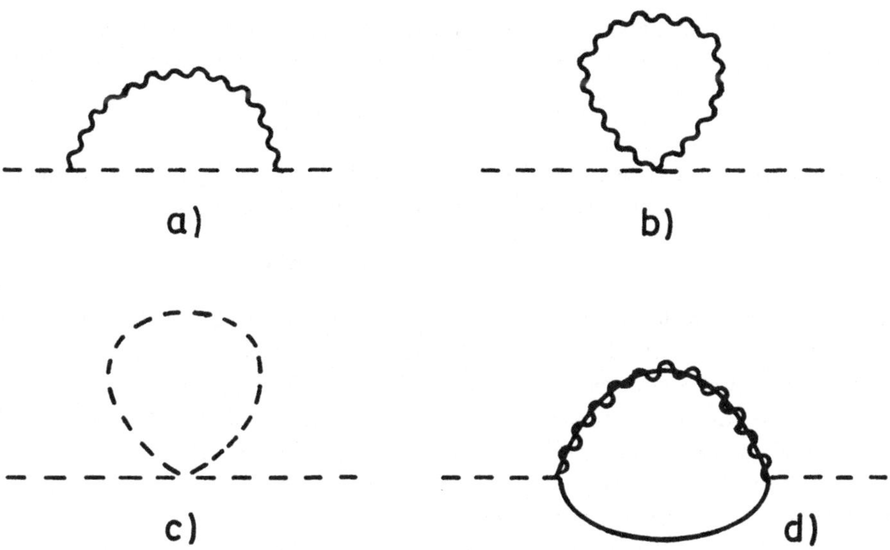

Fig. 7.1: Corrections to scalar masses. Wavy, dashed, solid and wavy-solid lines correspond to gauge bosons, scalars, fermions and gauginos respectively.

contributions as well as the scalar selfinteraction contribution are positive. In the supersymmetric limit these contributions are exactly cancelled by the remaining graph in Fig. 7.1 which contains a gauge fermion exchange. So if supersymmetry is exact nothing happens. Let us now suppose that supersymmetry is broken and for definiteness let us take m_0 (a gauge fermion mass) as the only source of supersymmetry breakdown. With this mass the contribution of the last graph will be suppressed and the cancellation will no longer be complete. Since a negative contribution is suppressed all scalar particles in the theory will receive a positive contribution to their m^2. For the partners of quarks the dominant contribution comes from the strong interactions

$$\delta m^2 \sim \alpha_3 m_0^2 \tag{7.15}$$

where α_3 denotes the $SU(3)$ coupling constant. Higges and partners of leptons will receive a smaller contribution

$$\delta m^2 \sim \alpha_W m_0^2 \tag{7.16}$$

where α_W denotes a combination of $SU(2) \times U(1)$ coupling constants. No indication for an induced $SU(2) \times U(1)$ breakdown whatsoever. But let us look more closely. With these corrections in particular the partners of quarks become heavy and this will reduce the contribution of the graph with the scalar quadrilinear interaction. This now also gives us an asymmetry between the mass corrections for Higgses and for the scalar partners of leptons. Going back to (7.7) and (7.9) we see that the partners of quarks couple to the Higgses but not to the partners of leptons. The suppression of this positive contribution then gives a negative contribution to the Higgs mass and in total we have

$$\delta m^2 \sim \alpha_W m_0^2 - \alpha_Y(\alpha_3 m_0^2) \tag{7.17}$$

where α_Y denotes the Yukawa coupling responsible for the Higgs-squark quartic interaction. The corresponding quantity for the partners of leptons is

$$\delta m^2 \sim \alpha_W m_0^2 - \alpha_Y(\alpha_W m_0^2) \tag{7.18}$$

and will under reasonable circumstances stay positive and the partner of quarks were already heavy enough in the first case. With a large enough Yukawa coupling (but still small enough to trust perturbation theory) the contributions in (7.17) could become negative and thus induce a breakdown of the weak interactions. A candidate for such a Yukawa coupling would be the one that is responsible for the mass of the heaviest particle in the model: the top quark. We can thus conclude two things from this simplified discussion of radiative corrections

1) $SU(2) \times U(1)$ breakdown can in principle be induced in a desirable way. Observe that it is nontrivial to have the situation that only the m^2 of a Higgs could become negative while other scalars keep positive m^2, i.e. everything could go wrong but it does not.
2) The mass of the top-quark $m_{\rm top}$, i.e. the top-quark Yukawa coupling is a crucial input parameter in the model (unlike in the standard model where it just parametrizes $m_{\rm top}$). Knowing $m_{\rm top}$ would tell us a lot more about the model and its predictions.

So far our quantitative discussion of these issues. We have now to go on and compute. As we said already, the parameters in (7.11) are defined at a large scale and we have to compute their evolution down to a scale of something like 100 GeV to discuss $SU(2) \times U(1)$ breakdown. For the whole model this then requires the integration of some 25 coupled renormalization group equations [40] and you can imagine that this cannot be done analytically. Let us first specify our boundary conditions. We should actually start our evolution at a scale $\tilde{\mu} = M \sim 2 \times 10^{18}$GeV, but we do not really know the spectrum of this theory at such large scales. There could be a grand unified sector and this could change the results. Let us therefore assume that the input parameters as given in (7.11) are valid at a scale $\tilde{\mu} = M_x \sim 3 \times 10^{16}$GeV where M_x is the grand unification scale in our model. If we assume that below this scale our model gives the complete spectrum we observe that the $SU(3) \times SU(2) \times U(1)$ couplings constants g_3, g_2 and g_1, once properly chosen at a scale $\tilde{\mu} \sim 100$ GeV meet at the scale $\tilde{\mu} = M_x$ with magnitude $\alpha(M_x) \sim 1/24$. This we will then take as our starting point. Let us now look at some of the renormalization group equations more closely and the most important ones are certainly those for the masses of the scalars. Here I will give you the equations for $\bar{h}$, φ_t and $\varphi_{\bar{t}}$ (the partners of t and $\bar{t}$-quark) in the approximation that only the top quark Yukawa coupling (g_t) is different from zero. In the calculation the effects of the other couplings have also been included but here we give just the simplified expressions

$$\begin{aligned} \tilde{\mu}\frac{\partial}{\partial\tilde{\mu}} m_{\bar{h}}^2 = & \frac{3}{8\pi^2} g_t^2 (m_{\bar{h}}^2 + m_{\varphi_t}^2 + m_{\varphi_{\bar{t}}}^2 + m_{3/2}^2 |A_t|^2 \\ & - \frac{1}{2\pi^2}[\frac{3}{4}|\tilde{m}_2|^2 g_2^2 + \frac{1}{4}|\tilde{m}_1|^2 g_1^2] \end{aligned} \tag{7.19}$$

$$\tilde{\mu}\frac{\partial}{\partial\tilde{\mu}}m^2_{\varphi_{\bar{t}}} = \frac{2}{8\pi^2}g_t^2(m^2_{\bar{h}} + m^2_{\varphi_t} + m^2_{\varphi_{\bar{t}}} + m^2_{3/2}|A_t|^2 - \frac{1}{2\pi^2}[\frac{4}{3}|\tilde{m}_3|^2 g_3^2 + \frac{4}{9}|\tilde{m}_1|^2 g_1^2] \tag{7.20}$$

$$\tilde{\mu}\frac{\partial}{\partial\tilde{\mu}}m^2_{\varphi_t} = \frac{1}{8\pi^2}g_t^2(m^2_{\bar{h}} + m^2_{\varphi_t} + m^2_{\varphi_{\bar{t}}} + m^2_{3/2}|A_t|^2 - \frac{1}{2\pi^2}[\frac{4}{3}|\tilde{m}_3|^2 g_3^2 + \frac{3}{4}|\tilde{m}_2|^2 g_2^2 + \frac{1}{36}|\tilde{m}_1|^2 g_1^2] \tag{7.21}$$

where the $\tilde{m}_i$ denote the $SU(3)\times SU(2)\times U(1)$ gaugino masses and A_t is the A parameter that comes with the term in the superpotential that contains the top-quark scalar. Observe that although we started with these A's to be the same they will no longer stay universal once we include the radiative corrections.

We can now look more closely at these equations and recover the qualitative behaviour we found in the simple example discussed above. The first term in (7.19) has a positive sign. This implies that the mass of $\bar{h}$ decreases if we lower $\tilde{\mu}$ from M_x down to 100 GeV. If the top quark mass (i.e. g_t) is large enough we could even imagine $m^2_{\bar{h}}$ to become negative, a sufficient (but not necessary) condition to have spontaneously broken $SU(2)\times U(1)$. But a lot of things could go wrong. The evolution equations of the partners of the top quark also have this first term with a positive coefficient and the m^2 of these particles should remain positive to keep $SU(3)_{\mathrm{colour}}\times U(1)_{\mathrm{e.m.}}$ unbroken. The reason why the model works is that the coefficients of these terms are $3:2:1$ in (7.19-21). Observe that these coefficients are not parameters which we can choose freely. They are an intrinsic property of the model and if they would have come out differently (like e.g. $1:2:3$) there would be no way for this model to be correct. The second terms in (7.19-21) depend on the effects of gaugino masses and our discussion up to now is only valid for $m_0 = 0$. Let us now include them. They have a negative coefficient and increase m^2 with decreasing $\tilde{\mu}$. At first sight they therefore do not favor an induced breakdown of $SU(2)\times U(1)$ (remember our simple example above). But indirectly they help. The terms in (7.20) and (7.21) contain the gluino contribution with the strong coupling constant g_3 which is not present in (7.19). When we now lower $\tilde{\mu}$ this could give a big contribution to $m^2_{\varphi_{t,\bar{t}}}$ but not to $m^2_{\bar{h}}$. The equations, however, are coupled and these large contributions enter the first term in (7.19) and speed up the evolution of $m^2_{\bar{h}}$ to small and possibly negative values. This is exactly the behaviour we had already guessed from our simplified discussion above. The discussion, however, also shows that it will be difficult (technically) to arrive at quantitative results. We have to solve all these coupled renormalization group equations numerically i.e. equations for the gauge couplings, gaugino masses, Yukawa couplings, scalar masses, A-parameters etc.. We have also to determine e.g. the value of the Yukawa couplings at M_x such that at $\tilde{\mu}\sim 100$GeV they have the correct values to parametrize the masses of quarks and leptons.

The parameters relevant for the breakdown of $SU(2)\times U(1)$ have been identified before. They are $m_{3/2}$, μ, m_0, A and g_t. They have to fulfill one constraint to give $SU(2)\times U(1)$ breakdown with the correct value for M_W and M_Z. To give you a feeling about this relation let me first discuss a simplified situation in which $\mu = m_0 = 0$ [41].

We know already from our discussion before that a model with $\mu = 0$ has problems but here we just want to exhibit the mechanism in a simplified case and also this case indicates what will hapen in models where μ is small.

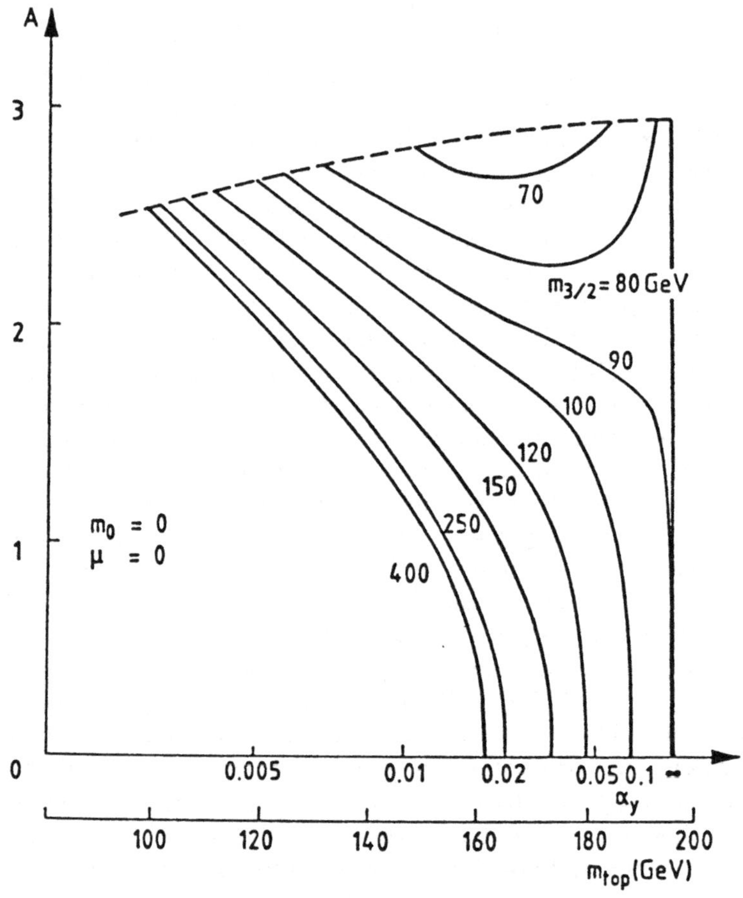

Fig. 7.2: Breakdown of $SU(2) \times U(1)$ for $\mu = m_0 = 0$ from ref. [41].

Fig. 7.2 shows the result of a numerical integration of the renormalization group equations in this simplified case $\mu = m_0 = 0$. Only a part of the parameter space in A, $m_{3/2}$ and m_{top} can lead to a breakdown of $SU(2) \times U(1)$, as we of course had to expect from our discussion above. Before we discuss the figure in detail, let me mention that with $\mu = m_0 = 0$ the sign of A is unphysical, only $|A|$ matters. The allowed region for $SU(2) \times U(1)$ breakdown is bounded at the left-hand side in Fig. 7.2, because m_{top} (i.e. g_t) is simply too small to drive $m_{\tilde{h}}^2$ negative enough to induce the breakdown. The actual value of this lower bound for m_{top} depends on A and in general large A allows smaller m_{top}. Nonetheless we obtain a large lower bound m_{top} for this configuration

$\mu = m_0 = 0$ of something like 100 GeV. We will later see that large m_{top} is usually required in models with small μ. The bound on the right and side of Fig. 7.2 cannot yet be understood from our discussion up to now. It simply comes from the evolution equation of the Yukawa coupling g_t

$$\tilde{\mu}\frac{\partial}{\partial\tilde{\mu}}g_t = \frac{3}{8\pi^2}g_t^3 - \frac{1}{8\pi^2}g_t\left(\frac{8}{3}g_3^2 + \frac{3}{2}g_2^2 + \frac{13}{18}g_1^2\right). \tag{7.22}$$

This shows that for large m_{top} we need a large Yukawa coupling at M_x. If g_t is large, however, the first term in (7.22) will be dominant. It has a positive coefficient and will reduce the magnitude of g_t when we lower $\tilde{\mu}$. This then gives us an upper bound on m_{top} of 200 GeV. Even if we choose g_t at M_x to be infinitely large the evolution (which goes with g_t^3) will reduce it to a "small" value at $\tilde{\mu} \approx 100$ GeV. Of course, the approximation on which (7.22) is based breaks down for large g_t but we see that in the minimal model it is hard and unlikely to have m_{top} to be larger than 200 GeV.

Last we have to discuss the bound in the upper part of Fig. 7.2 related to the parameter A. We have already seen that large A makes it easier to induce the breakdown of the weak interactions, i.e. the breakdown is possible for smaller values of m_{top}. This comes from the appearance of $|A|^2$ in the first term in (7.19). One could actually think that by increasing A sufficiently one could induce the $SU(2) \times U(1)$ breakdown for arbitrarily small m_{top} as long as g_tA_t stays large enough. This is true, but for large A other unpleasant things happen and we have to discuss this now in detail [42]. For this purpose consider again the potential (7.7) but for simplicity with just one Yukawa coupling:

$$V = \left|\frac{\partial g}{\partial y_a}\right|^2 + m_{3/2}^2\left(|h|^2 + |\varphi_e|^2 + |\varphi_{\bar{e}}|^2\right) + Am_{3/2}g_E\left(h\varphi_e\varphi_{\bar{e}} + \text{h.c.}\right). \tag{7.23}$$

We also dropped the $\frac{1}{2}D^2$ term because it is irrelevant for our discussion. For small A it is evident that the minimum of this potential will be at $h = \varphi_e = \varphi_{\bar{e}} = 0$ with $V = 0$. For large A the trilinear terms will dominate in a certain range and the minimum will be at $h = \varphi_e = \varphi_{\bar{e}} \neq 0$ breaking weak interactions but also electromagnetic and strong interactions and this has to be avoided. Thus A has to be small enough. In the special potential (7.23) (as well as in the general case (7.7) with universal A) the critical value is $A = 3$ because in this case the potential reads:

$$V = \left|\frac{\partial g}{\partial y_a} + m_{3/2}y_a^*\right|^2 \geq 0 \tag{7.24}$$

and we need $A \leq 3$. Including the radiative corrections in our model the A's will no longer stay universal and there will be separate bounds on A required by the absence of $SU(3) \times U(1)_{\text{e.m.}}$ breaking minima:

$$A_E^2 < 3\left(\frac{m_h^2 + m_{\varphi_e}^2 + m_{\varphi_{\bar{e}}}^2}{m_{3/2}^2}\right) \tag{7.25}$$

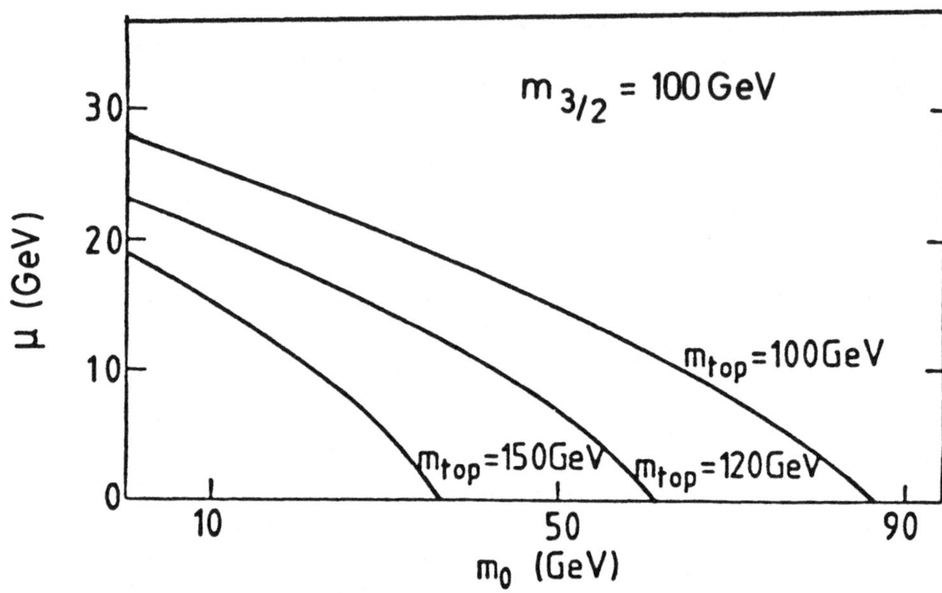

Fig. 7.3: The situation in the case of large top quark masses.

and similar expressions for A_D, A_U and all three families separately. These questions have to be carefully checked in all explicit models.

Curves similar to Fig. 7.2 can now be produced for arbitrary values of the parameters to explore the phenomenological consequences of the model. Apart from the parameters m_0, $m_{3/2}$, A and μ, we have seen that the value of m_{top} is very important. Since we know by now that m_{top} is quite heavy let us give another illustration of the behaviour of the model in this part of parameter space (fig. 7.3), where we have chosen $m_{3/2} = 100$ GeV. There is an approximate scaling law. If you want to know the behaviour for different values $am_{3/2}$, just scale m_0 and μ by the same factor a while A should be scaled by a^{-1} (at least in large parts of parameter space). The masses of the supersymmetric particles, of course, depend strongly on these parameters. Although they have to be computed in the explicit models under consideration one can use as a rule of thumb that the mass of the photino is $\sim \frac{1}{2}m_0$ while the mass of the gluinos is larger: approximately $3m_0$. The masses of the quark partners are given by

$$m_{\tilde{q}}^2 \approx m_{3/2}^2 + 7.4m_0^2 \tag{7.26}$$

while the lepton partners are less sensitive to the gaugino mass

$$m_{\tilde{l}}^2 \approx m_{3/2}^2 + 0.14m_0^2. \tag{7.27}$$

Of course, the masses can be computed exactly in a given model once the set of parameters is specified.

Let us now discuss the spectrum more thoroughly and start with the Higgs system [38]. The model contains five physical Higgs bosons, two charged and three neutral ones, one of which is a pseudoscalar. The mass of the charged Higgs bosons is given by

$$m_{\pm}^2 = m_1^2 + m_2^2 + M_W^2, \tag{7.28}$$

thus heavy since $m_1^2 + m_2^2$ has to be positive. The mass of the pseudoscalar is

$$m_{\rm PS}^2 = m_1^2 + m_2^2 \tag{7.29}$$

while the mass of the scalars is given by

$$m_S^2 = \frac{1}{2}\left(m_{\rm PS}^2 + M_Z^2\right) \pm \frac{1}{2}\sqrt{(m_{\rm PS}^2 + M_Z^2)^2 - 4m_{\rm PS}^2 M_Z^2 \frac{(v_1^2 - v_2^2)^2}{(v_1^2 + v_2^2)^2}} \tag{7.30}$$

which gives a tree-level prediction of the minimal supersymmetric extension of the standard model. One Higgs scalar is alway lighter than the Z boson. The bound is reached in the case where the ratio of the vev's of the Higgs fields is very large. If in contrast $v_1 = v_2$ we see that one of the scalars is massless. A crucial experimental test of the model will therefore come from Higgs searches. The above formulae and bounds are only valid at the tree level. Radiative corrections could change the results [43]. These corrections depend strongly on the mass of the top quark and become large [44] if $m_{\rm top}$ exceeds a value of 130 GeV. For such high values also the limit on the mass of the lightest Higgs-boson moves up beyond the Z-boson mass. These questions will be thoroughly discussed by S. Pokorski and R. Hempfling in this seminar series.

Let us next consider the gauginos. The gluinos only feel m_0 with $m_g \approx 3m_0$. The so-called charginos are combinations of charged gauginos and higgsinos with mass matrix

$$\begin{pmatrix} \tilde{m}_2 & \sqrt{2}M_W \\ \sqrt{2}M_W \frac{v_1}{v_2} & \mu \end{pmatrix} \tag{7.31}$$

and the spectrum depends strongly on m_0 and μ. Observe that one of the states is massless in the limit $\mu = 0$ since there also one finds $v_1 = 0$. The neutral gauginos $\tilde{W}^0$, $\tilde{B}^0$ mix with the neutral higgsinos leading to a complicated mass matrix which we shall not discuss here in detail [26,45]. Among these particles one usually expects to find the lightest supersymmetric particle (LSP) which is stable as long as R-parity remains unbroken. This could be (and is over a wide range of parameter space) the photino

$$\tilde{\gamma} = \sin\theta_W \tilde{W}^0 - \cos\theta_W \tilde{B}^0 \tag{7.32}$$

but there remain other possibilities, like a higgsino if μ is small or e.g. also a scalar partner of a neutrino in the case where $m_{3/2}$ is small compared to m_0 and μ.

The masses of the scalar partners of quarks and leptons are essentially determined by $m_{3/2}$ and m_0 (see (7.26-27)) with squarks feeling a stronger influence of m_0. One would then conclude that the sneutrino is the lightest of these particles, but this is not necessarily true. There could be an influence of quark masses m_q on the squark masses m_{sq} in case of a large A

$$\begin{pmatrix} m_{sq}^2 + m_q^2 & Am_q m_{sq} \\ Am_q m_{sq} & m_{sq}^2 + m_q^2 \end{pmatrix} \tag{7.33}$$

and it could very well be that the partner of the top-quark is the lightest squark. So far our first discussion of the minimal supersymetric extension of the standard model (also called the minimal low energy supergravity model). It depends on several parameters, those in the superpotential (μ and the Yukawa couplings) and those parametrizing the breakdown of supersymmetry ($m_{3/2}$, m_0 and A). The order of magnitude of the dimensionful quantities is supposed to lie in the 100 GeV to TeV region. The actual values, of course, are very important for a discussion of the supersymmetric particle spectra.

8. SUPERSYMMETRIC GRAND UNIFICATION

Grand unified models have been discussed in detail in chapter 2 and we shall concentrate here on the changes that occur in a supersymmetric environment. This concerns the scale M_x, a discussion of the superpotential, the question of the triplet-doublet splitting and proton decay via dimension 5 operators. We shall exclusively stay within the $SU(5)$ framework, with $\bar{5} + 10$ for a quark lepton family. The formulas for the evolution of the gauge coupling constants have been given in (2.4) and (2.5). If we assume a value of $\alpha_3 \sim 0.1$ and $\alpha \approx 1/128$ at a scale of 100 GeV we obtain in the nonsupersymmetric model a scale M_x of approximately 5×10^{14} GeV and desastrous proton decay. The supersymmetric model, however, has more light particles and as such the evolution of coupling constants changes [46]. The most important contribution comes from the gauginos implying a slow-down of the evolution. As a result we observe a larger $M_x \sim 3 \times 10^{16}$ GeV roughly 60 times larger than in the corresponding nonsupersymmetric model. Since proton decay is suppressed with the fourth inverse power of M_x there are no problems with proton stability in the supersymmetric $SU(5)$ model. In fact a precision analysis of electroweak data [47] has recently shown that the supersymmetric model (with two Higgs doublets and a supersymmetry breakdown scale in the TeV-region) gives, in constrast to nonsupersymmetric $SU(5)$ the correct prediction for $\sin^2 \theta_W(M_Z)$. The result is shown in Fig. 8.1. Thus supersymmetric grand unified models deserve further attention [48].

Let us first discuss the superpotential and the question of $SU(5)$ breakdown. We denote the quark superfields $X_i(10)$, $Y_i(\bar{5})$ $i = 1,2,3$ and the Higgs superfields $H(5)$, $\bar{H}(\bar{5})$ and $\Phi(24)$. The superpotential can then be writen as

$$g = g_{ij}X_iX_jH + h_{ij}X_iY_j\bar{H} + \lambda_1 H\Phi\bar{H} + \lambda_2\Phi^3 + M\Phi^2 + M'H\bar{H} \tag{8.1}$$

where g_{ij} determines the masses of up-type quarks and h_{ij} those of down-type quarks and leptons. The discussion of breakdown of $SU(5)$ is similar to the one in nonsupersymmetric $SU(5)$ models. The auxiliary fields read

$$\begin{aligned} -F_\Phi^* &= \lambda_1 H\bar{H} + 3\lambda_2\Phi^2 + 2M\Phi \\ -F_H^* &= \lambda\Phi\bar{H} + M'\bar{H} + g_{ij}X_iX_j \\ -F_{\bar{H}}^* &= \lambda_1\Phi H + M'H + h_{ij}X_iY_j \end{aligned} \tag{8.2}$$

and a minimum with $SU(5)$ broken to $SU(3) \times SU(2) \times U(1)$ can be found with

$$< H > = < \bar{H} > = < X_i > = < Y_i > = 0,$$

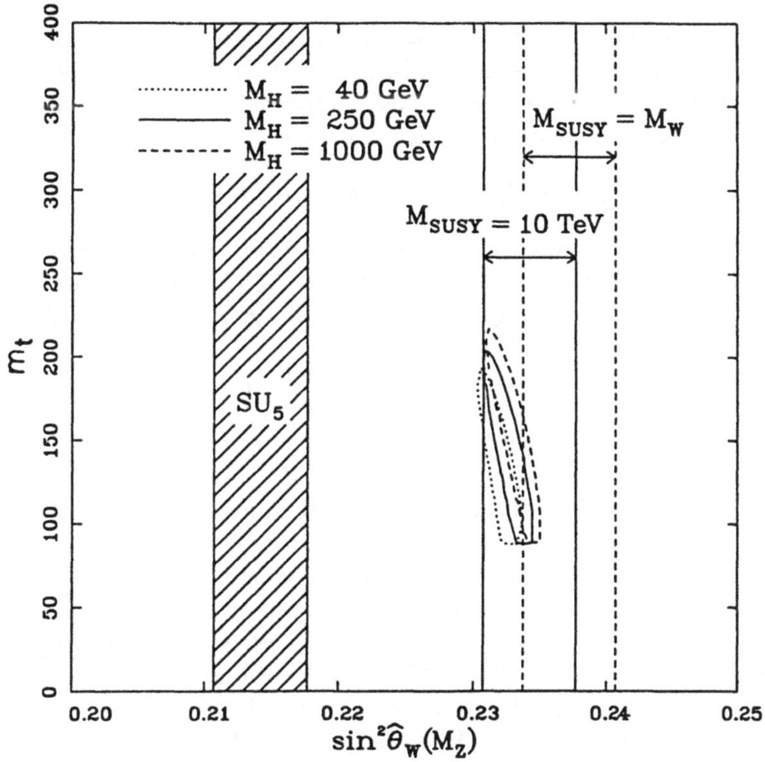

Fig.8.1: $\sin^2\theta$ versus m_t and the predictions from grand unified models [47].

and

$$< \Phi >= v \begin{pmatrix} 1 & 0 & 0 & 0 & 0 \\ 0 & 1 & 0 & 0 & 0 \\ 0 & 0 & 1 & 0 & 0 \\ 0 & 0 & 0 & -\frac{3}{2} & 0 \\ 0 & 0 & 0 & 0 & -\frac{3}{2} \end{pmatrix} \tag{8.3}$$

and vanishing vacuum energy. Since we have not discussed here the breakdown of supersymmetry there are degenerate minima with gauge group $SU(5)$ and $SU(4) \times U(1)$. Also the breakdown of $SU(2) \times U(1)$ has finally to be induced by the effects of supersymmetry breakdown along the lines discussed in the last chapter.

Again a fine-tuning has to be performed to keep the Higgs-doublets light. Here it amounts to

$$M' = \frac{3}{2} v \lambda_1 \tag{8.4}$$

This is similar to the nonsupersymmetric case but here we could argue that the fine-tuning concerns only parameters in the superpotential and is therefore not disturbed by radiative corrections. If we now would be able to find a reason why (8.4) should be valid at tree level we could claim to have solved the fine-tuning problem. There have been several interesting attempts in this direction. As a first we discuss the mechanism of a sliding singlet [49]. Take a gauge singlet superfield Z and add a term $\lambda H Z \bar{H}$ to the superpotential. The H auxiliary field reads now

$$-F_H^* = \bar{H}(\lambda_1 \Phi + \lambda Z + M'). \tag{8.5}$$

In the full theory, including supersymmetry breakdown, the doublet component of the scalar of $\bar{H}$ should receive a vev (in contrast to the $SU(3)$-triplet component). The vev of Z is undetermined and it can adjust its vev to have $F = 0$ for the doublet component, thus it slides to make

$$-\frac{3}{2}\lambda_1 v + \lambda z + M' = 0 \tag{8.6}$$

and the Higgs-doublet remains light. This looks nice, but also this mechanism has some problems. We do not understand why the allowed Z^2 and Z^3 terms are absent and also we cannot rule out the possibility that the absolute minimum of the potential occurs for large vev's of both the triplet and the doublet. Moreover, there are usually problems with a small supersymmetry breakdown scale in the presence of light singlets [50].

A second mechanism to be discussed here is the one of the missing partner [51]. H and $\bar{H}$ contain $(3,1) + (\bar{3},1)$ and $(1,2) + (1,\bar{2})$ of $SU(3)$ and $SU(2)$ respectively. Try to find now a new representation which only contains a (3,1) but not a (1,2). The former could then pair up with the $(\bar{3},1)$ in $\bar{H}$ while $(1,\bar{2})$ would remain massless. The simplest example is a 50 of $SU(5)$. It decomposes with respect to $SU(3) \times SU(2)$ as $(\bar{6},1) + (8,2) + (1,1) + (3,2) + (6,3) + (\bar{3},1)$ and as a cross term in the superpotential we could imagine $50 \times 5 \times 75$ with $75 = (1,1) + (3,1) + (3,2) + (\bar{3},1) + (\bar{3},2) + (\bar{6},2) + (6,2) + (8,1) + (8,3)$. Fortunately a vev of 75 can break $SU(5)$ to $SU(3) \times SU(2) \times U(1)$ thus avoiding the presence of Φ in (8.1). Instead we choose now for the superpotential

$$\begin{aligned} g = {} & \lambda 75 \times 75 \times 75 + M 75 \times 75 + \lambda_1 50 \times 75 \times \bar{50} \\ & + \lambda_2 50 \times 75 \times 5 + \lambda_3 \bar{50} \times 75 \times \bar{5} + \bar{M} 50 \times \bar{50} \end{aligned} \tag{8.7}$$

and as a mass matrix for the triplets we obtain

$$\begin{pmatrix} 0 & \lambda_2 v \\ \lambda_3 v & \bar{M} \end{pmatrix} \tag{8.8}$$

(where v is the vev of 75), while the doublets remain light. Of course, one still has to explain why we have omitted a direct $5 \times \bar{5}$ mass term in (8.7) and the question of a complete solution of the fine tuning problem remains open.

We had seen at the beginning of this chapter that M_x is quite large in supersymmetric grand unified models and that therefore proton decay via gauge boson exchange is sufficiently suppressed. This, however, is not the last word about proton decay in supersymmetric grand unified models. Remember, that in the supersymmetric version of the standard model we already had to suppress proton decay via dimension-4 operators by introducing an R-symmetry (see chapter 5). Here we have to worry about dimension five operators [52] leading to proton decay as shown in Fig. 8.2. The first step couples two fermions to two bosons (therefore the name dimension-5 operator) and has a propagator suppression of $1/M_x$ and the second step involves only light particles. Instead of

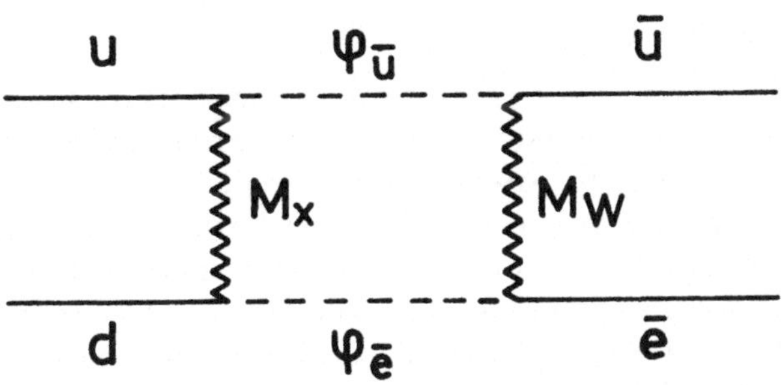

Fig. 8.2: Proton decay through dimension-5 operators.

$1/M_x^2$ in the amplitude we have now $1/M_x M_W$ and there is a potential danger of fast proton decay. A careful investigation of the dimension 5-operators has therefore to be performed. Out of the possible ones we need only consider those which are invariant under the R-symmetry discussed earlier and these are the two F-terms $(QQQL)_F$ and $(\bar{U}\bar{U}\bar{D}\bar{E})_F$. The latter reads in components

$$\bar{U}_{ia}\bar{U}_{jb}\bar{D}_{kc}\bar{E}_l\epsilon^{abc} \tag{8.9}$$

where a, b, c are $SU(3)$ indices and i, j, k, l are generation indices. All fields above are scalar superfields and should obey Bose-statistics. The two $\bar{U}$'s are antisymmetrized in a and b and therefore $i \neq j$ and one of the $\bar{U}$'s has to come from the second generation. Since the charmed quark is heavier than the proton the presence of the term in (8.9) does not constitute a problem. The other possibility reads

$$Q^a_{ir}Q^b_{js}Q^c_{kt}L_{lu}\epsilon_{abc}\epsilon^{rs}\epsilon^{tu} \tag{8.10}$$

where r, s, t, u are $SU(2)$-indices. Here we can have $i = j = 1$ but then we need $k = 2$ which leads to

$$\begin{pmatrix} c \\ s \end{pmatrix}_t \begin{pmatrix} \nu \\ e \end{pmatrix}_u \epsilon^{tu} \tag{8.11}$$

thus ce or $s\nu$. Proton decay therefore is only possible with the $(uds\nu)_F$ operator The dominant decay mode is proton to K^+ and antineutrino, a quite unique prediction of supersymmetric grand unified models. The rate is faster than the one from dimension-6 operators but it is not desastrously fast since $p \to K^+\bar{\nu}$ involves Yukawa couplings in graphs like Fig. 8.2 as compared to gauge couplings in the process with dimension-6 operators.

At the moment $p \to K^+\bar{\nu}$ seems to be at the border of observability. Supersymmetric grand unification deserves further attention; because of these predictions for proton decay but also because of the results from precision electroweak data as displayed in Fig. 8.1.

9. OUTLINE OF SEMINARS

So far our presentation of the theoretical construction of the minimal supersymmetric generalization of the standard model and its grand unified extension. We have now to investigate the apparent experimental consequences of this approach to stabilize the weak scale. We shall see in the following lectures, that present experimental results do already give constraints on the parameter space of the model consisting of the gravitino mass $m_{3/2}$, gaugino mass m_0, trilinear scalar couplings A as well as the Higgs-mass parameter μ. Of course, also the still unknown mass of the top-quark will play a crucial role in this discussion.

In the next lecture J. Kalinowski will present the experimental situation concerning the search for Higgs-bosons and the implications for the parameter space of the model. Then F. Borzumati will present a detailed overview of the mass spectra of supersymmetric particles and experimental bounds. S. Bertolini will represent implications of the supersymmetric models for flavour changing neutral currents, rare decays of Kaons and bottom-mesons as well as CP-violation. Strong constraints of the model can also come from cosmological considerations. They will be summarized by A. Klemm. W. Majerotto will present a complete overview of supersymmetric particle searches at high energy colliding beam facilities. Indirect searches and loop effects will be discussed by S. Pokorski, J. Sola and R. Hempfling. The search for supersymmetry has to go on and we might expect exciting new results in the near future.

REFERENCES

1. S. Glashow, Nucl. Phys. B22 (1961) 579;
   A. Salam, in Elementary Particle Theory, ed. N. Svartholm, (1968) 367;
   S. Weinberg, Phys. Rev. Lett. 419 (1967) 1264
2. M. Kobayashi and M. Maskawa, Progr. Theor. Phys. 49 (1973) 652
3. S. Adler, Phys. Rev. 177 (1969) 2426;
   J. Bell and R. Jackiw, Nuovo Cimento 51 A (1969) 47
4. E. Witten, Phys. Lett. 117B (1982) 324
5. L.E. Ibanez, Proceedings of ASI on High Energy Physics, Editor T. Ferbel, Plenum Press (1989)
6. C.Q. Geng and R.E. Marshak, Phys. Rev. D39 (1989) 693
7. R. Foot, G.C. Joshi, H. Lew and R.R. Volkas, Univ. of Melbourne preprint UM-P-89/39 (1989)
   K.S. Babu and R.N. Mohapatra, Univ. of Maryland preprint, MdDP-PP-89-189 (1989)
8. H. Georgi and S.L. Glashow, Phys. Rev. Lett. 32 (1974) 438
9. H. Georgi, H.R. Quinn and S. Weinberg, Phys. Rev. Lett. 33 (1974) 451
10. A.J. Buras, J. Ellis, M.K. Gaillard and D.V. Nanopoulos, Nucl. Phys. B135 (1978) 66
11. E. Gildener and S. Weinberg, Phys. Rev. D15 (1976) 3333
12. P. Langacker, Phys. Rep. 72C (1981) 185

13. M. Gell-Mann, P. Ramond and R. Slansky, in Supergravity, eds. P. van Nienwenhuizen and D.Z. Freedmann (North-Holland, 1979) p. 315
14. ALEPH Collab., D. Decamp et al., Phys. Lett. B246 (1990) 306;
OPAL Collab., M. Akrawy et al., Phys. Lett. B236 (1990) 224;
DELPHI Collab., P. Abreu et al., Nucl. Phys. B342 (1990) 1;
L3 Collab., B. Adeva et al., Phys. Lett. B248 (1990) 203
15. E. Farhi and L. Susskind, Phys. Rep. 74C (1981) 277 and references therein
16. J. Wess and B. Zumino, Nucl. Phys. B70 (1974) 39;
for a review see: J. Wess and J. Bagger, Princeton Series in Physics, Princeton Univ. Press (1983)
17. S. Ferrara, L. Girardello and F. Palumbo, Phys. Rev. D20 (1979) 403
18. P. Fayet and J. Iliopoulos, Phys. Lett. 51B (1974) 461
19. M.T. Grisaru, W. Siegel and M. Rocek, Nucl. Phys. B159 (1979) 429;
W. Fischler, H.P. Nilles, J. Polchinski, S. Raby and L. Susskind, Phys. Lett. 47 (1981) 757
20. P. West, Phys. Lett. B258 (1991) 375;
I. Jack, D.R.T. Jones and P. West, Phys. Lett. B258 (1991) 382;
P. West, Phys. Lett. B261 (1991) 396
21. D.C. Dunbar, I. Jack and D.R.T. Jones, CERN preprint TH-5996 (1991)
22. L. Girardello and M.T. Grisaru, Nucl. Phys. B194 (1982) 65
23. P. Fayet, Nucl. Phys. B90 (1975) 104;
A. Salam and J. Strathdee, Nucl. Phys. B87 (1975) 85
24. R.D. Peccei and H.R. Quinn, Phys. Rev. Lett. 38 (1977) 1440;
S. Weinberg, Phys. Rev. Lett. 40 (1978) 223;
F. Wilczek, Phys. Rev. Lett. 40 (1978) 229
25. J.E. Kim, Phys. Rev. Lett. 43 (1979) 103
26. H.P. Nilles, Phys. Rep. 110C (1984) 1
27. D.Z. Freedman, S. Ferrara and P. van Nieuwenhuizen, Phys. Rev. D13 (1976) 3214
28. For a review see: P. van Nieuwenhuizen, Phys. Rep. 68C (1981) 189
29. E. Cremmer, S. Ferrara, L. Girardello and A. van Proeyen, Nucl. Phys. B212 (1983) 413
30. H.P. Nilles, Phys. Lett. B115 (1982) 193 and Nucl. Phys. B217 (1983) 366
31. For a review see: H.P. Nilles, International Journal of Modern Physics A5 (1990) 4199
32. G. Veneziano and S. Yankielowicz, Phys. Lett. B113 (1982) 231
33. S. Ferrara, L. Girardello and H.P. Nilles, Phys. Lett. 125B (1983) 457
34. J.P. Derendinger, L.E. Ibanez and H.P. Nilles, Phys. Lett. B155 (1985) 65;
M. Dine. R. Rohm. N. Seiberg and E. Witten, Phys. Lett. B156 (1985) 55
35. J. Ellis, A.B. Lahanas, D.V. Nanopoulos and K. Tamvakis, Phys. Lett. 134B (1984) 429
36. H.P. Nilles, M. Srednicki and D. Wyler, Phys. Lett. 120B (1983) 346
37. J.E. Kim and H.P. Nilles, Phys. Lett. 138B (1984) 150
38. K. Inoue, A. Kakuto, H. Komatsu and S. Takeshita, Progr. Theor. Phys. 68 (1982) 927
39. L.E. Ibanez and G.G. Ross, Phys. Lett. 110B (1982) 215;
H.P. Nilles, Phys. Lett. 115 B (1982) 193

40. L.E. Ibanez and C. Lopez, Phys. Lett. 126B (1983) 54; Nucl. Phys. B233 (1984) 511
41. L. Alvarez-Gaume, J. Polchinski and M. Wise, Nucl. Phys. B221 (1983) 495
42. J.M. Frere, D.R.T. Jones and S. Raby, Nucl. Phys. B222 (1983) 11
43. S.P. Li and M. Sher, Phys. Lett. B140 (1984) 339
44. Y. Okada, M. Yamaguchi and T. Yamagido, Progr. Theor. Phys. 85 (1991) 1;
H. Haber and R. Hempfling, Phys. Rev. Lett. 66 (1991) 1815;
J. Ellis, G. Ridolfi and F. Zwirner, Phys. Lett. B257 (1991) 83;
R. Barbieri, F. Caravaglios and M. Frigeni, Phys. Lett. B258 (1991) 167;
P. Chankowski, S. Pokorski and J. Rosiek, MPI. München preprint Ph/91-57 (1991)
45. For a detailed description see: H.E. Haber and G.L. Kane, Phys. Rep. 117C (1985) 75
46. S. Dimopoulos, S. Raby and F. Wilczek, Phys. Rev. D24 (1981) 1681
47. P. Langacker, Precision Tests of the Standard Model, Univ. of Penn. preprint UPR-0435 T (1990)
48. J. Ellis, S. Kelley and D.V. Nanopoulos, Phys. Lett. B249 (1990) 441;
U. Amaldi, W. De Boer and H. Fürstenau, Phys. Lett. B260 (1991) 447;
P. Langacker and M.-X. Luo, Phys. Rev. D44 (1991) 817
49. E. Witten, Phys. Lett. 105B (1981) 267;
L.E. Ibanez and G.G.Ross, Phys. Lett. 110B (1982) 215
50. H.P. Nilles, M. Srednicki and D. Wyler, Phys. Lett. 124B (1983) 337
51. B. Grinstein, Nucl. Phys. B206 (1982) 387
52. S. Weinberg, Phys. Rev. D26 (1982) 287;
N. Sakai and T. Yanagida, Nucl. Phys. B197 (1982) 533

The Higgs sector of the minimal supersymmetric model

Jan Kalinowski
Institute for Theoretical Physics, University of Warsaw
00681 Warsaw, Poland

1 Introduction

Recent years have witnessed a spectacular success of the standard model (SM) of fundamental interactions - a gauge theory based on the SU(3)×SU(2)×U(1) symmetry group [1]. All the experimental results from LEP measurements are in perfect agreement with the SM predictions. The only missing pieces are the top quark and the spin zero elementary Higgs boson. The Higgs sector is one of the key points of the electroweak interactions needed for spontaneous symmetry breaking [2] which is responsible for the $W^{\pm}$, Z^0 and fermion masses. Although one could argue that it is only a matter of time until the missing pieces are discovered, some theoretical problems related to the Higgs boson suggest that it is necessary to look beyond the standard model.

There are three classes of problems which indicate that the standard model is incomplete. First, it has many arbitrary assumptions and parameters: gauge and Yukawa couplings, masses and mixing parameters, number and structure of fermion generations, handedness of fermionic representations etc. Second, the standard model is not asymptotically free which suggest that it is an effective low energy theory of a more fundamental one. The third class of problems is directly related to the presence of the elementary scalar in the theory. The scalar fields have nice properties. They can have nonzero vacuum expectation values (vev) without breaking Lorentz invariance. Therefore, they can trigger a spontaneous breakdown of gauge symmetries and thus provide masses to gauge bosons. Through Yukawa couplings to fermions they can also generate masses of quarks and leptons. However, they have also a bad property of aquiring quadratic divergences through radiative corrections. The correction to the mass of the Higgs boson is $\delta m^2 \sim g^2\Lambda^2$, where Λ is a physical scale beyond which the low energy theory no longer applies. To understand the physics at the Fermi scale $\sim 10^2$ GeV it would be inappropriate to have Λ of the order of the unification scale $\sim 10^{15}$ GeV or the Planck scale $\sim 10^{19}$ GeV but rather in the TeV region. This is so called the naturalness or hierarchy problem [3]. Therefore we are led to consider means to stabilize the Higgs sector below or at TeV scale. The only way to protect masses of elementary scalar particles is supersymmetry [4]. This symmetry relates bosons to fermions and thus makes the bosons to behave as well as fermions.

Supersymmetry (SUSY) imposes a new requirement on the Higgs sector. In the SM only one Higgs doublet is sufficient to give masses to quarks and leptons. In the SUSY models two Higgs doublets are needed to give masses to up-type and down-type quarks (and leptons) and to achieve the cancellation of trangle anomalies.

In this paper we discuss the Higgs sector of the minimal supersymmetric extension of the standard model. We study the mass spectrum of all the physical Higgs bosons and their interactions. We will not discuss the superpartners of the Higgs bosons nor Higgs boson decays into superparticles. These aspects of the SUSY model are dealt with in the lecture of A. Bartl [5]. Our discussion will be performed at the tree level without loop corrections. It is now well known that loop corrections in the case of a heavy top quark ($m_t \geq 120$ GeV) are very important [6]. Our disscussion will serve as a starting point for lectures by S. Pokorski [7] and R. Hempfling [8] where effects of loop corrections are considered, although ocasionally we will mention the role of loop corrections. Nevertheless taking conservatively $m_t = 90$ GeV (current experimental limit reads $m_t \geq 89$ GeV [9]) there exists a large domain in parameter space where these corrections are small and the tree level discussion is justified. Therefore we will discuss current experimental limits and prospects for discovering Higgs bosons in future colliders.

2 The Higgs sector in a minimal SUSY model

In a minimal supersymmetric standard model (MSSM) we employ two Higgs doublets $H_1 = (H_1^1, H_1^2) \equiv (\varphi_1^{0*}, -\varphi_1^-)$ and $H_2 = (H_2^1, H_2^2) \equiv (\varphi_2^+, \varphi_2^0)$ with opposite hypercharge $y(H_1) = -1$, $y(H_2) = 1$. The construction of the MSSM has beeen discussed in the lectures by H.P. Nilles [10]. Here we recall that the tree level scalar potential is given by

$$V = F_i^* F_i + \frac{1}{2}(D_a^2 + D'^2), \tag{1}$$

where the auxiliary fields F and D are as follows

$$\begin{aligned} D_a &= \frac{1}{2} g A_i^* \sigma_{ij}^a A_j, \\ D' &= \frac{1}{2} g' y_i A_i^* A_i + \xi, \\ F_i &= \frac{\partial W}{\partial A_i}. \end{aligned} \tag{2}$$

Here W is the superpotential with the superfields replaced by their scalar components, g and g' are the SU(2) and U(1) coupling constants, A_i is a generic notation for all scalar fields in the theory and ξ is the Fayet-Iliopoulos term [11]. In the context of many models it can be argued that ξ must be very small and we shall assume $\xi = 0$.

The superpotential W is a cubic gauge invariant function of the superfields containing the mass terms and Yukawa couplings. Its general form is as follows

$$W = \mu \epsilon_{ij} H_1^i H_2^j + W_F, \tag{3}$$

$$W_F = \epsilon_{ij}(g_E L^i H_1^j E + g_D Q^i H_1^j D + g_U Q^i H_2^j U), \tag{4}$$

with L and Q being left-handed lepton and quark doublets, E, U and D – right-handed electron, up- and down-type quark singlets and g_i - the corresponding Yukawa couplings. Unfortunately such a theory fails to break the SU(2)×U(1) gauge symmetry. There are two "minimal" solutions to achieve a spontaneous gauge symmetry breakdown. First, one can add a supersymmetric realization of a complex scalar Higgs field N. Then the superpotential takes the form

$$W = \epsilon_{ij}(\mu H_1^i H_2^j + \rho H_1^i H_2^j N) + rN + \Lambda N^3 + W_F. \tag{5}$$

A second approach, which we will discuss in more detail, is to add soft-supersymmetry breaking terms. They have been clasified by Girardello and Grisaru [12] and are of the form

$$m^2 \varphi\varphi^*, \quad m^2(\varphi^2 + \varphi^{*2}), \quad \mu(\varphi^3 + \varphi^{*3}), \quad \mu\lambda\lambda, \tag{6}$$

where φ is a complex scalar and λ a gauge fermion.

Let us now write explicitely the scalar potential as a function of H_1 and H_2. We assume that colour and lepton numbers are not spontaneously broken, *i.e.* squarks and sleptons do not aquire vacuum expectation values (vev). We can then ignore W_F and write (the N field not present)

$$V = \frac{1}{8}(g^2 + g'^2)(|H_1|^2 - |H_2|^2)^2 + \frac{1}{2}g^2|H_1|^2|H_2|^2 - \frac{1}{2}g^2|\epsilon_{ij}H_1^i H_2^j|^2 + |\mu|^2(|H_1|^2 + |H_2|^2) + V_{soft}, \tag{7}$$

$$V_{soft} = m_1^2|H_1|^2 + m_2^2|H_2|^2 + (m_3^2 \epsilon_{ij} H_1^i H_2^j + h.c.). \tag{8}$$

Furthermore, we assume that only neutral components of the Higgs fields aquire vev's

$$< H_1 >= \begin{pmatrix} v_1 \\ 0 \end{pmatrix}, \quad < H_2 >= \begin{pmatrix} 0 \\ v_2 \end{pmatrix} \tag{9}$$

and the Higgs potential for the neutral fields reads as follows

$$V = (m_1^2 + |\mu|^2)|\varphi_1^0|^2 + (m_2^2 + |\mu|^2)|\varphi_2^0|^2 + m_3^2(\varphi_1^{0*}\varphi_2^0 + \varphi_1^0\varphi_2^{0*}) + \frac{1}{8}(g^2 + g'^2)(|\varphi_1^0|^2 - |\varphi_2^0|^2)^2. \tag{10}$$

The potential consists of quadratic and quartic terms with the quartic coupling given in terms of the gauge couplings. To have the potential bounded from below one has to impose

$$m_1^2 + m_2^2 \geq 2|m_3|^2, \tag{11}$$

because the quartic term has a flat direction for $|\varphi_1^0|^2 = |\varphi_2^0|^2$. Since there are no trilinear terms, a stationary solution at $\varphi_1^0 = \varphi_2^0 = 0$ has to be unstable for SU(2)×U(1) breakdown. This can be achieved by requiring that

$$|m_3|^4 \geq m_1^2 m_2^2. \tag{12}$$

In a low-energy supergravity model such a scenario can be obtained as follows [13]. At a very large scale M the supersymmetry is softly broken and the SU(2)×U(1) gauge symmetry is exact. When the renormalization group equations are used to evolve the parameters of the model from M down to energies of the order 100 GeV one of m_i^2 becomes negative indicating the spontaneous breakdown of gauge symmetry.

3 Two-Higgs doublet model

Before we discuss explicitely the spectrum of the Higgs bosons in the SUSY model it is instructive to consider first a general two-Higgs doublet model. We shall then apply the results to the SUSY case.

Take two complex, $y = 1$, SU(2) doublet scalar fields $\phi_1 = (\varphi_1^+, \varphi_1^0)$ and $\phi_2 = (\varphi_2^+, \varphi_2^0)$. The most general Higgs potential that spontaneously breaks SU(2)×U(1) is of the form[1]

$$\begin{aligned} V &= \lambda_0 + \lambda_1(\phi_1^\dagger\phi_1 - v_1^2)^2 + \lambda_2(\phi_2^\dagger\phi_2 - v_2^2)^2 + \lambda_3[(\phi_1^\dagger\phi_1 - v_1^2) + (\phi_2^\dagger\phi_2 - v_2^2)]^2 \\ &\quad + \lambda_4[\phi_1^\dagger\phi_1\, \phi_2^\dagger\phi_2 - \phi_1^\dagger\phi_2\, \phi_2^\dagger\phi_1] + \lambda_5[\mathrm{Re}(\phi_1^\dagger\phi_2) - v_1 v_2\cos\chi]^2 \\ &\quad + \lambda_6[\mathrm{Im}(\phi_1^\dagger\phi_2) - v_1 v_2 \sin\chi]^2. \end{aligned} \tag{13}$$

Because of hermiticity λ_i are real. For $\lambda_i \geq 0$ the potential is positive semi-definite and the global minimum is at

$$<\phi_1>= \begin{pmatrix} v_1 \\ 0 \end{pmatrix}, \quad <\phi_2>= \begin{pmatrix} v_2 e^{i\chi} \\ 0 \end{pmatrix} \tag{14}$$

and breaks the electroweak symmetry. For $\lambda_5 \neq \lambda_6$ the potential violates CP in the Higgs sector. In the SUSY model we will see that $\lambda_5 = \lambda_6$. Then the last two terms can be rewritten as $\lambda_5|\phi_1^\dagger\phi_2 - v_1 v_2 e^{i\chi}|^2$ and by a redefinition $\phi_2 \to \phi_2 e^{i\chi}$ one can remove the phase χ. As a result v_1 and v_2 can be chosen to be real and positive. For a general discussion we

[1] We follow the notation of Gunion and Haber [14].

take $\lambda_5 \neq \lambda_6$ with $\chi = 0$. This corresponds to the most general CP-invariant two-Higgs doublet model.

The matrix for the squares of the Higgs boson masses can be obtained from

$$M_{ij}^2 = \frac{1}{2}\frac{\partial^2 V}{\partial\varphi_i \partial\varphi_j} \tag{15}$$

where φ_i is a generic notation for real and imaginary parts of the Higgs fields and the eq.(15) is evaluated at the minimum (14). With $\chi = 0$ the 8×8 mass matrix M^2 breaks into a set of 2×2 mass matrices and the diagonalization is straightforward.

3.1 Charged Higgs bosons

The positive φ_i^+ and negative $\varphi_i^- \equiv (\varphi_i^+)^*$ $(i = 1, 2)$ states decouple and have equal mass matrices of the form

$$\lambda_4 \begin{pmatrix} v_1^2 & -v_1 v_2 \\ -v_1 v_2 & v_2^2 \end{pmatrix}. \tag{16}$$

The eigenstates are easily found

$$\begin{aligned} H^\pm &= -\varphi_1^\pm \sin\beta + \varphi_2^\pm \cos\beta, \\ G^\pm &= \varphi_1^\pm \cos\beta + \varphi_2^\pm \sin\beta, \end{aligned} \tag{17}$$

with the masses

$$\begin{aligned} m_{G^\pm}^2 &= 0, \\ m_{H^\pm}^2 &= \lambda_4\,(v_1^2 + v_2^2), \end{aligned} \tag{18}$$

and where the mixing angle β is defined by

$$\tan\beta = v_2/v_1. \tag{19}$$

The massless would-be Goldstone bosons $G^\pm$ are eaten up by the $W^\pm$ and the $H^\pm$ are the physical charged Higgs bosons.

3.2 Imφ_1^0 and Imφ_2^0

The mass matrix is identical to the one in eq.(16) with λ_4 replaced by λ_6. Therefore the same mixing angle β diagonalizes it with the result

$$\begin{aligned} \frac{1}{\sqrt{2}}G^0 &= \mathrm{Im}\varphi_1^0 \cos\beta + \mathrm{Im}\varphi_2^0 \sin\beta, \\ \frac{1}{\sqrt{2}}A^0 &= -\mathrm{Im}\varphi_1^0 \sin\beta + \mathrm{Im}\varphi_2^0 \cos\beta, \end{aligned} \tag{20}$$

$$\begin{aligned} m_{G^0}^2 &= 0, \\ m_{A^0}^2 &= \lambda_6\,(v_1^2 + v_2^2). \end{aligned} \tag{21}$$

The massless would-be Goldstone boson G^0 is eaten up by the Z^0 and the A^0 is the physical neutral CP$= -1$ Higgs boson. Because of the nature of its couplings to fermions it is often called a pseudoscalar.

3.3 $\mathrm{Re}\varphi_1^0$ and $\mathrm{Re}\varphi_2^0$

The eigenstates of the mass matrix

$$\begin{pmatrix} A & B \\ B & C \end{pmatrix} = \begin{pmatrix} 4(\lambda_1+\lambda_3)v_1^2+\lambda_5 v_2^2 & (4\lambda_3+\lambda_5)v_1v_2 \\ (4\lambda_3+\lambda_5)v_1v_2 & \lambda_5 v_1^2+4(\lambda_2+\lambda_3)v_2^2 \end{pmatrix} \tag{22}$$

are as follows

$$\begin{aligned} \frac{1}{\sqrt{2}}H^0 &= (\mathrm{Re}\varphi_1^0 - v_1)\cos\alpha + (\mathrm{Re}\varphi_2^0 - v_2)\sin\alpha, \\ \frac{1}{\sqrt{2}}h^0 &= -(\mathrm{Re}\varphi_1^0 - v_1)\sin\alpha + (\mathrm{Re}\varphi_2^0 - v_2)\cos\alpha. \end{aligned} \tag{23}$$

The masses of the neutral scalar Higgs bosons and the mixing angle α are defined as

$$\begin{aligned} m^2_{H^0,h^0} &= \frac{1}{2}\left[A + C \pm \sqrt{(A-C)^2+4B^2}\right], \\ \tan 2\alpha &= \frac{2B}{A-C} \\ \sin 2\alpha &= \frac{2B}{\sqrt{(A-C)^2+4B^2}}. \end{aligned} \tag{24}$$

By convention the Higgs boson H^0 is chosen to be heavier than h^0.

Let us note that in the general CP-conserving two-Higgs doublet model there are 7 independent parameters: λ_1-λ_6, and one of the v_1 and v_2 (λ_0 plays no role and $v_1^2 + v_2^2$ is fixed by the gauge boson masses). Therefore there is a lot of freedom in fixing the parameters of the model and to test it experimentally. The Higgs bosons can have arbitrary masses. Since the couplings are proportional to the masses (as seen for example explicitely from eqs.(18) and (21)) the Higgs sector becomes strongly interacting for heavy Higgs bosons like in the standard model.

4 Implications of supersymmetry

Supersymmetry imposes strong restrictions on the Higgs sector. Comparing eqs (10) and (13) we find that

$$\begin{aligned} \lambda_1 &= \lambda_2, \\ \lambda_3 &= \frac{1}{8}(g^2+g'^2) - \lambda_1, \end{aligned}$$

$$\begin{aligned}
\lambda_4 &= 2\lambda_1 - \frac{1}{2}g'^2, \\
\lambda_5 &= \lambda_6 = 2\lambda_1 - \frac{1}{2}(g^2 + g'^2), \\
\lambda_0 &= -\frac{1}{8}(v_1^2 - v_2^2)^2(g^2 + g'^2), \\
m_1^2 &= -|\mu|^2 + 2\lambda_1 v_2^2 - \frac{1}{2}m_Z^2, \\
m_2^2 &= -|\mu|^2 + 2\lambda_1 v_1^2 - \frac{1}{2}m_Z^2, \\
m_3^2 &= \frac{1}{2}v_1 v_2(4\lambda_1 - g^2 - g'^2),
\end{aligned} \tag{25}$$

where as usual $m_Z^2 = (g^2 + g'^2)(v_1^2 + v_2^2)/2$.

4.1 The spectrum

Using the eqs. (18), (21), (24) and (25) we immediately obtain the spectrum of the physical Higgs bosons in the form

$$\begin{aligned}
m_{H^\pm}^2 &= m_{A^0}^2 + m_W^2, \\
m_{H^0,h^0}^2 &= \frac{1}{2}\left[m_{A^0}^2 + m_Z^2 \pm \sqrt{(m_{A^0}^2 + m_Z^2)^2 - 4m_Z^2 m_{A^0}^2 \cos^2 2\beta}\right], \\
\tan 2\alpha &= \frac{m_{A^0}^2 + m_Z^2}{m_{A^0}^2 - m_Z^2}\tan 2\beta.
\end{aligned} \tag{26}$$

We see that now the spectrum depends only on two free parameters for which $\tan\beta$ and m_{A^0} can be taken as independent. We have already noticed that v_1 and v_2 are real and non-negative, which means that one can choose $0 \leq \beta \leq \pi/2$. Furthermore from eqs.(24) and (25) follows that $\sin 2\alpha \leq 0$, so one can take $-\pi/2 \leq \alpha \leq 0$.

Before discussing the Higgs boson mass spectrum let us consider the supersymmetric limit. It is obtained by setting the soft-SUSY breaking parameters $m_1 = m_2 = m_3 = 0$. However, in this limit, eqs.(25) are inconsistent with $v_i \neq 0$ which proves the statement mentioned in Chapter 2 that in the MSSM with only two Higgs doublets soft-supersymmetry breaking terms are necessary for SU(2)×U(1) gauge symmetry to be spontaneously broken. With an additional Higgs field N the SUSY limit with broken gauge symmetry may exist [14].

In fig.1 the masses of the scalar H^0 and h^0 Higgs bosons as functions of m_{A^0} for fixed $\tan\beta$ are presented. We see that

$$\begin{aligned}
&m_{H^\pm} \geq m_W, \\
&m_{h^0} \leq m_{A^0} \leq m_{H^0}, \\
&m_{h^0} \leq m_Z \cos 2\beta \leq m_Z, \\
&m_{h^0}^2 + m_{H^0}^2 = m_{A^0}^2 + m_Z^2.
\end{aligned} \tag{27}$$

Note that the spectrum is symmetric under $\tan\beta \leftrightarrow 1/\tan\beta$ and one of the scalar Higgs bosons (h^0) is lighter than the Z boson. The bound is reached in the case when $\tan\beta$ is large. In fact model building prefers $\tan\beta = v_2/v_1 > 1$ which is in agreement with an old idea that the up-type and down-type quark masses are determined by two different v's with the Yukawa couplings roughly equal [15]. This is especially interesting now since we know that the top quark mass is large. Moreover in the limit $v_2/v_1 \geq 5$ the mass of either H^0 or h^0 (or A^0) is very close to m_Z. The presence of a Higgs boson roughly degenerate with the Z (irrespective of m_{A^0}) turns out to be a general feature of supersymmetric models with $v_2/v_1 > 1$ [16].

These nice predictions of the MSSM are modified by radiative corrections due to a heavy top quark [6]. In particular the relations (27) can be violated. However, for $m_t = 90$ GeV and $\tan\beta \geq 2$ the corrections are rather small, as can be seen in fig.2 where the mass m_{h^0} of the lightest Higgs boson is compared to the tree-level calculations [17]. They become more significant for smaller $\tan\beta$ or heavier top quark [7,8].

4.2 The Higgs boson couplings

The Higgs boson couplings to standard model fermions and gauge bosons are also fully determined in terms of $\tan\beta$ and m_{A^0}. The neutral Higgs boson couplings to the up (U) and down (D) fermions $\varphi f\bar{f}$ are as follows

$$g_{\varphi f\bar{f}} = -\frac{igm_f}{2m_W}\begin{cases} \frac{\sin\alpha}{\sin\beta}, & \frac{\cos\alpha}{\sin\beta}, \quad -i\gamma_5\cot\beta, \quad 1, \quad \text{for } f\text{=U}, \\ \frac{\cos\alpha}{\cos\beta}, & -\frac{\sin\alpha}{\cos\beta}, \quad -i\gamma_5\tan\beta, \quad 1, \quad \text{for } f\text{=D}, \end{cases} \tag{28}$$

and to gauge bosons φVV

$$g_{\varphi VV} = igm_V g^{\mu\nu}\left\{\cos(\beta-\alpha),\ \sin(\beta-\alpha),\ 0,\ 1\right\}, \tag{29}$$

for $\varphi = H^0, h^0, A^0$ and the standard model Higgs particle, respectively, and with $m_V = m_W$ for $V = W$ or $m_V = m_Z/\cos\theta_W$ for $V = Z$.

The charged Higgs boson coupling to fermions reads (all particles are incoming)

$$g_{H^+D\bar{U}} = \frac{ig}{2\sqrt{2}\,m_W}[m_D\tan\beta(1+\gamma_5) + m_U\cot\beta(1-\gamma_5)]. \tag{30}$$

The $V\varphi(p)\varphi(p')$ couplings are as follows (particles and momenta are incoming)

$$g_{Z\varphi A^0} = \frac{g(p-p')^\mu}{2\cos\theta_W}\begin{cases} \sin(\alpha-\beta), & \text{for } \varphi = H^0, \\ \cos(\alpha-\beta), & \text{for } \varphi = h^0, \end{cases} \tag{31}$$

$$g_{ZH^+H^-} = -\frac{ig\cos 2\theta_W}{2\cos\theta_W}(p-p')^\mu, \tag{32}$$

$$g_{W^+\varphi H^-} = -\frac{ig(p-p')^\mu}{2}\{\sin(\alpha-\beta),\ \cos(\alpha-\beta)\}, \tag{33}$$

for $\varphi = H^0$ and h^0, respectively, and finally

$$g_{W^+A^0H^-} = \frac{g(p-p')^\mu}{2}. \tag{34}$$

There are several comments in order.

- Note the presence of ZH^0A^0 and Zh^0A^0 couplings whereas Bose statistics forbid ZH^0H^0 and Zh^0h^0, and CP-invariance forbids ZH^0h^0.
- There is no tree-level ZW^+H^- vertex in a general model with doublet and singlet Higgs representations.
- The couplings for the vertices $W^+H^0H^-$ and $W^+h^0H^-$, ZH^0A^0 and Zh^0A^0, ZZH^0 and ZZh^0 are pairwise complementary in the sense that if one is suppressed by the combination of mixing angles the other is full-strength. Likewise, if for example ZZh^0 is suppressed the Zh^0A^0 is full-strength.
- As an illustration of the suppression factor, in fig.3 we show $\sin^2(\alpha-\beta)$, relevant for example for ZZh^0 vertex, as a function of m_{h^0} for several values of $\tan\beta$. We see that if the single Higgs boson production process (for instance the Bjorken process ZZh^0 in e^+e^- [18]) turns out to be suppressed then the Higgs pair production Zh^0A^0 may be an efficient source of Higgs bosons [19,20].
- Note that for large $\tan\beta$ the couplings of the neutral scalar Higgs boson which *is* degenerate with the Z boson are within a factor 2 equal to the corresponding couplings of the SM Higgs couplings. Therefore the experimental observation of a 90-GeV Higgs boson with standard couplings will not necesarily mean a confirmation of the standard model [16].
- Another interesting feature of the large $\tan\beta$ scenario is that the couplings of the other neutral scalar Higgs boson that is *not* degenerate with the Z and of the pseudoscalar A^0 to the down-type fermions are strongly enhanced. Therefore the bremsstrahlung of Higgs bosons from bottom quarks or τ leptons may turn out to be important sources of Higgs boson production [19,21].
- We do not discuss here couplings of supersymmetric partners. They can be found in the lecture by A. Bartl [5] and references therein.

Finally let us stress again that the above discussion is correct if radiative corrections can be neglected. If they are important, there are significant quantitative changes of the mass spectrum and couplings although some qualitative features remain similar. Moreover the phenomenology becomes more involved because the Higgs sector depends on other parameters, like m_t, SUSY breaking scale, mixing of squarks etc. For a discussion we refer to the lecture by S. Pokorski [7] and references therein.

5 Looking for SUSY Higgs bosons

Let us now discuss briefly the current experimental limits on Higgs boson masses and prospects for their discovery at present and future colliders. We will consider only e^+e^- and pp machines because ep machines with designed luminosities are not very promising for SUSY Higgs search[22]. Again we neglect radiative corrections and therefore our discussion will be rather limited.

5.1 e^+e^- colliders

Currently the most stringent experimental limits on the tree-level Higgs sector come from LEP I experiments. There are two main production mechanisms: single Higgs boson production via $Z \to Z^*h^0 \to f\bar{f}h^0$ important for relatively small $\tan\beta$, and the Higgs pair production $Z \to A^0h^0$ which is dominant for larger $\tan\beta(\geq 3)$. Since in the limit of large v_2/v_1 the mass of A^0 is close the mass of h^0 the results form LEP are rather restrictive [23]. They are shown in fig.4. Observe that these limits are valid in the MSSM and not in an arbitrary two-Higgs doublet model in which m_{A^0} is not related to m_{h^0} If the Higgs pair production is kinematically suppressed then the Higgs bremsstrahlung off heavy fermions gives a dominant contribution [19,21]. In fig.5 regions in the plane $(m_{h^0}, \tan\beta = v_2/v_1)$ corresponding to various dominant processes for $h^0b\bar{b}$ final state in the MSSM are shown [21]. The notation is as follows: (1) $e^+e^- \to Z \to Zh^0 \to h^0b\bar{b}$, (2) $e^+e^- \to Z \to A^0h^0 \to h^0b\bar{b}$ and (3) $e^+e^- \to Z \to b\bar{b} \to h^0b\bar{b}$. Therefore, to push the lower limit on m_{h^0} above $\sim$ 50 GeV the process (3) has to be taken into account. Impact of radiative corrections on the interplay of these processes has been also discussed in ref.[24]. The search for charged Higgs bosons at LEP gives a lower limit of approximately 43 GeV [25] but in the supersymmetric model one expects $m_{H^\pm} \geq m_W$.

Recently results of new analyses of experimental LEP I data including radiative corrections have been published [26]. OPAL collaboration excludes $m_{h^0} < 3.5$ GeV independently of the decay modes. The results of ALEPH collaboration are given in fig.6. They take $m_t = 140$ GeV, $m_{stop} = 1$ TeV and negligible mixing among stop squarks. For this choice of parameters $m_{h^0} < 41$ GeV and $m_{A^0} < 31$ GeV are excluded at 95% CL. Note that $\tan\beta = 1$ remains allowed when $m_{A^0} > 31$ GeV. If m_t and m_{stop} are varied in their allowed range the limits $m_{h^0} > 41$ GeV and $m_{A^0} > 20$ GeV remain valid. If these data are analysed with tree-level relations then $m_{h^0} = m_{A^0} < 44.4$ GeV for $\tan\beta \gg 1$ would be excluded.

At the tree-level the MSSM model predicts one of the scalar neutral Higgs bosons h^0 to be below and the other H^0 above m_Z with no restrictions on the pseudoscalar A^0. Therefore one could expect that LEP II would either discover h^0 or disprove the MMSM. However, the radiative corrections can push m_{h^0} above m_Z and h^0 may be outside the discovery limit of LEP II. Therefore more powerful e^+e^- colliders will be necessary to test completely the MSSM.

5.2 *pp* colliders

There are two dominant production processes: gluon-gluon fusion $gg \to \varphi$ and associated production with b-quarks $gg \to \varphi b\bar{b}$, where $\varphi = h^0, H^0, A^0$. For LHC they have been recently analysed by Kunszt and Zwirner [27]. The production cross sections for h^0 in the whole mass range and for A^0 up to 250-300 GeV assuming $\tan\beta > 3$ are larger than for the SM Higgs. On the other hand for the heavier neutral Higgs boson the enhancement of the production cross section is observed only at low values of m_{H^0}. In order to overcome the severe background the $\gamma\gamma$ decay modes have been investigated in detail. The results are summarized in fig.7. Discovery regions in $(m_{A^0}, \tan\beta)$ plane are divided into 'standard', 'good' and 'difficult'. The 'standard' regions (grey) are those where only one Higgs boson can be found in $\gamma\gamma$ decay mode with SM properties. The 'good' (hatched) are where the heavier Higgs boson can be found at LHC. In the 'difficult' (empty) region no measurable physics signal was found. Fortunately this region will be well covered by LEP II, and therefore *pp* machines appear as complementary to e^+e^- colliders.

Our discussion of the strategies for SUSY Higgs searches is far from complete. For more details we refer to proceedings of many workshops devoted to detailed studies of physics potentials of LEP, HERA, LHC and SSC colliders.

6 Conclusions

It has been obvious for many years that the Higgs sector of the SM is most sensitive to interactions at mass scales higher than actually probed experimentally. Therefore many theoretical problems concerning the Higgs sector have emerged. Only SUSY models are completely consistent internally, suffer no known phenomenological problems and allow to solve the naturalness/hierarchy problems with elementary Higgs bosons. We have reviewed the motivation for and properties of the MSSM. Supersymmetry requires the introduction of new degrees of freedom. Every particle of the SM receives a supersymmetric partner and at least two Higgs doublets are required to give masses to fermions. Simultaneously SUSY imposes severe constraints on the otherwise enormously model-dependent self-couplings of the two-Higgs doublet model. At the tree level the Higgs sector is fully determined in terms of only two parameters. With the heavy top quark new parameters make the Higgs sector more complicated. Nevertheless the MSSM suggest that some of the Higgs bosons are accesible to future colliders. If Higgs boson(s) are found their couplings have to be analysed to understand their origin. Thus, the Higgs sector may play a crucial role in revealing the nature of new physics as well as the secrets of the spontaneous symmetry breaking.

ACKNOWLEDGMENTS

I would like to thank A. Buras, P.H. Chankowski, B. Grzadkowski, W. Hollik, H.P. Nilles and S. Pokorski for many discussions on supersymmetric models. I am also grateful for warm hospitality to Physik-Department, TUM, Garching, where these lectures have been prepared.

References

[1] S. Glashow, *Nucl. Phys.* **B22** (1961) 579;
A. Salam, in *Elementary Particle Theory*, ed. N. Svartholm, (1968) 367;
S. Weinberg, *Phys. Rev. Lett.* **419** (1967) 1264.

[2] P. W. Higgs, *Phys. Lett.* **12**, 132 (1964);
F. Englert and R. Brout, *Phys. Rev. Lett.* **13**, 321 (1964);
G. S. Guralnik, C. R. Hagen and T. W. Kibble, *Phys. Rev. Lett.* **13**, 585 (1964).

[3] H. Georgi, H.R. Quinn and S. Weinberg, *Phys. Rev. Lett.* **33** (1974) 451;
G. 't Hooft, in *Recent Developments in Gauge Theories*, Proc. NATO ASI, Cargese, ed. G. 't Hooft (Plenum, New York, 1980).

[4] J. Wess and B. Zumino, *Phys. Lett.* **B49** (1974) 52;
A. Salam and J. Strathdee, *Phys. Rev.* **D11** (1975) 1521;
for a review see J. Wess and J. Bagger, *Supersymmetry and Supergravity*, Princeton University Press (1983).

[5] A. Bartl, these Proceedings.

[6] Y. Okada, M. Yamaguchi and T. Yamagido, *Prog. Theor. Phys.* **85** (1991) 1;
H. Haber and R. Hempfling, *Phys. Rev. Lett.* **66** (1991) 1815;
J. Ellis, G. Ridolfi and F. Zwirner, *Phys. Lett.* **B257** (1991) 83, CERN-TH.6002/91;
P.H. Chankowski, S. Pokorski and J. Rosiek, preprint MPI-Ph/91-57.

[7] S. Pokorski, these Proceedings.

[8] R. Hempfling, these Proceedings.

[9] K. Sliwa, CDF Collab., Proc. 25th Rencontres de Morionds, Les Arcs (1990).

[10] H.P. Nilles, these Proceedings;
see also for example H.P. Nilles *Phys. Rep.* **C110** (1984) 1;
H.E. Haber and G.L. Kane, *Phys. Rep.* **C117** (1985) 75.

[11] P. Fayet and J. Iliopoulos, *Phys. Lett.* **B51** (1974) 461.

[12] L. Girardello and M.T. Grisaru, *Nucl. Phys.* **B194** (1982) 65.

[13] K. Inoue, A. Kakuto, H. Komatsu and S. Takeshita, *Prog. Theor. Phys.* **67** (1982) 1889, **68** (1982) 927;
L.E. Ibanez and G.G. Ross, *Phys. Lett.* **B110** (1982) 215;
H.P. Nilles, *Phys. Lett.* **B115** (1982) 193.

[14] J.F. Gunion and H.E. Haber, *Nucl. Phys.* **B272** (1986) 1, **B278** (1986) 449;
S. Dawson, J.F. Gunion, H.E. Haber and G.L. Kane, *The Physics of Higgs Bosons: The Higgs Hunter's Guide* (Addison Wesley, Reading, MA 1990).

[15] P. Krawczyk and S. Pokorski, *Phys. Rev. Lett.* **60** (1988) 182;
P.Q. Hung and S. Pokorski, Fermilab-Pub-87/211-T.

[16] B. Grzadkowski, J. Kalinowski and S. Pokorski, Proc. XII Warsaw Conference, Kazimierz (1989), *Phys. Lett.* **B241** (1990) 534.

[17] P.H. Chankowski, private communication.

[18] J.D. Bjorken, in Proc. SLAC Summer Institute on Particle Physics, 1976, ed. M. Zipf.

[19] J. Kalinowski and S. Pokorski, *Phys. Lett.* **B219** (1989) 116.

[20] G.F. Giudice, *Phys. Lett.* **B208** (1988) 315.

[21] J. Kalinowski and H.P. Nilles, *Phys. Lett.* **B255** (1991) 134.

[22] K.J.F. Gaemers, R.M. Godbole and M. van der Horst, Proc. HERA Workshop, ed. R.D. Peccei, Hamburg (1987);
R. Rückl, Proc. Large Hadron Collider Workshop, ed. G. Jarlskog and D. Rein, Aachen (1990).

[23] ALEPH Collab., D. Decamp et al., *Phys. Lett.* **B246** (1990) 306;
OPAL Collab., M. Akrawy et al., *Phys. Lett.* **B236** (1990) 224;
DELPHI Collab., P. Abreu et al., *Nucl. Phys.* **B342** (1990) 1;
L3 Collab., B. Adeva et al., *Phys. Lett.* **B248** (1990) 203.

[24] J. Kalinowski, preprint TUM-T31-14/91, to appear in Proc. XIV Warsaw Conference, Warsaw (1991).

[25] ALEPH Collab., D. Decamp et al., *Phys. Lett.* **B236** (1990) 86.

[26] OPAL Collab., P.D. Acton et al., *Phys. Lett.* **B268** (1991) 122;
ALEPH Collab., D. Decamp et al., *Phys. Lett.* **B265** (1991) 475.

[27] Z. Kunszt and F. Zwirner, Proc. Large Hadron Collider Workshop, ed. G. Jarlskog and D. Rein, Aachen (1990).

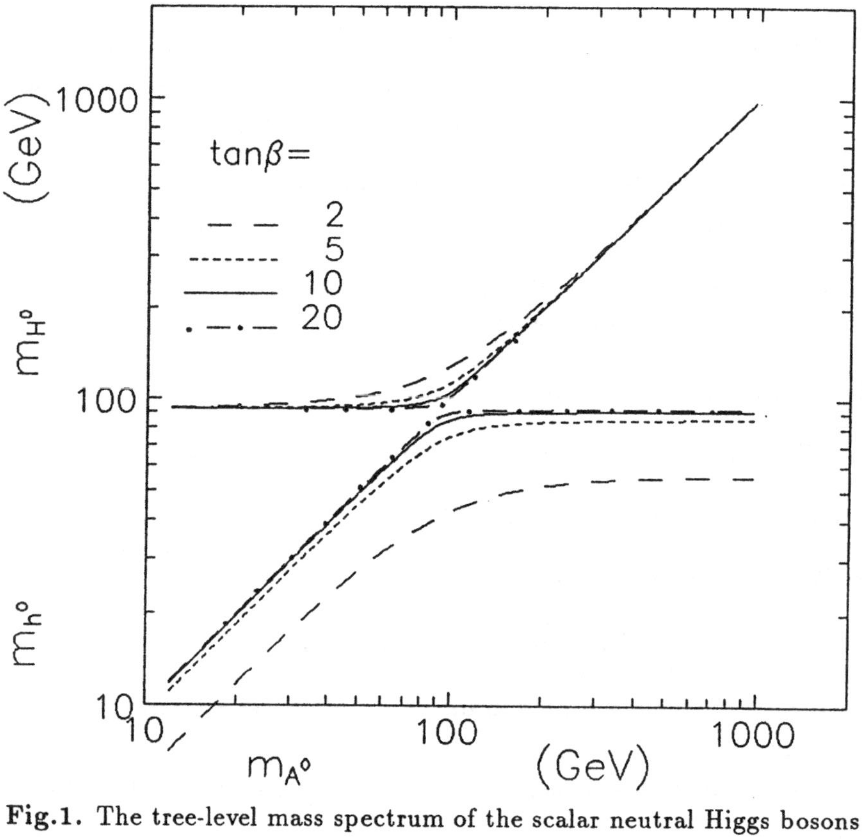

Fig.1. The tree-level mass spectrum of the scalar neutral Higgs bosons in the MSSM as a function of m_{A^0} for several values of $\tan\beta$. The set of curves below (above) the m_Z is for the h^0 (H^0) Higgs boson.

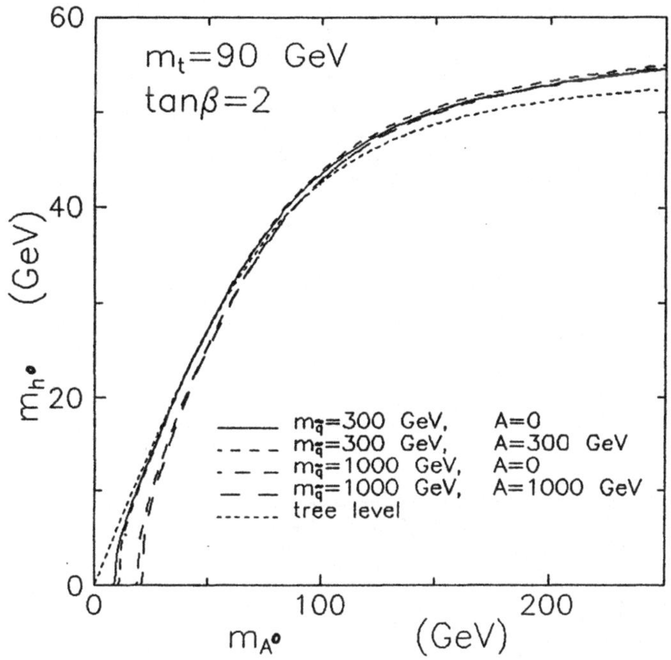

Fig.2. The mass of h^0 with one-loop radiative corrections for a set of parameters given in the figure compared to the tree-level result.

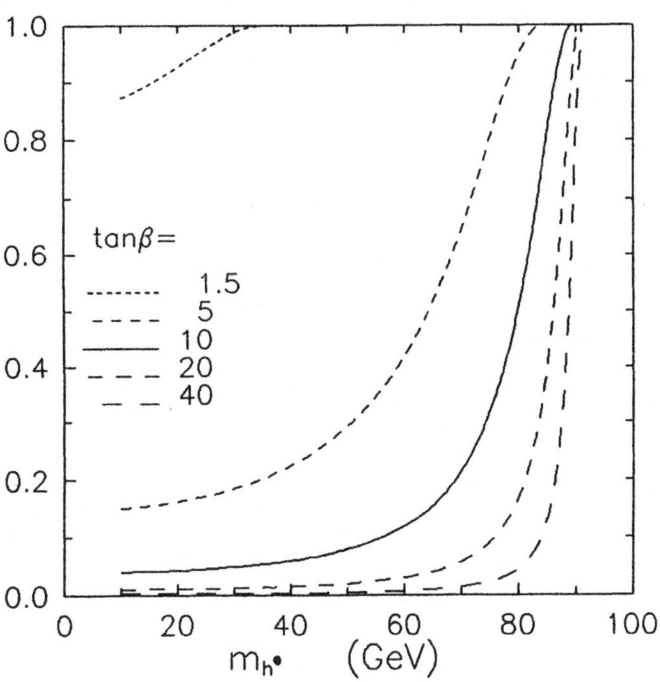

Fig.3. $\sin^2(\alpha - \beta)$ as a function of m_{h^0} for several values of $\tan\beta$ without radiative corrections.

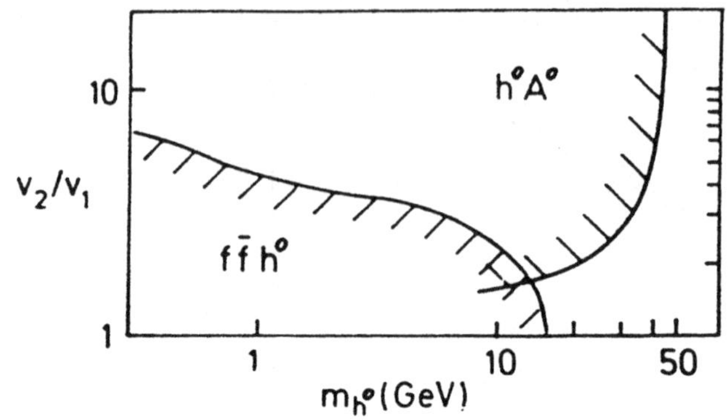

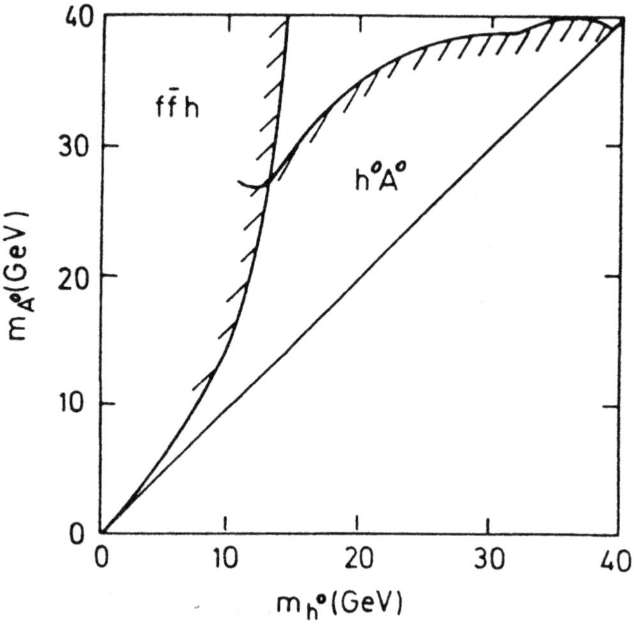

Fig.4. Limits on Higgs boson masses from ALEPH Collab. [23].

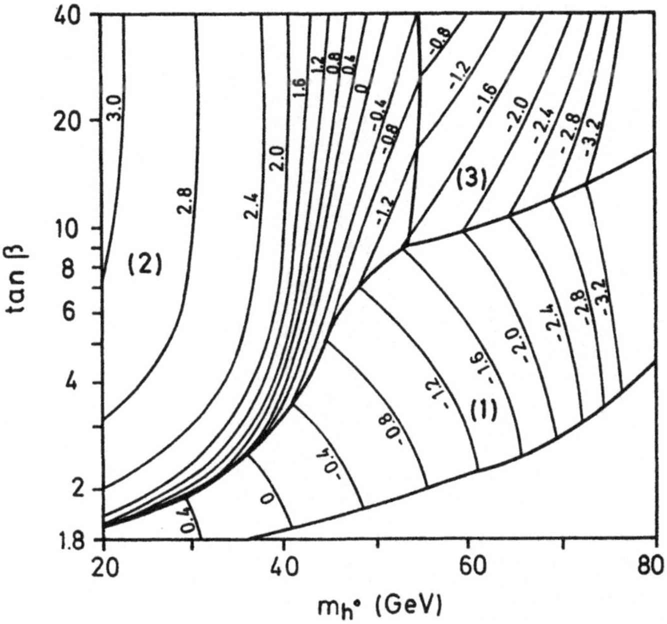

Fig.5. Regions in the plane $(m_{h^0}, v_2/v_1)$ corresponding to the dominant processes as explained in the text for the $h^0b\bar{b}$ final state. Contour lines are drawn for $\log_{10}(\sigma/1\,pb)$.

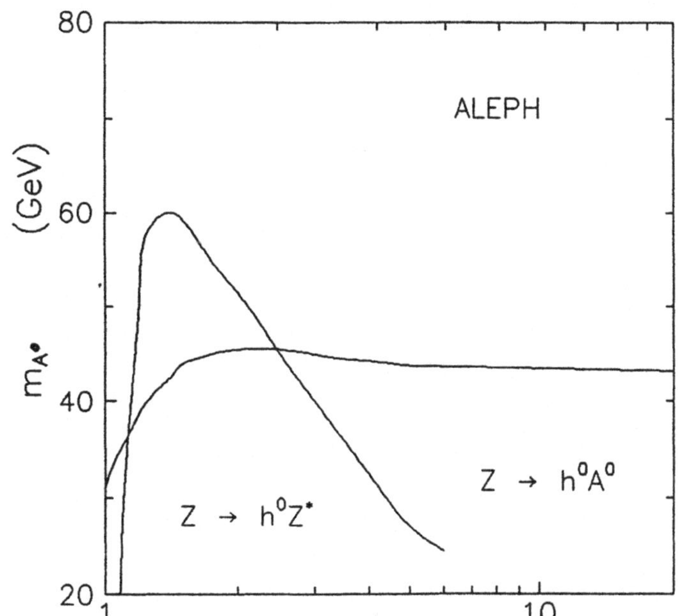

Fig.6. In the plane $(m_{A^0}, \tan\beta)$ domains excluded by ALEPH searches for $e^+e^- \to Z^*h^0$ and $\to A^0h^0$. One loop radiative corrections with $m_t = 140$ GeV and $m_{stop} = 1$ TeV are taken into account [26].

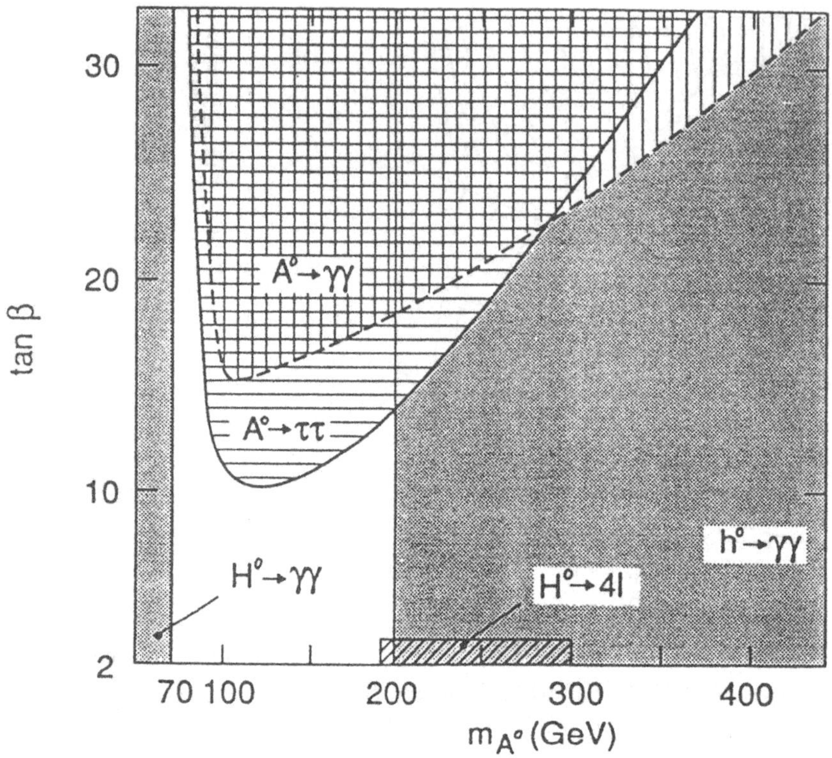

Fig.7. Standard (grey), good (hatched) and difficult (empty) discovery regions at LHC in the $(m_{A^0}, \tan\beta)$ plane [27].

MASS SPECTRA OF SUPERSYMMETRIC PARTICLES AND EXPERIMENTAL BOUNDS

Francesca M. Borzumati

II. Institut für Theoretische Physik ‡
der Universität Hamburg,
D-2000 Hamburg 50, Germany

ABSTRACT

In this lecture we analyze the Minimal Supersymmetric Standard Model with radiatively induced breaking of $SU(2)_L \times U(1)_Y$. An underlying Grand Unification of the strong and electroweak symmetries is assumed. The mass spectra of the supersymmetric particles are studied also in the case of heavy top quark masses and large values of v_2/v_1. In addition, a comparison with some of the existing lower bounds on these masses is given.

1. Introduction

The topic of this lecture is the analysis of the low-energy spectrum of an explicitly broken globally supersymmetric theory, as obtained after the spontaneous breaking of a local (N=1) Supersymmetry (SUSY) [1].

There is no need to give a motivation for the introduction of such a theory, already treated in great detail in the first two lectures of this workshop. It is enough only to recall that this theory fulfills the requirement of protecting the scalar fields from jumping to higher scales, like the Planck scale M_P or a possible Grand Unification scale M_X. The local Supersymmetry is able to cure the problem of quadratic divergences [2], and the spontaneous breaking of this symmetry [3] guarantees that this feature is not lost after the breaking. After decoupling of gravity, i.e. once the limit $M_P \to \infty$ is taken, we are left with a globally supersymmetric theory plus some terms of "soft breaking" of this symmetry. There is a very precise technical meaning of these words [4], which we will not investigate here.

The theory we are dealing with is the supersymmetrized version of the most *minimal* extension of the Standard Model (SM) compatible with Supersymmetry. That is, a two Higgs doublet model (2HDM) considered as embedded in an underlying Grand Unified Theory (GUT).

One of the reasons for the success of this theory is the observation that if the effective SUSY breaking is $\mathcal{O}(M_Z)$, then the logarithmic corrections to the mass parameters of the scalar field potential are able to induce the breaking of the electroweak gauge symmetry. The problem of the $SU(2)_L \times U(1)_Y$ symmetry breaking is then "solved" by being linked to the mechanism of breaking of the local Supersymmetry at the Planck scale. The dynamic of this mechanism is unknown, but it can be parametrized by a finite

‡Supported by the Bundesministerium für Forschung und Technologie, 05 4HH 92P/3, Bonn, FRG.

number of soft breaking terms, resulting in a finite number of new free parameters to be added to the ones already present in the globally supersymmetric Lagrangian, i.e. μ plus the usual SM parameters. The *minimal* choice of only three new parameters, m, A and M, can be justified by a criterion of economy and makes this model, called then Minimal Supersymmetric Standard Model, particularly suitable for phenomenological studies.

The rich range of possibilities offered by the supersymmetric mass spectrum has made this a vital model capable of evading experimental detection to date. Only a relatively small restriction of the allowed SUSY parameter space has been achieved so far through experimental searches. The recent precision measurements at LEP of the electroweak gauge coupling constants and the observation of their compatibility with the supersymmetric Grand Unification scenario [5], are far from providing any evidence for this model. Nevertheless, they may be suggestive that there is more to it than just theoretical speculation.

In this lecture we shall analyse in detail the mass spectrum of this model and we shall compare it, in a few cases, to the existing experimental lower bounds on supersymmetric masses. We shall give particular emphasis to the unification of the $SU(3)_{\rm C} \times SU(2)_{\rm L} \times U(1)_{\rm Y}$ gauge interactions, to the corrections of the mass parameters during the evolution from the GUT scale § to the electroweak scale and finally to the mechanism of breaking of $SU(2)_{\rm L} \times U(1)_{\rm Y}$.

Several analyses of these issues exist in the literature. In the earlier papers [6,7], not very high values of the top mass were considered and the supersymmetric parameter μ was taken to be small. In a more recent series of papers [8–12], the higher constraints on squark and gluino masses are implemented and the possibility of heavy top masses is considered. As a consequence, the increase of the value of $\tan\beta$ no longer allows the neglect of the Yukawa couplings of the bottom quark, h_b, and of the tau lepton, h_τ. In some of these papers the correct constraints on the GUT scale Yukawa couplings imposed by Grand Unified Theories are also implemented [10–12] and some uncertainty in the on-shell value of the bottom quark mass is allowed [11]. In some cases also the consequences of high values of $\tan\beta$ are studied [10,12]. In others, particular care is taken in the correct trading of low-energy parameters with high-energy ones when the radiative breaking of the electroweak sector is imposed [11,12]. The correspondence between high- and low-energy parameters is in general not one-to-one.

We shall discuss these issues in the following sections. The numerical results presented are taken from different sources in the literature and may be, at times, not in complete agreement because of different choices of α_S, $\sin^2\theta_W$ or m_b. They have therefore to be taken as indicative and some variation has to be considered possible due to changes of these input parameters.

The lecture is organized as follows. In section 2 we present the Lagrangian for this model. Section 3 explains the procedure followed in selecting from between all the possible points of the supersymmetric parameter space the ones allowing a radiatively induced breaking of $SU(2)_{\rm L} \times U(1)_{\rm Y}$. In sections 4, 5 and 6, a detailed description of the one-loop renormalization effects on different sectors of the model is given. Sections 7, 8

§ Renormalization effects between M_P and M_X are neglected here as well as in the existing literature.

and 9 deal with the theoretical predictions for the allowed range of supersymmetric masses and their comparison to the existing experimental lower bounds. Some conclusions will be finally drawn in section 10.

2. Lagrangian

The request of minimality mentioned in the introduction, strongly constrains the structure of the Lagrangian of our model. The requirements that only the fields due to the supersymmetrization of the SM are present and that matter parity is preserved, immediately dictate the expression of the $SU(3)_{\rm C} \times SU(2)_{\rm L} \times U(1)_{\rm Y}$ invariant superpotential:

$$W = h_U^{ij} Q_i U_j^c H_2 + h_D^{ij} Q_i D_j^c H_1 + h_E^{ij} L_i E_j^c H_1 + \mu H_1 H_2 \tag{1}$$

where the chiral matter superfields Q, U^c, D^c, L, E^c, H_1 and H_2 transform as follows under $SU(3)_{\rm C} \times SU(2)_{\rm L} \times U(1)_{\rm Y}$:

$$\begin{array}{llllll} Q & \equiv (3,2,1/6); & U^c & \equiv (\bar{3},1,-2/3); & D^c & \equiv (\bar{3},1,1/3); \\ L & \equiv (1,2,-1/2); & E^c & \equiv (1,1,1); & & \\ H_1 & \equiv (1,2,-1/2); & H_2 & \equiv (1,2,1/2)\,. & & \end{array} \tag{2}$$

Isospin and colour indices are contracted in the usual way. The couplings h_U, h_D and h_L are 3×3 matrices in the generation space $(i, j = 1, 2, 3)$.

The expression for the soft Supersymmetry breaking terms turns out to be quite simplified by the assumption of having a flat Kähler metric. At the Grand Unified scale M_X, they appear as:

- A cubic gauge invariant polynomial in the complex scalar fields:

$$S = mA \left[h_U \tilde{Q} \tilde{U}^c H_2 + h_D \tilde{Q} \widetilde{D}^c H_1 + h_E \tilde{L} \tilde{E}^c H_1 \right] + Bm\mu H_1 H_2 + h.c. \tag{3}$$

  where the tilde denotes the *scalar* component of the chiral matter superfields Q, U^c, D^c, L and E^c, while for simplicity we denote by H_1 and H_2 also the scalar components of the Higgs superfields H_1 and H_2. Later on, we shall often indicate with $\tilde{q}_L$ $(\tilde{u}_L, \tilde{d}_L, \ldots, \tilde{t}_L, \tilde{b}_L)$ the components of the three scalar $SU(2)_{\rm L}$ doublets $\tilde{Q}$ and similarly, with $\tilde{q}_L^c$ or $\tilde{q}_R$ the components of the three scalar $SU(2)_{\rm L}$ singlets $\tilde{U}^c$ and $\widetilde{D}^c$. We shall also (improperly) refer to $\tilde{q}_L$ and $\tilde{q}_R$ as the "left-handed" and "right-handed" component of the squark $\tilde{q}$. A similar convention will be adopted for the scalar leptons. The coefficients A and B in (3) are c-numbers and, in the presence of a flat Kähler metric, the equality $B = A - 1$ holds.

- A universal mass term for the scalar components y_i of the chiral superfields:

$$\mathcal{M}^2 \equiv m^2 \Sigma_i |y_i|^2 \,. \tag{4}$$

- Gaugino Majorana mass terms:

$$\widehat{M} \equiv \frac{M}{2}(\lambda_1\lambda_1 + \lambda_2\lambda_2 + \lambda_3\lambda_3) + h.c. \tag{5}$$

where λ_1, λ_2 and λ_3 denote the two-component gaugino fields of $U(1)_Y$, $SU(2)_L$ and $SU(3)_C$, respectively. Notice that in (5), the Grand Unification constraint of equal gaugino masses at M_X is added to the usual requirement of minimality.

Hence, the Minimal Supersymmetric Standard Model considered here is described at the Grand Unification scale by the Lagrangian

$$L = W + S + \mathcal{M}^2 + \widehat{M} + \text{kinetic terms} \,. \tag{6}$$

In order to discuss the physical implications of this Lagrangian at low-energy (i.e. at the electroweak scale), we need to renormalize the relevant parameters from M_X down to M_Z. A detailed discussion of this procedure will be given in the following sections. For the time being, for the sake of setting our notation, we shall focus on the expression of the scalar potential at M_Z. The correct $SU(2)_L \times U(1)_Y$ breaking down to $U(1)_{em}$ is achieved by the vacuum

$$\langle H_1 \rangle = \begin{pmatrix} v_1 \\ 0 \end{pmatrix}; \quad \langle H_2 \rangle = \begin{pmatrix} 0 \\ v_2 \end{pmatrix}; \quad \langle \tilde{q} \rangle = \langle \tilde{\ell} \rangle = 0, \tag{7}$$

where the last two equalities have to be satisfied by all the scalar quarks and leptons of the model. It is possible to redefine the phases of H_1 and H_2 so that v_1 and v_2 turn out to be real and non-negative. The low-energy Higgs potential along the neutral direction is then

$$V = \mu_1^2 |H_1^0|^2 + \mu_2^2 |H_2^0|^2 - \mu_3^2 (H_1^0 H_2^0 + h.c.) + \frac{1}{8}(g^2 + g'^2)(|H_1^0|^2 - |H_2^0|^2)^2, \tag{8}$$

where g and g' denote the $SU(2)_L$ and $U(1)_Y$ gauge coupling constants, respectively, and $\mu_{1,2,3}$ are running mass parameters. At M_X they read

$$\mu_1^2 = \mu_2^2 = m^2 + \mu^2 \tag{9}$$

$$\mu_3^2 = -Bm\mu \,. \tag{10}$$

The minimization of the Higgs potential (8) yields the two constraints

$$v_1^2 + v_2^2 = \frac{2\left[\mu_2^2 - \mu_1^2 - (\mu_1^2 + \mu_2^2)\cos 2\beta\right]}{(g^2 + g'^2)\cos 2\beta} \tag{11}$$

$$\sin 2\beta = \frac{2 v_2 v_1}{v_1^2 + v_2^2} = \frac{2\mu_3^2}{\mu_1^2 + \mu_2^2} \,. \tag{12}$$

By recalling that $2M_Z^2 = (g^2 + g'^2)(v_1^2 + v_2^2)$, one obtains for (11) the more convenient form

$$\tan^2 \beta = \left(\frac{v_2}{v_1}\right)^2 = \frac{\mu_1^2 + M_Z^2/2}{\mu_2^2 + M_Z^2/2} \,. \tag{13}$$

For the above desired minimum (7) to occur we must enforce a stability condition which ensures that the potential is bounded from below together with the condition that the origin $\langle H_1 \rangle = \langle H_2 \rangle = 0$ is a local maximum. They read:

$$\mathcal{S} \equiv \mu_1^2 + \mu_2^2 - 2|\mu_3^2| > 0 \tag{14}$$

$$\mathcal{B} \equiv \mu_1^2 \cdot \mu_2^2 - \mu_3^4 < 0 \,, \tag{15}$$

respectively. From these last two conditions it is clear that the tree-level potential, where μ_1, μ_2, μ_3 take the GUT scale values provided in (9) and (10), cannot yield the desired minimum. The renormalization effects are however big enough to ensure that the stability and breaking conditions (14) and (15) can be simultaneously satisfied together with the two minimization conditions (12) and (13).

3. Numerical Procedure

The Lagrangian described in the previous section has still to undergo renormalization down to the weak scale. These evolution effects will be analysed in the following sections for each sector of the theory. In this section, following the analysis of [11], we shall present the kind of procedure one uses to restrict the SUSY parameter space to those regions where the requirement of radiative breaking of the electroweak sector can be correctly implemented.

The free parameters present in the Lagrangian (6) are the SUSY breaking parameters m, M, A plus the supersymmetric parameter μ. Gauge couplings and Yukawa couplings appear also in (6). The former ones are quite well known at low energy. As far as the latter ones are concerned, only the third generation fermions are considered here as massive and we take the limit of Cabibbo-Kobayashi-Maskawa matrix as being equal to the unit matrix $\mathbb{1}$ ¶. The matrices h_U, h_D and h_L are diagonal and reduce at low-energy to the elements $h_t(M_Z)$, $h_b(M_Z)$ and $h_\tau(M_Z)$ which can be linked to $m_t(M_Z)$, $m_b(M_Z)$ and $m_\tau(M_Z)$ through the vacuum expectation values v_1 and v_2. The low-energy unknowns are then $m_t(M_Z)$ and $\tan\beta$. In principle, there is some uncertainty also in the value of m_b. As we shall see in section 5, the limited freedom for this parameter is partially cancelled by a further Grand Unification condition which also simultaneously reduces the range of allowed values of m_t and $\tan\beta$. For the time being, in this section, we shall neglect this complication and consider m_b as known.

Contact has to be made between the low-energy and high-energy parameters of the theory. This will be achieved through the renormalization group equations (RGE) and the two minimization conditions (12) and (13). These two conditions will enable us to fix two of the high-energy parameters, for example A and μ ‖.

In practice we shall proceed as follows. We shall evolve the low-energy gauge couplings $\alpha_S(M_Z)$, $\sin^2\theta_W$ and $\alpha_{em}(M_Z)$ to higher values of energy until a unification point is found. For an effective SUSY mass at the weak scale ($M_{SUSY} = M_Z$) and for

$$\alpha_S(M_Z) = 0.114, \qquad \sin^2\theta_W = 0.233, \qquad \alpha_{em}(M_Z) = 1/128 \tag{16}$$

we obtain the following values for the unification mass and for the common gauge coupling constant at the unification point:

$$M_X = 1.5 \times 10^{16}, \qquad \alpha_X^{-1}(M_X) = 24.4\ . \tag{17}$$

¶The results presented in Fig. 1 include also some generational mixing. This and some of the following figures were obtained in [11] with the aim of studying the effects of flavour change induced by SUSY. These results are obviously still valid in the approximation considered here. The values of the squark masses used for these figures, in fact, are not sizeably affected by the small inter-generational mixing elements in the squark mass matrices.

‖A different procedure of choosing the value of A and solving for $\tan\beta$ is used in [12].

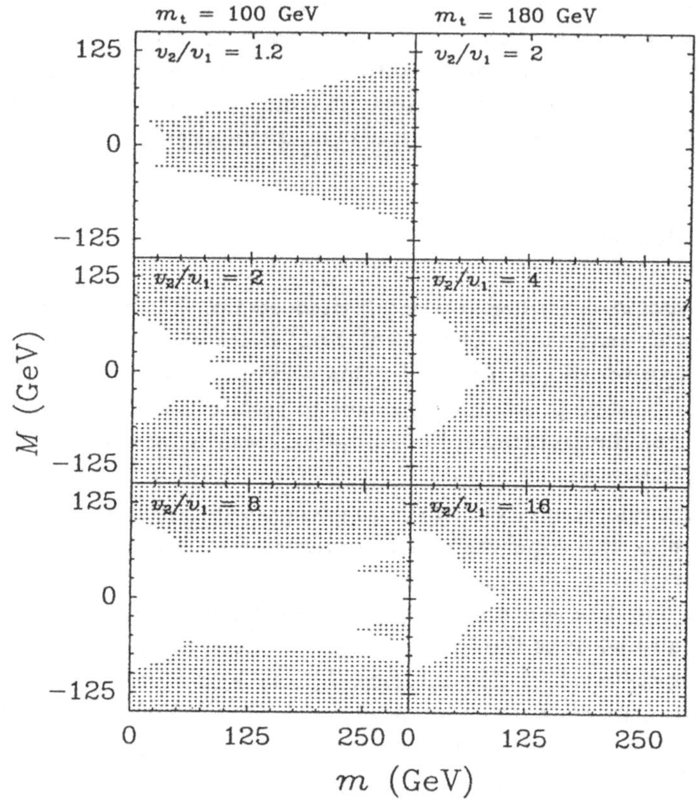

Figure 1: **Possible configurations in the (m, M) space allowed by the conditions of radiative breaking of the electroweak sector, for different values of m_t and $\tan\beta$. The range considered here is $|m|$ and $m < 150\,\mathrm{GeV}$. Each allowed point in this plot can have up to a fourfold degeneracy in (A, μ).**

A change of M_{SUSY} up to the TeV range affects mildly these results. For $M_{\mathrm{SUSY}} > M_Z$, the gauge couplings have to be evolved in two steps, according to the 2HDM-RGE up to the M_{SUSY} scale and then up again according to the SUSY-RGE. This type of analysis is obviously a simplified one since a sharp transition between the two regions is assumed. No threshold effects are included when the energy becomes high enough to allow the appearence of a new degree of freedom in the evolution of the gauge coupling constants and of all the other parameters in the Lagrangian (6). The same kind of sharp transition is assumed at the scale M_X.

Once M_X and α_X are known, for a fixed choice of $(m_t, \tan\beta)$, the Yukawa couplings can be evolved up to M_X in the same way. The values obtained will be used as high-scale input in the RGE for the remaining parameters whose high-scale boundary conditions are given by (10) and by:

$$\begin{aligned} &m^2_{\tilde{Q}} = m^2_{\tilde{U}^c} = m^2_{\tilde{D}^c} = m^2_{\tilde{L}} = m^2_{\tilde{E}^c} = m^2 \\ &A_t = A_b = A_\tau = A \ . \end{aligned} \tag{18}$$

A formal solution for all the low-energy parameters, in terms of m, A, M and μ and their products, can be easily obtained by inspection of the relevant RGE. Once a pair (m, M) is chosen, the two minimization conditions (12) and (13) will provide solutions for A and μ. These are in general not unique solutions. Therefore a choice of the four parameters $(m, M, m_t, \tan\beta)$ can correspond to different possible supersymmetric realizations as observed in [11] and [12].

Now, among the "multiple" points of the four-dimensional space (m, M, m_t, $\tan\beta$) the ones giving the correct realization of SUSY have to be selected. For each point, it has to be checked for example, that the conditions (14) and (15) are verified, that no parameter of the theory is growing too much during the evolution from M_Z to M_X, and so on. Moreover, it has especially to be checked that no further unphysical vacua breaking charge and/or colour are accidentally generated. This can still happen, even after the removal of the points (m, M, m_t, $\tan\beta$) where negative values for the squared masses of squarks and sleptons are obtained. The minimization of the Higgs potential (8) is, in fact, not sufficient to guarantee that the vacuum (7) is the absolute minimum of the full scalar low-energy potential. In practice, the minimization of the full potential cannot be achieved through simple means. In order to avoid dangerous charge and colour breaking absolute minima, one imposes further constraints on the allowed range of the SUSY parameters [13–15]. These conditions are only necessary to avoid such minima [14]. They are given by:

$$A_t m^2 < 3(m^2_{\tilde{t}_L} + m^2_{\tilde{t}^c_L} + \mu^2_2) \tag{19}$$

$$A_b m^2 < 3(m^2_{\tilde{b}_L} + m^2_{\tilde{b}^c_L} + \mu^2_1) \tag{20}$$

$$A_\tau m^2 < 3(m^2_{\tilde{\tau}_L} + m^2_{\tilde{\tau}^c_L} + \mu^2_1) \tag{21}$$

In [15] one finds a detailed discussion on the conditions for which (19)-(21) may become sufficient. In general, no necessary and sufficient conditions can be derived analytically, and the absence of stable colour and charge breaking vacua has to be checked numerically point by point in the parameter space. The unphysical vacua become stable in the direction of those charged and/or coloured scalars whose fermion partners have small Yukawa couplings (compared to the $SU(2)_{\rm L}$ gauge constant). This is certainly the case for the bottom and tau fields, when one considers not too large values of $\tan\beta$. On the other hand, precisely in the limit of small Yukawa couplings, (19)-(21) become also sufficient. This does not apply to (19), which remains only necessary. However, for large enough Yukawa couplings it is possible to show that the unwanted minima lie above the "correct" ones.

Finally, the existing experimental lower bounds on the supersymmetric masses have to be considered and an analysis of the cuts that these impose on the SUSY parameter space has to be made. Care has to be taken to use bounds as model independent as possible not conflicting with the main assumptions of the model we consider. These bounds are not taken into account in Figs. 1, 6, 7 and 8. The aim there is mainly to show the kind of restrictions imposed by the condition of radiatively induced electroweak breaking in the region $m < 300\,\text{GeV}$ and $|M| < 150\,\text{GeV}$ for several values of m_t and $\tan\beta$.

We close this section with a comment. It has been observed that one-loop corrections to the Higgs potential can unexpectedly play a major role [16–18]. Only for certain ranges of the scale at which one stops the renormalization of the parameters are we allowed to neglect the one-loop corrections and use the *RG improved* tree-level potential [16]. For the results presented here we have not tried to determine, case by case, the correct range of the renormalization scale. We have, however, a posteriori verified the stability of our results for different choices of the effective low-energy scale in the $M_Z - 2M_Z$ range. A more careful procedure was followed only in the particular case of $m_t = 150\,\mathrm{GeV}$ and $\tan\beta = 40$ described in section 6.

4. Gauge Couplings

We shall verify in this section that SUSY does allow the unification of the three gauge interactions without predicting too short a lifetime for the proton. To this end, we start by normalizing the hypercharge Y, in order to guarantee that the generators of $SU(3)_\mathrm{C}$, $SU(2)_\mathrm{L}$ and $U(1)_\mathrm{Y}$ have the same normalization once embedded in a Grand Unifying group. The one-loop evolution equations for coupling constants, g_3, g_2, g_1 (or g_S, g and $\sqrt{5/3}\ g_Y$) look like [19]

$$\frac{\mathrm{d}\,\tilde{\alpha}_i}{\mathrm{d}\,t} = -b_i\,\tilde{\alpha}_i^2 \tag{22}$$

where $\tilde{\alpha}_i$ is defined as $\tilde{\alpha}_i \equiv \alpha_i/(4\pi) = g_i^2/(16\pi^2)$, the variable t is given by $t = \log(\mu/M_Z)$ and b_i are the coefficients of the beta function of the couplings g_i. These equations can be easily integrated to give, in terms of the low-energy parameters $\alpha_S(M_Z)$, $\alpha_{\mathrm{em}}(M_Z)$ and the Weinberg angle θ_W, the following relations:

$$\begin{aligned} \tilde{\alpha}_S^{-1}(M_Z) &= \tilde{\alpha}_3^{-1}(\mu) - b_3\,t \\ \sin^2\theta_W\ \tilde{\alpha}_{\mathrm{em}}^{-1}(M_Z) &= \tilde{\alpha}_2^{-1}(\mu) - b_2\,t \\ \frac{3}{5}\cos^2\theta_W\ \tilde{\alpha}_{\mathrm{em}}^{-1}(M_Z) &= \tilde{\alpha}_1^{-1}(\mu) - b_1\,t\,. \end{aligned} \tag{23}$$

The unification of the three interactions is then possible if there exists a scale $\mu = M_X$, such that

$$\tilde{\alpha}_3(M_X) = \tilde{\alpha}_2(M_X) = \tilde{\alpha}_1(M_X) = \tilde{\alpha}_X(M_X)\,. \tag{24}$$

The system (23) is a system of three equations and two unknowns, $\tilde{\alpha}_X(M_X)$ and $t_X \equiv \log(M_X/M_Z)$ which may not necessarily be solvable. Moreover, even if a solution is found, one still has to verify that it is not in conflict with the existing lower bound for the proton lifetime. It is easy to see from (23) that a unifying point for the three couplings can be found if $\alpha_S(M_Z)$, $\alpha_{\mathrm{em}}(M_Z)$ and $\sin^2\theta_W$ satisfy the consistency condition

$$\frac{5}{3}\,\frac{\sin^2\theta_W\alpha_S(M_Z) - \alpha_{\mathrm{em}}(M_Z)}{\left(1-\frac{8}{3}\sin^2\theta_W\right)\alpha_S(M_Z)} = \frac{b_2-b_3}{b_1-b_2}\,. \tag{25}$$

The SM does not fulfill this requirement. We recall that the one-loop expression for the coefficients b_i is given by

$$(b_3, b_2, b_1) = \left(-11+\frac{4}{3}\mathrm{N_G},\ -\frac{22}{3}+\frac{4}{3}\mathrm{N_G}+\frac{1}{6}\mathrm{N_H},\ \frac{4}{3}\mathrm{N_G}+\frac{1}{10}\mathrm{N_H}\right) \tag{26}$$

where N_G is the number of generations and N_H is the number of light Higgs doublets. For $N_G = 3$ and $N_H = 1$, the right hand side of (25) gives a value which is not possible to match on the left hand side, not even when the experimental uncertainty on $\alpha_S(M_Z)$ and $\sin^2\theta_W$ is taken into account. A careful analysis of these issues is contained in[20] and in [5] from where the Fig. 2 is borrowed. The results shown in Fig. 2a are based on the two-loop evolution equations for the running coupling constants. The same qualitative features can be obtained also by simply using the coefficients in (26).

A quick inspection of (26) shows that the unification of the electroweak and strong couplings cannot be achieved in the two simplest possible extensions of the SM:

- An increase in the number of generations, for example, gives the same result as the SM with $N_G = 3$. The reason is that N_G enters with the same weight in all the coefficients b_i and cancels on the right hand side of (25).

- An increase in the number of Higgs doublets is in principle enough to achieve unification. By plugging (26) in (25) one gets in fact a solvable equation in N_H with possible solution $N_H = 7$. The two-loop analysis lowers this value to $N_H = 6$ [5]. Unfortunately these values do not give viable unification points, as it can be seen in Fig. 2b. A generic increase of N_H, in fact, while not touching the slope of the strong coupling, slows down the increase of α_2^{-1} more than it speeds up the falling of α_1^{-1}. If one keeps increasing the value of N_H, eventually the unification becomes possible, but for values of energy which give too fast a proton decay.

In contrast, SUSY introduces a drastic change in the values of the coefficients b_i, here compared with the ones for the SM and the two Higgs doublet model (2HDM):

$$(b_3, b_2, b_1) = \begin{cases} (-7,\ -19/6,\ 41/10) & \text{SM} \\ (-7,\ -3,\ 21/5) & \text{2HDM} \\ (-3,\ 1,\ 33/5) & \text{SUSY.} \end{cases} \tag{27}$$

In particular, it manages to slow down the strong coupling constant evolution and to achieve unification at the scale given in (17) if $M_{SUSY} = M_Z$ and the input values (16) are used. The results obtained in [5] are displayed in Fig. 2c.

Before closing this session, we should mention that there still exist other non-supersymmetric possibilities to achieve unification, by allowing larger groups than $SU(5)$ to break to the SM in at least two stages [20]. An ordinary $SO(10)$, for example, can break first to a left-right symmetric $SU(3)_C \times SU(2)_L \times SU(2)_R \times U_{B-L}$ at a scale M_X and then to the SM at M_R. The unification cannot be achieved though if one tries to fix M_R in the TeV range. On the contrary, by leaving M_R as a free parameter, one obtains solvable equations with viable unification point for $M_R = 10^{10}$ GeV. While SUSY still allows a one-step unification and can fill the so called desert between M_Z and M_X in the lower corner, these models seem to be offering a complementary alternative.

5. Yukawa Couplings

The Yukawa couplings form the other sector of the model where the influence of SUSY is simply felt through the increased number of degrees of freedom and the value

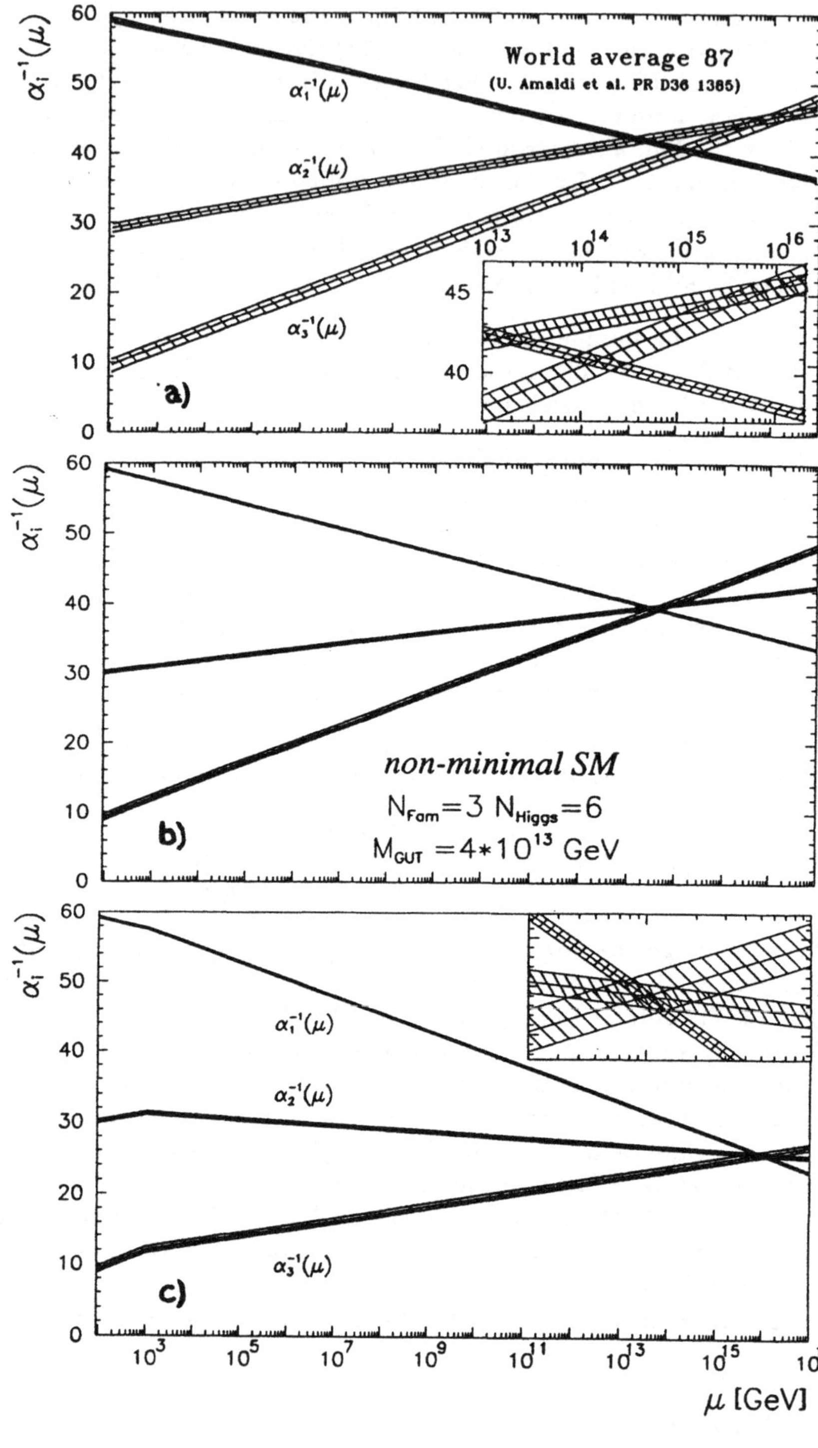

Figure 2: Evolution of the strong and electroweak gauge couplings in the Standard Model, an extension of the Standard Model with six Higgs doublets and the minimal supersymmetric model. The input values of $\alpha_S(M_Z) = 0.108 \pm 0.005$ and $\sin^2\theta_W(\overline{MS}) = 0.2336 \pm 0.0018$ are used here.

of M_{SUSY}. The evolution equations for these couplings are

$$\begin{aligned} 2\dot{\tilde{h}}_t &= \left[(GY)_t - 6\tilde{h}_t^2 - \tilde{h}_b^2\right]\tilde{h}_t & & (GY)_t &= \tfrac{16}{3}\tilde{\alpha}_3 + 3\tilde{\alpha}_2 + \tfrac{13}{15}\tilde{\alpha}_1 \\ 2\dot{\tilde{h}}_b &= \left[(GY)_b - 6\tilde{h}_b^2 - \tilde{h}_t^2 - \tilde{h}_\tau^2\right]\tilde{h}_b & \text{with} \quad & (GY)_b &= \tfrac{16}{3}\tilde{\alpha}_3 + 3\tilde{\alpha}_2 + \tfrac{7}{15}\tilde{\alpha}_1 \\ 2\dot{\tilde{h}}_\tau &= \left[(GY)_\tau - 4\tilde{h}_\tau^2 - 3\tilde{h}_b^2\right]\tilde{h}_\tau & & (GY)_\tau &= 3\tilde{\alpha}_2 + \tfrac{9}{5}\tilde{\alpha}_1 \end{aligned} \tag{28}$$

and with $\tilde{h}_i \equiv h_i/4\pi$. The convention $\dot{\tilde{h}}_i = \mathrm{d}\tilde{h}_i/\mathrm{d}t$ has been adopted in order to have a more compact form of the equations. Since the 2HDM-RGE have exactly the same structure as the supersymmetric ones, we give for comparison only the evolution equations for the Yukawa couplings in the SM

$$\begin{aligned} 2\dot{\tilde{h}}_t &= \left[(GY)_t - \tfrac{9}{2}\tilde{h}_t^2 - \tfrac{3}{2}\tilde{h}_b^2 - \tilde{h}_tau^2\right]\tilde{h}_t & & (GY)_t &= 8\tilde{\alpha}_3 + \tfrac{9}{4}\tilde{\alpha}_2 + \tfrac{17}{20}\tilde{\alpha}_1 \\ 2\dot{\tilde{h}}_b &= \left[(GY)_b - \tfrac{9}{2}\tilde{h}_b^2 - \tfrac{3}{2}\tilde{h}_t^2 - \tilde{h}_\tau^2\right]\tilde{h}_b & \text{with} \quad & (GY)_b &= 8\tilde{\alpha}_3 + \tfrac{9}{4}\tilde{\alpha}_2 + \tfrac{1}{4}\tilde{\alpha}_1 \\ 2\dot{\tilde{h}}_\tau &= \left[(GY)_\tau - 3\tilde{h}_t^2 - 3\tilde{h}_b^2 - \tfrac{5}{2}\tilde{h}_\tau^2\right]\tilde{h}_\tau & & (GY)_\tau &= \tfrac{9}{4}\tilde{\alpha}_2 + \tfrac{9}{4}\tilde{\alpha}_1 \,. \end{aligned} \tag{29}$$

Due to the presence of only one Higgs doublet, the equations (29) differ from (28) in the contribution of the top quark to the renormalization of the tau Yukawa coupling.

The nonlinearity of the equations (28) makes the increase of the Yukawa couplings with energy quite fast. This feature, present also in the SM and the 2HDM equations, is in (28) quantitatively enhanced by the presence of new degrees of freedom. Given the value of the coefficients of $\tilde{\alpha}_3$ and $\tilde{h}_t$ in (28), the Yukawa coupling relative to the existing lower bound on the top mass of 89 GeV [21],

$$h_t = \frac{m_t}{v_2} > \frac{m_t}{v} \sim 0.5 \tag{30}$$

is still not quite competitive with the strong coupling g_3, but it becomes dangerously so as soon as m_t approaches $180 - 190$ GeV. The requirement of applicability of (28) throughout the evolution of the low-energy parameters up to the GUT scale, is then the origin of the existence of an upper bound on the top mass in supersymmetric models. The approximate value of 196 GeV is obtained in the limit $\tan\beta \gg 1$, but lower upper bounds are obtained for intermediate values of $\tan\beta$. The too fast growth of h_t while approaching the GUT scale M_X is also the reason why sufficiently high values of m_t may not give acceptable solutions of the equations (28) for $\tan\beta$ too close to 1 **.

A few more observations are in order regarding the boundary conditions of the evolution equations (28):

- The Yukawa couplings entering in (28) are related to the third generation fermion masses and to $\tan\beta$ by

$$\frac{m_t(M_Z)}{m_b(M_Z)} = \frac{h_t(M_Z)}{h_b(M_Z)}\tan\beta, \qquad \frac{m_\tau(M_Z)}{m_b(M_Z)} = \frac{h_\tau(M_Z)}{h_b(M_Z)} \,. \tag{31}$$

**We restrict ourselves here to values of $\tan\beta > 1$ as predicted by SUSY before the introduction of radiative corrections to the Higgs potential. This modification, while allowing smaller values of $\tan\beta$ would not modify the discussion for values greater than 1.

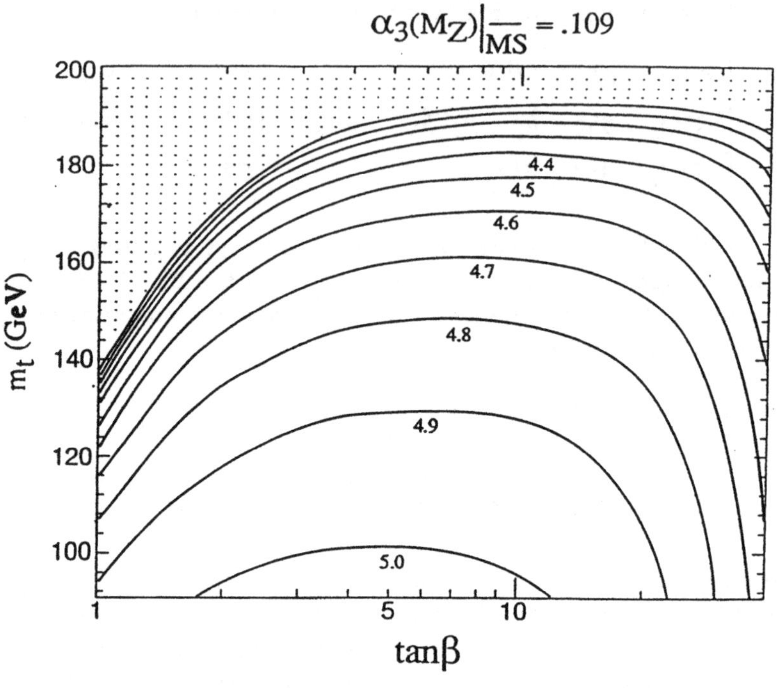

Figure 3: Restrictions imposed upon the values of m_t, m_b and $\tan\beta$ by the condition $h_b(M_X) = h_\tau(M_X)$ and the request of validity of perturbative calculations.

Besides m_t and $\tan\beta$ being unknown, also the value of m_b is plagued by uncertainties. We can only restrict ourselves to a range of values $4.2 - 5.2$ for $m_b(2m_b)$. The values of $m_b(M_Z)$ can then be obtained by QCD evolution.

- The GUT scale values of the Yukawa couplings, while unknown, are nevertheless linked to each other by relations depending on the particular Grand Unified model considered. For an underlying $SU(5)$ and $SO(10)$ symmetry we have respectively

$$\tilde{h}_b(M_X) = \tilde{h}_\tau(M_X) \qquad (SU(5)) \tag{32}$$
$$\tilde{h}_t(M_X) = \tilde{h}_b(M_X) = \tilde{h}_\tau(M_X) \qquad (SO(10)) \; . \tag{33}$$

Both conditions reduce the region of parameter space, $\tan\beta$, m_t, m_b [11,22,23].

We choose to impose the condition (32). For fixed values of m_t, for the input parameters (16) and for the GUT scale parameters (17), we scan the allowed values of m_b and $\tan\beta$ until solutions of the evolution equations (28) satisfying the unification conditions (32) are found. The results shown in Fig. 3 (taken from [22]) are quite obvious:

- For fixed values of $\tan\beta$ an increase in the top mass favours smaller values of m_b in the phenomenologically allowed range. Higher values of h_b would further increase the growth of h_t, as it can be seen from (28), spoiling (32) first and then giving values of h_t not acceptable in the perturbative regime. This explains why in Fig. 3 the curves with smaller values of m_b are always above the ones with higher m_b.

- On the other hand, for fixed values of m_t and for small values of $\tan\beta$, an increase in $\tan\beta$ gives naturally higher values of h_b and of m_b for $h_b \ll h_t$. Further increases of $\tan\beta$ require still higher values of h_b without influencing very much m_b. All the curves display in fact a central plateau. Moreover, smaller values of h_t have to be used in order to mantain m_t unchanged. Eventually, the region $h_b \sim h_t$ is approached. Further increases of $\tan\beta$ can start affecting the condition (32) explaining therefore the sharp drop of the curves for high values of $\tan\beta$.

- The lower possible value of $m_b(M_Z)$, compatible with a fixed value of $m_t(M_Z)$ gives a top dependent upper bound of $\tan\beta$

$$(\tan\beta)_{\max} \sim \frac{m_t(M_Z)}{m_b(M_Z)} \,. \tag{34}$$

This value is obviously obtained in the limit $h_t \sim h_b$ ††. It was also shown in a different context [12] that values exceeding this bound would be in disagreement with (14).

These features are also implicitly studied in [11] and used for Figs. 1, 6, 7 and 8. It can be observed in Fig. 1, for example, that for $m_t = 100\,\text{GeV}$, a numerical solution of (28) is found for $\tan\beta$ as small as 1.2, while for $m_t = 180\,\text{GeV}$, a solution is obtained only for $\tan\beta > 2$. For this same m_t, values of $\tan\beta$ in the range $2 - 2.5$ violate the condition (32).

We conclude this section by observing that it is quite hard to satisfy the condition (32) in the SM. The presence of the top Yukawa coupling in the evolution equation for h_τ and the lack of α_3, makes the ratio h_b/h_τ too big for a top mass $\sim 140\,\text{GeV}$ [24].

6. Higgs Potential Parameters and $SU(2)_{\text{L}} \times U(1)_{\text{Y}}$ Breaking

As already mentioned, one of the reasons for the success of this theory was due to the realization that the logarithmic radiative corrections to the mass parameters of the scalar potential were big enough to induce the breaking of the electroweak sector. This result, initially obtained for an effective SUSY breaking of $\mathcal{O}(M_Z)$, remains unchanged if the SUSY breaking scale moves up to the TeV range. This mechanism, besides being appealing, has the clear advantage of avoiding additional Higgs fields and therefore of keeping the number of new free parameters of the theory quite small.

It was observed in section 2 that the GUT scale values of μ_1, μ_2, μ_3 are quite far from satisfying the stability and breaking conditions (14) and (15) and the two

††Some attempts have been made in the past to obtain an absolute upper bound on $\tan\beta$. It was argued in [6] and [8] that for increasing values of $h_t(m_Z) \sim h_b(M_Z)$, h_b would eventually evolve to non-perturbative values. Some non-admissible value of $h_b(M_Z)$ was then used to constrain the value of $\tan\beta$ independently of m_t.

minimization conditions (12) and (13). Their low-energy values can be obtained from the evolution equations

$$
\begin{aligned}
(\dot{\bar{\mu}}_1^2) &= (GH)_H - 3\tilde{h}_b\,(SS)_b - \tilde{h}_\tau\,(SS)_\tau \\
(\dot{\bar{\mu}}_2^2) &= (GH)_H - 3\tilde{h}_t\,(SS)_t
\end{aligned}
\tag{35}
$$

$$
\begin{aligned}
(\dot{\mu}_{\mathrm{R}}^2) &= \left[(GH)_\mu - \left(3\tilde{h}_t^2 + 3\tilde{h}_b^2 + \tilde{h}_\tau^2\right)\right]\mu_{\mathrm{R}}^2 \\
m\dot{B}_{\mathrm{R}} &= -\left[(GH)_B - \left(3\tilde{h}_t^2 A_t + 3\tilde{h}_b^2 A_b + \tilde{h}_\tau^2 A_\tau\right) m\right]
\end{aligned}
\tag{36}
$$

where the definition $\bar{\mu}_{1,2}^2 \equiv \mu_{1,2}^2 - \mu_{\mathrm{R}}^2$ has been used. The subscript R in μ_{R}^2 and B_{R} distinguishes these renormalized parameters from the high-scale ones μ and B. The relation (10) at low-energy then reads $\mu_3^2 = -B_{\mathrm{R}} m \mu_{\mathrm{R}}$. From this and (36) one can easily obain the evolution equation for μ_3^2. The gaugino and scalar masses contribution to (35) and (36) are

$$
\begin{aligned}
(GH)_H &= 3\tilde{\alpha}_2 M_2^2 + \tfrac{3}{5}\tilde{\alpha}_1 M_1^2 & \qquad (SS)_t &= (m_{\tilde{Q}}^2 + m_{\tilde{U}}^2)_{33} + \bar{\mu}_2^2 + A_t^2 m^2 \\
(GH)_\mu &= 3\tilde{\alpha}_2 + \tfrac{3}{5}\tilde{\alpha}_1 & \text{and} \qquad (SS)_b &= (m_{\tilde{Q}}^2 + m_{\tilde{D}}^2)_{33} + \bar{\mu}_1^2 + A_b^2 m^2 \\
(GH)_B &= 3\tilde{\alpha}_2 M_2 + \tfrac{3}{5}\tilde{\alpha}_1 M_1 & \qquad (SS)_\tau &= (m_{\tilde{L}}^2 + m_{\tilde{E}}^2)_{33} + \bar{\mu}_1^2 + A_\tau^2 m^2 \,.
\end{aligned}
\tag{37}
$$

It is easy to see from the previous equations that μ_2^2 decreases faster than μ_1^2 since its evolution is driven by the top quark, and both μ_1^2 and μ_2^2 decrease faster than μ_3^2. These three quantities start evolving down from the initial conditions (9) and (10) to a situation where

$$
\left(\mu_3^2\right)^2 < \left(\mu_2^2\right)^2 < \mu_1^2 \cdot \mu_2^2 < \left(\frac{\mu_1^2 + \mu_2^2}{2}\right)^2 < \left(\mu_1^2\right)^2 \,. \tag{38}
$$

It is still $B > 0$, which also implies $S > 0$. Afterwards, μ_2^2 decreases even further, while evolving down to lower energies and eventually equals μ_3^2. The $SU(2)_{\mathrm{L}} \times U(1)_{\mathrm{Y}}$ breaking becomes possible when μ_3^2 is located between the geometrical and arithmetical mean of μ_1^2 and μ_2^2 without μ_1^2 and μ_2^2 having necessarily to be negative. An example is given in Fig. 4 where suitable values of A and μ are taken after having tested that they do satisfy the correct breaking pattern.

A few more considerations can be drawn here on the possible values of $\tan\beta$ with respect to the SUSY parameters:

- If no correction to the Higgs potential (8) are added, at the electroweak scale one generally has $\mu_2^2 < \mu_1^2$. The minimization condition (13) then implies $\tan\beta > 1$.

- For top masses above the existing experimental lower bound, values of $\tan\beta$ too close to one (i.e. $h_t \gg h_b$) but still compatible with the requirements of perturbation theory, may admit only a very limited region of SUSY parameters. This is the case of $\tan\beta = 1.2$ for $m_t = 100\,\mathrm{GeV}$ depicted in Fig. 1. We see there in fact, that too big values of the gaugino parameter M would bring μ_2^2 too far from μ_1^2 through the indirect effect of the term $(SS)_t$ in (35).

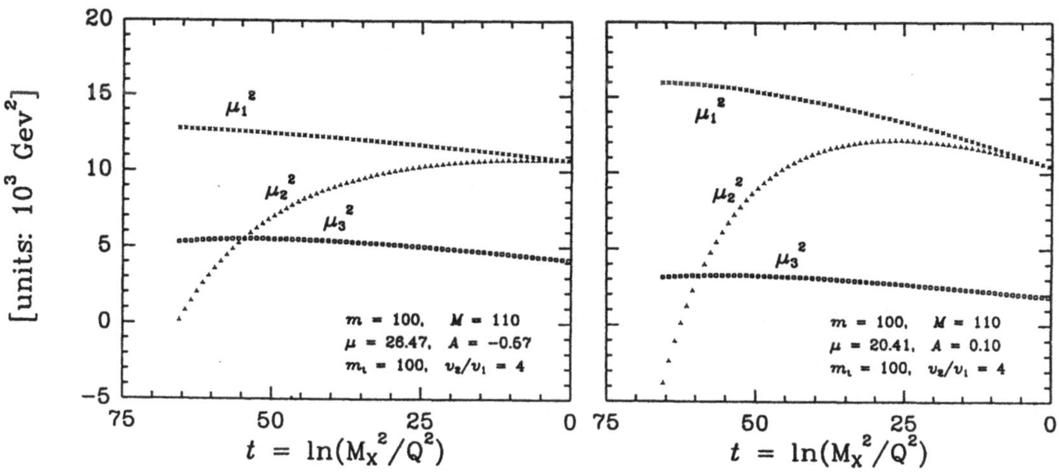

Figure 4: **Evolution of μ_1^2, μ_2^2 and μ_3^2, from the unification scale to the weak scale. The two figures correspond to the same value (m, M), but different solutions of A and μ.**

- Heavy enough values of m_t cannot allow an arbitrary increase of $\tan\beta$. Since it is in this case $h_b \sim h_t$, the evolution of μ_1^2 and μ_2^2 becomes more and more similar limiting therefore the value of $\tan\beta$ in (13). As already mentioned, the upper bound (34) was also derived in [12] making use of the boundedness of the Higgs potential and the conditions imposed by radiative breaking.

- For $h_b \sim h_t$ the value of μ_3^2 giving the correct electroweak breaking, forced to lie between μ_1^2 and μ_2^2, becomes smaller and smaller the more h_b approaches h_t. A small value of μ_3^2 implies either a small value of $\mu_{\rm R}$ or a small value of $B_{\rm R}$. We find this second possibility for $m_t = 150\,\text{GeV}$ and $\tan\beta = 40$ throughout the full range (m, M) considered, i.e. $0 < m < 250\,\text{GeV}$ and $0 < M < 470\,\text{GeV}$. This solution is shown in Fig. 5 where for each point in the two-dimensional space (m, M), also the values of $\mu_{\rm R}$ and $B_{\rm R}$ compatible with the radiative breaking of $SU(2)_{\rm L} \times U(1)_{\rm Y}$ are displayed. Negative values of M, also allowed, are not shown for simplicity. This situation is quite "extreme" since it requires a value of m_b at the lower limit of the range we allow. It is nevertheless quite instructive. The narrow distribution of the values of $B_{\rm R}$ around zero shows, however, that some fine-tuning among all the parameters is needed. For this reason, the approximation $M_{\rm SUSY} = \mathcal{O}(M_Z)$ is not possible here. This solution was obtained by stopping the downward renormalization of squarks and gluinos at the average value of their masses obtained for each pair (m, M).

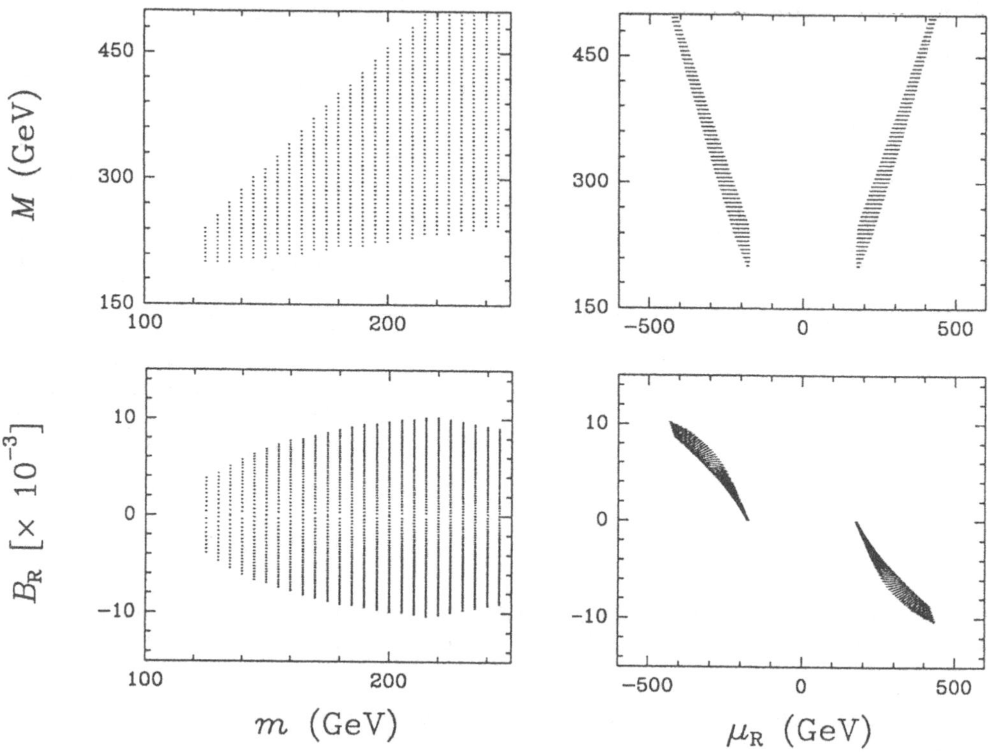

Figure 5: Possible supersymmetric realizations for the "extreme" choice of $m_t = 150$ GeV and $\tan\beta = 40$. For simplicity, only positive values of M are plotted here.

7. Gaugino Masses

Starting from this section we shall discuss the type of supersymmetric particle spectra one can obtain in the Minimal Supersymmetric Standard Model. We start here with the gaugino masses determined by the parameter M. This common parameter in (5) renormalizes down to three distinct values M_1, M_2 and M_3 according to the equations

$$\frac{\mathrm{d}\,M_i}{\mathrm{d}\,t} = -b_i\;\tilde{\alpha}_i M_i \tag{39}$$

with the coefficients b_i already given in (27). For $M_{\mathrm{SUSY}} = M_Z$ and the usual input parameters (16), the ratios M_3/M, M_2/M and M_1/M are 2.79, 0.82 and 0.42, respectively. M_1 and M_2 enter in the chargino and neutralino mass matrices. M_3 gives the gluino mass $m_{\tilde{g}}$, once a possible negative sign allowed for the parameter M is removed. Except for some particular values of m_t and $\tan\beta$ and only in a limited region of m (see the cases $m_t = 100$ GeV and $\tan\beta = 1.2$ and 8), no lower limit on $m_{\tilde{g}}$ can be inferred in this model. A comparison with the experimental lower bound for $m_{\tilde{g}}$ is given in section 9.

8. Chargino, Neutralino and Higgs Masses

The chargino and neutralino sector is entirely determined by M, the value of parameter μ at the weak scale, which we indicate with μ_R, $\tan\beta$, the Yukawa couplings and the electroweak gauge couplings. We shall examine these three sectors in the following. The results are shown in Figs. 6 and 7.

Chargino Masses:

The relevant terms in the low-energy Lagrangian are [25]

$$\frac{1}{2}\left\{(\widetilde{W}^- \widetilde{H}_1^-)\ M_C \begin{pmatrix} \widetilde{W}^+ \\ \widetilde{H}_2^+ \end{pmatrix} + h.c.\right\} \quad \text{with} \quad M_C = \begin{pmatrix} M_2 & gv_2 \\ gv_1 & \mu_R \end{pmatrix} \tag{40}$$

where $\widetilde{W}^\pm \equiv -i\lambda^\pm = -i(\lambda_1 \mp i\lambda_2)/\sqrt{2}$. We denote by $\tilde{\chi}^+_{1,2}$ the mass eigenstates for the Lagrangian terms in (40). By using the relation $2M_W^2 = g^2(v_1^2 + v_2^2)$ and by assuming M and μ to be real, it is easy to see that the chargino mass eigenvalues can be written as

$$m^2_{\tilde{\chi}^+_{1,2}} = \frac{M_2^2 + \mu_R^2 + 2M_W^2}{2} \pm \frac{1}{2}\sqrt{(M_2^2 + \mu_R^2 + 2M_W^2)^2 - 4\left(M_2\mu_R - M_W^2 \sin 2\beta\right)^2}\ . \tag{41}$$

For $\tan\beta \gg 1$ one may expect to obtain the characteristic gaugino-higgsino spectrum. In this limit and for small μ_R in fact, the two eigenvalues are:

$$m^2_{\tilde{\chi}^+_2} \sim \frac{M_2^2 \mu_R^2}{M_2^2 + 2M_W^2}\ , \qquad m^2_{\tilde{\chi}^+_1} \sim M_2^2 + 2M_W^2\ . \tag{42}$$

It was argued that high values of $\tan\beta$, favoured by a heavy top mass, should have led to the detection of a light chargino at LEP. In practice the situation is far from being so simple. High values of $\tan\beta$ do imply small values of μ_3^2, but this does not allow us to draw a definite conclusion on μ (see Fig. 5). The interplay between μ_R, M_2, $\tan\beta$ and indirectly also of all the other independent parameters of the theory, is more complex for $\tan\beta \geq 1$. In principle, the full range of masses from approximately zero up to the rough upper bound given by the chosen value of M is allowed for the lightest chargino $\tilde{\chi}^+_2$.

Neutralino Masses:

A similar situation holds also for the neutralino masses. The relevant term of the low-energy Lagrangian is in this case:

$$\frac{1}{2}(\widetilde{B}, \widetilde{W}_3, \widetilde{H}_1^0, \widetilde{H}_2^0) \begin{pmatrix} M_1 & 0 & -g'v_1/\sqrt{2} & g'v_2/\sqrt{2} \\ 0 & M_2 & gv_1/\sqrt{2} & -gv_2/\sqrt{2} \\ -g'v_1/\sqrt{2} & gv_1/\sqrt{2} & 0 & -\mu_R \\ g'v_2/\sqrt{2} & -gv_2/\sqrt{2} & -\mu_R & 0 \end{pmatrix} \begin{pmatrix} \widetilde{B} \\ \widetilde{W}_3 \\ \widetilde{H}_1^0 \\ \widetilde{H}_2^0 \end{pmatrix} + \ h.c. \tag{43}$$

We indicate by $\tilde{\chi}^0_i$ the four-component (Majorana) spinors obtained in terms of the two-components mass eigenstates χ^0_i $(i = 1, 4)$, $(\tilde{\chi}^0_i \equiv (\chi^0_i, \bar{\chi}^0_i))$. As shown in Fig. 6, the lightest of these states, $\tilde{\chi}^0_1$ can have a mass from zero up to a maximum value in

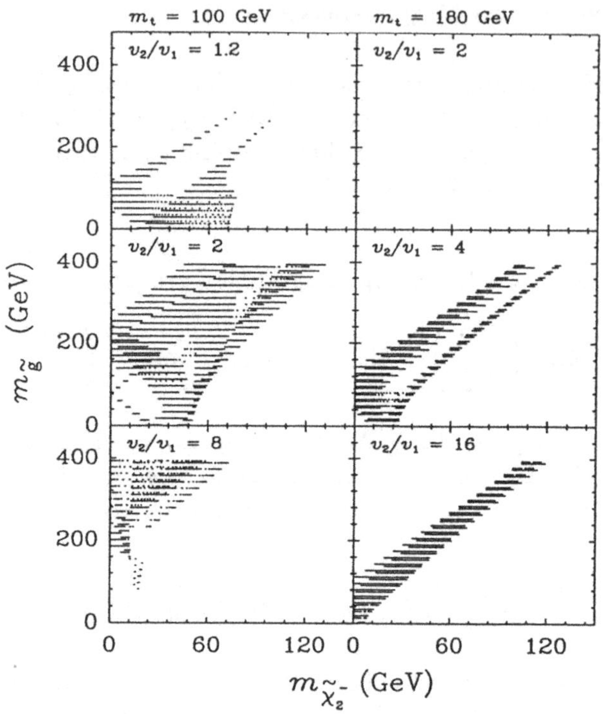

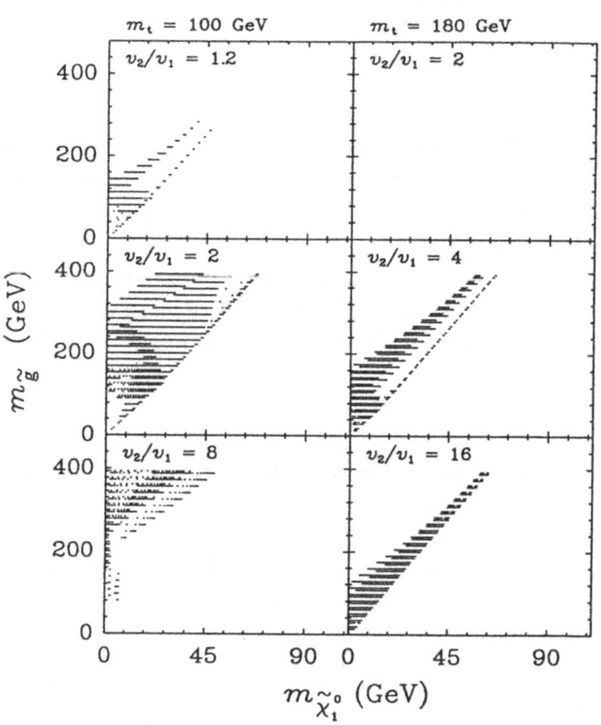

Figure 6: Lightest chargino and lightest neutralino masses obtained for the same values of the SUSY parameter space considered in Fig. 1.

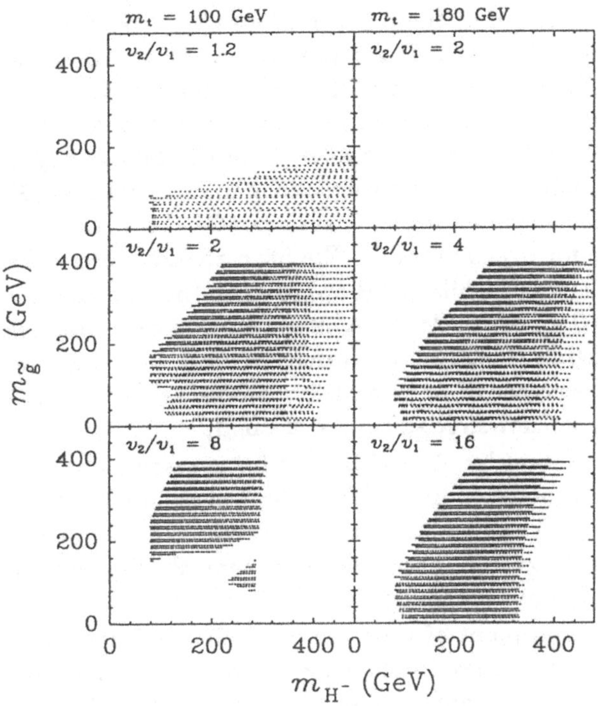

Figure 7: Charged Higgs masses obtained for the same values of the SUSY parameter space considered in Fig. 1.

general quite smaller than the corresponding maximum value allowed for $m_{\tilde{\chi}_2^+}$ in the same region of SUSY parameter space.

Higgs Masses:

The Higgs sector is the one which has received the greatest attention lately. Five physical scalars appear in this sector after the spontaneous breaking of $SU(2)_L \times U(1)_Y$. We denote by $H_{1,2}^0$ the two CP-even mass eigenstates, by H_3^0 the CP-odd state, and by $H^\pm$ the two charged ones. The tree-level effective potential (8) allows the following mass spectrum for these scalars:

$$m_{H_3^0}^2 = \mu_1^2 + \mu_2^2 = \frac{2\mu_3^2}{\sin 2\beta} \tag{44}$$

$$m_{H^\pm}^2 = M_W^2 + m_{H_3^0}^2 \tag{45}$$

and

$$m_{H_1^0,H_2^0}^2 = \frac{1}{2}\left[m_{H_3^0}^2 + M_Z^2 \pm \sqrt{(m_{H_3^0}^2 + M_Z^2)^2 - 4m_{H_3^0}^2 M_Z^2 \cos^2 2\beta}\,\right] . \tag{46}$$

At the tree-level then, the following relations hold independently of the value of $\tan\beta$

$$m_{H_2^0} < m_{H_3^0} < m_{H_1^0} , \qquad m_{H_2^0} < M_Z . \tag{47}$$

Until recently, the second of these two relations was considered the main test of SUSY. The lack of observation of a Higgs particle below the Z mass was considered enough to completely rule out the model. The situation has changed drastically after the radiative corrections to the potential (8) have been added (more details will be given in the following lectures). The most affected of all the five Higgs is of course H_2^0 whose mass can be as small as zero if no corrections to the potential (8) are considered. Smaller is the effect on the remaining four scalars. Of these, the pseudoscalar is the one which sets the scale for $m_{H_1^0}$ and $m_{H^\pm}$. We have already observed that in the limit $h_t \sim h_b$, μ_3^2 can be quite small while $\tan\beta$ approaches the maximum value allowed by m_t and m_b (34) i.e. $\tan\beta_{\max} \sim m_t(M_Z)/m_b(M_Z)$. In this case, $m_{H_3^0}$ can be very light or nearly massless bringing therefore the values of $m_{H_1^0}$ and $m_{H^\pm}$ down to M_Z and M_W, respectively. The decrease of $m_{H^\pm}$ with increasing values of $\tan\beta$ can already be observed in Fig. 7, even for not so high values of $\tan\beta$.

We conclude this section by observing that for average values of $\tan\beta$, as shown in Fig. 6 and Fig. 7, this model indicates that the neutralino has the lightest possible spectrum among the particles considered in this section. For the same choice of SUSY parameters, the charged Higgs is *on average* heavier than the lightest chargino, although the situation may be different for $\tan\beta \gg 1$.

From the experimental point of view the possibility of imposing lower bounds on the masses of these particles is far from being obvious or assumption-free. Absolute lower bounds exist only for the charged particles, i.e. chargino [26] and charged Higgs [27]. They come from LEP and they are about 45 GeV.

As for the neutral Higgses H_0^2 and H_0^3, the situation has been considerably complicated by the addition of the one-loop corrections to the potential (8). The prediction $m_{H_2^0} < m_{H_3^0}$ is not valid anymore and the new decay $H_2^0 \to H_3^0 H_3^0$ becomes possible. Moreover, the search for neutral Higgses at LEP in the two channels $e^+e^- \to Z \to H_2^0 Z^*$ and $e^+e^- \to Z \to H_2^0 H_3^0$ is now sensitive to the top mass and the scalar quark masses, in particular the scalar partner of the top quark. The regions $m_{H_2^0} < 41\,\mathrm{GeV}$ and $m_{H_3^0} < 31\,\mathrm{GeV}$ have been excluded at 95% CL [28] for $m_t = 140\,\mathrm{GeV}$, $m_{\tilde{t}} = 1\,\mathrm{TeV}$ and under the assumption that no mixing exists between the left-handed and right-handed component of the stop. As we shall see in the next section, the scalar partner of the top quark plays a particular role in the model we discuss here and these assumptions are quite far from being theoretically justified in this particular framework.

9. Scalar –Quark and –Lepton Masses

We start by considering the squark mass matrices. The 6×6 matrix of the $Q = 2/3$ sector is formally written in terms of the 3×3 submatrices $M^2_{U_{LL}}$, $M^2_{U_{RR}}$ and $M^2_{U_{LR}}$ as follows:

$$\widetilde{M}_U^2 = \begin{pmatrix} M^2_{U_{LL}} & M^2_{U_{LR}} \\ \\ M^2_{U_{LR}} & M^2_{U_{RR}} \end{pmatrix} . \tag{48}$$

$M^2_{U_{LL}}$ and $M^2_{U_{RR}}$ are the mass matrices of the left- and right-handed component of the up-type squarks, while $M^2_{U_{LR}}$ contains the mass terms mixing the two components. It holds that:

$$
\begin{aligned}
M^2_{\tilde{U}_{LL}} &= \mathrm{diag}\left(m^2_{\tilde{Q}_{11}}, m^2_{\tilde{Q}_{22}}, m^2_{\tilde{Q}_{33}} + m_t^2\right) - |DT^L_{\tilde{U}}|\ \mathbb{1} \\
M^2_{\tilde{U}_{RR}} &= \mathrm{diag}\left(m^2_{\tilde{U}_{11}}, m^2_{\tilde{U}_{22}}, m^2_{\tilde{U}_{33}} + m_t^2\right) - |DT^R_{\tilde{U}}|\ \mathbb{1} \\
M^2_{\tilde{U}_{LR}} &= \mathrm{diag}\left(0, 0, (A_t m + \mu_{\mathrm{R}} \cot\beta)\, m_t\right) \qquad (49)
\end{aligned}
$$

where $DT^{L,R}_{\tilde{U}}$ are the so-called "D-term" contributions given by

$$
DT^{L,R}_{\tilde{U}} = M_Z^2 \cos 2\beta \left(T^3_{\tilde{U}} - Q_U \sin\theta_W\right) . \qquad (50)
$$

Similarly, the corresponding 3×3 submatrices $M_{D_{LL}}$, $M_{D_{RR}}$ and $M^2_{\tilde{D}_{LR}}$ for the $Q = -1/3$ sector are

$$
\begin{aligned}
M^2_{\tilde{D}_{LL}} &= \mathrm{diag}\left(m^2_{\tilde{Q}_{11}}, m^2_{\tilde{Q}_{22}}, m^2_{\tilde{Q}_{33}} + m_b^2\right) + |DT^L_{\tilde{D}}|\ \mathbb{1} \\
M^2_{\tilde{D}_{RR}} &= \mathrm{diag}\left(m^2_{\tilde{D}_{11}}, m^2_{\tilde{D}_{22}}, m^2_{\tilde{D}_{33}} + m_b^2\right) + |DT^R_{\tilde{D}}|\ \mathbb{1} \\
M^2_{\tilde{D}_{LR}} &= \mathrm{diag}\left(0, 0, (A_b m + \mu_{\mathrm{R}} \tan\beta)\, m_b\right) . \qquad (51)
\end{aligned}
$$

The "D-term" contributions for the down-sector can be obtained from (50), once the correct quantum numbers are substituted for the ones for the up-sector. The same rules apply for building up the mass matrices of the lepton superpartners. The scalar neutrino mass matrix is only a 3×3 matrix, since neutrinos are here considered massless and do not have right-handed components.

A simple reshuffling of rows and columns in the matrices (49) and (51) gives a block-diagonal matrix with 2×2 diagonal submatrices for the first and second generation and 2 × 2 non-diagonal ones for the third generation. Moreover, since we neglect here the masses of the first two generations of quarks and leptons, the evolution of the masses of the first two generations of sfermions is determined by the same equations. Once we introduce the notation $m^2_{\tilde{Q}_{11}} = m^2_{\tilde{Q}_{22}} = m^2_{\tilde{u}_L} = m^2_{\tilde{d}_L}$, $m^2_{\tilde{U}_{11}} = m^2_{\tilde{U}_{22}} = m^2_{\tilde{u}_R}$, $m^2_{\tilde{D}_{11}} = m^2_{\tilde{D}_{22}} = m^2_{\tilde{d}_R}$ and similar definitions for the leptonic sector, these equations have the simple form

$$
\begin{aligned}
(\dot{m^2_{\tilde{u}_L}}) = (\dot{m^2_{\tilde{d}_L}}) &= (GG)_Q = \frac{16}{3}\tilde{\alpha}_3 M_3^2 + 3\tilde{\alpha}_2 M_2^2 + \frac{1}{5}\tilde{\alpha}_1 M_1^2 \\
(\dot{m^2_{\tilde{u}_R}}) &= (GG)_U = \frac{16}{3}\tilde{\alpha}_3 M_3^2 + \frac{16}{15}\tilde{\alpha}_1 M_1^2 \\
(\dot{m^2_{\tilde{d}_R}}) &= (GG)_D = \frac{16}{3}\tilde{\alpha}_3 M_3^2 + \frac{4}{15}\tilde{\alpha}_1 M_1^2 \\
(\dot{m^2_{\tilde{e}_L}}) = (\dot{m^2_{\tilde{\nu}_L}}) &= (GG)_L = \tilde{\alpha}_2 M_2^2 + \frac{3}{5}\tilde{\alpha}_1 M_1^2 \\
(\dot{m^2_{\tilde{e}_R}}) &= (GG)_E = \frac{12}{5}\tilde{\alpha}_1 M_1^2 . \qquad (52)
\end{aligned}
$$

The evolution of the first two generations of sfermion masses is driven only by the gaugino masses and their low-energy value can be expressed as $m^2_{\tilde{q}} = m^2 + C_{\mathrm{M}}(\tilde{q})M^2 + DT(\tilde{q})$. In particular, for the input values (16) and for $M_{\mathrm{SUSY}} = M_Z$, it holds that:

$$
m^2_{\tilde{u}_L} = m^2 + 6.51\ M^2 - 0.35\ M_Z^2\,|\cos 2\beta|
$$

$$\begin{aligned}
m^2_{\tilde{u}_R} &= m^2 + 6.09\ M^2 - 0.16\ M_Z^2\,|\cos 2\beta| \\
m^2_{\tilde{d}_L} &= m^2 + 6.51\ M^2 + 0.42\ M_Z^2\,|\cos 2\beta| \\
m^2_{\tilde{d}_R} &= m^2 + 6.04\ M^2 + 0.08\ M_Z^2\,|\cos 2\beta| \\
m^2_{\tilde{e}_L} &= m^2 + 0.52\ M^2 + 0.27\ M_Z^2\,|\cos 2\beta| \\
m^2_{\tilde{e}_R} &= m^2 + 0.15\ M^2 + 0.23\ M_Z^2\,|\cos 2\beta| \\
m^2_{\tilde{\nu}_{eL}} &= m^2 + 0.52\ M^2 - 0.50\ M_Z^2\,|\cos 2\beta|\ .
\end{aligned} \qquad (53)$$

We observe here:

- The corrections to the initial value m^2 are quite substantial.
- The values of the coefficients $C_{\rm M}(\tilde{q})$ have to be considered as indicative. A change of $\alpha_{\rm S}$ and $\sin^2\theta_W$ in (16), within the experimental error, induces small changes in in the gauge couplings contribution to the equations (52) which are only partially neutralized by the variations also induced in the unification scale M_X. Moreover, a change in $M_{\rm SUSY}$ from M_Z to $2M_Z$ can affect these coefficients up to a factor of 15%. Notice that for the same variation of the SUSY scale and for the initial value of $\alpha_{\rm S}$ given in (16), the ratio $m_{\tilde{g}}/M$ changes from 2.79 to 2.61.
- The "D-term" contributions tend to increase the values of the down-type squarks and slepton masses, but to decrease the value of up-type ones. The effect is not very big for squarks, but is quite sizeable for sleptons. For $m \sim M \sim M_Z$ in fact, the decrease/increase for up/down squarks is about 5%, but the correction can be up to 20 and 30 % for scalar electrons and sneutrinos.

The spectrum of the third generation of squarks and sleptons is more complicated. To begin with, the 2×2 submatrices which can be obtained for this generation from (48), (49) and (51), are not diagonal. We give explicitly the ones for the squarks of the third generation

$$M_{\tilde{t}}^2 = \begin{pmatrix} m^2_{\tilde{Q}_{33}} + m_t^2 - |DT_U^L|\,\mathbb{1} & (A_t m + \mu_{\rm R}\cot\beta)\, m_t \\ (A_t m + \mu_{\rm R}\cot\beta)\, m_t & m^2_{\tilde{U}_{33}} + m_t^2 - |DT_U^R|\,\mathbb{1} \end{pmatrix} \qquad (54)$$

$$M_{\tilde{b}}^2 = \begin{pmatrix} m^2_{\tilde{Q}_{33}} + m_b^2 - |DT_D^L|\,\mathbb{1} & (A_t m + \mu_{\rm R}\tan\beta)\, m_b \\ (A_t m + \mu_{\rm R}\tan\beta)\, m_b & m^2_{\tilde{U}_{33}} + m_b^2 - |DT_D^R|\,\mathbb{1} \end{pmatrix}\ . \qquad (55)$$

As we can see, the non-diagonal pieces can have a quite substantial size. Moreover, the renormalization group equations for the diagonal terms are also different from the equations (52). Again, by using the notation $m^2_{\tilde{Q}_{33}} = m^2_{\tilde{t}_L} = m^2_{\tilde{b}_L}$, $m^2_{\tilde{U}_{33}} = m^2_{\tilde{t}_R}$, $m^2_{\tilde{D}_{33}} = m^2_{\tilde{b}_R}$ and similar definitions for the leptonic sector, we can write them as

$$(\dot{m}^2_{\tilde{t}_L}) = (\dot{m}^2_{\tilde{b}_L}) \;=\; (GG)_Q - \;\tilde{h}_t^2\,(SS)_t - \;\tilde{h}_b^2\,(SS)_b$$

$$
\begin{aligned}
(m^2_{\tilde{t}_R})^{\cdot} &= (GG)_U - 2\,\tilde{h}_t^2\,(SS)_t \\
(m^2_{\tilde{b}_R})^{\cdot} &= (GG)_D - 2\,\tilde{h}_b^2\,(SS)_b \\
(m^2_{\tilde{\tau}_L})^{\cdot} = (m^2_{\tilde{\nu}_{\tau L}})^{\cdot} &= (GG)_L - \,\tilde{h}_\tau^2\,(SS)_\tau \\
(m^2_{\tilde{\tau}_R})^{\cdot} &= (GG)_E - 2\,\tilde{h}_\tau^2\,(SS)_\tau \ .
\end{aligned} \tag{56}
$$

The gauge terms $(GG)_i$ are given in (52) and the scalar field terms $(SS)_t$, $(SS)_b$ and $(SS)_\tau$ in (37). The evolution down to the weak scale is driven not only by the gaugino masses, but also by the third generation's Yukawa couplings, i.e. mainly by the top mass and the value of $\tan\beta$. The effect is that the values of the coefficients $C_{\rm M}(\tilde{q})$ for the third generation's sfermion masses are smaller than the ones listed in (53). We shall distinguish here the two situations $h_b \ll h_t$ and $h_b \sim h_t$.

$h_b \ll h_t$

In the specific case of $m_t = 150\,\text{GeV}$ and $\tan\beta = 3$ (the condition $h_b \ll h_t$ is here certainly verified), the decrease of the coefficients $C_{\rm M}(\tilde{q})$ with respect to the ones for the first two generations is about 15 % and 30 % for $m^2_{\tilde{Q}_{33}} = m^2_{\tilde{t}_L} = m^2_{\tilde{b}_L}$ and $m^2_{\tilde{t}_R}$, respectively. It is negligible for $m^2_{\tilde{b}_R}$, $m^2_{\tilde{\tau}_L}$ and $m^2_{\tilde{\tau}_R}$. The conclusions which can be drawn for the stop, sbottom and stau mass eigenstates are therefore the following:

- Both left-handed and right-handed diagonal terms in the stop matrix (54), are below the corresponding values of the up-type squarks of the first two generations. As already observed, the decrease of the coefficient $C_{\rm M}(\tilde{t}_R)$ is twice as much as the decrease of $C_{\rm M}(\tilde{t}_L)$ (notice the factor of two multiplying h_t^2 in the evolution equation for $m^2_{\tilde{t}_R}$). Moreover, the off-diagonal terms in (54) are in this case quite big. The effect is that one of the two stop mass eigenstates, which we call $\tilde{t}_1$ has a mass smaller than the entry $m_{\tilde{t}_R}$ in (54). On the contrary, a partial/full compensation of the difference between the value of $m_{\tilde{t}_L}$ in (54) and the masses of the remaining four up-type squarks is obtained for the second mass eigenstate $\tilde{t}_2$. One mass eigenstate is then either comparable to the almost degenerate four up-type squarks, or moderately below them, according to the value of m_t and the region of SUSY parameter space considered. The second one $\tilde{t}_1$, can have a much smaller mass, in principle even compatible with zero as shown in Fig. 8 for $m_t = 100$. For some of the points already excluded in Fig. 8, the off-diagonal terms of the matrix (54) can be big enough to drive $m^2_{\tilde{t}_1}$ to negative values. The contribution of m_t to the diagonal entries of (54), up until now neglected, has the effect of reducing the splitting between the two stop mass eigenstates. For increasing values of m_t, but still such that $h_b \ll h_t$, the diagonal terms in the matrix (54) grow faster than the off-diagonal ones. Notice in fact, that no values close to zero are present in Fig. 8 for $m_t = 180$.

- As far as the sbottom mass matrix is concerned, practically no splitting is introduced by the off-diagonal terms in the matrix (55). Since $h_b \ll h_t$, the heavier of the two mass eigenstates, $\tilde{b}_2$, is roughly the right-handed component $\tilde{b}_R$, nearly degenerate with the other four down-type squarks of the first two generations. The other one, which we indicate with $\tilde{b}_1$, is smaller. The size of the splitting of

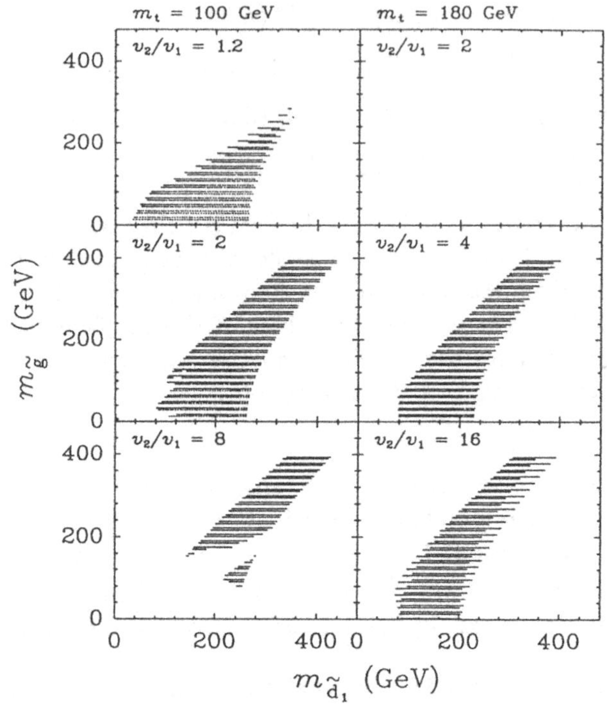

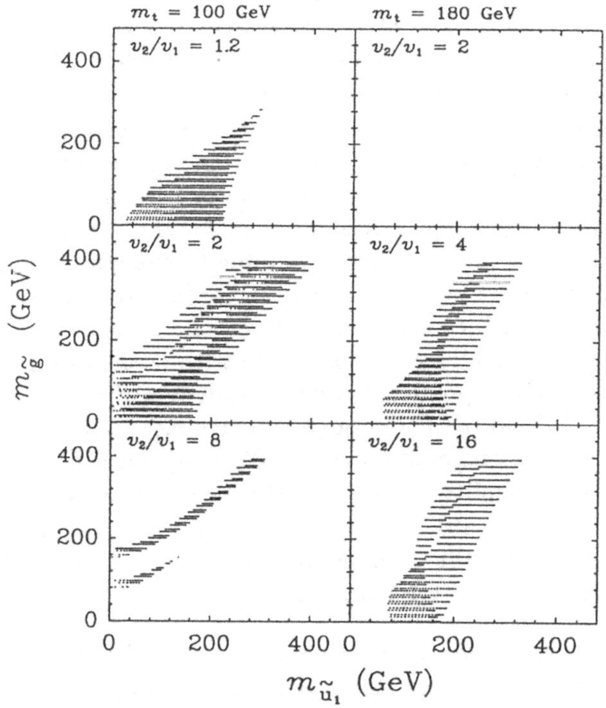

Figure 8: Masses of the lightest stop and sbottom quark mass eigenstates, here indicated as $\tilde{u}_1$ and $\tilde{d}_1$, obtained for the same values of the SUSY parameter space considered in Fig. 1.

these two eigenstates depends of course on the value of m and M. By inspection of Fig. 8a and 8b, we can conclude that as far as $h_b \ll h_t$, $\tilde{t}_1$ is *on average* lighter than $\tilde{b}_1$.

- The third generation of sleptons is a replica of the first two. The effect of the Yukawa coupling h_τ in the RGE for $m_{\tilde{L}}^2$ and $m_{\tilde{R}}^2$, is in fact almost negligible compared to the gaugino contribution. The lightest mass eigenstate $\tilde{\tau}_1$ is practically the $SU(2)_L$-singlet component $\tilde{\tau}_R$.

$h_b \sim h_t$

In this case, the decrease of all diagonal entries of both stop and sbottom mass matrices is similar. For $m_t = 150$ and $\tan\beta = 40$, the gaugino contribution to these masses decreases with respect to the contribution it gives to the first two generation's squarks masses of about 25-30 %. In the leptonic case this decrease is sizeable only for $m_{\tilde{\tau}_R}^2$ (roughly 20 %) but negligible for $m_{\tilde{\tau}_L}^2$. The observations one can make in this case are:

- The off-diagonal terms are now quite relevant in both stop and sbottom mass matrices, either for the presence of m_t or for the presence of $\tan\beta$. The same considerations made before for the $\tilde{t}$ hold here also for $\tilde{b}$. Moreover, since the sbottom matrix does not have a direct dependence on m_t in the diagonal terms, $\tilde{b}_1$ can be in principle smaller than $\tilde{t}_1$ for heavy top masses and for big enough $\tan\beta$.

- Also in the case of $\tilde{\tau}$, a large splitting between the two mass eigenstates can be introduced by big off-diagonal terms in the mass matrix. The different renormalization pattern present in this case can further enlarge this splitting. Moreover, due to the smaller sensitivity to the value of M, the lightest stop $\tilde{\tau}_1$, can be lighter than $\tilde{t}_1$ and $\tilde{b}_1$ [12] in some regions of the SUSY parameter space

Some interesting phenomenological implications which can be obtained for large values of $\tan\beta$, are discussed in [12].

As for the case of chargino, neutralino and Higgs bosons, no experimental search has brought so far any evidence for the existence of squarks and sleptons. The best lower limits on the slepton masses coming from LEP [26] exclude values of about 43 GeV for photino masses up to $20 - 30$ GeV. The assumption made in these searches that the sleptons decay directly to the lightest supersymmetric particle (LSP), assumed to be the photino, are obviously acceptable for these values of masses. Moreover, a possible splitting between the left- and right-handed components $\tilde{l}_L$ and $\tilde{l}_R$ is also considered in these analyses. The situation is more complex as far as the bounds on squarks and gluino masses are concerned. The recent CDF limits [29] $m_{\tilde{g}} > 150$ GeV (independently of $m_{\tilde{q}}$) and $m_{\tilde{q}} > 150$ GeV (for $m_{\tilde{g}} < 400$ GeV), rely on at least two assumptions not supported by the Minimal Supersymmetric Standar Model. One of them is that all the squarks are considered as degenerate, the second is that squarks and gluinos are supposed to decay directly to the LSP withouth intermediate decays to charginos or neutralinos [30]. Obviously a more complete analysis which is able to relax these two assumptions, cannot be performed without specifying a particular theoretical framework. An attempt in this direction is made in [31] where the study of these cascade decays, as predicted by the Minimal Supersymmetric Standard Model, allows one to estimate that the CDF limits should be lowered by about 30 GeV.

10. Conclusions

In closing this lecture we would like to point out that the predictive power of this model strongly relies on the minimal choice of parameters made at some high-scale and on the fact the electroweak breaking can be induced by radiative effects. The possibility of reducing to only four the number of new parameters to be introduced in the theory, is certainly not theoretically motivated and may be considered as a drawback of this model. On the other hand, we have shown through a few examples that the experimental searches for supersymmetric particles are far from being assumption-free and are in many cases already inspired by this model. It might also become harder and harder to perform searches of supersymmetric particles heavier than the ones excluded up to now in a model-independent way. Moreover, in spite of its small number of parameters, this model still offers a very rich spectrum of masses and only a serious threat from experimental side could convince us to dismiss it.

Acknowledgments

I wish to thank S. Bertolini, A. Masiero and G. Ridolfi for a fruitful collaboration, M. Olechowski with whom I worked out many of the issues discussed in this lecture and G. Blair for a careful reading of the manuscript. Finally I thank W. Hollik, R. Rückl and J. Wess for their efforts in organizing this workshop and most of all for their patience in waiting for this manuscript.

References

[1] J. Bagger and J. Wess, *Supersymmetry and Supergravity*, Princeton University Press, Princeton, NJ, 1983;
H.P. Nilles, *Phys. Rep.* 110 (1984) 1;
H.E. Haber and G.L. Kane, *Phys. Rep.* 117 (1985) 75;
A.B. Lahanas and D.V. Nanopoulos, *Phys. Rep.* 145 (1987) 1.

[2] E. Witten, *Nucl. Phys.* B188 (1981) 513;
N. Sakai, *Z. Phys.* C11 (1981) 153;
S. Dimopoulos and H. Georgi, *Nucl. Phys.* B193 (1981) 150.

[3] E. Cremmer, B. Julia, J. Scherk, P. van Nieuwenhuizen, S. Ferrara and L. Girardello, *Nucl. Phys.* B147 (1979) 105.

[4] L. Girardello, M.T. Grisaru, *Nucl. Phys.* B194 (1982) 65.

[5] U. Amaldi, W. de Boer and H. Fürstenau, *Phys. Lett.* B260 (1991) 447.

[6] L.E. Ibáñez and C. Lopez, *Phys. Lett.* B126 (1983) 54 and *Nucl. Phys.* B233 (1984) 511;
L.E. Ibáñez, C. Lopez and C. Muñoz, *Nucl. Phys.* B256 (1985) 218.

[7] A. Bouquet, J. Kaplan and C.A. Savoy, *Phys. Lett.* B148 (1984) 69 and *Nucl. Phys.* B262 (1985) 218;
C. Kounnas, A.B. Lahanas, D.V. Nanopoulos and M. Quirós, *Nucl. Phys.* B236 (1984) 438.

[8] G.F. Giudice and G. Ridolfi, *Z. Phys.* C41 (1988) 447.

[9] W. Majerotto and B. Mösslacher, *Z. Phys.* C48 (1990) 273.

[10] M. Olechowski and S. Pokorski, *Phys. Lett.* B214 (1988) 239.

[11] S. Bertolini, F. Borzumati, A. Masiero and G. Ridolfi, *Nucl. Phys.* B353 (1991) 591.

[12] M. Drees and M.M. Nojiro, *Nucl. Phys.* B369 (1992) 54.

[13] J.M. Frère, D.R.T. Jones and S. Raby, *Nucl. Phys.* B222 (1983) 11;
M. Claudson, L.J. Hall and I. Hinchliffe, *Nucl. Phys.* B228 (1983) 501;
M. Drees, M. Glück and K. Grassie, *Phys. Lett.* B57 (1985) 164;
H. Komatsu, *Phys. Lett.* B15 (1988) 323.

[14] J.-P. Derendinger and C.A. Savoy, *Nucl. Phys.* B237 (1984) 307.

[15] J.F. Gunion, H.E. Haber and M. Sher, *Nucl. Phys.* B306 (1988) 1.

[16] G. Gamberini, G. Ridolfi and F. Zwirner, *Nucl. Phys.* B331 (1990) 331;
P.H. Chankowski, *Phys. Rev.* D41 (1990) 2877.

[17] J. Ellis, G. Ridolfi and F. Zwirner, *Phys. Lett.* B257 (1991) 83.

[18] Y. Okada, M. Yamaguchi and T. Yanagida, *Prog. Theor. Phys.* 85 (1991) 1;
R. Barbieri, M. Frigeni and F. Caravaglio, *Phys. Lett.* B258 (1991) 167;
H.E. Haber and R. Hempfling, *Phys. Rev. Lett.* 66 (1991) 1815;
A. Yamada, *Phys. Lett.* B263 (1991) 233;
P.H. Chankowski, Univ. of Warsaw preprint, IFT/7/91 (1991);
P.H. Chankowski, S. Pokorski and J. Rosiek *Phys. Lett.* B274 (1992) 191.

[19] for the RGE used in this text, see:
M.E. Machacek and M.T. Vaughn, *Phys. Lett.* B103 (1981) 427;
C.T. Hill, *Phys. Rev.* D24 (1981) 691;
M.B. Einhorn and D.R.T. Jones, *Nucl. Phys.* B196 (1982) 475;
K. Inoue, A. Kakuto, H. Komatsu and S. Takeshita, *Prog. Theor. Phys.* 68 (1982) 927, *ibid.* 71 (1984) 413;
L.E. Ibáñez and C. Lopez, and L.E. Ibáñez, C. Lopez and C. Muñoz, in ref. [6];
J. Bagger, S. Dimopoulos and E. Massó, *Phys. Lett.* B156 (1985) 357;
N.K. Falck, *Z. Phys.* C30 (1986) 247.

[20] P. Langacker and M. Luo, *Phys. Rev.* D44 (1991) 817.

[21] CDF Collab., F. Abe et al., *Phys. Rev. Lett.* 64 (1990) 142 and *Phys. Rev.* D43 (1991) 664.

[22] J. Ellis, S. Kelley, D.V. Nanopoulos, CERN preprint, CERN-TH.6140/91, CTP-TAMU-40/91, ACT-40 (1991).

[23] H. Arason, D. Castaño, b. Keszthelyi, S. Mikaelian, E. Piard, P. Ramond and B. Wright *Phys. Rev. Lett.* 67 (1991) 2933.

[24] S. Dimopoulos, L. Hall and S. Raby, Ohio State University preprint, OSU-DOE-ER-01545-567 (1991).

[25] J.F. Gunion and H.E. Haber, *Nucl. Phys.* B272 (1986) 1.

[26] ALEPH Coll., D. Decamp et al., *Phys. Lett.* B236 (1990) 86;
L3 Coll., B. Adeva et al., *Phys. Lett.* B233 (1989) 530;
OPAL Coll., M.Z. Akrawy et al., *Phys. Lett.* B240 (1990) 261.

[27] L3 Coll., B. Adeva, et al., *Phys. Lett.* B252 (1990) 511.

[28] ALEPH Collab., D. Decamp et al., *Phys. Lett.* B265 (1991) 475.

[29] S. Kuhlmann, talk presented at the *Second International Symposium on Particles, Strings and Cosmology*, Northeastern University, Boston, Massachusetts, March 25-30, 1991.

[30] H. Baer, J. Ellis, G.B. Gelmini, D.V. Nanopoulos and X. Tata, *Phys. Lett.* B161 (1985) 175;
G. Gamberini, *Z. Phys.* C30 (1986) 605;
R.M. Barnett, H.E. Haber and G.L. Kane, *Nucl. Phys.* B267 (1986) 625;
H. Baer, D. Karatas and X. Tata, *Phys. Lett.* B183 (1987) 220;
G. Gamberini, G.F. Giudice, B. Mele and G. Ridolfi, *Phys. Lett.* B203 (1988) 453.

[31] H. Baer, X. Tata and J. Woodside, *Phys. Rev. Lett.* 63 (1989) 352.

Cosmological constraints on supersymmetric models

Albrecht Klemm

Physik Department, Technische Universität München,
D-8046 Garching

Abstract

Supersymmetric models with R-symmetry predict a stable lightest supersymmetric particle (LSP). The relic mass density of the LSP in the standard big bang model is discussed. For the minimal supersymmetric model the constraints on the mass spectrum from the new LEP data imply that this density is negligible, unless the LSP would be heavy.

1 Introduction

The main achievements of the standard big bang model of cosmology (see [1-2] for reviews) are the explanation of the microwave background radiation and the abundances of the light elements. The theoretical description of both phenomena starts with an equilibrium state in an energy range where the particle spectrum and the interactions are well understood and traces the development of the particle densities in the expanding and cooling universe by rate equations. It is quite natural to extend this method in order to find the cosmological implications of *hypothetical* particles and interactions. If the strength of the latter one are comparable at least with the weak interaction, the particles will be in equilibrium down to temperatures of $O(1\ \text{MeV})$-$O(1\ \text{GeV})$. The initial conditions for the calculations are therefore not affected by earlier cosmological events, such as the weak phase transition or details of inflationary scenarios, which are less understood. The bounds obtained by these calculations are therefore *reliable*. The main uncertainty in the above mentioned temperature interval is due to the unknown details of the presumed quark-hadron transition at temperatures between $(150 - 400)$ MeV.

Such investigations have especially been carried out for supersymmetric theories since they predict a particle spectrum and interactions which are sufficiently constrained by the symmetry.

In the minimal supersymmetric extension of the standard model MSSM [3] it is the most economical way to introduce an R-parity in order to rule out baryon and lepton number violating dimension-four operators which would lead to an unacceptably high proton decay rate. The particles of the MSSM have a multiplicative quantum number R under this discrete symmetry, which can be written in terms of the baryon number B, the lepton number L and the spin J as

$$R = (-1)^{2J+3B+L}. \tag{1.1}$$

Since all supersymmetric partner particles have $R = -1$ and all standard model particles have $R = 1$, R parity forbids the decay of the lightest supersymmetric particle (LSP). In the standard big bang scenario, the LSP was in equilibrium in early epochs of our cosmos. A reduction of the number of LSPs can only proceed via pair annihilation, which however becomes more and more unlikely in the expanding universe. The calculation of the present energy density can be performed by integrating the detailed rate equation for the LSP [5-8]. Since this energy density is constraint by the lower bounds on the age of the universe, which come e.g. from radioactive dating methods, we can get constraints on the supersymmetric model. On the other hand, a neutral LSP with a sensible contribution to the energy density of the universe could account for the so called dark matter problem. Also galaxy formation models with weakly interacting massive particles (WIMPs) are nowadays preferred.

If the LSP would be charged or strongly interacting it would condensate with ordinary matter and super-heavy isotopes should be found. The constraints on the latter ones [15] exclude this possibility. Unbroken R parity therefore implies that the LSP is weakly interacting and should be detected in collider experiments by missing energy signatures. One should keep in mind that the introduction of R parity is not compelling. Other discrete symmetries which avoid too fast proton decays do not imply a LSP [4]. Relaxing R symmetry would invalidate the above chain of argumentation.

Also the gravitinos can have cosmological implications even if they are not the LSPs. Since they couple only gravitationally, they went out of equilibrium at a very early time when the temperature (cf. sec. 3.4) was of order $O(10^{17}$ GeV), i.e. without being suppressed by a Boltzmann factor like in (4.3). Most likely the gravitino number density was in the following diluted by an inflationary expansion and recreated hereinafter during the reheating period, which is caused by the thermalisation of the vacuum energy. That is the initial abundance of the gravitino density depends on the reheating temperature of the assumed inflationary scenario. Because of this uncertainty in the contraints coming from the gravitino, we like to sketch the corresponding arguments only briefly in the following and devote the main part of the time to the contraint on the relic density of the WIMP.

Due to its relatively long lifetime $\tau = 4.2 \times 10^3 \left(1 \text{ TeV}/\text{m}_{3/2}\right)^3$ sec the energy density of the gravitino or its decay products, respectively, can invalidate the successful nucleosynthesis predictions and/or distort the distribution of the background radiation, if the gravitinos where copiously reproduced after the inflation.

One possibility to avoid these difficulties is to make the gravitino so heavy that it decays before nucleosynthesis, which implies $m_{3/2} \geq O(10 \text{ TeV})$ [9]. The first effect of the presence of the gravitino mass density on nucleosynthesis is to enhance the Hubble expansion roughly at the *cosmic time* $t \simeq .7$ sec, which leads to a higher 'freeze out' temperature (cf. sec 3.4) of the neutron to proton ratio $(N_n/N_p) = \exp(-(m_n - m_p)/T)$. This will lead to more neutrons at the decoupling temperature. Since practically all neutrons which did not freely decay in the following are incorporated into ^4He this leads to an higher ^4He abundance. In [7] the restriction on the latter one was used in order to refine the bound of [9] to $m_{3/2} \geq 19$ TeV. In this scenario the gravitino decays when it dominates the energy density as a non-relativistic particle. The corresponding entropy increase will wash out any baryon asymmetry, which was produced before.

If the gravitinos are lighter $m_{3/2} = O(10 \text{ GeV}) - \text{O}(1 \text{ TeV})$ they will survive until the synthesis of the light elements starts roughly at $t \simeq 150$ sec and beyond that. In this case the authors of [10] have found that the strongest bound on the gravitino number density comes from the photo-dissociation of of ^4He into deuterium and ^3He, by cascades of energetic photons triggered by the decay products of the gravitino. This bound implies that the reheating temperature has to be less then $T_R < 2.5 \times 10^8 \text{ GeV} \left(\text{m}_{3/2}/100 \text{ GeV}\right)^{-1}$. This low reheating temperature did not allow for the production of a sufficient baryon asymmetry in the framework of the corresponding supersymmetric GUT. However since it was noticed that non-perturbative baryon, lepton and CP violating processes can generate the baryon asymmetry below the weak phase transition ([11] and ref. therein), this bound on T_R became doubtful.

Finally there exists the possibility that the gravitino is the LSP. If there were no suppression by inflation the gravitinos would have to be rather light $m_{3/2} \leq 1$ KeV [12], otherwise they would contribute to much to the present day energy density. The next lightest SUSY particle will decay into the gravitino with a lifetime comparable to the one above. But this causes not as serious problems with nucleosynthesis and background radiation because the number density of this WIMP is sufficiently Boltzmann suppressed (cf. sec. 4.1).

Problems similar to the gravitino problem but harder to avoid occur in supergravity theories, where the supersymmetry breaking is performed spontaneously in the hidden sector. This is because the hidden sector contains light fields which couple only through gravity. When the latter one become non-relativistic they develop coherent field oscillations like the 'invisible axion' [13]. Their energy density then falls like $\rho \propto R^{-3}$. This occurs when the energy density of the matter is still radiation dominated $\rho_m \propto R^{-4}$, such that the energy density of the 'hidden field(s)' will overcome the one of the ordinary matter and contribute too much to the present day density. This so called Polonyi problem has up to now no convincing solution see ([14] and ref. therein).

We begin our discussion with a short review of the dark matter problem. Especially we will see whether a weakly interacting particle WIMP can play a role as a dark matter constituent in the universe. In section three we explain the basic ingredients of the standard big bang cosmology. We will discuss in some detail the age constraint on the density of the LSP. The methods to calculate the relic density of the LSP are described in section four. Recent results on the relic density of the LSP in view of the new LEP data are given at the end of section three.

2 Dark matter in the Universe

When astronomers study the dynamics of systems on scales greater then a few hundred pc's they are regularly led to the conclusion that either Newtons law of gravitation is violated in the regime of low acceleration [16] or that there must be a large amount of matter, which is not associated with light [17]. Adopting the latter interpretation one observes that the larger the amounts of *dark matter* is the larger the scale of the investigated system, see figure 1.

Oort reported in 1932 [18] that roughly half of the matter in the galactic disk in the vicinity (kpc) of the sun must be hidden. This fact was obtained by fitting gravitational potentials in order to explain the observed density distribution and the velocity dispersion perpendicular to the galactic disk of so-called tracer stars. More recent investigations [19] along the same lines confirm that within layers of scale heights less than 0.7 kpc above the galactic plane the ratio between visible and dark matter is $0.5 \leq \rho_{dark}/\rho_{vis} \leq 1.5$. Estimates for the dark matter in the *elliptical galaxies* and the *bulges of spiral galaxies* using stellar dynamics have also been performed [20] and references therein.

On the scale of tens of kpc's one observes [21] that the orbital velocities V_0 of radial motions around spiral galaxies are roughly constant from the edges of the galaxies, where the light effectively ceases, up to the far outer regions, where the brightness is only one percent of the central region. These rotation curves suggest by applying simply Keplers 3rd law $GM(r) = v^2 r$ that the galaxies are surrounded by a roughly spherical halo of dark matter with a radial mass density distribution $M(r) \propto r$. To calculate the total mass contribution of the dark matter one has to estimate the radius of the halo. This can be done using the upper observational bound on the the local *escape* velocity. Likewise the kinematics of the *Magellanic cloud*, local group timing or the kinematics of *satellite galaxies* [20] can be used to estimate this radius, see figure 1. Using an averaged number density for the galaxies one gets for the mass contribution of the dark matter in the whole universe ($\Omega_i = \rho_i/\rho_{crit}$), cf. (3.8) and (3.7),

$$\Omega_{dark} \simeq 0.1 \geq 10\, \Omega_{vis},$$

i.e. its mass represents 10 % of the critical mass of the universe, while the visible mass contributes only one percent to the critical mass.

On the next scale, 100 kpc's to 1 Mpc, one estimates the mass of groups of galaxies using the virial theorem [22], which states that one has for the time average of the kinetic energy K and the potential energy W : $2\langle K\rangle_{time} + \langle W\rangle_{time} = 0$. In practice one can measure by blue-red-shift determination only the momentary velocity in the line of sight. If the system is well relaxed one can use $2K + W \approx 0$. Assuming the masses of galaxies in groups are typical, one gets from this observations mass contributions from the galaxies in the range of

$$0.1 < \Omega < 0.3\ . \tag{2.1}$$

On the scales of ten's of Mpc's, distortions of the Hubble *flow* [24], e.g. due to the mass of the *Virgo* cluster, can be measured and used to determine the averaged mass to light ratio of the system (see Figure 1).

On cosmological scales one uses the relation between the number of galaxies N_{gal} and the red-shift ($z = (\lambda(t_0)/(\lambda(t_1)) - 1$, cf. (3.11))

$$\frac{1}{z^2}\frac{dN_{gal}}{dzd\Omega} = (H_0R_0)^{-3}n_c(t)\left[1 - 2(\frac{\Omega_0}{2} + 1)z + \ldots\right], \tag{2.2}$$

in order to determine Ω_0. This way Loh and Spillar [42] have obtained $\Omega_0 = 0.9^{+1.1}_{-0.7}$. The problem here are the assumptions about the galaxy development at very large red-shifts. In [42] $n_c(t) \approx const.$ was used, but see [43].

Likewise the relation between true luminosity distance d_L and the red-shift [1]

$$d_L = H_0^{-1}[z + \frac{1}{2}(1 - \frac{\Omega_0}{2})z^2 + \ldots] \tag{2.3}$$

could in principle be used to determine[1] Ω_0. However the size of the errors here allow only to state that Ω_0 is roughly of order 1.

Equation (2.3) is used to determine the present value of the Hubble constant $H_0 := (\dot{R}(t)/R(t))|_{t=t_0}$. Due to the difficulties in fixing the true luminosity distance d_L of the cosmological objects it is likewise badly known [30]

$$H_0 = 100\ h \text{ km sec}^{-1}\text{ Mpc}^{-1}, \tag{2.4}$$

[1]Here and in the following the subscript $_0$ indicates the present value of the corresponding quantity.

Figure 1: A survey of dark matter in the universe at different scales. The mass to light ratio M/L is normalized to the mass to light ratio of the sun. Using the averaged luminosity density in the visible spectrum $j_0 = 1.7 \times 10^8 h (L_{sun} \mathrm{Mpc}^{-3})_{vis}$, the present density can be expressed as $\Omega_0 = (M/L)(j_0/\rho_c) \approx (1/1600h)(M/L)$. The different systems are sketched in the text.

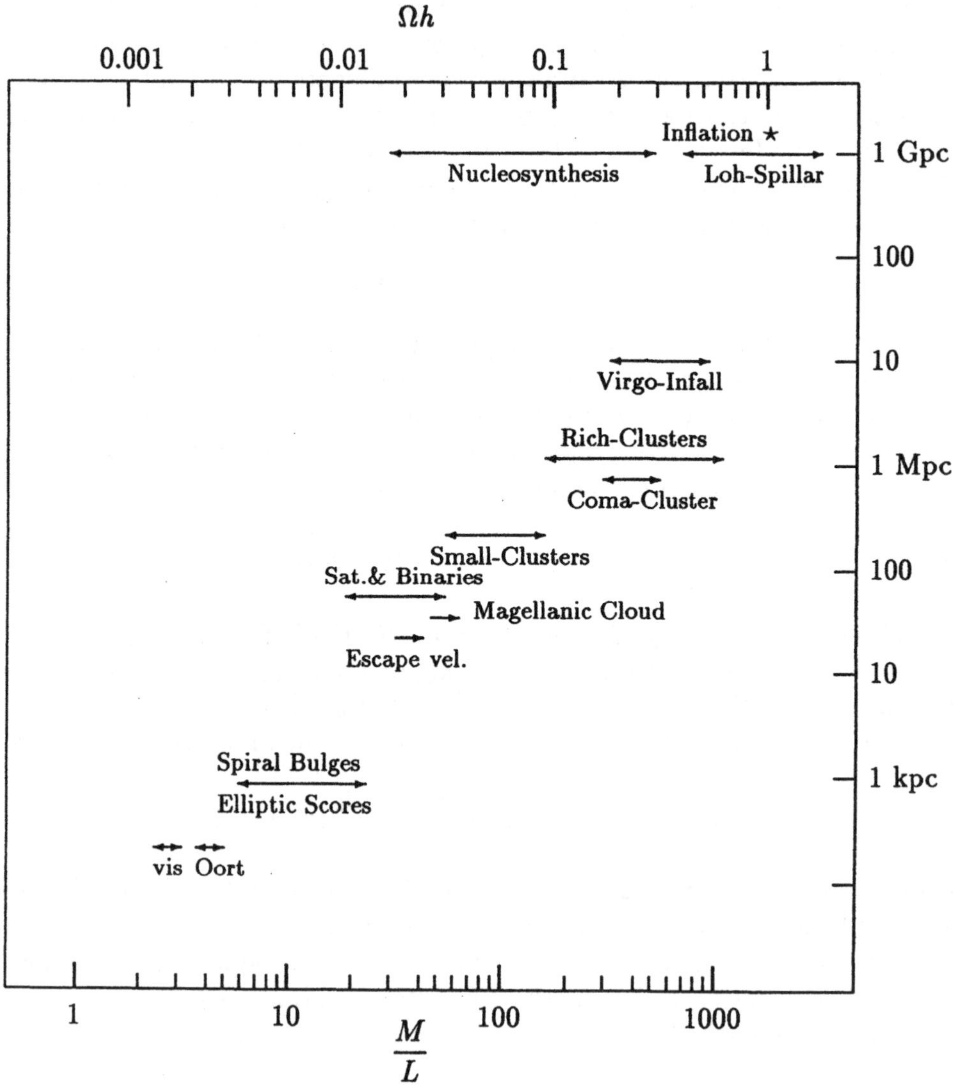

where

$$0.4 \leq h \leq 1.0. \tag{2.5}$$

The baryon to photon ratio η is a crucial parameter for the formation of the light nuclei. From the comparison [25] between the theoretical *nucleosynthesis* calculations [33] and the observed abundance [27] of light nuclei one can infer a constraint on the baryon contribution to Ω_0:

$$0.015 \leq \Omega_B h^2 \leq 0.026. \tag{2.6}$$

The bounds for the various contributions to the density of the universe on different scales are depicted in figure 1.

As far as baryonic matter is concerned we see two things. There must be some amount of baryonic dark matter in order to explain the difference between the lower bound of (2.6) and the amount of visible matter. At the scales of galaxies *all* the dark matter *can* be baryonic. However there exist arguments pro [28] and contra [29] this possibility. If we follow [29], then the LSP would be an interesting candidate for halo dark matter. It has also turned out that structure formation models can better explain small structures in presence of massive weakly interacting particles so called cold dark matter [31].

A weakly interacting *massive* particle like the LSP would cluster at scales less then 10 Mpc, where it could explain the halo dark matter, if it's energy density is roughly *one quarter* of the critical density. It can not make up for the mass difference between the mass estimates below this scale and higher mass estimates on the cosmological scale. Especially if one believes for theoretical reasons that $\Omega_0 = 1$ one needs rather so called hot dark matter like light $m_\nu \simeq 90h^2$ eV neutrinos which are relativistic during structure formation and do not cluster with galaxies. Alternatively one could introduce a cosmological constant, which may also be attractive in order to solve the age constraint, see section (3.2).

If the LSP particles were present in our galaxy there would be indirect methods to detect them. The most common scenario is that LSP's annihilate in the sun, which would lead to an significant 250 MeV neutrino flux [32]. If the $s-$neutrino would make the halo mass then the limits on this flux from proton decay detectors imply $m < O(3-5)$ GeV.

Furthermore, direct detection experiments which limit neutrinos also limit $s-$ *neutrinos* to be lighter then $O(12$ GeV$)$ [34]. The LEP data on the decay width of the Z imply for $s-$neutrinos $m > O(30)$ GeV which *rules out* the $s-$neutrino as a dark matter candidate.

3 Basics of the theoretical description of big bang cosmology

3.1 The Einstein equations

The *cosmological principle* is an assumption — the validity of which is confirmed by the highly isotropic microwave background radiation and the radio galaxy distribution on very large scales — that the three dimensional spatial subspace of our cosmos is at all times *homogeneous* and *isotropic*, i. e. it has the maximal number $(6 = 3(3+1)/2)$ of possible isometries. These symmetries restrict the freedom of the metric which determines the intrinsic geometry of our cosmos. So one can write it e.g. in the Robertson-Walker form[2]

$$ds^2 = g_{\mu\nu}dx^\mu dx^\nu = dt^2 - R^2(t)\left\{\frac{dr^2}{1-kr^2} + r^2 d\theta^2 + r^2 \sin^2\theta d\phi^2\right\}. \tag{3.1}$$

Here the only free parameters are the time dependent scale factor $R(t)$ and k, which take the values $(-1, 0, 1)$, indicating whether the scalar curvature of the spatial three space $\mathcal{R}^{(3)} = 6k/R^2(t)$ (constant all over the spatial subspace) is negative, vanishing or positive, respectively.

The Einstein equations can be written in the form

$$G_{\mu\nu} := R_{\mu\nu} - \frac{1}{2}g_{\mu\nu}\mathcal{R} = 8\pi G T^m_{\mu\nu} + \Lambda g_{\mu\nu}, \tag{3.2}$$

where $R_{\mu\nu}$ is the Ricci tensor, $\mathcal{R}$ is the scalar curvature, $T^m_{\mu\nu}$ is the energy momentum tensor for the matter, and we have allowed for a cosmological constant Λ. A natural candidate for the energy momentum tensor of the matter which is compatible with the symmetries of the metric is the one of an incompressible, frictionless fluid. In comoving coordinates it reads $T^m_{\mu\nu} = \text{diag}(\rho^m, p^m, p^m, p^m)$. Here ρ^m is the energy density and p^m is the pressure of the matter, respectively. Inserting this into (3.2) one obtains for the $(time, time)$, $(space, space)$ components

$$\frac{\dot{R}^2}{R^2} + \frac{k}{R^2} = \frac{8\pi G}{3}\rho^m + \frac{\Lambda}{3}, \tag{3.3}$$

$$2\frac{\ddot{R}}{R} + \frac{\dot{R}^2}{R^2} + \frac{k}{R^2} = -8\pi G p^m + \Lambda, \tag{3.4}$$

while the other components vanish identically. We may also use the conservation of the *full* stress energy tensor, including the vacuum contribution, $T^{\mu\nu}_{;\mu} = 0$, which reads

$$d(\rho R^3) = -p d(R^3). \tag{3.5}$$

[2]We use the signature of the metric $g_{\mu\nu} = \text{diag}(+,-,-,-)$. In the book of S. Weinberg [1] the signature is $g_{\mu\nu} = \text{diag}(-,+,+,+)$.

Only two equations among (3.3),(3.4) and (3.5) are independent, because contracting two times (λ, ν and κ, μ) the second Bianchi identity $\sum_{cyclic(\mu\nu\rho)} R_{\kappa\lambda\mu\nu;\rho} = 0$ yields $G^{\mu\rho}{}_{;\mu} = 0$, i.e. the energy momentum conservation is a consequence of (3.2) and the Bianchi identity. Usually one chooses the energy conservation (3.5) and the so-called Friedman equation (3.3) as the independent equations. We define the critical density by requiring $k = 0$ in (3.3) this yielding[3]

$$\rho_c = \frac{3H_0^2}{8\pi G} = 0.81 \times 10^{-47} h^2 \text{ GeV}^4 = 1.88 \times 10^{-29} \text{h}^2 \frac{\text{g}}{\text{cm}^3}, \tag{3.7}$$

where we have used (2.4). Introducing the ratio

$$\Omega := \frac{\rho}{\rho_c}, \tag{3.8}$$

(3.3) is often rewritten as

$$\frac{k}{H^2 R^2} = \Omega - 1 + \frac{\Lambda}{3H_0^3}. \tag{3.9}$$

A useful auxiliary equation is obtained by solving (3.3) and (3.4) for $\ddot{R}$:

$$\frac{\ddot{R}}{R} = -4\pi G(\rho^m + 3p^m) + \Lambda. \tag{3.10}$$

The scale factor $R(t)$ is positive by definition. Also the energy density and the pressure of the matter is positive. Neglecting the cosmological constant for the moment, we learn from (3.10) that $R(t)$ is a convex function. Considering the kinematics of particle motion in the background of (3.1) one obtains [37] for massless particles emitted at time t_1 and detected at time t_0 that the wavelength behaves like

$$\frac{\lambda(t_1)}{\lambda(t_0)} = \frac{R(t_1)}{R(t_0)}. \tag{3.11}$$

Since today we observe a red shifted ($\lambda(t_0) > \lambda(t_1)$) spectrum in the light of distant galaxies we have $\dot{R}(t) > 0$. So $R(t)$ is a convex curve with a nowadays positive slope, which means that the age $t_0 - t_1$ of the Universe counted from the initial singularity $R(t_1) = 0$ (one defines $t_1 = 0$) is finite and bounded from above by

$$t_0 < \frac{R(t)}{\dot{R}(t)}\Big|_{t=t_0} =: \frac{1}{H_0}. \tag{3.12}$$

[3]Here we use so called natural units by setting $k_{Boltzmann} = h/(2\pi) = c = 1$ and keeping as fundamental unit only the energy in GeV,MeV,... Conventional units can be restored by remembering

$$\begin{aligned} 1\ \text{GeV} &\hat{=} 1.1605 \times 10^{13} \text{Kelvin} \\ 1\ \text{GeV} &\hat{=} 1.7827 \times 10^{-24} \text{g} \\ 1\ \text{GeV}^{-1} &\hat{=} 6.5822 \times 10^{-25} \text{sec} \\ 1\ \text{GeV}^{-1} &\hat{=} 1.9733 \times 10^{13} \text{cm}. \end{aligned} \tag{3.6}$$

3.2 Matter and vacuum contribution to the energy momentum tensor

An epoch of thermodynamic equilibrium can establish an equation of state $p = p(\rho)$, by which one can solve (3.5) for $\rho(R)$. Explicit solutions $R(t)$ can then be obtained by integrating (3.3). E. g. for ideal gases of bosons ($-$) or fermions ($+$) in *kinetic equilibrium* the Bose-Einstein or Fermi-Dirac phase distributions

$$f(E) = \frac{1}{\exp[(E-\mu)/T] \pm 1} \tag{3.13}$$

provides us with explicit expressions [38] for the number density $n(T)$, the energy density $\rho(T)$, the pressure $p(T)$ and the entropy density $s(T)$:

$$n(T) = \frac{g}{2\pi^2} \int_m^\infty f(E)|\vec{p}|E dE, \tag{3.14}$$

$$\rho(T) = \frac{g}{2\pi^2} \int_m^\infty f(E)|\vec{p}|E^2 dE, \tag{3.15}$$

$$p(T) = \frac{g}{6\pi^2} \int_m^\infty f(E)|\vec{p}|^{3/2} dE, \tag{3.16}$$

$$s(T) = \frac{\rho(T) + p(T) - \mu n(T)}{T}. \tag{3.17}$$

Here μ is the chemical potential. The necessary conditions for such an epoch to occur is that the interaction rates Γ driving the particles towards equilibrium are large compared with the Hubble expansion rate $H = \frac{\dot{R}}{R}$. Whether this is a realistic assumption will be discussed in some detail in section (3.4).

Simple examples of matter content can be obtained by using the high relativistic and non-relativistic limits of (3.15) and (3.16):

$$\text{Radiation } p_R = (1/3)\rho_R \quad \Rightarrow \quad \rho_R \propto R^{-4}, \tag{3.18}$$

$$\text{Non-relativistic Matter } p_{nr} = 0 \quad \Rightarrow \quad \rho_{nr} \propto R^{-3}, \tag{3.19}$$

$$\text{Vacuum energy } p_{vac} = -\rho_{vac} = \frac{\Lambda}{8\pi G} \quad \Rightarrow \quad \rho_{vac} \propto const. \tag{3.20}$$

The last case corresponds to an universe without matter but a non-vanishing cosmological constant. Inserting (3.20) into (3.3) leads to $H = (8\pi G\rho_{vac}/3)^{(1/2)} = const$ and hence an exponential growth of the scale factor $R(t) \propto \exp(Ht)$. The space defined by (3.1) with this scale factor has a ten-parameter group of isometries which is the maximal number in four dimensions ($10 = 4(4+1)/2$). It is called de Sitter space. Such an epoch with a large value of Λ is probably realised during inflation

[39]. It solves the horizon problem and leads in a natural way to a flat universe [39]. This explanation for $\Omega = 1$ is especially desired, because the present day universe seems to rather flat $|\Omega - 1| = O(1)$, while $\Omega = 1$ is an unstable fix point of (3.9).

Today we live as far as the *known* contribution to the energy momentum tensor is concerned in an universe which is dominated by non-relativistic matter, i.e. $p^m = 0$. The original motivation for introducing the cosmological constant was to obtain a static solution in the case of a matter dominated universe with $k = 1$. This is done by adjusting the cosmological constant such that the right-hand-side of (3.10) vanishes $\rho^m = (\Lambda/(4\pi G)) = 2\rho_{vac}$ and by adjusting the radius so that (3.3) states that $(\dot{R}/R) = 0$ vanishes, i.e. $R = (1/\sqrt{\Lambda})$, one gets the desired static solution.

The observed Hubble expansion contradicts a static solution. Nevertheless a small cosmological constant can be tolerated. See (Weinberg 89) [40] for a review of the cosmological constant problem. From direct observations in the solar system the strictest bound comes from the observed advance of Mercury's perihelion [41]. It states $\rho_{vac} := (\Lambda/8\pi G) = 2.31 \times 10^{-36}$ GeV^4 or $\Omega_{vac} = h^{-2}\ 2.8 \times 10^{10}$. Bounds from the global dynamics are ten orders of magnitudes more restrictive. By relating the number of faint galaxies to their redshift Loh and Spillar [42] have found a very stringent bound $\Omega_{vac} \leq 0.1$. This was reconsidered in [43], proposing new models for the galaxy evolution at very high redshifts with the result that the best fit to the data was at $\Omega_{matter} \leq .1$ and $\Omega_{vac} \simeq 0.5 - 1$; $\Omega_{vac} = 0$ is also consistent with their data. In [44] Durrer and Straumann concluded from the development of fluctuations in a $k = 0$ cold dark matter universe that $\Omega_{vac} < .7$. From the statistics of gravitational lenses the authors of [45] inferred that $\Omega_{vac} < .9$.

3.3 The age constraint on Ω_0

As mentioned in the introduction, the upper bound on the relic density comes from a lower constraint on the age of the universe. The dating methods which concern us here come from estimating the age of the oldest stars by stellar evolution models and from radioactive dating methods. The initial conditions for the ^4He and metallicity abundances are predicted by nucleosynthesis calculations. These methods predict an age bound of the universe ranging from $10 \times 10^9\ yrs \leq t_0 \leq 20 \times 10^9\ yrs$ [46], which has to be compared with the dynamical age of the universe in order to constraint Ω_0.

Using $\rho/\rho_0 = (R/R_0)^{-3}$ for non-relativistic matter (3.19) or $\rho/\rho_0 = (R/R_0)^{-4}$ for radiation (3.18), respectively, we can rewrite (3.3) as

$$\left(\frac{\dot{R}}{R_0}\right)^2 + \frac{k}{R_0^2} = \begin{cases} \frac{8\pi G}{3}\rho_0\left(\frac{R_0}{R}\right) + \left(\frac{R}{R_0}\right)^2 \frac{\Lambda}{3} & \text{Non-relativistic matter} \\ \frac{8\pi G}{3}\rho_0\left(\frac{R_0}{R}\right)^2 & \text{radiation.} \end{cases} \qquad (3.21)$$

Here we include the possiblity of a cosmological constant only in the matter dominated area, since in the radiation dominated phase the energy densities are anyway so high that one can neglect Λ. Using now the definition of the critical density ρ_c (3.7) and of $\Omega_0 = (\rho_0/\rho_c)$ we can rewrite this as

$$\left(\frac{\dot{R}}{R_0}\right) = \begin{cases} H_0\sqrt{1-\Omega_0-\Omega_{vac}+\Omega_0(R_0/R)+\Omega_{vac}(R/R_0)^2} & \text{Non-relativistic matter} \\ H_0\sqrt{1-\Omega_0+\Omega_0(R_0/R)^2} & \text{radiation,} \end{cases} \tag{3.22}$$

where $\Omega_{vac} = \Lambda/(3H_0^2)$. The equation for the non-relativistic matter case for $\Omega_{vac} = 0$ can be intergrated in a closed form from $R = 0$ to $R = R_0$

$$t_0 = \begin{cases} H_0^{-1}\frac{\Omega_0}{2(\Omega-1)^{3/2}}\left[\arccos(2\,\Omega_0^{-1}-1)-\frac{2}{\Omega_0}\sqrt{\Omega_0-1}\right], & \Omega_0 < 1\,, \\ \frac{2}{3}H_0^{-1}, & \Omega_0 = 1, \\ H_0^{-1}\frac{\Omega_0}{2(1-\Omega)^{3/2}}\left[\frac{2}{\Omega_0}\sqrt{1-\Omega_0}-\operatorname{arccosh}(2\,\Omega_0^{-1}-1)\right], & \Omega_0 > 1. \end{cases} \tag{3.23}$$

The same calculation for a radiation dominated universe yields

$$t_0 = H_0^{-1}\frac{\sqrt{\Omega_0}-1}{\Omega_0-1}. \tag{3.24}$$

For fixed H_0 the present age of the universe decreases in all cases as Ω_0 increases. A conservative lower bound is $t_0 \geq 12\times10^9$ yrs. The one which is favoured by star development models states $t_0 \geq 15\times10^9$ yrs. Assuming both theoretical prejudices $\Omega_0 = 1$ and $\Lambda = 0$ to hold, these age contraints imply $h \leq 0.54$ and $h \leq 0.43$, respectively (cf. (2.5)). Reconciling larger values of h with a flat universe seems only possible if one introduces a cosmological constant. In table 1 we list constraints on Ω_0 assuming these bounds for the actual range of the Hubble parameter h, taking into account the possibility of a nonvanishing cosmological constant with values which were discussed at the end of section (3.2). The latter cases are known as Lemaıtre models.

Integrating the matter dominated case in (3.22) from $R(0) = 0$ to $R(t)$ and expanding the result for $(R_0/R(t)) >> \Omega_0^{-1}$ one obtains

$$t \simeq \frac{2}{3}\left(\frac{R}{R_0}\right)^{3/2} H_0^{-1}\Omega_0^{-1/2}. \tag{3.25}$$

Today the energy contribution of the radiation, assuming the photons and three massless neutrino species to contribute, can be calculated from the temperature of the microwave background radiation $T_{\gamma_0} = 2.75$ K to be $\rho_r^0 = (\pi^2/30)g_*T_{\gamma_0}^4 = 8.09\times10^{-34}$ g cm^{-3} (compare (3.35) for g_*). Parameterizing the non-relativistic

Table 1: Bounds on Ω_0 from the age constraint for $t_0 \geq 12 \times 10^9$ yrs and $t_0 \geq 15 \times 10^9$ yrs for different values of Ω_{vac}. The — sign means that age constraint cannot be fullfilled.

h	$t_0 \geq 12 \times 10^9$ yrs			$t_0 \geq 15 \times 10^9$ yrs		
	$\Omega_{vac} = 0$	$\Omega_{vac} = .7$	$\Omega_{vac} = .9$	$\Omega_{vac} = 0$	$\Omega_{vac} = .7$	$\Omega_{vac} = .9$
.4	$\Omega_0 \leq 3.51$	$\Omega_0 \leq 4.12$	$\Omega_0 \leq 4.30$	$\Omega_0 \leq 1.48$	$\Omega_0 \leq 2.01$	$\Omega_0 \leq 2.16$
.5	$\Omega_0 \leq 1.48$	$\Omega_0 \leq 2.00$	$\Omega_0 \leq 2.16$	$\Omega_0 \leq 0.45$	$\Omega_0 \leq 0.87$	$\Omega_0 \leq 1.00$
.6	$\Omega_0 \leq 0.58$	$\Omega_0 \leq 1.02$	$\Omega_0 \leq 1.16$	$\Omega_0 \leq 0.068$	$\Omega_0 \leq 0.38$	$\Omega_0 \leq 0.49$
.7	$\Omega_0 \leq 0.17$	$\Omega_0 \leq 0.53$	$\Omega_0 \leq 0.65$	—	$\Omega_0 \leq 0.16$	$\Omega_0 \leq 0.25$
.8	—	$\Omega_0 \leq 0.27$	$\Omega_0 \leq 0.37$	—	$\Omega_0 \leq 0.05$	$\Omega_0 \leq 0.12$
.9	—	$\Omega_0 \leq 0.13$	$\Omega_0 \leq 0.21$	—	$\Omega_0 \leq 0.007$	$\Omega_0 \leq 0.06$
1	—	$\Omega_0 \leq 0.05$	$\Omega_0 \leq 0.13$	—	—	$\Omega_0 \leq 0.027$

matter contribution as $\rho^0_{nrM} = 1.88 \times 10^{-29}\ \Omega_0\ h^2$ g cm^{-3}, we get using (3.18) and (3.19) $(\rho_{nrM}/\rho_r) = (R/R_0) \times (2.32 \times 10^4 \Omega_0 h^2)$. The time t_{EQ} when the non-relativistic matter contribution and the radiation contribution were equal follows from (3.25) to be

$$t_{EQ} = \frac{2}{3}\left(\frac{R_{EQ}}{R_0}\right)^{3/2} H_0^{-1}\Omega_0^{-1/2} = 1.4 \times 10^3 (\Omega_0\ h^2)^{-2}\ \text{yrs}. \tag{3.26}$$

Comparing this time for all sensible Ω_0 and h values with the age constraint $t_0 = 12 \times 10^9$ yrs we see that the duration of the radiation dominated area can be neglected.

3.4 The early radiation dominated epoch

We want now to investigate very early epochs of the universe. From (3.18) and (3.19) we have already seen that the radiation contribution dominates the energy density as one goes back in time towards the initial singularity $R \to 0$. Inserting the energy density of the radiation ρ_R in (3.3) and neglecting k/R^2 against $(8\pi G)/(3R^4)$ at early epochs we learn that

$$R(t) \propto \sqrt{t}. \tag{3.27}$$

Now we use the high relativistic expansion of the equilibrium expression for the energy density (3.15). For a radiation gas with several species of bosons and fermions we get

$$\rho_R = \frac{\pi^2}{30} g_* T^4, \quad \text{where } g_* = \sum_{i=boson} g_i \left(\frac{T_i}{T}\right)^4 + \frac{7}{8} \sum_{i=fermion} g_i \left(\frac{T_i}{T}\right)^4. \tag{3.28}$$

Here the g_i are the degrees of freedom of species i. The factor $(\frac{T_i}{T})^4$ (cf. (3.35)) is appropriate for species which are dropped out of equilibrium. These may still have the equilibrium distributions but their 'temperature' can differ from that of the termalized species, as it is explained in some detail below for the neutrino. Using (3.27) and (3.31) in (3.3) we get

$$t = 0.301 g_*^{-1/2} \frac{m_{pl}}{T^2} \simeq \left(\frac{T}{1 \text{ MeV}}\right)^{-2} \text{sec.} \tag{3.29}$$

The approximate sign is valid in the 100 MeV $\geq$ T $>$ 1 MeV range, where the photon, the three neutrino species and electrons-positrons can be considered as relativistic particles in equilibrium. Hence we have for $g_*(100 \text{ MeV} \geq T > 1 \text{ MeV}) = 2 + \frac{7}{8}[(3 \times 2) + 4] = 10.75$. Above temperatures of 300 GeV according to the standard model, we would have the following spectrum: the Gauge bosons γ, Z_0, $W^{\pm}$, 8 Gluons, the Higgs douplett $\binom{\Phi_+}{\Phi_0}$ and three Matter generations $\binom{\nu_e}{e_L}$, e_R, $\binom{\bar{u}_L}{d_l}$, u_R, d_R such that $g_* = 106.75$.

According to (3.29) we have at the initial singularity $t \rightarrow 0$, $R \rightarrow 0$ but $T \rightarrow \infty$. Note that in the nonrelativistic range the two-point interaction rate i.e. $\Gamma = n_{eq}\langle v_{relative}\sigma\rangle$, increases with the temperature as $\Gamma = n_{eq}\langle v_{relative}\sigma\rangle \propto (m/T)^{\frac{3}{2}} \exp(-m/T)\langle v_{relative}\sigma\rangle$, while the Hubble expansion (3.3) goes like

$$H = 1.66 g_*^{1/2} \frac{T^2}{m_{pl}}, \tag{3.30}$$

such that the interaction rate will generally overcome the Hubble expansion rate.

For relativistic particles we have $n \propto T^3$. By dimensional considerations we have $n_{eq}\langle v_{relative}\sigma\rangle \propto T^3 \frac{T}{T} g^2 (T^{-2-2\text{dimg}})$. E. g. for gravitational interactions or Fermi interactions (below the weak scale) the dimension of the coupling is dimg $= -2$ such that $\Gamma \propto T^5$. Comparing the weak interaction rate $n_{eq}\langle v_{relative}\sigma\rangle = G_F^2 T^5$ with (3.30) one gets as the temperature when the neutrinos just went out of equilibrium $T \approx 1$ MeV.

For dimension-less couplings we have $\Gamma \propto T$, but for reasonable temperatures ($T < m_{pl}$ where these considerations make sense at all) Γ is greater than H because of the suppression by m_{pl} in (3.30). Hence we assume that a generic particle species start with its equilibrium value of the energy density at early times. Comparing the interaction rates with the Hubble expansion rate one can calculate the development of $g_*(T)$ assuming the standard model spectrum and interactions. The result is depicted in figure 2.

The sudden fall at 200 MeV is due to the quark-hadron transition. The exact shape in this range is not known. Since it can be sensible for the freeze out of the LSP, different possibilities have to be considered. Figure 3 shows the behaviour of $g_*(T)$ with two hypotheses about the quark-hadron transition temperature.

Figure 2: The evolution of $g_*(T)$ as a function of temperature in the $SU(3)_C \otimes SU(2)_L \otimes U(1)_Y$ standard model, from [2]

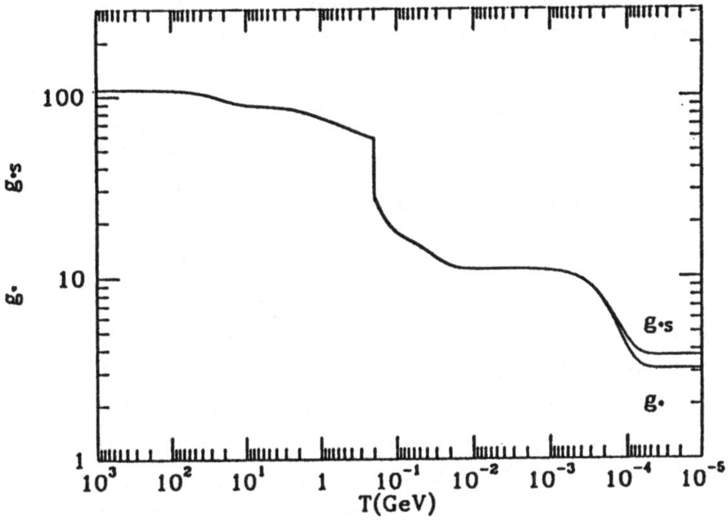

Figure 3: The evolution of $g_*(T)$ according to two assumptions about the quark-hadron transition temperature (150 MeV and 400 MeV, respectively). The dashed lines indicate the curve for free quarks using current algebra masses and the corresponding curve for free hadrons, from [8]

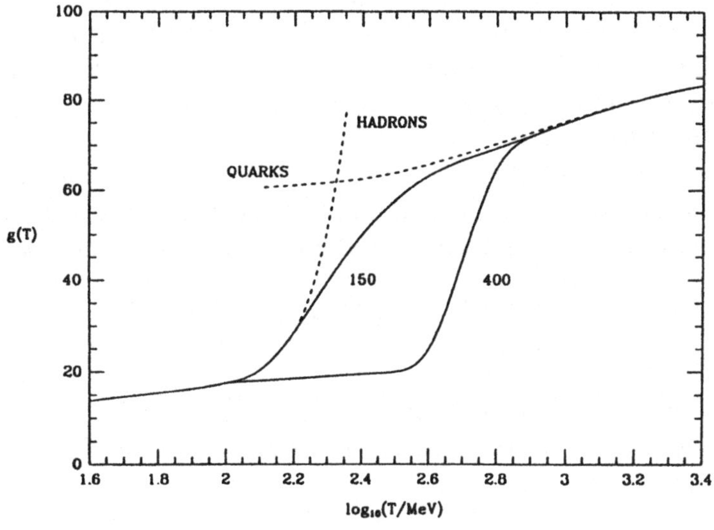

If the interaction rates are great enough to maintain equilibrium, the Hubble expansion can be considered as adiabatic. Let $S = sR^3$ be the entropy in a comoving volume, then one has $dS = 0$. Furthermore one sees from (3.17), the relativistic and non-relativistic expressions for $\rho(T)$, $p(T)$ and $n(T)$, that the radiation contribition dominates the entropy density, which is therefore given by

$$s = \frac{2\pi^2}{45} g_*^{(s)} T^3, \quad \text{where } g_* = \sum_{i=boson} g_i \left(\frac{T_i}{T}\right)^3 + \frac{7}{8} \sum_{i=fermion} g_i \left(\frac{T_i}{T}\right)^3. \tag{3.31}$$

By the conservation of entropy we have

$$gT^3R^3 = const., \tag{3.32}$$

where we understand by g the wheigted degrees of freedom $g = \sum_{i=boson} g_i + \frac{7}{8}\sum_{i=fermion} g_i$ for particles which are in equilibrium. This statement can easily be used in order to relate the present day number density of a particle, which went out of equilibrium at a temperature T_{freeze} to the present day number density of photons, which is related to the measured temperature of the background radiation. Let us explain the mechanism with the neutrinos as an example. For $T > 1$ MeV the three species of neutrinos were kept in chemical and kinetical equilibrium with the photons, electrons and positrons by the weak interaction, especially we have $T_\nu = T_{e^\pm} = T_\gamma$. Below the 'freeze out' temperature of $T_{freeze} \simeq 1$ MeV the weak interaction rates compared with the Hubble expansion are not fast enough to keep the neutrinos in equilibrium, so they decouple and the effective degrees of freedom for interacting radiation drop to from 10.75 (see above) to $g(1 \text{ MeV} > \text{T} > m_e) = 2 + 4 \times (7/8) = 11/2$. As long as $g = const.$ the photon temperature cools down like the neutrino temperature since we have $T_\gamma \propto R^{-1}$ by (3.32) and $T_\nu \propto R^{-1}$ by (3.11). Since the number density of radiation is $n \propto T^3$, the latter fact simply states that the total amount of neutrinos in a comoving volume $N = nR^3$ is constant. This is trivially also true for any non interacting particle.

When T drops below $T = m_e$, the $e^\pm$ pairs become nonrelativistic and do no longer contribute to g such that $g(T^a << m_e) = 2$. The corresponding domination of pair-annihilation will heat the photons but not the decoupled neutrinos. The photon temperature *before* and *after* $e^\pm$ annihilation is according to (3.32) related by

$$\frac{(R^aT_\gamma^a)^3}{(R^bT_\gamma^b)^3} = \frac{g(T^b > m_e)}{g(T^a << m_e)} = \frac{11}{4}, \tag{3.33}$$

while for the neutrinos we have from (3.11) $(R^bT_\gamma^b)^3 = (R^bT_\nu^b)^3 = (R^aT_\nu^a)^3$. Since $g = 2$ has not passed any further threshold we have from (3.33)

$$T_\gamma/T_\nu = (11/4)^{1/3} \tag{3.34}$$

still today. Note the spreading of $g_*^{(s)}$ and g_* in figure 1 due the difference between photon temperature and neutrino temperature after pair-annihilation starts. The effective degree of freedom g_* after the decoupling is given by

$$g_* = 2 + \frac{7}{8} \times 2 \times 3 \times \left(\frac{4}{11}\right)^{4/3} \simeq 3.36. \tag{3.35}$$

In order to calculate the number density of the LSP (the particle is denoted by χ in the following) it is convenient to normalize the number density n_χ to the number density of the equilibrium radiation, i.e. to introduce $f := (n_\chi/T_{eq}^3)$. Having obtained the value for f at a temperature where χ is out of chemical equilibrium, say $f(T_{freeze})$, by solving the rate equation (4.2) the present day density can readily related to the present day photon temperature. As $n_\chi^{freeze} R_{freeze}^3 = n_\chi^0 R_0^3$, we have from (3.32)

$$n_\chi^0 = f(T_{freeze}) \frac{g(T_0)}{g(T_{freeze})} T_{\gamma 0}^3. \tag{3.36}$$

4 The freeze out of the LSP

4.1 Rate equation for the decoupling of the LSP

In this section we want to discuss the freeze out of the LSP. An up to date detailed technical treatment can be found in [8]. The developement of the number density of the LSP is governed by the rate equation[4]

$$\frac{dn_\chi}{dt} = -3\frac{\dot{R}}{R} n_\chi - \langle \sigma v_{rel} \rangle_T (n_\chi^2 - n_{eq}^2). \tag{4.1}$$

Here v_{rel} is the relative velocity of two particles, σ is the cross section and $\langle\ \rangle_T$ indicates thermal weighting. The first term on the right-handside of (4.1) describes the dilution of the number density due to the Hubble expansion. The term $-\langle v_{rel}\sigma\rangle_T n_\chi^2$ comes from the pair-annihilation of the χ's. Note that $-\langle v_{rel}\sigma\rangle_T n_\chi \cdot$dt is the number dN of antiparticles which can be reached by χ in a unit time intervall dt. Likewise the term $\langle v_{rel}\sigma\rangle_T n_{eq}^2$ describes the pair-production rate of χ's by particles which are in equilibrium. In describing by (4.1) the decouple from *chemical* equilibrium one assumes that the particles are still in *kinetical* equilibrium, hence the same $\langle v_{rel}\sigma\rangle_T$ for both processes. It is suitable to introduce new variables $x := T_{eq}/m_\chi$ and $f(x) := n_\chi/T_{eq}^3$ to get the rescaled equation

$$\frac{df}{dx} = \frac{m_\chi}{k^3} \left(\frac{4\pi^2 g_* G}{45}\right)^{-1/2} \langle v_{rel}\sigma\rangle_T (f_\chi^2 - f_{eq}^2). \tag{4.2}$$

[4] Here one makes the natural assumption that no particle anti-particle asymmetry is present.

Here we have used (3.29). Furthermore (3.32) was used in order to replace $-\frac{\dot{R}}{R}$ by $\frac{\dot{T}}{T}$, which of course is only true as long as $g(T)$ is constant. If the temperature depence of $g(T)$ is taken into account, as it is done in [8], one has $-\frac{\dot{R}}{R} = \frac{\dot{T}}{T} + \frac{1}{3g}\frac{dg}{dT}\dot{T}$. For most purposes, especially our qualitative discussion (4.2) is sufficient.

For particles χ which are high relativistic $n \propto T^3$, so $f = const.$ is a solution of (4.2). Here $const. = g_\chi/2$ for bosons and $const. = (7/4)g_\chi/2$ for fermions, as can be seen from the $T >> m_\chi$ limit of (3.14). If these particles decouple before they become non-relativistic, their present number density can be related to the present photon temperature by (3.36) i.e. $n_\chi^0 = const.(g(T_0)/g(T_{freeze}))T_{\gamma\,0}^3$.

If the particles remain in equilibrium when x drops below 1 their normalized number density will start to fall like

$$f \propto \exp(-m_\chi/T) \tag{4.3}$$

according to the non-relativitic limit of (3.14). The *longer* a heavy particle will stay in equilibrium the *stronger* will its number density be Boltzmann suppressed against the photon number density.

The elementary particle physics input is encoded in $\langle\sigma v_{rel}\rangle_T$. Let p_1 and p_2 be the incoming momenta of the two χ particles. Then the quantity $\langle\sigma v_{rel}\rangle_T$ is given by

$$\langle\sigma v_{rel}\rangle_T := \frac{1}{n_{eq}^2}\int d^3p_1 d^3p_2 f(E_1)f(E_2)\frac{1}{E_1E_2}w(s), \tag{4.4}$$

where $f(E)$ is given by (3.13), n_{eq} is given by by (3.14) and one has defined

$$w(s) = \frac{1}{4}\int |\mathcal{M}|^2(2\pi^4)\delta^4\left(p_1 + p_2 - \sum_j p_j\right)\prod_i \frac{dp^3}{(2\pi)^3 2p_i^0}. \tag{4.5}$$

Here $\mathcal{M}$ is the reduced transition matrix element for the annihilation of the two χ particles, summed over the final spins and averaged over the initial spins. Since for the couplings of the LSP one has typically $x \approx \frac{1}{20}$, at the decoupling temperature one can use the non-relativistic limit of (3.13) $f(E) = \kappa(2\pi)^{-3}\exp(-E/T)$ ($\kappa =$ 4,2 for Dirac, Majorana particles respectively) and expand $\langle\sigma v_{rel}\rangle_T$ in a power series in x. See [8] for the explicite expressions up to $O(x^4)$. The corresponding rate equation can be easily integrated numerically with the equilibrium value as initial conditions in order to get $f(x_{freeze})$, where x_{freeze} is the rescaled temperature at which the χ particles are out of chemical equilibrium. By (3.36) this is readily converted in the present energy density of the LSP.

Heavy neutrinos were the first particles which were investigated along these lines [5]. Generally, for particles which couple to heavy vector mesons one gets the following picture. As mentioned in the last section the weak interaction went out of equilibrium at 1 MeV. If the mass of the neutrino is below this value the neutrinos

Figure 4: The relic density $\Omega_X h_0^2$ for weakly interacting particles of mass M_X shows three crossings with $\Omega h_0^2 = 1$. Also indicated are densities for particles which interact more strongly or weakly then the neutrino via the Z^0. Extreme strong couplings reach the unitarity bound at $M_X \approx 340$ Tev, from [35].

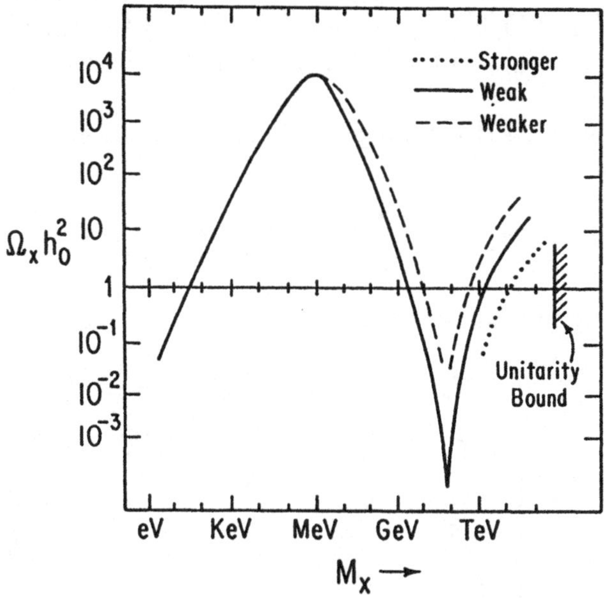

decouple as relativistic particles so the present day density rises with the mass in this region. At $m_\nu \approx 1$ MeV this curve turns over and falls since the Boltzmann suppression (4.3) will become significant. However for mass values larger then the vector boson mass the cross section falls as $(1/E^2)$, so the decoupling temperature rises as the mass rises, which leads again to higher relic densities. These trends are depicted in figure 4.

4.2 The relic density for the LSP

As argued in the introduction the LSP has to be among the neutral states of the minimal supersymmetric standard model. Since the s−neutrinos are already ruled out by the combination of LEP results and cosmic neutrino experiments cf. section 2, we concentrate on the neutralino as LSP candidate. The latter one is in general a mass eigenstate combination of the Higgsinos $\tilde{H}_{1,2}$, superpartners of the Higgses as well as of the wino $\tilde{W}^3$ and the bino $\tilde{B}$ which are superpartners of the 3rd $SU(2)$ gauge boson and the $U(1)$ gauge boson, respectively.

The masses depend on the soft supersymmetry breaking gaugino masses M_1, M_2 as well as on the higgs mixing ϵ which is introduced in order to avoid a phenomenological unaccepatible axion

$$L_{mass} = \epsilon\varepsilon_{ij}\tilde{H}_1^i\tilde{H}_1^j - M_2\tilde{W}^i\tilde{W}^i - M_1\tilde{B}\tilde{B}. \tag{4.6}$$

If one assumes a grand unification scenario one has the following relation between the gaugino masses $M_1 = (5/3)(\alpha_1/\alpha_2)M_2$. The ratio of the Higgs vacuum expectation values $\tan\beta = (v_1/v_2)$ is a further free parameter. Recent LEP results imply that $\tan\beta \geq 1.2$.

Combined with the usual Higgs gauge field couplings one get the following mass matrix for the charginos

$$(\tilde{W}^+, \tilde{H}_1^+)\begin{pmatrix} M_2 & g_2v_2 \\ g_2v_2 & -\epsilon \end{pmatrix}\begin{pmatrix} \tilde{W}^- \\ \tilde{H}_2^- \end{pmatrix}. \tag{4.7}$$

Likewise one has in the neutral sector the following mass matrix

$$(\tilde{W}^3, \tilde{B}, \tilde{H}_1^0, \tilde{H}_2^0)\begin{pmatrix} M_2 & 0 & \frac{-g_2v_1}{\sqrt{2}} & \frac{g_2v_2}{\sqrt{2}} \\ 0 & M_1 & \frac{g_1v_1}{\sqrt{2}} & \frac{-g_1v_2}{\sqrt{2}} \\ \frac{-g_2v_1}{\sqrt{2}} & \frac{g_1v_1}{\sqrt{2}} & 0 & \epsilon \\ \frac{g_2v_2}{\sqrt{2}} & \frac{-g_1v_2}{\sqrt{2}} & \epsilon & 0 \end{pmatrix}\begin{pmatrix} \tilde{W}^3 \\ \tilde{B} \\ \tilde{H}_1^0 \\ \tilde{H}_2^0 \end{pmatrix}. \tag{4.8}$$

The parameter space (ϵ, M_2) is restricted from the new LEP results either because of the bound on the charginos which have to be heavier then

$$m_{char} \geq 45 \text{ GeV} \tag{4.9}$$

[47] or because the of the limit on the gluino mass [48]

$$m_{gluino} \geq 22 \text{ GeV}. \tag{4.10}$$

Especially these bounds eliminate all the regions were the neutralinos contribuite considerable to the present mass density of the universe in the analysis of [6].

The situation for the abundances for the LSP after LEP can be inferred from figures 5 and 6 which are taken from reference [49]. In figure 5 the contours in the (ϵ, M_2) plane are depicted were the neutralino is according to (4.8) a pure (99%) photino $\tilde{\gamma} := \sin\Theta_w\tilde{W}^3 + \cos\Theta_W\tilde{B}$, a pure higssino $\tilde{H}_{(12)} := (\tilde{H}_1^0 + \tilde{H}_2^0)/\sqrt{2}$, $\tilde{H}_{[12]} := (\tilde{H}_1^0 - \tilde{H}_2^0)/\sqrt{2}$ a $\tilde{S}$ state for $\epsilon \to 0$ or a pure bino. The plot is for $\tan\beta = 2$. The light shaded area is forbidden by (4.9) and the heavy shaded area is forbidden by (4.10).

The present mass density can be calculated along the lines of section (4.1). Of course also the $s-$fermion masses, the higgs mass and the top quark mass must

Figure 5: Mass contours for the neutralino in the (ϵ, M_2) plane with the bounds from LEP on the charginos and the gluinos, from [50].

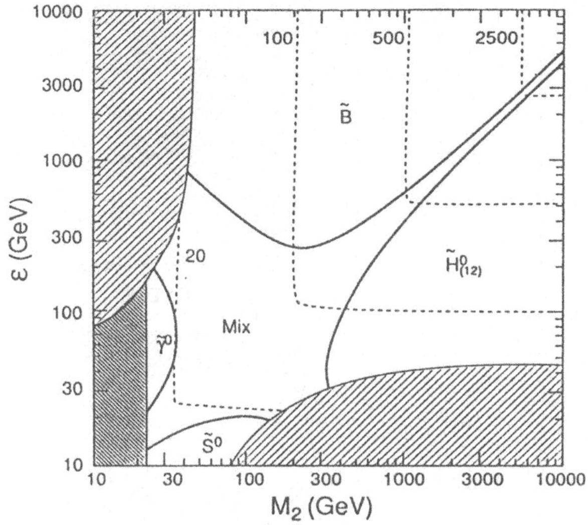

be fixed to obtain $\langle\sigma_{rel}\rangle_T$. In figure 6 the present day relic density in fractions of the critical density is plotted for $\tan\beta = 2$, $m_{\tilde{f}} = 74$ GeV, $m_0 = 37$ GeV and $m_{top} = 74$ GeV. Detailed investigations exhausting sensible values of $m_{\tilde{f}}$, m_0 and m_{top} can be found in [50]. The conclusions are that by the LEP results the photino can be the LSP only in a very tiny parameter region: 23 GeV $\geq$ M$_2$ $\geq$ 50 GeV, $|\epsilon| \leq 300$ GeV and $\tan\beta \leq 3$. Here it is also possible to adjust $m_{\tilde{f}}$ to obtain the preferred relic density about $\Omega h^2 \approx 1/4$. In the dominante part of the parameter space the LSP is either a bino or a higgsino. One can get values $\Omega h^2 \approx 1/4$ by adjusting the s-fermion masses in a rather broad mass range for the bino 30 GeV $\leq \mathrm{m}_{\tilde{B}} \leq$ 350 GeV, which has therefore the best prospects as LSP dark matter candidate.

The bino or at least one $s-$lepton or $s - quark$ must be lighter then ≈ 350 GeV in order to obtain $\Omega h^2 \leq 1/4$. The corresponding number for the higgsinos is 1 TeV. This bound comes essentially from the the rising tail in figure 4.

5 Conclusions

Combined data from cosmic neutrino detectors and collider experiments have ruled out the $s-$neutrino as dark matter candidate. The new data from LEP shrink the parameter space (ϵ, M_2) for the photino in the (MSSM) so strongly that this previously preferred dark matter candidate seems to be rather unlikely. The remaining

Figure 6: The relic densities in fractions of Ωh_0^2 for the bino $\tilde{B}$ and the higgsino combination $H_{[12]}$, from [50].

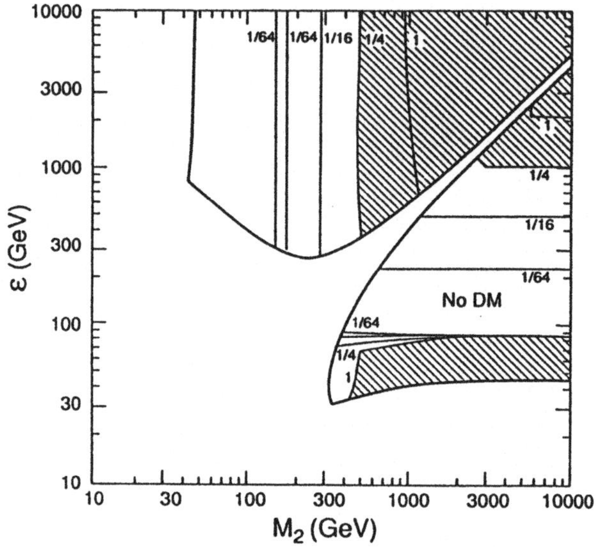

possibilities are the bino or the higgsino with masses above 20 − 30 GeV. The latter ones can be seen in dark matter detectors mainly by their spin dependent interactions which lead to counting rates which are three orders of magnitude below the current sensivity. For masses higher then the weak scale the relic density bounds on the LSP can set useful upper bounds on the masses which are complementary to the collider bounds.

References

[1] S. Weinberg *Gravitation and Cosmology: Principles and Applications of the General Theory of Relativity* J. Wiley & Sons, New York(1972)

[2] E. W. Kolb & M. S. Turner: *The Early Universe* Addison Wesley (1990)

[3] H. P. Nilles: Phys. Rep. 110 (1984) 1; H. E. Haber and G. L. Kane: Phys. Rep. 117 (1985) 75; M. F. Sohnius: Phys. Rep. 128 (1985) 41

[4] L. J. Hull and M. Suzuki: Nucl.Phys. B **231** (1984) 419; L. J. Hull: Mod.Phys.Lett. A **5** (1990) 467; L. E. Ibáñez and G. G. Ross: *Should Discrete Symmetries Be Anomaly-Free ?*, CERN-TH.6000/91

[5] Ya. Zeldovich Adv. Astron. and Astrophys. **3**, (1965) 241; B. W. Lee and S. Weinberg: Phys. Rev. Lett. **39** (1977) 183

[6] H. Goldberg: Phys. Rev. Lett. **50** (1983) 219; J. Ellis, J. S. Hagelin, D. V. Nanopolous, K. Olive and M. Srednicki: Nucl.Phys. B **B238** (1984) 453;

[7] A. Klemm and M. G. Schmidt: Z. f. Phys. C **40** (1988) 591

[8] M. Srednicki K. Olive and R. Watkins: Nucl.Phys. B **B310** (1988) 693;

[9] S. Weinberg: Phys. Rev. Lett. 48 (1982) 1303

[10] J. Ellis, D. V. Nanopolous and S. Sarkhar: Nucl.Phys. B 259 (1985) 175

[11] J. Kripfganz and A. Ringwald: Z. f. Phys. 44 (1989) 213, A. Ringwald Nucl.Phys. B 330 (1990) 19

[12] H. Pagels and J. R. Primack: Phys. Rev. Lett. 48 (1982) 223

[13] L. Abbott and P. Sikivie: Phys. Lett. 120 B (1983) 133, J. Preskill, M. Wise and F. Wilczek: Phys. Lett. 120 B (1983) 127, M. Dine and W. Fischler: Phys. Lett. 120 B 137

[14] O. Bertolami: Phys. Lett. 209 B (1988) 277

[15] P. F. Smith et al.: Nucl.Phys. B **206** (1982) 333

[16] J. Bekenstein and M. Milgrom: *Does the missing mass problem signal the breakdown of Newtons gravity*, Astrophys.J. **286** (1984) 7, see also ref. 17

[17] *Dark matter in the Universe*, Proceedings of the Jerusalem Winter School for theor. Phys. (86-87) Editors: J. Bahcall, T. Piran and S. Weinberg

[18] J. H. Oort: Bull. astr. Insts, Neth. **6** (1932) 249

[19] J. N. Bahcall: Astrophys.J. **276** (1984) 156, Astrophys.J. **276** (1984) 169, Astrophys.J. **287** (1984) 926

[20] S. Tremaine and H. M. Lee: *Dark Matter in Galaxies and Galaxy Systems* in [17].

[21] R. Sancisi and T. S. van Albada: in *Dark matter in the Universe*, Eds. J. Kormendy and G. Knapp (Reidel, Dordrecht, 1987)

[22] F. Zwicky: Helvetica Physica Acta **6** (1933) 110; J. P. Huchra and M. J. Geller: Astrophys.J. **257** 432; D. Merritt: Astrophys.J. **331** (1987) 121;

[23] *The early universe and cosmic structures*, Eds.: J. M. Alimi, A. Blanchard, A. Bouquet, F. Martin de Volnay and J. Trân Thanh Vân, Edition Frontières (1990)

[24] E. Bertschinger: *Large-scale motions in the universe* in [23]

[25] J. Yang, M. S. Turner, G. Steigman, D. N. Schramm and K. A. Olive: Astrophys.J. **281** (1984) 493

[26] G. Gamov, Phys. Rev. **70** (1946) 527; R. A. Alpher, *H. Bethe* and G. Gamov: Phys. Rev. **73** (1948) 803; R. V. Wagoner, Astrophys.J. **179** (1973) 343

[27] D. N. Schramm and R. V. Wagoner, Ann. Rev Nucl. Part. Sci. **27** (1979) 37

[28] J. N. Bahcall and S. Casertano: *Some possible regularities in Missing Mass* in [17]

[29] D. J. Hegyi and K. A. Olive: Phys.Lett. **126B** (1983) 34

[30] G.A Tammann / R. B. Tully in *The Extragalactic Distance Scale, ASP Conference Series No. 4* Eds. S. v. d. Bergh and C. J. Pritchet, Provo: Brigham Young University Press, (1988); R. B. Tully in [46] (1989)

[31] M. Davis, G. Efstathiou, C. S. Frenk and S. D. M. White: Astrophys.J. **292** (1985) 371, Astrophys.J. **313** (1987) 505

[32] W. H. Press and D. N. Spergel: Astrophys.J. **296** (1985) 679; J. Silk, K. Olive and M. Srednicki: Phys. Rev. Lett. **55** (1985) 257; K. A. Olive and M. Srednicki: Phys.Lett. **205** (1988) 315

[33] M. W. Goodman and E. Witten: Phys. Rev. D **31** (1985) 3059; I. Wasserman Phys. Rev D **33** (1985) 2071; K. Griest Phys. Rev D **38** (1988) 2357 (erratum Fermilab-Pub-89/139-A

[34] S. Ahlen et al. Phys.Lett. **B195** (1987) 603 D.D. Caldwell et al. Phys. Rev. Lett. 61 (1988) 510

[35] D. N. Schramm: *Cosmology and Experimental Particle Physics* in [23].

[36] *The Implications of Z Width Measurement For The Search of Dark Matter* in [23].

[37] L. D. Landau & E. M. Lifshitz: Lehrbuch der Theor. Physik. Band II *Klassische Feldtheorie* Kapitel II & XI, Berlin (1965)

[38] L. D. Landau & E. M. Lifshitz: Lehrbuch der Theor. Physik. Band V *Statistische Physik* Kapitel V, Berlin (1966)

[39] A. H. Guth, Phys. Rev. **D23** (1981) 347; A. D. Linde, Phys.Lett. **108B** (1982) 389; A. Albrecht and P. J. Steinhardt, Phys. Rev. Lett. **48** (1982) 1220; for a review see A. D. Linde, *Particle Physics and Inflationary Cosmology*, Gordon & Breech, N. Y. (1990)

[40] S. Weinberg: Rev. of Mod. Phys. Lett **61** (1989) 1

[41] W. Rindler, *Essential Relativity* Springer Heidelberg (1977)

[42] Loh E. D. and E. J. Spillar: Astrophys. J. **303** (1986) 154

[43] M. Fukugita, F. Takahara, K. Yamashita and Y. Yoshi: Astrophys. J. **361** (1990) L1

[44] R. Durrer and N. Straumann: Mon. Not. R. astr. Soc. **242** (1990) 221

[45] M. Fukugita, T. Futamasec, M. Kasai and E. L. Turner: *Statistical Properties of Gravitational Lenses with a Non-Zero Cosmological Constant* Preprint (1991) POP-422 sub to Astrophys. J. **242** (1990) 221

[46] *Astrophysical Ages and Dating Methods*, Proceedings of the Fifth IAP Workshop Paris (1989), Editors: E. Vangioni-Flam, M. Cassé, J. Audouze and J. Trân Thanh Vân, Editons Frontiéres (1990)

[47] M. Z. Akrawy et al. Phys.Lett. 252 B (1990) 290 ; D. Decamp et al. Phys.Lett. 255 B (1990) 623

[48] J. Allati et al. , Phys.Lett. 235 B (1990) 363

[49] K. A. Olive: *Dark matter canditates in the supersymmetric standard model* in [23]

[50] K. A. Olive and M. Srednicki: Nucl.Phys. B **355** (1991) 208

Production and Decay of Supersymmetric Particles at Future Colliders

A. Bartl
Institut für Theoretische Physik
Universität Wien, A-1090 Vienna

W. Majerotto and B. Mößlacher
Institut für Hochenergiephysik
Österreichische Akademie der Wissenschaften, A-1050 Vienna

1 Introduction

Despite its impressive success it is generally believed that the Standard Model is still not the final theory of fundamental particle interactions. At present, one of the most attractive possibilities of extending the Standard Model is provided by supersymmetry (SUSY). A supersymmetric extension of the Standard Model allows us to incorporate in a natural way elementary scalar particles such as the Higgs boson into a quantum field theoretical framework. Moreover, it may enable us to relate the weak symmetry breaking scale to the grand unification scale [1, 2, 3].

If these ideas about 'low-energy' supersymmetry are true, one expects the supersymmetric particles to have masses below or about 1 TeV. The possibility that some of the SUSY particles may have masses even much lower than this value has tremendously stimulated the interest in experimental searches at the present accelerators (CERN-S$p\bar{p}$S-Collider, FNAL-Tevatron, TRISTAN, LEP100). The search for supersymmetry will also play an important rôle in the future, when experiments at higher energies can be performed.

In this paper we shall describe how supersymmetric particles could be detected at the new colliders HERA, LEP200, LHC, SSC, and at a possible future linear e^+e^- collider. We shall present theoretical predictions for production cross sections and decay probabilities, as well as for the important signatures. Our calculations will be based on the Minimal Supersymmetric Standard Model (MSSM) which is the simplest supersymmetric extension of the Standard Model.

Like the Standard Model, the MSSM is based on the $SU(3) \times SU(2) \times U(1)$ gauge symmetry group, spontaneously broken down to $SU(3) \times U(1)_{em}$. It has the minimal particle content, i.e. it contains the known particles, the gauge bosons, quarks and leptons, plus their superpartners, the gauginos ($\tilde{g}, \tilde{W}^{\pm}, \tilde{Z}, \tilde{\gamma}$), squarks $\tilde{q}$ and sleptons $\tilde{\ell}$. As is well known, two Higgs doublets are necessary, $H_1 = (H_1^0, H_1^-)$ and $H_2 = (H_2^+, H_2^0)$, together with their superpartners, the higgsinos ($\tilde{H}_{1,2}^0, \tilde{H}^{\pm}$).

The non-strongly interacting gauginos mix with the higgsinos to form corresponding mass eigenstates. There are two pairs of charginos $\tilde{\chi}_i^\pm$, $i = 1, 2$, and four neutralinos $\tilde{\chi}_i^0$, $i = 1, \ldots, 4$, in order of their mass eigenvalues. The masses and couplings of the charginos and neutralinos are determined by the corresponding mass matrices [2, 4]. These depend on the parameters M, M', μ, and $\tan\beta = v_2/v_1$. M and M' are the $SU(2)$ and $U(1)$ gaugino masses, sometimes also called M_2 (or $m_{1/2}$) and M_1, respectively. μ is the mass parameter introduced in the superpotential term $\mu H_1 H_2$. v_1 and v_2 are the vacuum expectation values of the two Higgs doublets. In the following we shall use the relations $M/m_{\tilde{g}} = (\alpha_2/\alpha_3) \approx 0.3$ and, assuming GUT, $M'/M = (5/3)\tan^2\theta_W \approx 0.5$, where $m_{\tilde{g}}$ is the gluino mass. Without loss of generality we take $M \geq 0$.

To the two chirality states of the charged leptons and quarks correspond the right and left scalar particles $\tilde{\ell}_R$, $\tilde{\ell}_L$ and $\tilde{q}_R$, $\tilde{q}_L$. The masses of the scalar particles follow from renormalization group equations [5]

$$m^2_{\tilde{f}_{L,R}} = m_f^2 + m_0^2 + C(\tilde{f})M^2 \pm m_Z^2 \cos 2\beta (T_3^f - Q^f \sin^2\theta_W) \qquad (1.1)$$

where T_3^f and Q^f are the third component of the weak isospin and charge of the corresponding left- or right-handed fermion, m_0 is the common scalar mass at the unification point, and $C(\tilde{\ell}_R) \approx 0.23$, $C(\tilde{\ell}_L) = C(\tilde{\nu}_e) \approx 0.79$, $C(\tilde{q}_L) \approx 10.8$, $C(\tilde{q}_R) \approx 10.1$. Usually, $\tilde{f}_L$ and $\tilde{f}_R$ are to a good approximation also mass eigenstates. An exception may be the scalar top quark where there is $\tilde{t}_L - \tilde{t}_R$ mixing proportional to the large top quark mass.

The MSSM contains five physical Higgs particles: h^0, H^0, $A^0(0^-)$ and $H^\pm$. In lowest order the masses of the Higgs particles are given in terms of two parameters (commonly used m_A and $\tan\beta$), implying the bounds $m_{h^0} \leq m_Z \leq m_{H^0}$, $m_{h^0} \leq m_A \leq m_{H^0}$, $m_{H^\pm} \geq m_W$ [6]. Radiative corrections can change these tree level masses appreciably due to the large top quark mass [7, 8]. It is possible that $m_{h^0} > m_Z$ and/or $m_{h^0} > m_A$.

Moreover, an analysis of the renormalization group equations based on radiative electroweak symmetry breaking suggests $\tan\beta$ to be in the range $1 \lesssim \tan\beta \lesssim m_t/m_b$.

In the MSSM the multiplicative quantum number R-parity is conserved ($R = +1$ for standard model particles including Higgs bosons, $R = -1$ for the supersymmetric particles). This implies that there is a lightest supersymmetric particle (LSP) which is stable and into which all supersymmetric particles eventually decay. A further consequence is that supersymmetric particles can only be produced in pairs. Usually, the lightest neutralino $\tilde{\chi}_1^0$ is assumed to be the LSP.

To summarize, the following four basic parameters are necessary to specify the SUSY particle sector of the MSSM: m_0, M, μ and $\tan\beta$. For the description of the Higgs sector, at least one further parameter is needed (apart from the top quark mass m_t), for which usually m_A is taken. Production cross sections as well as decay rates depend in a characteristic way on these parameters. Here a systematic study of this dependence in the whole parameter space relevant for this energy will be performed.

In Sect. 2 we shortly review the present experimental bounds and give the expectations for SUSY particle production at future colliders. In Sect. 3, 4, and 5 we treat SUSY particle production and decay in e^+e^-, pp, and ep collisions, respectively. Sect. 6 contains a summary. In the Appendix we give formulae for the e^+e^- production cross sections and for the three-body decays of charginos and neutralinos.

2 Present Experimental Bounds and Expectations

For the charged scalar partners of the fermions the LEP experiments have led to the bound $m_{\tilde{f}} \gtrsim 45$ GeV, and similarly for charginos $m_{\tilde{\chi}_1^\pm} \gtrsim 46$ GeV [9]. From the invisible width

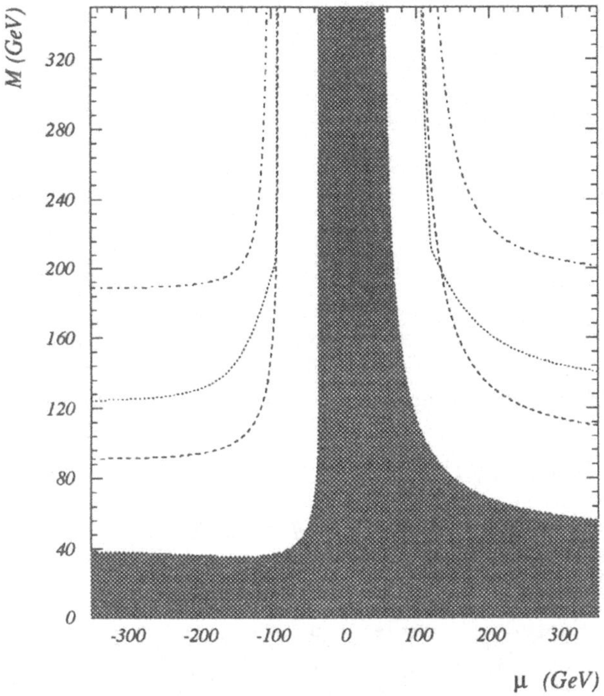

Fig. 1: Parameter region of M, μ, for $\tan\beta = 4$, explorable at LEP200 by chargino and neutralino searches. Shown are border lines for: $\tilde{\chi}_1^0\tilde{\chi}_1^0$ $(- \cdot - - \cdot -)$, $\tilde{\chi}_1^0\tilde{\chi}_2^0$ $(\cdots\cdots)$, $\tilde{\chi}_1^+\tilde{\chi}_1^-$ $(- - -)$ production at $\sqrt{s} = 190$ GeV. Also shown is the region excluded by chargino searches at the Z^0.

of the Z^0 one can derive $m_{\tilde{\nu}} \geq 42$ GeV. An analysis [10] of the CDF data, taking into account cascade decays, has given the lower bound $m_{\tilde{g}} \gtrsim 135$ GeV, whereas for squarks the limits $m_{\tilde{q}} \gtrsim 170$ GeV (for $m_{\tilde{g}} < m_{\tilde{q}}$) and $m_{\tilde{q}} \gtrsim 130$ GeV (for $m_{\tilde{q}} < m_{\tilde{g}} \leq 400$ GeV) are obtained, assuming that all squark masses are equal. However, due to $\tilde{t}_L - \tilde{t}_R$ mixing the stop mass could be much smaller, even below 100 GeV [11]. With $M \approx 0.3 m_{\tilde{g}}$ the bound on the gluino mass implies $M \gtrsim 40$ GeV. From this and the negative neutralino searches at the Z^0 one can derive $m_{\tilde{\chi}_1^0} \gtrsim 19$ GeV for the lightest neutralino.

The present experimental bounds on the neutral Higgs particles in the MSSM, including one-loop radiative corrections are $m_h \geq 41$ GeV and $m_A \geq 20$ GeV [9]. At present the experimental data do not restrict $\tan\beta$.

In the LEP energy range ($\sqrt{s} \leq 190$ GeV) charged sleptons and charginos can be observed if their masses are below 90 GeV. The supersymmetry parameter region accessible is very roughly $m_0 \leq 90$ GeV, and $M \lesssim 90$ GeV with $|\mu|$ arbitrary, or $|\mu| \lesssim 90$ GeV with M arbitrary. At the LHC and SSC gluinos and squarks with masses up to around 1 TeV can be detected. This implies that the region $m_0 \lesssim 1$ TeV and $M \lesssim 300$ GeV can be explored.

An e^+e^- collider at $\sqrt{s} = 500$ GeV allows chargino and selectron searches up to a mass of approximately 250 GeV. This yields an accessible parameter region of roughly $m_0 \lesssim 250$ GeV, and $M \lesssim 250$ GeV with $|\mu|$ arbitrary, or $|\mu| \lesssim 250$ GeV with M arbitrary.

3 e^+e^- Collisions

In e^+e^- collisions the following SUSY particle processes are the most important ones and would give direct evidence for supersymmetry:

$$e^+e^- \rightarrow \tilde{\chi}_i^+ + \tilde{\chi}_j^-, \quad i,j = 1,2 \tag{3.1}$$
$$e^+e^- \rightarrow \tilde{\chi}_i^0 + \tilde{\chi}_j^0, \quad i,j = 1,\dots 4 \tag{3.2}$$
$$e^+e^- \rightarrow \tilde{e}_{L,R}^+ + \tilde{e}_{L,R}^- \tag{3.3}$$
$$e^+e^- \rightarrow \tilde{\mu}_{L,R}^+ + \tilde{\mu}_{L,R}^- \tag{3.4}$$
$$e^+e^- \rightarrow \tilde{\nu} + \bar{\tilde{\nu}} \tag{3.5}$$
$$e^+e^- \rightarrow \tilde{q}_{L,R} + \bar{\tilde{q}}_{L,R}. \tag{3.6}$$

In order to work out suitable signatures it is also necessary to study the decays of these particles.

The three-body decays of charginos and neutralinos are:

$$\tilde{\chi}_i^\pm \rightarrow l^+ + l^- + \tilde{\chi}_k^\pm,\ q + \bar{q} + \tilde{\chi}_k^\pm,\ l^\pm + \overset{(-)}{\nu} + \tilde{\chi}_k^0,\ q + \bar{q}' + \tilde{\chi}_k^0 \tag{3.7}$$
$$\tilde{\chi}_i^0 \rightarrow l^+ + l^- + \tilde{\chi}_k^0,\ q + \bar{q} + \tilde{\chi}_k^0,\ l^\pm + \overset{(-)}{\nu} + \tilde{\chi}_k^\mp,\ q + \bar{q}' + \tilde{\chi}_k^\pm. \tag{3.8}$$

If kinematically allowed, the two-body decays will be the dominant ones:

$$\tilde{\chi}_i^\pm \rightarrow Z^0 + \tilde{\chi}_k^\pm,\ h^0(A^0 \,\mathrm{or}\, H^0) + \tilde{\chi}_k^\pm,\ W^\pm + \tilde{\chi}_k^0,\ H^\pm + \tilde{\chi}_k^0 \tag{3.9}$$
$$\tilde{\chi}_i^0 \rightarrow Z^0 + \tilde{\chi}_k^0,\ W^\pm + \tilde{\chi}_k^\mp,\ h^0(A^0 \,\mathrm{or}\, H^0) + \tilde{\chi}_k^0,\ H^\pm + \tilde{\chi}_k^\mp. \tag{3.10}$$

The produced sfermions $\tilde{f}_{L,R}$ (slepton or squark) decay into a fermion f (lepton or quark) and a neutralino $\tilde{\chi}_i^0$ or chargino $\tilde{\chi}_i^\pm$:

$$\tilde{e}_{L,R} \rightarrow e + \tilde{\chi}_i^0,\ \tilde{e}_L \rightarrow \nu + \tilde{\chi}_i^- \tag{3.11}$$
$$\tilde{\nu}_L \rightarrow \nu + \tilde{\chi}_i^0,\ e^- + \tilde{\chi}_i^+ \tag{3.12}$$
$$\tilde{q}_{L,R} \rightarrow q + \tilde{\chi}_i^0, \tilde{q}_L \rightarrow q' + \tilde{\chi}_i^\pm,\ \tilde{u}_L \rightarrow d + \tilde{\chi}_i^+,\ \tilde{d}_L \rightarrow u + \tilde{\chi}_i^-. \tag{3.13}$$

In this section we consider only squarks corresponding to light quark flavours. If the gluino is lighter than the squark, then the squark would first decay into a gluino:

$$\tilde{q}_{L,R} \rightarrow q + \tilde{g}. \tag{3.14}$$

As the lightest neutralino $\tilde{\chi}_1^0$ is supposed to be the LSP, all SUSY particles will eventually decay into $\tilde{\chi}_1^0$. At high energies and masses in general cascade decays of SUSY particles occur leading to a rather complex decay pattern. This has a strong influence on the experimental signature of SUSY particles, in particular on the classical 'missing energy' signature. (In the following missing momentum will be denoted by $\not{p}$.) According to eqs. (3.9) and (3.10) some of the SUSY particles may also decay into Higgs particles. Therefore, the actual masses and couplings of the Higgs particles will influence some of the signatures. As is well known radiative corrections may appreciably change the Higgs masses compared with their tree level values [7, 8].

The formulae for the various cross sections and decays have been worked out by several authors, see [12], where a rather complete list of references is given. In the Appendix we present the formulae for the cross sections of the reactions (3.1 – 3.6), and for the three-body decays (3.7) and (3.8). The formulae for the two-body decays (3.9) and (3.10) can be found in [12, 13], and for the sfermion decays (3.11 – 3.13) in [12].

3.1 LEP 200

The predictions for cross sections will be given for $\sqrt{s} = 190$ GeV, assuming an integrated luminosity of 500 pb^{-1} per year. For earlier studies we refer to [14].

3.1.1 Chargino production

In Fig. 1 we show the region in the (M, μ) plane (for $\tan\beta = 4$) which can be explored at LEP200 by searching for charginos and neutralinos. The region already excluded by unsuccessful search at the Z^0 is also indicated. For other values of $\tan\beta$ similar plots are obtained.

The production mechanism for $e^+e^- \to \tilde{\chi}_i^+ + \tilde{\chi}_j^-$, $i, j = 1, 2$, is γ, Z exchange in the s channel and $\tilde{\nu}$ exchange in the t channel. The total production cross section for $e^+e^- \to \tilde{\chi}_1^+ + \tilde{\chi}_1^-$ is shown in Fig. 2 for $m_{\tilde{\nu}} = 300$ GeV and $\tan\beta = 4$. It can reach 6 fb. If both $M \gtrsim 90$ GeV and $|\mu| \gtrsim 90$ GeV, the light chargino becomes too heavy to be produced at this energy. If $|\mu| > M$ then $\tilde{\chi}_1^+$ is more gaugino-like, and if $M > |\mu|$ it is more higgsino-like. Therefore, sneutrino exchange can have a substantial influence for $|\mu| > M$, especially if $m_{\tilde{\nu}} \simeq m_Z$. The $\tan\beta$ dependence is in general weak.

The decay of the chargino, $\tilde{\chi}_1^+ \to \tilde{\chi}_1^0 + l + \nu$, $\tilde{\chi}_1^0 + q + \bar{q}$, lead to 'two sided' events with acollinear, acoplanar hard leptons or jets and large missing energy: $l^+ + l^- + \not{p}$ or $l^\pm + \text{jets} + \not{p}$ or $\text{jets} + \text{jets} + \not{p}$. For $m_{\tilde{e}} = m_{\tilde{\nu}} = m_{\tilde{q}} = 300$ GeV and $\tan\beta = 4$ we obtain for these signatures a rate of 0.07 pb, 0.5 pb, 3.5 pb, respectively.

A priori, it cannot be excluded that the sneutrino and/or the left slepton $\tilde{l}_L$ are lighter than the chargino. In this unlikely case the chargino would first decay into $\tilde{\nu}$ or $\tilde{l}_L$. This would drastically enhance the leptonic signal.

A recent detailed study of chargino production at LEP200 is given in [15].

3.1.2 Neutralino production

As can be seen in Fig. 1 the SUSY parameter region for the reaction $e^+e^- \to \tilde{\chi}_1^0\tilde{\chi}_2^0$ is larger than that for chargino production. Here the reaction mechanism is Z exchange in the s channel and $\tilde{e}_{L,R}$ exchange in the t and u channel. The Z^0 couples only to the Higgsino components of the neutralinos, whereas the selectron couples only to the gaugino-component. For $M < |\mu|$ the $\tilde{\chi}_1^0$ is gaugino-like, whereas for $M > |\mu|$ it has large higgsino-components. Therefore, one expects sufficiently large cross sections for $M > |\mu|$. Only if $m_{\tilde{e}} \simeq m_Z$ we get sizeable cross sections also in the region $M < |\mu|$. The total cross section can go up to 1.7 pb (for $m_{\tilde{e}} = 300$ GeV).

The $\tilde{\chi}_2^0$ has the following decay modes: $\tilde{\chi}_2^0 \to \tilde{\chi}_1^0 f\bar{f}$, and $\chi_1^\pm f\bar{f}'$ if kinematically allowed. This leads to one-sided events. In Fig. 3 we show the cross section for $e^+e^- \to \tilde{\chi}_1^0\tilde{\chi}_2^0 \to (e^+e^-) + \not{p}$ as a function of M and μ, for $m_{\tilde{e}} = m_{\tilde{\nu}} = m_{\tilde{q}} = 300$ GeV and $\tan\beta = 4$. The cross section for $e^+e^- \to \tilde{\chi}_1^0\tilde{\chi}_2^0 \to (\text{jet}\,\text{jet} + \not{p}$ is approximately a factor 20 higher and can reach 1.3 pb.

If the sneutrino is lighter than the $\tilde{\chi}_2^0$, and the $\tilde{\chi}_2^0$ has a non-vanishing Z-ino component, then the decay $\tilde{\chi}_2^0 \to \tilde{\nu} + \bar{\nu}$ would dominate, where the sneutrino is invisible. Then the $\tilde{\chi}_2^0$ would be produced undetectably.

In the case of a light slepton, $m_{\tilde{l}} < m_{\tilde{\chi}_2^0}$, the $\tilde{\chi}_2^0$ would first decay into $\tilde{l}^\pm + l^\mp$, yielding strongly enhanced leptonic rates.

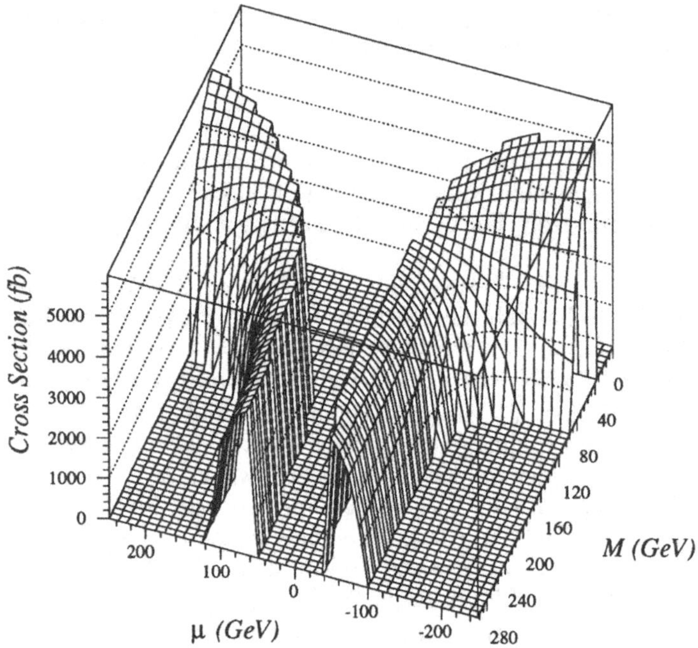

Fig. 2: Cross section for $e^+e^- \to \tilde{\chi}_1^+\tilde{\chi}_1^-$, as a function of M and μ, for $\sqrt{s} = 190$ GeV, $m_{\tilde{\nu}} = 300$ GeV, $\tan\beta = 4$.

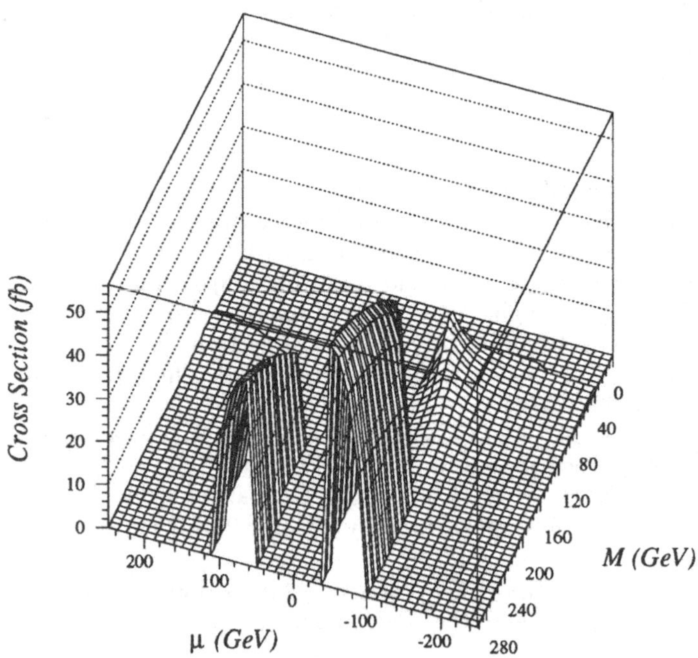

Fig. 3: Cross section for $e^+e^- \to \tilde{\chi}_1^0\tilde{\chi}_2^0 \to (e^+e^-) + \not{p}$ (one-sided), as a function of M and μ, for $\sqrt{s} = 190$ GeV, $m_{\tilde{e}} = m_{\tilde{\nu}} = m_{\tilde{q}} = 300$ GeV, $\tan\beta = 4$.

3.1.3 Selectron production

Here the reactions $e^+e^- \to \tilde{e}_R^+ + \tilde{e}_R^-, \tilde{e}_L^+ + \tilde{e}_L^-, \tilde{e}_L^\pm + \tilde{e}_R^\mp$ can occur. The reaction mechanism for selectron pair production $e^+e^- \to \tilde{e}_L^+ + \tilde{e}_L^-, \tilde{e}_R^+ + \tilde{e}_R^-$, is (γ, Z) exchange in the s channel and $\tilde{\chi}_i^0$ exchange in the t channel. The process $e^+e^- \to \tilde{e}_L^\pm + \tilde{e}_R^\mp$ receives only t-channel contributions.

The total cross section $\sigma(\tilde{e}_L\tilde{e}_L + \tilde{e}_R\tilde{e}_R + 2(\tilde{e}_L\tilde{e}_R))$ varies between 0.75 and 1.75 pb for $m_{\tilde{e}} = 80$ GeV, depending mainly on the SUSY parameter M and only weakly on μ and tanβ .

$\tilde{e}_L$ and $\tilde{e}_R$ decay differently: $\tilde{e}_L \to e\tilde{\chi}_1^0, e\tilde{\chi}_2^0, \nu\tilde{\chi}_1^-$, $\tilde{e}_R \to e\tilde{\chi}_1^0, e\tilde{\chi}_2^0$, where $\tilde{\chi}_2^0$ and $\tilde{\chi}_1^-$ further decay until $\tilde{\chi}_1^0$ is produced. The decay pattern of $\tilde{e}_R$ is simpler because the direct decay into the LSP dominates. For illustration we show in Fig. 4 the branching ratios for $\tilde{e}_L$ decays as a function of μ for $m_{\tilde{e}_L} = 80$ GeV, $M = 80$ GeV, $\tan\beta = 4$. As can be seen, cascade decays are possible if approximately $M < m_{\tilde{e}_L}$ or $|\mu| < m_{\tilde{e}_L}$.

In Fig. 5 we show the cross section for the 'classical' signature $e^+e^- \to e^+ + e^- + \not{p}$, as a function of M and μ, for $m_{\tilde{e}} = 80$ GeV, tan$\beta = 4$. Notice that the parameter region covered by LEP at the Z^0 is left out. The region $|\mu| \gtrsim 120$ GeV for $M \gtrsim 160$ GeV is excluded because the selectron would be heavier than the LSP. The cross section for this signature can reach a value of 1.5 pb. It can also be seen that the rate is much smaller where cascade decays are possible.

In $e^+e^- \to \tilde{\mu}_{L,R}^+\tilde{\mu}_{L,R}^-$ there is only (Z,γ) exchange. Hence the total cross section is smaller, $\sigma = 0.19$ pb for $m_{\tilde{\mu}} = 80$ GeV.

3.1.4 Sneutrino Production

In the MSSM there is only a left-sneutrino $\tilde{\nu} \equiv \tilde{\nu}_L$. The reaction $e^+e^- \to \tilde{\nu}_e + \bar{\tilde{\nu}}_e$ has contributions from Z^0 exchange in the s channel and $\tilde{\chi}_i^\pm$ exchange in the t channel.

The total cross section can go up to 3 pb for $m_{\tilde{\nu}} = 80$ GeV. The sneutrino can, however, only be seen if it decays in a cascade, $\tilde{\nu} \to e^-\tilde{\chi}_1^+ \to e^-\tilde{\chi}_1^0 f\bar{f}'$, or $\tilde{\nu} \to \nu\tilde{\chi}_2^0 \to \nu\tilde{\chi}_1^0 f\bar{f}$, because the direct decay into the LSP, $\tilde{\nu} \to \nu\tilde{\chi}_1^0$, is invisible. Such visible decays can occur only if $M < m_{\tilde{\nu}_e}$ or $|\mu| < m_{\tilde{\nu}_e}$. Possible signatures are the one-sided events: e^+e^- (or $e^-\mu^+$) $+ \not{p}$, e^-jets $+ \not{p}$, jets $+ \not{p}$, and the two-sided events: (e jets) + (e jets) $+ \not{p}$, etc. Fig. 6 shows the rate for $e^+e^- \to \tilde{\nu}\bar{\tilde{\nu}} \to (e^+e^-) + \not{p}$ (one-sided), as a function of M and μ, for $m_{\tilde{\nu}} = 80$ GeV und tan$\beta = 4$. The cross section for $e^+e^- \to \tilde{\nu}\bar{\tilde{\nu}} \to$ (e jet $+ \not{p}$ is roughly a factor of 7 higher.

3.2 e^+e^- collisions at 500 GeV

It is conceivable that SUSY particles only show up at energies of several 100 GeV. In this case an e^+e^- collider with an energy well above 200 GeV would be well suited to test the idea of supersymmetry. Presently, several options for a linear e^+e^- collider in the energy range $\sqrt{s} = 0.5 - 2$ TeV are being discussed [16]. Most recently, detailed studies on the physics potential of such a collider at $\sqrt{s} = 500$ GeV were performed [17]. This step in energy would considerably enlarge the explorable parameter range. Experiments at such a collider would allow a crucial test of the MSSM. In particular, if no Higgs particles were found, the MSSM would be ruled out [18]. It is therefore appropriate to study the SUSY particle production reactions (3.1 – 3.6) and the decays of these particles also at this energy. A detailed study is given in [19]. A new feature is that also the higher mass states of the charginos and the neutralinos may be accessible, and cascade decays play a

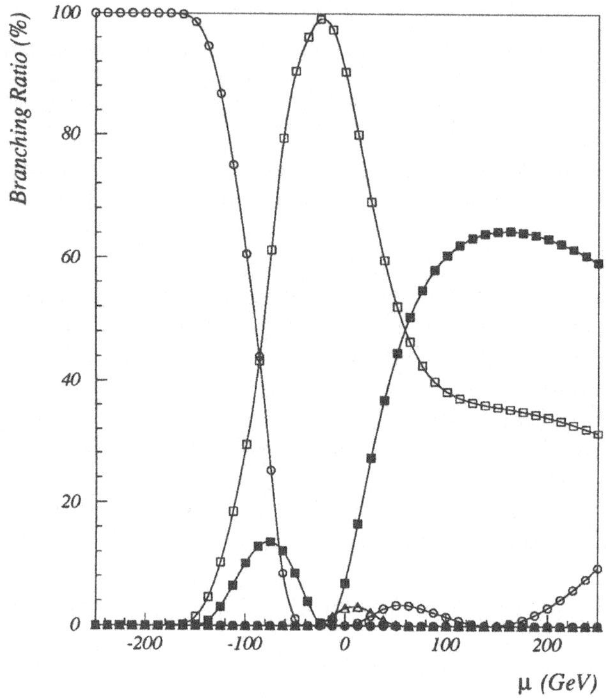

Fig. 4: Branching ratios for $\tilde{e}_L$ decays into charginos and neutralinos, as a function of μ, for $m_{\tilde{e}_{L,R}} = 80$ GeV, $M = 80$ GeV, $\tan\beta = 4$, $\sin^2\theta_W = 0.23$. The curves correspond to the following transitions:

- ■ into the light chargino $\tilde{\chi}_1^\pm$
- ● into the heavy chargino $\tilde{\chi}_2^\pm$
- ○ into $\tilde{\chi}_1^0$
- □ into $\tilde{\chi}_2^0$
- △ into $\tilde{\chi}_3^0$

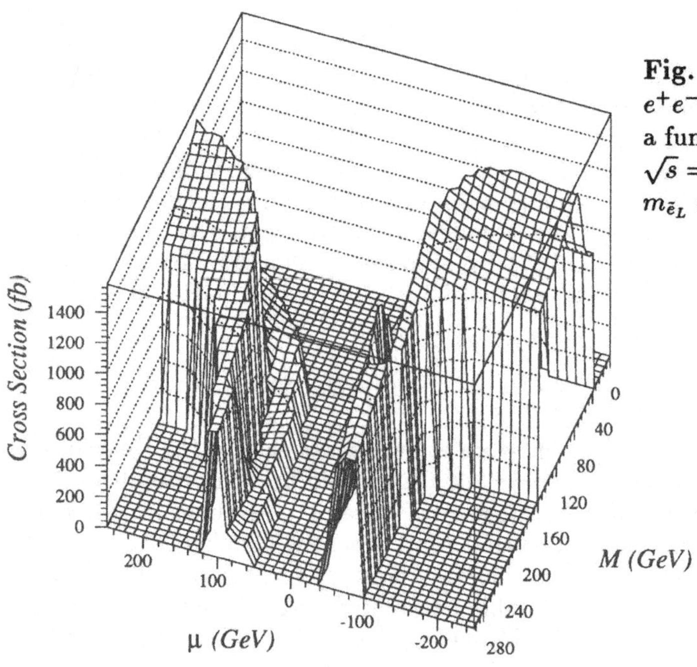

Fig. 5: Cross section for $e^+e^- \to \tilde{e}_{L,R} + \tilde{e}_{L,R} \to e^+e^- + \not{p}$, as a function of M and μ, for $\sqrt{s} = 190$ GeV, $m_{\tilde{e}_L} = m_{\tilde{e}_R} = 80$ GeV, $\tan\beta = 4$.

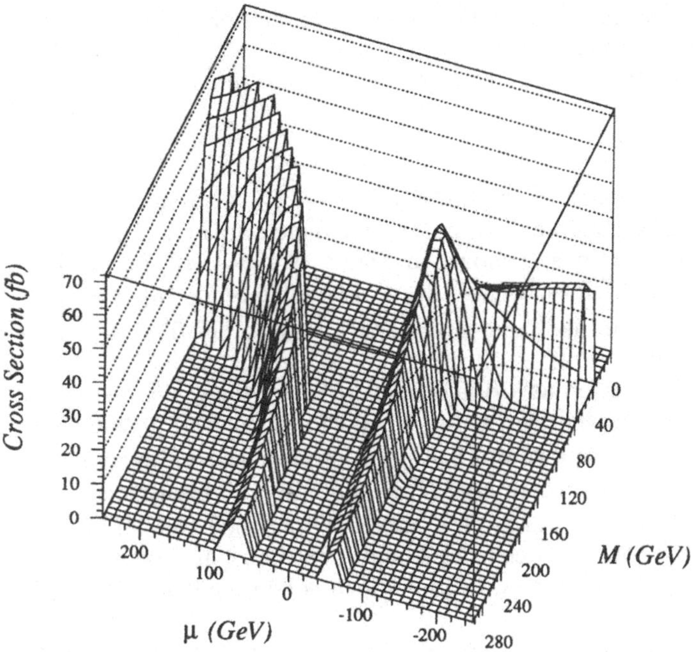

Fig. 6: Cross section for $e^+e^- \to \tilde{\nu}\tilde{\nu} \to (e^+e^-) + \not{p}$ (one-sided), as a function of M and μ, for $\sqrt{s} = 190$ GeV, $m_{\tilde{\nu}} = 80$ GeV and $\tan\beta = 4$.

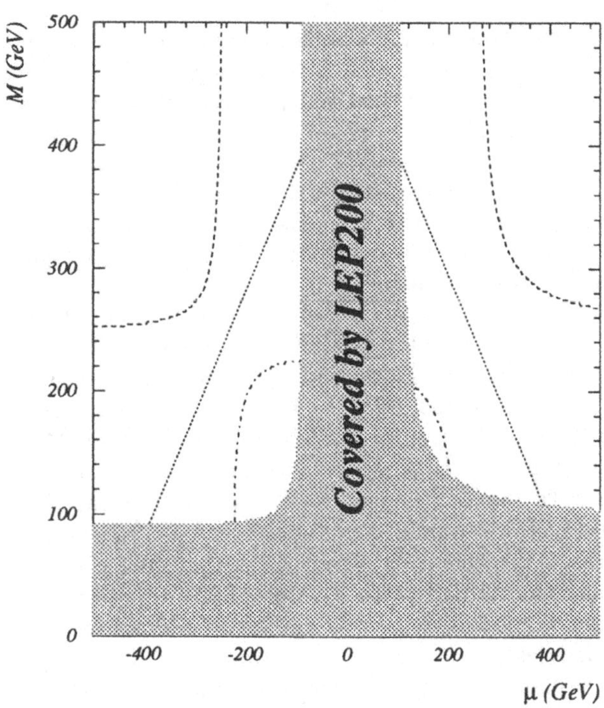

Fig. 7: Supersymmetry parameter space (M, μ) for chargino production at $\sqrt{s} = 500$ GeV, $\tan\beta = 4$. Shown are borderlines for $e^+e^- \to \tilde{\chi}_1^+ + \tilde{\chi}_1^-$ (– – –), $e^+e^- \to \tilde{\chi}_1^\pm + \tilde{\chi}_2^\mp$ (· · ·), $e^+e^- \to \tilde{\chi}_2^+ + \tilde{\chi}_2^-$ (– · – · –). Also shown the region explored by LEP200.

more important rôle than at lower energies. For the following predictions of event rates we shall assume an integrated luminosity of 20 fb^{-1} per year.

A study of SUSY particle production in e^+e^- collisions at $\sqrt{s} = 2$ TeV is given in [12].

3.2.1 Chargino production

Besides the pair production of the lightest chargino, $e^+e^- \to \tilde{\chi}_1^+\tilde{\chi}_1^-$, also the reactions $e^+e^- \to \tilde{\chi}_1^\pm\tilde{\chi}_2^\mp$ and $e^+e^- \to \tilde{\chi}_2^+\tilde{\chi}_2^-$ are possible. The parameter region in the (M, μ) plane (for $\tan\beta = 4$) accessible by these reactions is shown in Fig. 7. Notice that the contour lines $m_{\tilde{\chi}_1^\pm} + m_{\tilde{\chi}_2^\pm} =$ const are straight lines in the (M, μ) plane. If $M > |\mu|$, then $\tilde{\chi}_1^\pm$ is more higgsino-like and $\tilde{\chi}_2^\pm$ more wino-like, and vice versa for $M < |\mu|$. The chargino masses are approximately M and $|\mu|$.

In Fig. 8 we show the total cross section as a function of M and μ for $\sqrt{s} = 500$ GeV, $\tan\beta = 4$, $m_{\tilde{\nu}} = 600$ GeV. The cross section can go up to 550 fb. If both $M \gtrsim 250$ GeV and $|\mu| \gtrsim 250$ GeV the light chargino becomes too heavy to be produced at this energy. It should be noted that for $M \lesssim 250$ GeV there is a dependence on the sneutrino mass. The cross section is smaller by 50 % for $m_{\tilde{\nu}} = 100$ GeV than for $m_{\tilde{\nu}} = 600$ GeV. This can be explained by a destructive interference between the Z and $\tilde{\nu}$ exchange and because the $\tilde{\nu}$ couples only to the gaugino component of the chargino which is dominant for $M \leq |\mu|$. The $\tan\beta$ dependence is in general weak.

In a restricted range of parameters the chargino $\tilde{\chi}_1^\pm$ can also decay first into $\tilde{\chi}_2^0$ (see eq. (3.7)) , although the direct decay into $\tilde{\chi}_1^0$ is still the dominant one. In a certain region of parameters the $\tilde{\chi}_1^\pm$ can decay into real $W^\pm$'s, $\tilde{\chi}_1^\pm \to \tilde{\chi}_1^0 + W^\pm$.

The signatures are 'two-sided' events with acollinear, acoplanar, hard leptons or jets and large missing energy. The rate for $e^+e^- \to \tilde{\chi}_1^+\tilde{\chi}_1^- \to e^+ + e^- + \not{p}$ can go up to 7 fb (for $m_{\tilde{\nu}} = m_{\tilde{e}} = m_{\tilde{q}} = 600$ GeV, $\tan\beta = 4$) leading to 140 events/year, assuming an integrated luminosity of 20 fb^{-1}. The rates for the other signatures, $e^+e^- \to \tilde{\chi}_1^+ + \tilde{\chi}_1^- \to e + \text{jets} + \not{p}$ and $e^+e^- \to \tilde{\chi}_1^+ + \tilde{\chi}_1^- \to \text{jets} + \text{jets} + \not{p}$ are larger than the rate for electrons by approximately a factor 14 and 50, respectively, if the masses of the exchanged scalar particles are much larger than m_W. If they are around m_W, then these ratios are different for $M \leq 250$ GeV.

In the case that the sneutrino and/or the left-slepton are lighter than the chargino, the leptonic decay rates would again be strongly enhanced.

A Monte Carlo study of $e^+e^- \to \tilde{\chi}_1^+\tilde{\chi}_1^-$ taking into account beamstrahlung, background processes, and detector simulation, has been performed in [20]. The most severe Standard Model background is $W^\pm$ pair production, which can be reduced by appropriate cuts.

We now discuss the production of a heavy chargino, $e^+e^- \to \tilde{\chi}_1^\pm + \tilde{\chi}_2^\mp$. The production cross section is shown in Fig. 9 as a function of M and μ, for $\tan\beta = 4$, and $m_{\tilde{\nu}} = 600$ GeV. It can go up to 50 fb. The heavy chargino may have quite a complex decay pattern because in general it decays in cascades until the LSP is produced. Two-body decays into real $W^\pm$'s, Z's, and Higgs particles, eq. (3.9), are possible in a large region of parameter space. The $\tilde{\chi}_k^0$ and $\tilde{\chi}_1^\pm$ then further decay. An interesting signature is therefore given by $e^+e^- \to \tilde{\chi}_1^\pm\tilde{\chi}_2^\mp \to Z + X$. The corresponding rate can reach a value of 20 fb yielding 400 events per year. However, this rate depends more sensititvely on the SUSY parameters, especially on the masses of the Higgs particles.

In a smaller region of the parameter space (see Fig. 7) pair production of the heavy chargino is possible, $e^+e^- \to \tilde{\chi}_2^+ + \tilde{\chi}_2^-$. Characteristic signatures would be given by the decays of $\tilde{\chi}_2^\pm$ into real W's and Z's, $e^+e^- \to \tilde{\chi}_2^+\tilde{\chi}_2^- \to Z + Z + X$, $Z + W^\pm + X$, $W^+ + W^- + X$.

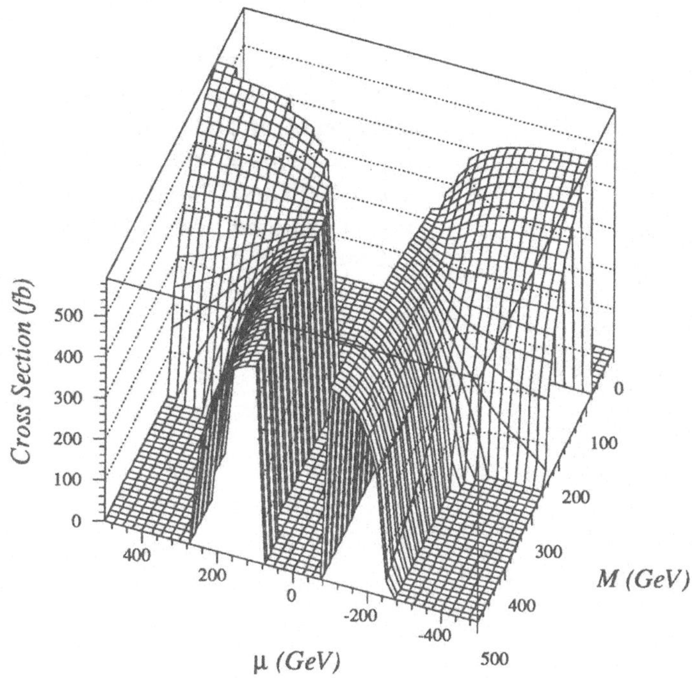

Fig. 8: Total cross section for $e^+e^- \to \tilde{\chi}_1^+ + \tilde{\chi}_1^-$ as a function of M and μ, for $m_{\tilde{\nu}} = 600$ GeV. The parameter region explored by LEP200 is cut out.

3.2.2 Neutralino production

At $\sqrt{s} = 500$ also higher neutralino states can be produced, $e^+e^- \to \tilde{\chi}_i^0 + \tilde{\chi}_j^0$. The parameter region in the (M, μ) plane accessible by the reaction $e^+e^- \to \tilde{\chi}_1^0\tilde{\chi}_2^0$ is again larger than that by $e^+e^- \to \tilde{\chi}_i^+\tilde{\chi}_1^-$. If $m_Z < M < |\mu|$, then $\tilde{\chi}_1^0$ is approximately B-ino-like, $\tilde{\chi}_2^0$ is W^3-ino-like, whereas $\tilde{\chi}_3^0$ and $\tilde{\chi}_4^0$ are higgsino-like. On the other hand, if $M > |\mu| > m_Z$, $\tilde{\chi}_1^0$ and $\tilde{\chi}_2^0$ have large higgsino components, whereas $\tilde{\chi}_3^0$ and $\tilde{\chi}_4^0$ are B-ino and W^3-ino-like, respectively. The neutralino masses are approximately $M/2$, M, $|\mu|$, and $|\mu|$ [13, 21].

In the paramter domain also accessible for pair production of light charginos, the cross section for $e^+e^- \to \tilde{\chi}_1^0\tilde{\chi}_2^0$ can go up to 200 fb (for $m_{\tilde{e}} = 600$ GeV). Outside this domain this cross section is, unfortunately at most a few fb, unless the mass of the exchanged selectron is very small (~ 100 GeV).

Again, the most interesting signatures for $\tilde{\chi}_1^0\tilde{\chi}_2^0$ production are one-sided events. The cross section for $e^+e^- \to \tilde{\chi}_1^0\tilde{\chi}_2^0 \to$ (jet jet + $\not{p}$ can reach 130 fb (for $m_{\tilde{e}} = m_{\tilde{q}} = m_{\tilde{\nu}} =$ 600 GeV). The rate for $e^+e^- \to \tilde{\chi}_1^0 + \tilde{\chi}_2^0 \to (e^+e^-) + \not{p}$ would be smaller by a factor of approximately 20.

It turns out that some of the production cross sections for the higher neutralino states are comparable to that of $e^+e^- \to \tilde{\chi}_1^0\tilde{\chi}_2^0$. For the higher neutralino states more decay channels are open, and, therefore, cascade decays are more likely. This will lead to a more complex decay pattern and in general reduce the $\not{p}$ signal. On the other hand, more decays into Z's and $W^{\pm}$'s are possible leading to multileptons and/or multijets in the final state. In addition, the rates will depend more sensitively on the properties of the Higgs sector. For a more detailed discussion we refer to [19].

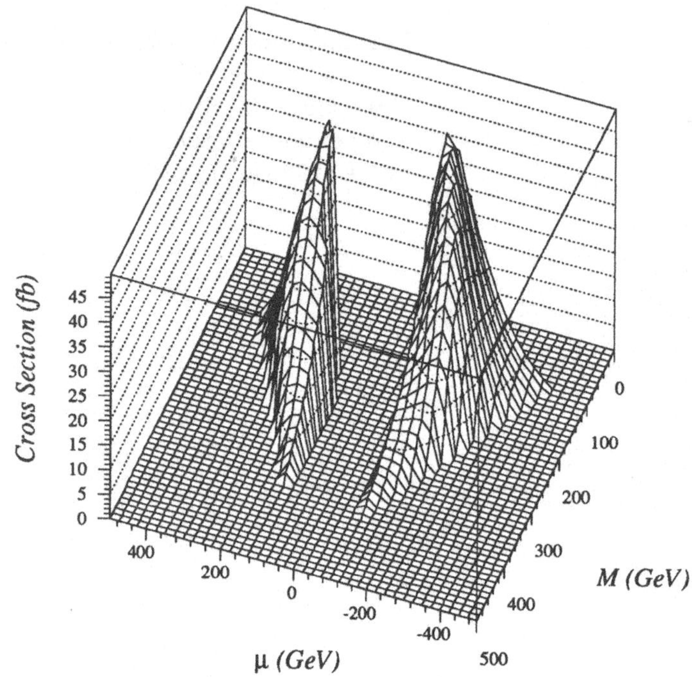

Fig. 9: Total cross section for $e^+e^- \to \tilde{\chi}_1^+ + \tilde{\chi}_2^- + c.c.$ as a function of M and μ for $\tan\beta = 4$, $m_{\tilde{\nu}} = 600$ GeV, $\sqrt{s} = 500$ GeV. The parameter region explored by LEP200 is cut out.

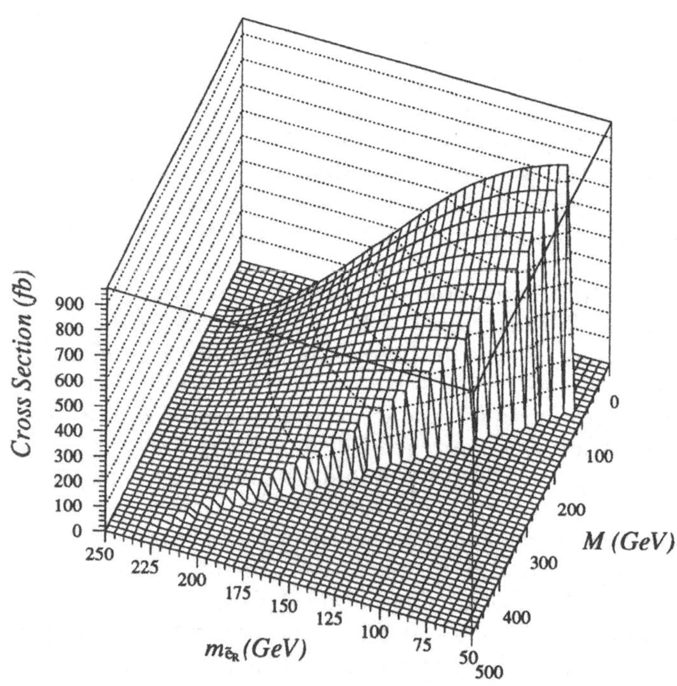

Fig. 10: Total cross section for selectron pair production at $\sqrt{s} = 500$ GeV, $\tan\beta = 4$, $\mu = 200$ GeV $e^+e^- \to \tilde{e}_R^+ + \tilde{e}_R^-$, as a function of M and $m_{\tilde{e}_R}$. The parameter region explored by LEP200 is cut out.

3.2.3 Selectron production

In the energy range considered the mass difference between $\tilde{e}_R$ and $\tilde{e}_L$ according to eq. (1.1) can be quite sizeable. Therefore, in our calculations of selectron production cross sections we shall always use eq. (1.1) to relate $m_{\tilde{e}_R}$ and $m_{\tilde{e}_L}$ to the SUSY parameters m_0, M, $\tan\beta$. Quite generally $\tilde{e}_R$ is lighter than $\tilde{e}_L$.

Since the production cross sections depend only weakly on $\tan\beta$ and μ (apart from the region $|\mu| \lesssim 100$ GeV), we present in Fig. 10 the cross section for $e^+e^- \to \tilde{e}_R^+ + \tilde{e}_R^-$ as a function of M and $m_{\tilde{e}_R}$ for $\mu = -200$ GeV and $\tan\beta = 4$. For $m_{\tilde{e}_R} = 200$ GeV the cross section varies between 50 and 150 fb depending on M. Notice that according to the mass relations, eq. (1.1), there exists a maximal value of M for a given value of $m_{\tilde{e}_L}$ and $m_{\tilde{e}_R}$. It is $M = 410$ GeV for $m_{\tilde{e}_R} = 200$ GeV, and $M = 240$ GeV for $m_{\tilde{e}_L} = 200$ GeV, for $\tan\beta = 4$.

The cross section for $e^+e^- \to \tilde{e}_L^+\tilde{e}_L^-$ is similar in magnitude to that for $\tilde{e}_R\tilde{e}_R$ production. It should be noted that also the cross section for $e^+e^- \to \tilde{e}_L^\pm + \tilde{e}_R^\mp$, which has only t channel contributions, has a comparable cross section, depending on the SUSY parameters.

In the mass region considered, the difference in the decay pattern of $\tilde{e}_R$ and $\tilde{e}_L$ is more pronounced, and cascade decays are expected to play a more important rôle. We show in Figs. 11a and b the branching ratios for the $\tilde{e}_R$ and $\tilde{e}_L$ decays into neutralinos and charginos, eq. (3.11), as function of μ for $M = 150$ GeV, $\tan\beta = 4$, for $m_{\tilde{e}_R} = 200$ GeV and $m_{\tilde{e}_L} = 200$ GeV.

It is noticeable that in a large region of μ, more precisely $|\mu| > M$, the $\tilde{e}_R$ decays to 100 % into $\tilde{\chi}_1^0$. This can be understood because the $\tilde{e}_R$ couples only to the B-ino component of the neutralinos, and $\tilde{\chi}_1^0$ is essentially a B-ino in this parameter region. For $|\mu| < M$, the decays into $\tilde{\chi}_2^0$ and $\tilde{\chi}_3^0$ are significant because here these particles have large B-ino components.

Concerning the $\tilde{e}_L$ decays, the direct decay into $\tilde{\chi}_1^0$ is not the most important one. For $|\mu| \gtrsim M$ the decay into the light chargino dominates, the branching ratio being between 40 % and 50 % . For $|\mu| \lesssim M$ the decays into $\tilde{\chi}_2^0$ and $\tilde{\chi}_3^0$ can have large branching ratios because in this region the heavy neutralinos have large gaugino components.

Despite the possibility of cascade decays the most interesting signature for pair production of $\tilde{e}_R$ or $\tilde{e}_L$ is again provided by the two-sided events with acollinear, acoplanar, hard electrons plus missing energy. The rate for $e^+e^- \to \tilde{e}_R^+ e_R^- \to e^+ + e^- + \not{p}$ can go up to 180 fb for $m_{\tilde{e}_R} = 200$ GeV and $\tan\beta = 4$, whereas the corresponding one for $e^+e^- \to \tilde{e}_L^+ e_L^- \to e^+ + e^- + \not{p}$ reaches 60 fb.

A Monte Carlo study of selectron production is given in [22].

3.2.4 Sneutrino production

The cross section for $e^+e^- \to \tilde{\nu}_e + \bar{\tilde{\nu}}_e$ is in general higher than that for selectron production because the Z^0 and the charginos couple more strongly. For $m_{\tilde{\nu}_e} = 200$ GeV the cross section can reach 700 fb.

The sneutrino can only be observed if it decays first into a chargino or a higher mass neutralino (see eq. (3.12), which then decays into leptons or quarks. Only if $M < m_{\tilde{\nu}_e}$ or $|\mu| < m_{\tilde{\nu}_e}$, the sneutrino is visible. It turns out that in this region the decay into the light chargino has the largest branching ratio. Therefore, the important signatures are again one-sided and two-sided events with hard leptons and jets, as already discussed for LEP200 (see Sect. 3.1.4). More details on pair production of sneutrinos as well as of the other sfermions ($\tilde{\mu}_{L,R}, \tilde{\nu}_\mu, \tilde{u}_{L,R}, \tilde{d}_{L,R}$ etc.) can be found in [19]. A Monte Carlo study of smuon production is given in [23].

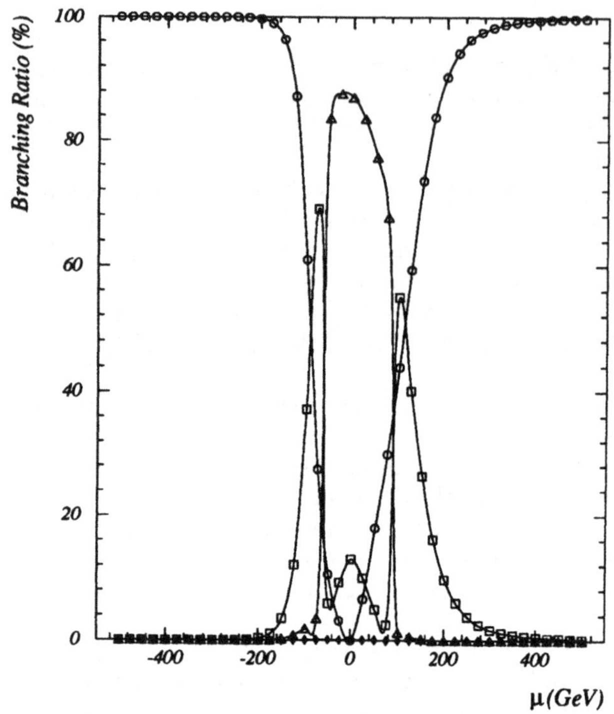

Fig. 11a: Branching ratios for $\tilde{e}_R^- \to e^-\tilde{\chi}_i^0$, as a function of μ, for $m_{\tilde{e}_{L,R}} = 200$ GeV, $M = 150$ GeV, $\tan\beta = 4$, $\sin^2\theta_W = 0.23$. The curves correspond to the following transitions:

- ○ into $\tilde{\chi}_1^0$
- □ into $\tilde{\chi}_2^0$
- △ into $\tilde{\chi}_3^0$
- ◇ into $\tilde{\chi}_4^0$

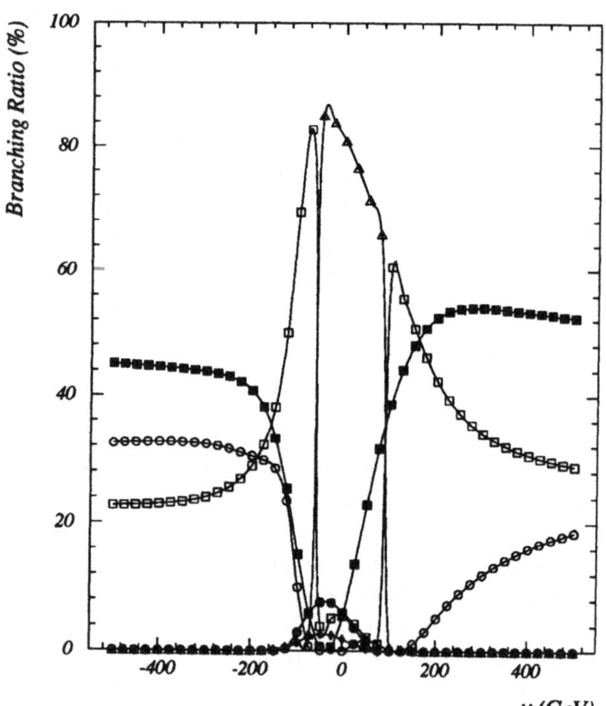

Fig. 11b: Branching ratios for $\tilde{e}_L^- \to e^-\tilde{\chi}_i^0, \nu\tilde{\chi}_i^-$, as a function of μ, for $m_{\tilde{e}_{L,R}} = 200$ GeV, $M = 150$ GeV, $\tan\beta = 4$, $\sin^2\theta_W = 0.23$. The curves correspond to the following transitions:

- ■ into the light chargino $\tilde{\chi}_1^\pm$
- ● into the heavy chargino $\tilde{\chi}_2^\pm$
- ○ into $\tilde{\chi}_1^0$
- □ into $\tilde{\chi}_2^0$
- △ into $\tilde{\chi}_3^0$
- ◇ into $\tilde{\chi}_4^0$.

4 *pp* Collisions

In *pp* collisions one expects the strongly interacting SUSY particles, gluinos and squarks, to be produced with the largest cross section of all supersymmetric particles [24, 25].

The basic processes for gluino and squark production in hadronic collisions are [26]

$$g + g \rightarrow \tilde{g} + \tilde{g} \tag{4.1}$$
$$q + \bar{q} \rightarrow \tilde{g} + \tilde{g} \tag{4.2}$$
$$g + g \rightarrow \tilde{q} + \bar{\tilde{q}} \tag{4.3}$$
$$q + \overset{(-)}{q} \rightarrow \tilde{q} + \overset{(-)}{\tilde{q}} \tag{4.4}$$
$$g + \overset{(-)}{q} \rightarrow \tilde{g} + \overset{(-)}{\tilde{q}}\ . \tag{4.5}$$

For example, for a gluino mass $m_{\tilde{g}} = 500$ GeV one obtains at $\sqrt{s} = 16$ TeV, the design c.m. energy of the LHC, a production cross section of ~ 50 pb for $pp \rightarrow \tilde{g}\tilde{g} + X$. Similarly, for a squark mass $M_{\tilde{q}} = 500$ GeV one gets a cross section of ~ 15 pb for $pp \rightarrow \tilde{q}\bar{\tilde{q}} + X$. Assuming an integrated luminosity of 10^5 pb^{-1} this would mean a rate of 5×10^6 gluino pairs and 1.5×10^6 squark pairs per year. For the SSC one obtains a cross section $\sigma \approx 10$ pb for $pp \rightarrow \tilde{g}\tilde{g} + X$ at $\sqrt{s} = 40$ TeV and $m_{\tilde{g}} = 1000$ GeV. Taking an integrated luminosity of 10^4 pb^{-1} yields 10^5 such events per year.

If squarks are heavier than gluinos then the gluinos produced decay into

$$\tilde{g} \rightarrow q + \bar{q} + \tilde{\chi}_i^0 \tag{4.6}$$
$$\rightarrow t + \bar{t} + \tilde{\chi}_i^0 \tag{4.7}$$
$$\rightarrow q + \bar{q}' + \tilde{\chi}_i^{\pm}, \tag{4.8}$$
$$\rightarrow t + \bar{b} + \tilde{\chi}_i^-,\ \bar{t} + b + \tilde{\chi}_i^+ \tag{4.9}$$
$$\rightarrow g + \tilde{\chi}_i^0. \tag{4.10}$$

Here we distinguish between decays into light quarks q and those which involve the top quark because they show a different decay pattern. This is due to the fact that in the decays (4.7) and (4.9) the top Yukawa coupling and additional quark mass terms are important. The decays (4.6 – 4.9) proceed via virtual squark exchange. The decay into a gluon, eq. (4.10), is possible via quark-squark loops, the $t - \tilde{t}$ contribution being the most important one. It is not negligible in certain regions of the SUSY parameter space.

In the mass range for gluinos and squarks we shall consider, also the charginos and neutralinos will be heavy. Therefore, in most cases they will decay into $W^{\pm}, Z^0$ or Higgs particles and lower mass states of charginos or neutralinos according to eqs. (3.9) and (3.10). If these two-body decays are kinematically not allowed, the three-body decays, eqs. (3.7) and (3.8), may also become important. Thus in general one has cascade decays of gluinos. They end when the lightest supersymmetric particle $\tilde{\chi}_1^0$ is reached. The final states can therefore contain Z^0-bosons, $W^{\pm}$-bosons, neutral and charged Higgs-bosons, leptons and jets and missing momentum from the lightest neutralino and the neutrinos.

If the gluinos are heavier than the squarks, then squark production will be more important. In the case of light quark flavours the squarks then decay according to eq. (3.13). For the following decays of the top squark and the bottom squark

$$\tilde{t}_{L,R} \rightarrow t + \tilde{\chi}_i^0 \tag{4.11}$$
$$\tilde{t}_R \rightarrow b + \tilde{\chi}_i^+ \tag{4.12}$$
$$\tilde{b}_L \rightarrow t + \tilde{\chi}_i^- \tag{4.13}$$

it is again necessary to take into account the top Yukawa coupling. (Notice that the Yukawa coupling terms in $\tilde{t}_L \to b + \tilde{\chi}_i^+$ and $\tilde{b}_R \to t + \tilde{\chi}_i^-$ are proportional to the bottom quark mass which we shall neglect here.)

The neutralinos and charginos again decay as described above. Also here a variety of cascade decays is possible until the lightest neutralino is produced.

For illustration we show in Fig. 12 the domains in the (M, μ) plane (for $\tan\beta = 4$) where for a squark with a mass of 500 GeV cascade decays are kinematically possible (domain I), or where only transitions into the lightest neutralino are allowed (domain II). For a gluino with a mass of 500 GeV the same picture as in Fig. 12 is valid taking $M \simeq 0.3\, m_{\tilde{g}}$.

In pp collisions sleptons, charginos and neutralinos could also be produced by the Drell-Yan mechanism, but their cross sections are smaller [27] and are not considered here.

4.1 Gluino decays and signatures

In order to understand the complex decay pattern of gluinos and squarks it is necessary to consider the individual steps in the cascade. For the decays we have used the formulae as given in [28] where a complete list of these is presented.

Considering first only the gluino decays into light quarks, $\tilde{g} \to q + \bar{q} + \tilde{\chi}_i^{0,\pm}$, and leaving out the decay $\tilde{g} \to g + \tilde{\chi}_i^0$, the overall picture of gluino decays would be rather simple [29] because then only the gaugino components of the neutralinos couple: For $|\mu| \lesssim m_{\tilde{g}}/3$ mainly the heavy chargino and the heaviest neutralino are produced, the light chargino and the two lightest neutralinos being mainly higgsinos. For $|\mu| \gtrsim m_{\tilde{g}}/3$ the decays into the light chargino and the second lightest neutralino would dominate, because here the heavy particles are higgsino like. This is illustrated in Fig. 13 for $m_{\tilde{g}} = 750$ GeV and $\tan\beta = 4$. (Furthermore, we have taken $M_{\tilde{q}_L} = M_{\tilde{q}_R} = 2\, m_{\tilde{g}}$, $\alpha_s = 0.1$, and $\sin^2\theta_W = 0.23$.)

This simple picture is changed if in the decays $\tilde{g} \to t + \bar{t} + \tilde{\chi}_i^0$, $\tilde{g} \to t + \bar{b} + \tilde{\chi}_i^-$, $\tilde{g} \to b + \bar{t} + \tilde{\chi}_i^+$ the top mass is taken into account [28, 30]. (We take $m_t = 150$ GeV.) In the parameter domain $|\mu| \lesssim m_{\tilde{g}}/3$ the dominant transition of these decays is now that into the light chargino, contrary to the simple picture described above. For $|\mu| > m_{\tilde{g}}/3$ the influence of the Yukawa coupling terms is smaller because the higgsino-like particles are now heavy and, therefore, reduce the phase space, a feature which is even more pronounced for smaller gluino masses.

The decays $\tilde{g} \to g + \tilde{\chi}_i^0$ play a minor rôle. In the cases considered the sum of the branching ratios for $\tilde{g} \to g + \tilde{\chi}_i^0$ is a few percent. It can, however, be larger for smaller gluino masses and smaller $\tan\beta$ [28, 30].

The branching ratios for $\tilde{g} \to \tilde{\chi}_i^0 + q + \bar{q}$ (or g) and $\tilde{g} \to \tilde{\chi}_i^{\pm} + q + \bar{q}'$, summed over all quark flavours including the top quark, are shown in Fig. 14 as a function of μ for $m_{\tilde{g}} = 750$ GeV, $\tan\beta = 4$. For $|\mu| \lesssim m_{\tilde{g}}/3$ there are substantial transition rates into the light chargino and the light neutralinos. For $|\mu| > m_{\tilde{g}}/3$ the transitions into the light chargino and the second lightest neutralino are the most important ones. Notice that the branching ratio for the decay into the lightest neutralino is only about 15 percent. It is smaller than that of the decay into $\tilde{\chi}_2^0$ because for larger $|\mu|$ the $\tilde{\chi}_1^0$ is mainly a B-ino and the $\tilde{\chi}_2^0$ a W^3-ino. Notice that the decay into $\tilde{\chi}_3^0$ appears due to $\tilde{g} \to g + \tilde{\chi}_3^0$ and $\tilde{g} \to t + \bar{t} + \tilde{\chi}_3^0$.

Next we discuss some interesting signatures for gluino production $p + p \to \tilde{g} + \tilde{g} + X$ (assuming that the squarks are heavier than gluinos). The 'classical' signature would be that both gluinos go directly into the lightest supersymmetric particle $\tilde{\chi}_1^0$, leading to events

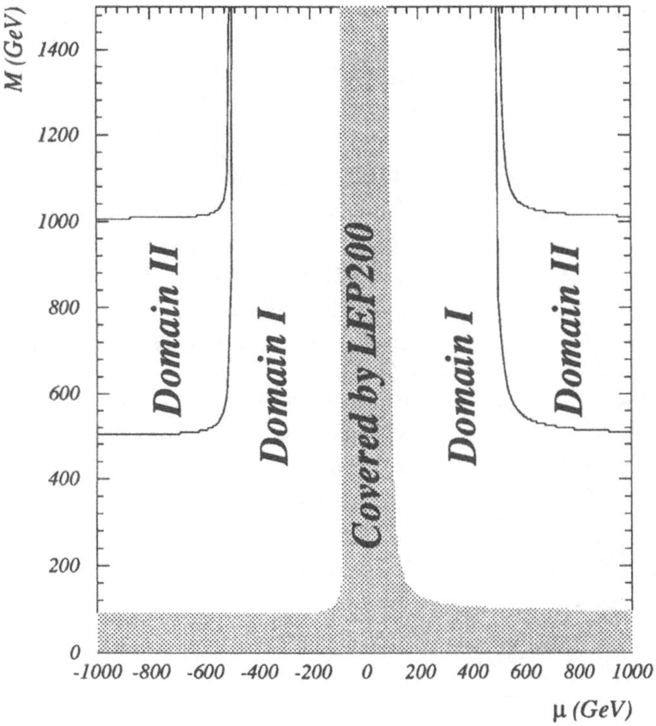

Fig. 12: Supersymmetry parameter space (M, μ) for decays of a squark with $M_{\tilde{q}} = 500$ GeV, $\tan\beta = 4$.
Domain I: Cascade decays possible.
Domain II: Only transitions into $\tilde{\chi}_1^0$ possible.
Also shown the region covered by LEP200. For a gluino with $m_{\tilde{g}} = 500$ GeV the same figure is valid taking $M \simeq 0.3\, m_{\tilde{g}}$. For another squark mass an analogous figure holds if M and μ are scaled by the ratio of the squark masses.

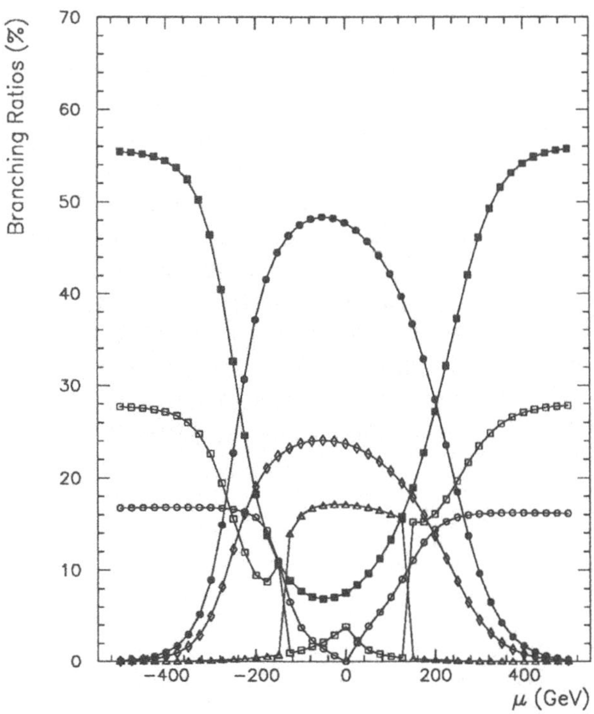

Fig. 13: Branching ratios for the decays $\tilde{g} \to q + \bar{q} + \tilde{\chi}_i^0$, $\tilde{g} \to u + \bar{d} + \tilde{\chi}_i^-$ plus $d + \bar{u} + \tilde{\chi}_i^+$, neglecting all quark masses, as a function of μ for $m_{\tilde{g}} = 750$ GeV and $\tan\beta = 4$. Here the decays $\tilde{g} \to g + \tilde{\chi}_i^0$ are left out. We have taken $M_{\tilde{q}_L} = M_{\tilde{q}_R} = 2\, m_{\tilde{g}}$, $\alpha_s = 0.1$, and $\sin^2\theta_W = 0.23$. The curves correspond to the following transitions:

- ■ into the light chargino $\tilde{\chi}_1^{\pm}$
- ● into the heavy chargino $\tilde{\chi}_2^{\pm}$
- ○ into $\tilde{\chi}_1^0$
- □ into $\tilde{\chi}_2^0$
- △ into $\tilde{\chi}_3^0$
- ◇ into $\tilde{\chi}_4^0$.

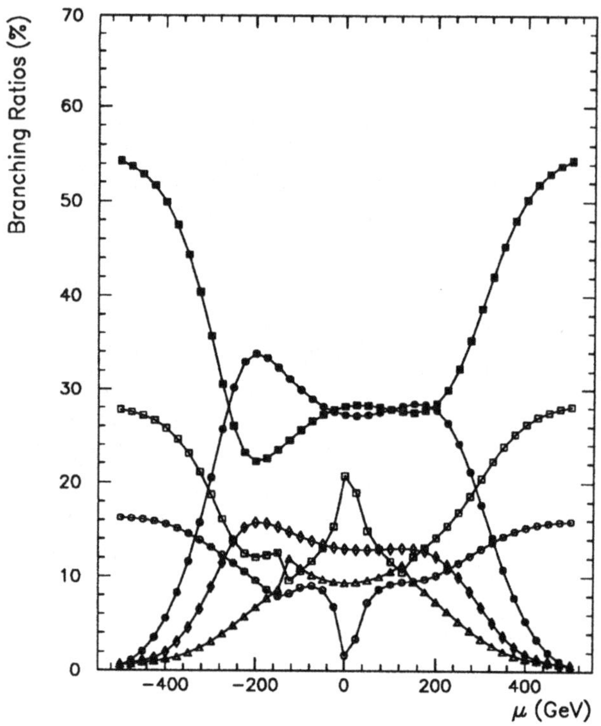

Fig. 14: Branching ratios for $\tilde{g} \to \tilde{\chi}_i^0 + q + \bar{q}$, $\tilde{\chi}_i^0 + g$ and $\tilde{g} \to \tilde{\chi}_j^+ + q + \bar{q}'$ plus $\tilde{\chi}_j^- + \bar{q} + q'$, summed over all quark flavours, as a function of μ, for $m_{\tilde{g}} = 750$ GeV and $\tan\beta = 4$. We have taken $m_t = 150$ GeV, $M_{\tilde{q}_L} = M_{\tilde{q}_R} = 2\, m_{\tilde{g}}$, $\alpha_s = 0.1$, and $\sin^2\theta_W = 0.23$. The curves correspond to the following transitions:

- ■ into the light chargino $\tilde{\chi}_1^\pm$
- ● into the heavy chargino $\tilde{\chi}_2^\pm$
- ○ into $\tilde{\chi}_1^0$
- □ into $\tilde{\chi}_2^0$
- △ into $\tilde{\chi}_3^0$
- ◇ into $\tilde{\chi}_4^0$.

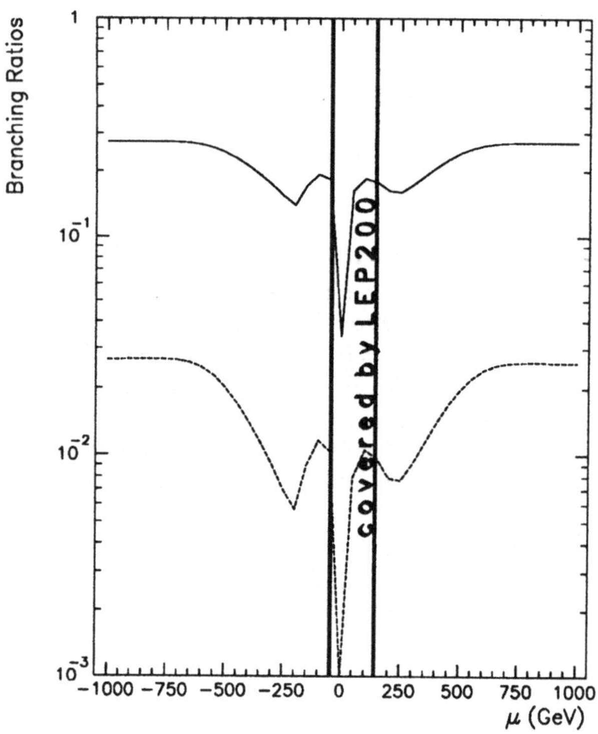

Fig. 15: The branching ratios for the two produced gluinos to go directly into $\tilde{\chi}_1^0$ and $2\,\tilde{\chi}_1^0$; i.e. $\tilde{g} + \tilde{g} \to \tilde{\chi}_1^0 + X$ (——), and $\tilde{g} + \tilde{g} \to \tilde{\chi}_1^0 + \tilde{\chi}_1^0 + X$ (- - -), (where X does not contain another $\tilde{\chi}_1^0$) for $m_{\tilde{g}} = 1000$ GeV as a function of μ, for $\tan\beta = 4$. We have taken $m_t = 150$ GeV, $M_{\tilde{q}_L} = M_{\tilde{q}_R} = 2\, m_{\tilde{g}}$, $\alpha_s = 0.1$, and $\sin^2\theta_W = 0.23$. Also shown the region covered by LEP200.

with large missing energy. This is shown in Fig. 15 for $m_{\tilde{g}} = 1000$ GeV. The branching ratio for both gluinos going directly into the LSP is at most a few percent. Thus with an integrated luminosity of 10^5 pb^{-1} one could expect $\sim 10^3$ such events at the LHC. Similarly at the SSC one would have 2×10^3 such events. The probability that only one gluino decays directly into $\tilde{\chi}_1^0$ is higher and can go up to 30 percent, as also shown in Fig. 15. The region in μ which will be explored by LEP 200 is also indicated.

A further interesting signature would be two Z^0's in the final state which originate from the cascade decays. Fig. 16 shows the branching ratio for $\tilde{g} + \tilde{g} \to 2\,Z^0 + X$ for a gluino mass of 1000 GeV. (We have taken the mass of the Higgs particle h^0, $m_{h^0} = 80$ GeV, and the masses of all other Higgs particles much heavier.) The branching ratio can go up to 10 % , but there is a strong dependence on the SUSY parameter μ.

Fig. 16 also shows the branching ratios for $\tilde{g} + \tilde{g} \to 4\,\mu$'s, $5\,\mu$'s, and $6\,\mu$'s. Here it is summed over all events with muons coming from Z^0, $W^\pm$, t-quark decays into $b\,W$, and three-body decays of charginos and neutralinos. Over a large range of μ the four muon rate is about 10^{-4}.

We want to emphasize that some details of the $2\,Z^0$ and multimuon rates depend on the mass spectrum in the Higgs sector. If the decay of a chargino or a neutralino into a Higgs particle is possible the other decay rates are reduced.

A more detailed discussion of gluino signatures is given in [31]. Monte Carlo studies taking into account the standard model background ($t\bar{t}$, Z^0Z^0 production, etc.) were performed and have shown that at the LHC gluinos in the mass range 300 GeV $\leq m_{\tilde{g}} \leq$ 1000 GeV can be detected [25, 27].

4.2 Squark decays and signatures

As to the squark decays, one has again to distinguish between those into light quarks and heavy quarks, and of course between left and right squarks. All cases were discussed in detail in [28] where also the corresponding formulae are given.

If the quark masses can be neglected the pattern of squark decays according to eq. (3.13) is again rather simple. The branching ratios for right squarks are then the same (independent of flavour) as for right selectrons of the same mass. These are already shown in Fig. 11a. For $|\mu| > M$ right squarks decay practically 100 % into $\tilde{\chi}_1^0$, and for $|\mu| < M$ to more than 90 % into $\tilde{\chi}_3^0$.

Left squarks decay dominantly into charginos as can be seen in Fig. 17 where we plot the branching ratios of $\tilde{u}_L \to d + \tilde{\chi}_i^+$, $u + \tilde{\chi}_i^0$, for $M_{\tilde{q}} = 1000$ GeV. For $|\mu| > M$ ($|\mu| < M$) the transition into the lighter (heavier) chargino has the largest branching ratio due to its gaugino nature in this region of μ. Among the decays into neutralinos, for $|\mu| > M$ ($|\mu| < M$) the transitions into $\tilde{\chi}_2^0$ ($\tilde{\chi}_4^0$) have also big rates as the $\tilde{\chi}_2^0$ ($\tilde{\chi}_4^0$) is mainly a W^3-ino in this region of μ. A qualitatively similar pattern holds for $\tilde{d}_L$ decays.

The influence of the top quark mass terms in the decays of $\tilde{t}_R$ is seen in Fig. 18 where we plot the branching ratios into charginos and neutralinos as a function of μ for $M_{\tilde{q}} = 1000$ GeV, $M = 500$ GeV, $\tan\beta = 4$. Now the decays into charginos become very important, for $|\mu| \lesssim M$ the decay into $\tilde{\chi}_1^+$ being the dominant one. In the $\tilde{t}_L$ decays the top Yukawa coupling only changes the width of $\tilde{t}_L \to t + \tilde{\chi}_i^0$.

As to the $\tilde{b}_L$ decays, the main differences between these and the decays of $\tilde{d}_L$ appear again in the region $|\mu| \lesssim M$. Due to the top Yukawa coupling the width for $\tilde{b}_L \to t + \tilde{\chi}_1^-$ becomes the largest one.

Concerning the signatures for squark production $p + p \to \tilde{q} + \bar{\tilde{q}} + X$, one expects the direct transition of a squark into the lightest supersymmetric particle $\tilde{\chi}_1^0$ to be more

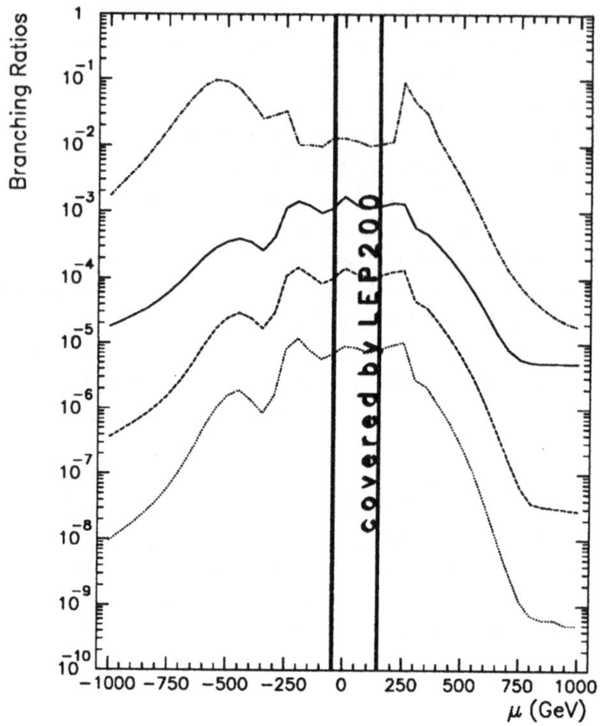

Fig. 16: The branching ratios for $\tilde{g}+\tilde{g}\to 2Z^0+X$ (- · - · -), $\tilde{g}+\tilde{g}\to 4\mu+X$ (——), $\tilde{g}+\tilde{g}\to 5\mu+X$ (- - -), $\tilde{g}+\tilde{g}\to 6\mu+X$ (· · · ·) for $m_{\tilde{g}}=1000$ GeV, as a function of the parameter μ, for $\tan\beta=4$. We have summed over all events with muons coming from Z^0, $W^\pm$, and three body decays of charginos and neutralinos, and from $t\to Wb$. We have taken $m_t=150$ GeV, $M_{\tilde{q}_L}=M_{\tilde{q}_R}=2\,m_{\tilde{g}}$, $\alpha_s=0.1$, $\sin^2\theta_W=0.23$, $m_{h^0}=80$ GeV and all other Higgs particles heavy. Also shown the region covered by LEP200.

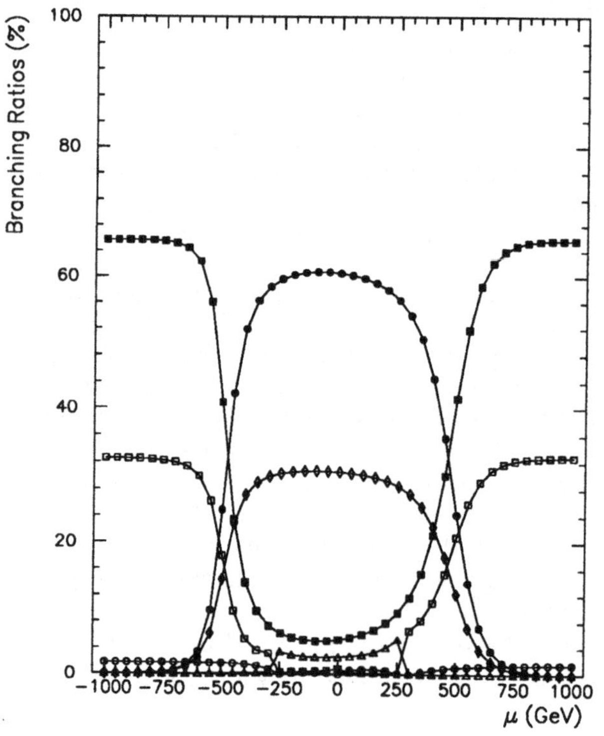

Fig. 17: Branching ratios for $\tilde{u}_L$ decays into charginos and neutralinos for $u_L\to d+\tilde{\chi}_i^+$, $u+\tilde{\chi}_i^0$, as a function of μ, for $M_{\tilde{q}}=1000$ GeV, $M=500$ GeV, $\tan\beta=4$. We have taken $m_t=150$ GeV, $\alpha_s=0.1$, $\sin^2\theta_W=0.23$. The curves correspond to the following transitions:

- ■ into the light chargino $\tilde{\chi}_1^\pm$
- ● into the heavy chargino $\tilde{\chi}_2^\pm$
- ○ into $\tilde{\chi}_1^0$
- □ into $\tilde{\chi}_2^0$
- △ into $\tilde{\chi}_3^0$
- ◇ into $\tilde{\chi}_4^0$.

important than for gluinos because right squarks of light flavours decay into $\tilde{\chi}_1^0$ to more than 90 percent (for $|\mu| \gtrsim M$). Fig. 19 exhibits the branching ratios for $\tilde{q} + \bar{\tilde{q}} \to \tilde{\chi}_1^0 + X$ and $\tilde{q} + \bar{\tilde{q}} \to \tilde{\chi}_1^0 + \tilde{\chi}_1^0 + X$, i.e. only one of the squarks or both directly decay into $\tilde{\chi}_1^0$, for $M_{\tilde{q}} = 1000$ GeV, $\tan\beta = 4$, and $M = 500$ GeV. We have averaged over left and right squarks and all flavours. It is interesting to note that in a large range of μ the branching ratio into two $\tilde{\chi}_1^0$ is an order of magnitude larger than that into one $\tilde{\chi}_1^0$. As a consequence, about 40 % of the events where a $\tilde{q}\bar{\tilde{q}}$ pair is produced will have the 'classical' signature of large missing energy. This feature is quite independent of $\tan\beta$ and holds in a large range of squark masses, provided $|\mu| \gtrsim M$.

Other signatures would be $\tilde{q}+\bar{\tilde{q}} \to ZZ+X, 4\mu+X, 5\mu+X$, etc. They are discussed in [31]. Monte Carlo studies [25, 27] have shown that also for squarks a mass reach of about 1 TeV is possible at LHC.

5 ep Collisions

The processes most favourable for finding supersymmetric particles in *ep* collisions are the reactions [32, 33, 34]

$$e + p \to \tilde{e}_{L,R} + \tilde{q}_{L,R} + X \tag{5.1}$$

$$e + p \to \tilde{\nu}_L + \tilde{q}_L + X, \tag{5.2}$$

The basic reaction mechanism is $eq \to \tilde{e}_{L,R}\ \tilde{q}_{L,R}$ or $eq \to \tilde{\nu}\tilde{q}_L$ scattering via t channel neutralino or chargino exchange. The sea quarks and the antiquarks in the proton also contribute, though their contribution is smaller.

The produced slepton or squark decays into the corresponding fermion and a neutralino $\tilde{\chi}_i^0$ or chargino $\tilde{\chi}_i^{\pm}$, eqs. (3.11 – 3.13). (Here we assume that the gluino is heavier than the squarks.) The higher mass neutralinos and charginos then further decay until the LSP is reached, see eqs. (3.7 – 3.10). Therefore, also here cascade decays are possible.

The production cross sections for both reactions (5.1 – 5.2) depend most strongly on the sum of the masses of the slepton and squark produced, and of course on the centre of mass energy.

5.1 HERA

At HERA *ep* collisions at $\sqrt{s} = 314$ GeV will be studied. The design luminosity is 1.5×10^{31} cm^{-2}s^{-1}. The reactions (5.1 – 5.2) have been studied for this energy in [33, 34]. Cross sections larger than 0.1 pb have been found for $m_{\tilde{e},\tilde{\nu}}+m_{\tilde{q}} \lesssim 180$ GeV. If cross sections down to 5×10^{-2} pb can be measured, the observable mass range extends to $m_{\tilde{e},\tilde{\nu}}+m_{\tilde{q}} \lesssim 200$ GeV. With the exisiting experimental bounds on the squark and selectron masses from CDF and LEP (see Sect. 2) and within the MSSM, the possibilities of detecting SUSY particles at HERA are therefore limited. At a higher c.m. energy, e.g. $\sqrt{s} = 450$ GeV, the mass range $m_{\tilde{e},\tilde{\nu}} + m_{\tilde{q}} \leq 300$ GeV could be explored.

Relaxing the MSSM by including R-parity violating terms would allow single SUSY particle production and thus extend the explorable mass region for SUSY particles. For further studies we refer to [35].

5.2 LEP/LHC

The LHC will offer the unique possibility of being operated also in the *ep* collider mode at $\sqrt{s} = 1.4$ TeV with a luminosity of about 10^{32} cm^{-2}s^{-1}. Also here the search for SUSY

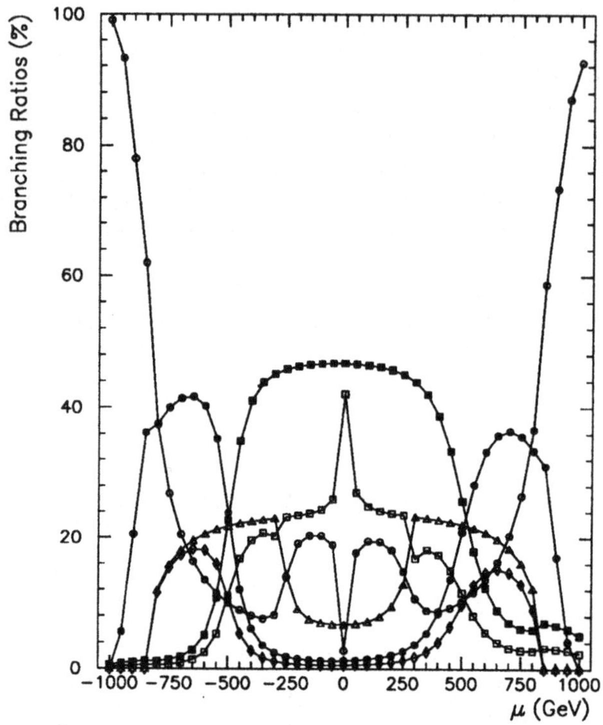

Fig. 18: Branching ratios for $\tilde{t}_R$ decays into charginos and neutralinos for $t_R \to b + \tilde{\chi}_i^+, t + \tilde{\chi}_i^0$, as a function of μ, for $M_{\tilde{q}} = 1000$ GeV, $M = 500$ GeV, $\tan\beta = 4$. We have taken $m_t = 150$ GeV, $\alpha_s = 0.1$, $\sin^2\theta_W = 0.23$. The curves correspond to the following transitions:

- ■ into the light chargino $\tilde{\chi}_1^\pm$
- ● into the heavy chargino $\tilde{\chi}_2^\pm$
- ○ into $\tilde{\chi}_1^0$
- □ into $\tilde{\chi}_2^0$
- △ into $\tilde{\chi}_3^0$
- ◇ into $\tilde{\chi}_4^0$.

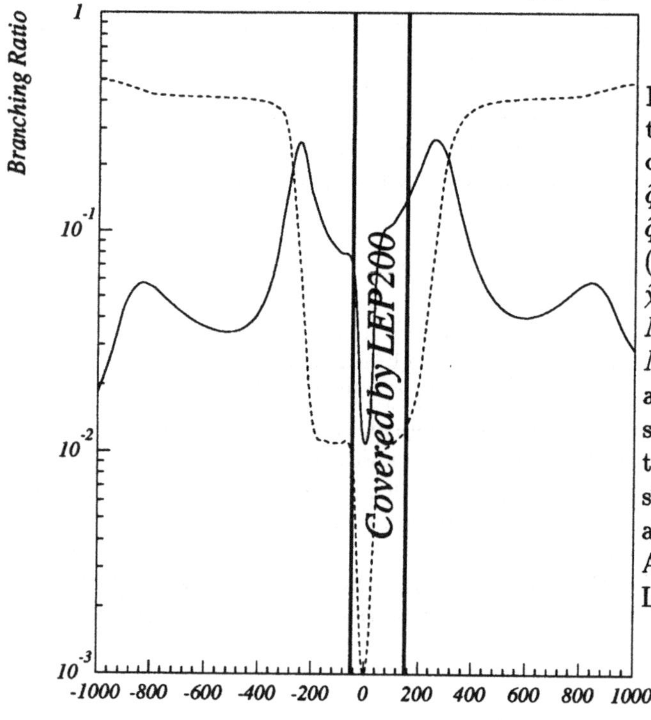

Fig. 19: The branching ratios for the two produced squarks to go directly into $\tilde{\chi}_1^0$ and $2\,\tilde{\chi}_1^0$, i.e. $\tilde{q} + \bar{\tilde{q}} \to \tilde{\chi}_1^0 + X$ (——) and $\tilde{q} + \bar{\tilde{q}} \to \tilde{\chi}_1^0 + \tilde{\chi}_1^0 + X$ (- - -), (where X does not contain another $\tilde{\chi}_1^0$) as a function of μ for $M_{\tilde{q}_L} = M_{\tilde{q}_R} = 1000$ GeV, $M = 500$ GeV, $\tan\beta = 4$. We have averaged over left and right squarks and all flavours. We have taken $m_t = 150$ GeV, $\alpha_s = 0.1$, $\sin^2\theta_W = 0.23$, $m_{h^0} = 80$ GeV and all other Higgs particles heavy. Also shown the region covered by LEP200.

particles will be an interesting subject. Such an *ep* collider will substantially extend the explorable mass region, especially for selectrons and sneutrinos [36, 37].

In Fig. 20 we show the cross section for $e^- p \to \tilde{e}_{L,R} + \tilde{q}_{L,R} + X$ as a function of the sfermion mass, summed over all possible pairs of selectrons and squarks, taking $m_{\tilde{e}} = m_{\tilde{q}}$, for $\mu = -200$ GeV and various values of M. These cross sections are larger than 0.1 pb for $m_{\tilde{f}} \lesssim 300$ GeV. More generally, the cross section is larger than 0.1 pb for $m_{\tilde{e}} + m_{\tilde{q}} \leq$ 600 GeV. Therefore, sfermions in this mass range would give between 50 and 1000 events per year (assuming an integrated luminosity of 500 pb^{-1}). The production cross section for sneutrinos is in general larger than that for selectrons because the chargino couplings are stronger than the neutralino couplings.

In Fig. 20 also the dependence on the SUSY paramter M can be seen. The cross section varies approximately within a factor of 2 for $m_{\tilde{e}} = m_{\tilde{q}} = 250$ GeV and in the parameter range 200 GeV $\leq M \leq$ 800 GeV. The dependence on μ and $\tan\beta$ is rather weak.

The individual production cross sections for different slepton and squark species are very different from each other. $\sigma(e^- p \to \tilde{\nu}_L + \tilde{d}_L + X)$ is in general the largest cross section due to the larger couplings of charginos to sleptons and squarks. In selectron production $ep \to \tilde{e}_L + \tilde{u}_L + X$ has the largest cross section.

To find suitable signatures of SUSY particle production in *ep* collisions a proper treatment of all the decays is necessary. Selectron, sneutrino and squark decays have already been discussed in detail in Sect. 3.2 and 4.2.

The signatures substantially depend on the supersymmetry parameters M and μ, or more precisely, whether cascade decays are possible or not. A good signature for selectron production is one isolated electron + missing energy on the lepton side. For $m_{\tilde{f}} = 250$ GeV we get a rate of ~ 0.1 pb. A signature for sneutrino production is a lepton pair e^+e^- (or μ^+e^-) + missing energy on the lepton side. If cascade decays of the sneutrino are possible the cross section for such events is ~ 0.01 pb. More details can be found in [38].

Monte Carlo studies taking into account the background have shown that detection of selectron and squark production in the mass range $m_{\tilde{e}} + m_{\tilde{q}} \leq 600$ GeV should be possible [39].

6 Summary

Here we have presented a systematic and rather complete study of supersymmetric particle production and decay within the Minimal Supersymmetric Standard Model in e^+e^-, *pp*, and *ep* collisions at high energies. The prospects of searching for supersymmetry in experiments to be performed at HERA, LEP200, LHC, SSC, LEP/LHC and at a 500 GeV e^+e^- collider are investigated in the corresponding supersymmetry parameter domain. Production and in particular the decays of supersymmetric particles depend in a characteristic way on the supersymmetry parameters. A general feature is that cascade decays become more important with increasing particle mass, leading to a complex decay pattern. Furthermore, this strongly influences the signatures for supersymmtry. Quite generally, the "classical" signature, large missing momentum, will be weakened. Instead of it, new signatures as $W^\pm$, Z^0, Higgs particles in the final state will appear.

Within the next years, experiments at FNAL will extend the mass range for gluinos and squarks beyond 200 GeV. At HERA sleptons and squarks are detectable if approximately $m_{\tilde{e},\tilde{\nu}} + m_{\tilde{q}} \leq 180$ GeV. Non-strongly interacting supersymmetric particles, in particular sleptons and charginos, will be observed at LEP200 if their mass is below 90 GeV.

A big step forward in the search for supersymmetry is possible by the *pp* colliders LHC and SSC. A mass range for gluinos and squarks up to the TeV region will be explored.

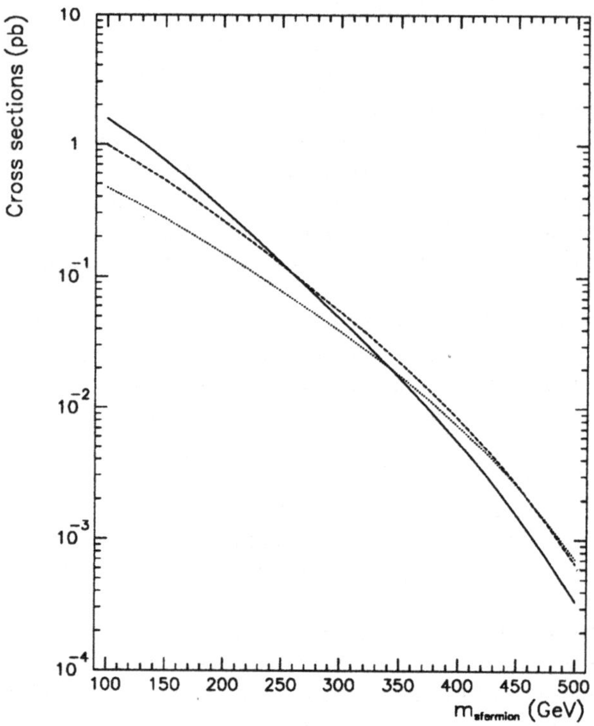

Fig. 20: Cross section for $e + p \to \tilde{e} + \tilde{q} + X$, summed over all left and right selectron and squark flavours, as a function of the sfermion mass taking $m_{\tilde{e}} = m_{\tilde{q}}$, for $\tan\beta = 4$, for $\mu = -200$ GeV and $M = 200$ GeV (——), 400 GeV (- - -), 800 GeV (· · · ·)..

Missing momentum as well as Z^0's and multileptons in the final state serve as good signatures. If LEP/LHC will also be operated in the *ep* collider mode, then it will be possible to cover a mass range for selectrons and sneutrinos according to $m_{\tilde{e},\tilde{\nu}} + m_{\tilde{q}} \leq 600$ GeV. A linear e^+e^- collider with c.m. energy of several hundred GeV up to 2 TeV would offer the unique possibility of exploring the properties of the non-strongly supersymmetric particles. We have studied in detail supersymmetric particle production at a 500 GeV e^+e^- collider. Such a collider would allow us to rule out the Minimal Supersymmetric Standard Model if no Higgs particle will be found. Moreover, it would extend the mass range of charged supersymmetric particles up to almost 250 GeV.

Concerning the explorable SUSY parameter range, a 500 GeV e^+e^- collider is roughly comparable to the LHC. Whereas at the LHC the strongly interacting SUSY particles up to a high mass can be detected, an e^+e^- collider is best suited for a detailed study of the non-strongly interacting SUSY particles. This would be necessary for fixing the SUSY model and its parameters.

A Appendix

A.1 Chargino and neutralino mixing

The chargino mass matrix as defined in [2, 4] is diagonalized by two unitary 2×2 matrices U and V,

$$\chi_i^+ = V_{ij}\psi_j^+, \qquad \chi_i^- = U_{ij}\psi_j^-, \qquad i = 1, 2 \tag{A.1}$$

where

$$\psi_j^+ = (-i\lambda^+, \psi_{H_2}^1), \qquad \psi_j^- = (-i\lambda^-, \psi_{H_1}^2) \tag{A.2}$$

with $\lambda^\pm$, and $\psi_{H_2}^1$, $\psi_{H_1}^2$ the two-component spinors of the $W^\pm$-ino, and the charged higgsinos $\tilde{H}_1^+$, $\tilde{H}_2^-$, respectively. The two-component spinors χ_i^+ and χ_i^- are combined to the four-component Dirac spinors of the charginos:

$$\tilde{\chi}_i^+ = \begin{pmatrix} \chi_i^+ \\ \bar{\chi}_i^- \end{pmatrix} \qquad i = 1,2. \tag{A.3}$$

Explicit formulae for the elements of U and V can be found in [15].

The 4×4 neutralino mass matrix as defined in [2, 4] is diagonalized by the 4×4 unitary matrix N,

$$\chi_i^0 = N_{ij}\psi_j^0 \qquad j = 1,4 \tag{A.4}$$

in the basis

$$\psi_j^0 = (-i\lambda_\gamma, -i\lambda_Z, \psi_{H_1}^1 \cos\beta - \psi_{H_2}^2 \sin\beta, \psi_{H_1}^1 \sin\beta + \psi_{H_2}^2 \cos\beta) \tag{A.5}$$

with λ_γ, λ_Z, $\psi_{H_1}^1$ and $\psi_{H_2}^2$ denoting the two-component spinor-fields of the photino, Z-ino, and the neutral higgsinos, respectively. The four-component Majorana spinors of the neutralinos $\tilde{\chi}_i^0$ are then constructed from the two-component spinors χ_i^0.

A.2 Chargino production

The cross section for $e^+e^- \to \tilde{\chi}_i^+\tilde{\chi}_j^-$ is given by [40]:

$$\frac{d\sigma}{dt} = \frac{d\sigma_\gamma}{dt} + \frac{d\sigma_Z}{dt} + \frac{d\sigma_{\tilde{\nu}}}{dt} + \frac{d\sigma_{\gamma Z}}{dt} + \frac{d\sigma_{\gamma\tilde{\nu}}}{dt} + \frac{d\sigma_{Z\tilde{\nu}}}{dt} \tag{A.6}$$

$$\begin{aligned}
\frac{d\sigma_\gamma}{dt} &= \frac{e^4}{8\pi s^4}\delta_{ij}\left((M_i^2-u)(M_j^2-u)+(M_i^2-t)(M_j^2-t)+2\,M_iM_js\right) && \text{(A.7)}\\
\frac{d\sigma_Z}{dt} &= \frac{g^4|D_Z(s)|^2}{32\pi s^2\cos^4\theta_W}\Big((L_e^2+R_e^2)((O'^L_{ij})^2+(O'^R_{ij})^2)\\
&\quad \left[(M_i^2-u)(M_j^2-u)+(M_i^2-t)(M_j^2-t)\right]+\\
&\quad 4\,(L_e^2+R_e^2)O'^L_{ij}O'^R_{ij}\eta_i\eta_jM_iM_js-(L_e^2-R_e^2)((O'^L_{ij})^2-(O'^R_{ij})^2)\\
&\quad \left[(M_i^2-u)(M_j^2-u)-(M_i^2-t)(M_j^2-t)\right]\Big) && \text{(A.8)}\\
\frac{d\sigma_{\tilde{\nu}}}{dt} &= \frac{g^4|D_{\tilde{\nu}}(t)|^2}{64\pi s^2}(V_{i1})^2(V_{j1})^2(M_i^2-t)(M_j^2-t) && \text{(A.9)}\\
\frac{d\sigma_{\gamma Z}}{dt} &= \frac{e^2g^2\,\mathrm{Re}(D_Z(s))}{16\pi s^3\cos^2\theta_W}\delta_{ij}\Big((L_e+R_e)(O'^L_{ij}+O'^R_{ij})\\
&\quad \left[(M_i^2-u)(M_j^2-u)+(M_i^2-t)(M_j^2-t)+2\,M_iM_js\right]- && \text{(A.10)}\\
&\quad (L_e-R_e)(O'^L_{ij}-O'^R_{ij})\left[(M_i^2-u)(M_j^2-u)-(M_i^2-t)(M_j^2-t)\right]\Big)
\end{aligned}$$

channel	$\tilde{q}_H\bar{\tilde{q}}_K$		
s	γ, Z		
coeff.	LL	LR	RR
A^{HK}_{γ}	$\frac{3\,e^4 e_q^2}{8\pi}$	–	$\frac{3\,e^4 e_q^2}{8\pi}$
A^{HK}_{Z}	$\frac{3\,g^4 L_q^2(L_e^2+R_e^2)}{16\,\pi\cos^4\theta_W}$	–	$\frac{3\,g^4 R_q^2(L_e^2+R_e^2)}{16\,\pi\cos^4\theta_W}$
$A^{HK}_{\gamma Z}$	$\frac{3\,e^2 e_q g^2 L_q(L_e+R_e)}{8\,\pi\cos^2\theta_W}$	–	$\frac{3\,e^2 e_q g^2 R_q(L_e+R_e)}{8\,\pi\cos^2\theta_W}$

Tab. 1: The coefficients as appearing in eqs. (A.18 – A.23) for the cross section $e^+e^- \to \tilde{q}_H\bar{\tilde{q}}_K$. Here $e = g\sin\theta_W$, $e > 0$, and e_q is the charge of the quark in units of e. The couplings are given in Appendix A.6. Notice that $A^{HK}_{\tilde{\chi}} = B^{HK}_{\tilde{\chi}} = A^{HK}_{\gamma\tilde{\chi}} = A^{HK}_{Z\tilde{\chi}} = 0$.

$$\frac{d\sigma_{\gamma\tilde{\nu}}}{dt} = \frac{e^2 g^2 D_{\tilde{\nu}}(t)(V_{i1})^2}{16\pi s^3}\delta_{ij}\left((M_i^2 - t)(M_j^2 - t) + M_i M_j s\right) \tag{A.11}$$

$$\frac{d\sigma_{Z\tilde{\nu}}}{dt} = \frac{g^4\,\mathrm{Re}(D_{\tilde{\nu}}(t)D_Z^*(s))}{16\pi s^2\cos^2\theta_W}L_e V_{i1}V_{j1} \left[O'^{L}_{ij}(M_i^2 - t)(M_j^2 - t) + O'^{R}_{ij}\eta_i\eta_j M_i M_j s\right] \tag{A.12}$$

where $D_Z = (s - m_Z^2 + im_Z\Gamma_Z)^{-1}$, $D_{\tilde{\nu}} = (t - m_{\tilde{\nu}}^2)^{-1}$. M_i is the (positive valued) mass of the chargino $\tilde{\chi}_i^{\pm}$, and $\eta_i = \pm 1$ is the sign of the corresponding mass eigenvalue. The couplings L_e, R_e, $O'^{L,R}_{ij}$ are given in Appendix A.6.

A.3 Neutralino production

The cross section for $e^+e^- \to \tilde{\chi}_i^0\tilde{\chi}_j^0$ is given by [40]:

$$\frac{d\sigma}{dt} = \frac{d\sigma_Z}{dt} + \frac{d\sigma_{\tilde{e}}}{dt} + \frac{d\sigma_{Z\tilde{e}}}{dt} \tag{A.13}$$

$$\frac{d\sigma_Z}{dt} = \frac{g^4}{16\pi s^2\cos^4\theta_W}|D_Z(s)|^2(O''^{L}_{ij})^2(L_e^2 + R_e^2) \left[(m_i^2 - u)(m_j^2 - u) + (m_i^2 - t)(m_j^2 - t) - 2\eta_i\eta_j m_i m_j s\right] \tag{A.14}$$

$$\frac{d\sigma_{\tilde{e}}}{dt} = \frac{g^4}{64\pi s^2}\Bigg\{\left(f^L_{ei}f^L_{ej}\right)^2\Big[|D_{\tilde{e}_L}(t)|^2(m_i^2 - t)(m_j^2 - t) +$$

$$|D_{\tilde{e}_L}(u)|^2(m_i^2 - u)(m_j^2 - u) - 2\,\mathrm{Re}(D_{\tilde{e}_L}(t)D^*_{\tilde{e}_L}(u))\eta_i\eta_j m_i m_j s\Big] +$$

$$\left(f^R_{ei}f^R_{ej}\right)^2\Big[|D_{\tilde{e}_R}(t)|^2(m_i^2 - t)(m_j^2 - t) +$$

$$|D_{\tilde{e}_R}(u)|^2(m_i^2-u)(m_j^2-u) - 2\,\mathrm{Re}(D_{\tilde{e}_R}(t)D^*_{\tilde{e}_R}(u))\eta_i\eta_j m_i m_j s\Big]\Big\} \quad \text{(A.15)}$$

$$\begin{aligned}\frac{d\sigma_{Z\tilde{e}}}{dt} &= \frac{g^4}{16\pi s^2\cos^2\theta_W}\mathrm{Re}(D_Z(s))O''^{L}_{ij} \\ &\Big\{L_e f^L_{ei} f^L_{ej}\Big[D_{\tilde{e}_L}(t)\left((m_i^2-t)(m_j^2-t)-\eta_i\eta_j m_i m_j s\right) + \\ &D_{\tilde{e}_L}(u)\left((m_i^2-u)(m_j^2-u)-\eta_i\eta_j m_i m_j s\right)\Big] - \\ &R_e f^R_{ei} f^R_{ej}\Big[D_{\tilde{e}_R}(t)\left((m_i^2-t)(m_j^2-t)-\eta_i\eta_j m_i m_j s\right) + \\ &D_{\tilde{e}_R}(u)\left((m_i^2-u)(m_j^2-u)-\eta_i\eta_j m_i m_j s\right)\Big]\Big\}, \end{aligned} \quad \text{(A.16)}$$

where $D_Z = (s - m_Z^2 + i m_Z \Gamma_Z)^{-1}$, $D_{\tilde{e}_{L,R}}(x) = (x - m^2_{\tilde{e}_{L,R}})^{-1}$. m_i is the (positive valued) mass of the neutralino $\tilde{\chi}^0_i$, and $\eta_i = \pm 1$ is the sign of the corresponding mass eigenvalue. The couplings $O''^{L,R}_{ij}$, $f^{L,R}_{ei}$ are given in Appendix A.6.

A.4 Sfermionproduction

The cross section for $e^+e^- \to \tilde{f}_H \bar{\tilde{f}}_K$, where $\tilde{f}$ is a slepton or a squark (H, $K = L$ or R), is given by [40]:

$$\frac{d\sigma}{dt} = \frac{d\sigma_\gamma}{dt} + \frac{d\sigma_Z}{dt} + \frac{d\sigma_{\tilde{\chi}}}{dt} + \frac{d\sigma_{\gamma Z}}{dt} + \frac{d\sigma_{\gamma\tilde{\chi}}}{dt} + \frac{d\sigma_{Z\tilde{\chi}}}{dt}, \quad \text{(A.17)}$$

$$\frac{d\sigma_\gamma}{dt} = A^{HK}_\gamma \frac{\left(ut - M^2_{\tilde{f}_H} M^2_{\tilde{f}_K}\right)}{s^4} \quad \text{(A.18)}$$

$$\frac{d\sigma_Z}{dt} = A^{HK}_Z \frac{\left(ut - M^2_{\tilde{f}_H} M^2_{\tilde{f}_K}\right)}{s^2}|D_Z(s)|^2 \quad \text{(A.19)}$$

$$\frac{d\sigma_{\tilde{\chi}}}{dt} = \frac{\left(ut - M^2_{\tilde{f}_H} M^2_{\tilde{f}_K}\right)}{s^2}\left(\sum_i A^{HK}_{\tilde{\chi}} D_i(t)\right)^2 + \sum_{i,j}\frac{\eta_i\eta_j m_i m_j B^{HK}_{\tilde{\chi}} D_i(t) D_j(t)}{s} \quad \text{(A.20)}$$

$$\frac{d\sigma_{\gamma Z}}{dt} = A^{HK}_{\gamma Z} \frac{\left(ut - M^2_{\tilde{f}_H} M^2_{\tilde{f}_K}\right)}{s^3}\mathrm{Re}(D_Z(s)) \quad \text{(A.21)}$$

$$\frac{d\sigma_{\gamma\tilde{\chi}}}{dt} = \frac{\left(ut - M^2_{\tilde{f}_H} M^2_{\tilde{f}_K}\right)}{s^3}\sum_i A^{HK}_{\gamma\tilde{\chi}} D_i(t) \quad \text{(A.22)}$$

$$\frac{d\sigma_{Z\tilde{\chi}}}{dt} = \frac{\left(ut - M^2_{\tilde{f}_H} M^2_{\tilde{f}_K}\right)}{s^2}\mathrm{Re}(D_Z(s))\sum_i A^{HK}_{Z\tilde{\chi}} D_i(t), \quad \text{(A.23)}$$

where $D_Z = (s - m_Z^2 + i m_Z\Gamma_Z)^{-1}$, $D_i(t) = (t - m_i^2)^{-1}$. m_i is the (positive valued) mass of the exchanged neutralino $\tilde{\chi}^0_i$ or chargino $\tilde{\chi}^\pm_i$, and $\eta_i = \pm 1$ is the sign of the corresponding

channel	$\tilde{e}^+_H\tilde{e}^-_K$			$\tilde{\nu}_e\bar{\tilde{\nu}}_e$
s	γ, Z			Z
t	$\tilde{\chi}^0_i$			$\tilde{\chi}^\pm_i$
coeff.	LL	LR	RR	LL
A^{HK}_γ	$\frac{e^4}{8\pi}$	–	$\frac{e^4}{8\pi}$	–
A^{HK}_Z	$\frac{g^4 L_e^2(L_e^2+R_e^2)}{16\pi\cos^4\theta_W}$	–	$\frac{g^4 R_e^2(L_e^2+R_e^2)}{16\pi\cos^4\theta_W}$	$\frac{g^4(L_e^2+R_e^2)}{64\pi\cos^4\theta_W}$
$A^{HK}_{\tilde{\chi}}$	$\frac{g^2\left\|f^L_{ei}\right\|^2}{8\sqrt{\pi}}$	–	$\frac{g^2\left\|f^R_{ei}\right\|^2}{8\sqrt{\pi}}$	$\frac{g^2(V_{i1})^2}{8\sqrt{\pi}}$
$B^{HK}_{\tilde{\chi}}$	–	$\frac{g^4}{64\pi} f^R_{ei} f^L_{ei} f^R_{ej} f^L_{ej}$	–	–
$A^{HK}_{\gamma Z}$	$\frac{e^2 g^2 L_e(L_e+R_e)}{8\pi\cos^2\theta_W}$	–	$\frac{e^2 g^2 R_e(L_e+R_e)}{8\pi\cos^2\theta_W}$	–
$A^{HK}_{\gamma\tilde{\chi}}$	$\frac{e^2 g^2\left\|f^L_{ei}\right\|^2}{16\pi}$	–	$\frac{e^2 g^2\left\|f^R_{ei}\right\|^2}{16\pi}$	–
$A^{HK}_{Z\tilde{\chi}}$	$\frac{g^4 L_e^2\left\|f^L_{ei}\right\|^2}{16\pi\cos^2\theta_W}$	–	$\frac{g^4 R_e^2\left\|f^R_{ei}\right\|^2}{16\pi\cos^2\theta_W}$	$\frac{g^4 L_e(V_{i1})^2}{32\pi\cos^2\theta_W}$

channel	$\tilde{l}^+_H\tilde{l}^-_K$, $\tilde{l}=\tilde{\mu},\tilde{\tau}$			$\tilde{\nu}_l\bar{\tilde{\nu}}_l$
s	γ, Z			Z
t	–			–
coeff.	LL	LR	RR	LL
A^{HK}_γ	$\frac{e^4}{8\pi}$	–	$\frac{e^4}{8\pi}$	–
A^{HK}_Z	$\frac{g^4 L_l^2(L_e^2+R_e^2)}{16\pi\cos^4\theta_W}$	–	$\frac{g^4 R_l^2(L_e^2+R_e^2)}{16\pi\cos^4\theta_W}$	$\frac{g^4(L_e^2+R_e^2)}{64\pi\cos^4\theta_W}$
$A^{HK}_{\tilde{\chi}}$	–	–	–	–
$B^{HK}_{\tilde{\chi}}$	–	–	–	–
$A^{HK}_{\gamma Z}$	$\frac{e^2 g^2 L_l(L_e+R_e)}{8\pi\cos^2\theta_W}$	–	$\frac{e^2 g^2 R_l(L_e+R_e)}{8\pi\cos^2\theta_W}$	–
$A^{HK}_{\gamma\tilde{\chi}}$	–	–	–	–
$A^{HK}_{Z\tilde{\chi}}$	–	–	–	–

Tab. 2: The coefficients as appearing in eqs. (A.18 – A.23) for the cross section $e^+e^- \to \tilde{f}_H\bar{\tilde{f}}_K$, $\tilde{f} = \tilde{e}, \tilde{\mu}, \tilde{\tau}, \tilde{\nu}_e, \tilde{\nu}_\mu, \tilde{\nu}_\tau$. The couplings are given in Appendix A.6.

$\tilde{\chi}^+_i \to$	$\tilde{\chi}^0_k u\bar{d}$	$\tilde{\chi}^+_k u\bar{u}$	$\tilde{\chi}^+_k d\bar{d}$
A_s	$6\,(O^L_{ki})^2$	$12\,\frac{(L_u O'^L_{ki})^2+(R_u O'^R_{ki})^2}{\cos^4\theta_W}$	$12\,\frac{(L_d O'^L_{ki})^2+(R_d O'^R_{ki})^2}{\cos^4\theta_W}$
B_s	$6\,(O^R_{ki})^2$	$12\,\frac{(L_u O'^R_{ki})^2+(R_u O'^L_{ki})^2}{\cos^4\theta_W}$	$12\,\frac{(L_d O'^R_{ki})^2+(R_d O'^L_{ki})^2}{\cos^4\theta_W}$
C_s	$-6\,O^L_{ki}O^R_{ki}$	$-12\,\frac{(L_u^2+R_u^2)O'^L_{ki}O'^R_{ki}}{\cos^4\theta_W}$	$-12\,\frac{(L_d^2+R_d^2)O'^L_{ki}O'^R_{ki}}{\cos^4\theta_W}$
A^L_t	$3\,(f^L_{uk}V_{i1})^2$	0	$3\,(V_{k1}V_{i1})^2$
A^L_u	$3\,(f^L_{dk}U_{i1})^2$	$3\,(U_{k1}U_{i1})^2$	0
A^L_{tu}	$3\,f^L_{uk}f^L_{dk}V_{i1}U_{i1}$	0	0
A^L_{st}	$-3\sqrt{2}f^L_{uk}V_{i1}O^L_{ki}$	0	$6\,\frac{V_{k1}V_{i1}O'^L_{ki}L_d}{\cos^2\theta_W}$
B^L_{st}	$3\sqrt{2}f^L_{uk}V_{i1}O^R_{ki}$	0	$-6\,\frac{V_{k1}V_{i1}O'^R_{ki}L_d}{\cos^2\theta_W}$
A^L_{su}	$3\sqrt{2}f^L_{dk}U_{i1}O^R_{ki}$	$-6\,\frac{U_{k1}U_{i1}O'^R_{ki}L_u}{\cos^2\theta_W}$	0
B^L_{su}	$-3\sqrt{2}f^L_{dk}U_{i1}O^L_{ki}$	$6\,\frac{U_{k1}U_{i1}O'^L_{ki}L_u}{\cos^2\theta_W}$	0

Tab. 3: Coefficients in eqs. (A.25 – A.30) for hadronic chargino decays. The couplings are given in the Appendix A.6. Notice that $A^R_t = A^R_u = A^R_{ut} = A^R_{st} = A^R_{su} = B^R_{st} = B^R_{su} = 0$.

mass eigenvalue. $M_{\tilde{f}_H}$ is the mass of the produced sfermion $\tilde{f}_H$. Here $m_f = 0$ has been taken for the corresponding fermion f. The coefficients $A^{H,K}_\gamma$, $A^{H,K}_Z$ etc. are given in Tab. 1 and 2.

A.5 Three-body decay

The width for the decay of a chargino or neutralino into a lighter chargino or neutralino and a fermion pair is given by [34]:

$$\Gamma(\tilde{\chi}_i \to \tilde{\chi}_k + f\bar{f}') = \frac{\alpha^2}{32\pi \sin^4\theta_W m_i^3} \int d\bar{s}\, d\bar{t}\, (W_s + W_t + W_u + W_{tu} + W_{st} + W_{su}) \quad \text{(A.24)}$$

$$\begin{aligned} W_s &= \frac{1}{(\bar{s}-M_s^2)^2}\Big(A_s(m_i^2-\bar{t})(\bar{t}-m_k^2) + \\ &\quad B_s(m_i^2-\bar{u})(\bar{u}-m_k^2) + 2\,C_s\eta_i\eta_k m_i m_k \bar{s}\Big) && \text{(A.25)} \\ W_t &= A^L_t\,\frac{(m_i^2-\bar{t})(\bar{t}-m_k^2)}{(\bar{t}-M_L^2)^2} + A^R_t\,\frac{(m_i^2-\bar{t})(\bar{t}-m_k^2)}{(\bar{t}-M_R^2)^2} && \text{(A.26)} \\ W_u &= A^L_u\,\frac{(m_i^2-\bar{u})(\bar{u}-m_k^2)}{(\bar{u}-M_L^2)^2} + A^R_u\,\frac{(m_i^2-\bar{u})(\bar{u}-m_k^2)}{(\bar{u}-M_R^2)^2} && \text{(A.27)} \end{aligned}$$

$\tilde{\chi}_i^+ \to$	$\tilde{\chi}_k^0 l^+ \nu$	$\tilde{\chi}_k^+ l^+ l^-$	$\tilde{\chi}_k^+ \nu\bar{\nu}$
A_s	$2\,(O^L_{ki})^2$	$4\,\frac{(L_l O'^L_{ki})^2+(R_l O'^R_{ki})^2}{\cos^4\theta_W}$	$\frac{(O'^L_{ki})^2}{\cos^4\theta_W}$
B_s	$2\,(O^R_{ki})^2$	$4\,\frac{(L_l O'^R_{ki})^2+(R_l O'^L_{ki})^2}{\cos^4\theta_W}$	$\frac{(O'^R_{ki})^2}{\cos^4\theta_W}$
C_s	$-2\,O^L_{ki}O^R_{ki}$	$-4\,\frac{(L_l^2+R_l^2)O'^L_{ki}O'^R_{ki}}{\cos^4\theta_W}$	$-\frac{O'^L_{ki}O'^R_{ki}}{\cos^4\theta_W}$
A_t^L	$(f^L_{\nu k}V_{i1})^2$	$(V_{k1}V_{i1})^2$	0
A_u^L	$(f^L_{lk}U_{i1})^2$	0	$(U_{k1}U_{i1})^2$
A_{tu}^L	$f^L_{\nu k}f^L_{dk}V_{i1}U_{i1}$	0	0
A_{st}^L	$-\sqrt{2}f^L_{\nu k}V_{i1}O^L_{ki}$	$2\,\frac{V_{k1}V_{i1}O'^L_{ki}L_l}{\cos^2\theta_W}$	0
B_{st}^L	$\sqrt{2}f^L_{\nu k}V_{i1}O^R_{ki}$	$-2\,\frac{V_{k1}V_{i1}O'^R_{ki}L_l}{\cos^2\theta_W}$	0
A_{su}^L	$\sqrt{2}f^L_{lk}U_{i1}O^R_{ki}$	0	$-\frac{U_{k1}U_{i1}O'^R_{ki}}{\cos^2\theta_W}$
B_{su}^L	$-\sqrt{2}f^L_{lk}U_{i1}O^L_{ki}$	0	$\frac{U_{k1}U_{i1}O'^L_{ki}}{\cos^2\theta_W}$

Tab. 4: Coefficients in eqs. (A.25 – A.30) for leptonic chargino decays. The couplings are given in the Appendix A.6. Notice that $A_t^R = A_u^R = A_{ut}^R = A_{st}^R = A_{su}^R = B_{st}^R = B_{su}^R = 0$.

$$W_{tu} = A^L_{tu}\frac{2\eta_i\eta_k m_i m_k\bar{s}}{(\bar{t}-M_L^2)(\bar{u}-M_L^2)} + A^R_{tu}\frac{2\eta_i\eta_k m_i m_k\bar{s}}{(\bar{t}-M_R^2)(\bar{u}-M_R^2)} \tag{A.28}$$

$$W_{st} = \frac{2}{(\bar{t}-M_L^2)(\bar{s}-M_s^2)}\left(A^L_{st}(m_i^2-\bar{t})(\bar{t}-m_k^2) + B^L_{st}\eta_i\eta_k m_i m_k\bar{s}\right) + (L \leftrightarrow R) \tag{A.29}$$

$$W_{su} = \frac{2}{(\bar{u}-M_L^2)(\bar{s}-M_s^2)}\left(A^L_{su}(m_i^2-\bar{u})(\bar{u}-m_k^2) + B^L_{su}\eta_i\eta_k m_i m_k\bar{s}\right) + (L \leftrightarrow R). \tag{A.30}$$

m_i is the (positive valued) mass of the chargino or neutralino $\tilde{\chi}_i$, and $\eta_i = \pm 1$ is the sign of the corresponding mass eigenvalue. M_s is the mass of the $W^\pm$, Z, exchanged in the $\bar{s}$ channel. M_L, M_R denote the masses of the left and right sfermion exchanged in the $\bar{t}$- and $\bar{u}$ channel. The range of integration for $\bar{s}$ and $\bar{t}$ is given by the bounds, $\bar{s}_{min} = 0$, $\bar{s}_{max} = (m_i - m_k)^2$, $\bar{t}_{min,max} = 1/2(m_i^2 + m_k^2 - \bar{s} \mp \sqrt{\lambda(m_i^2, m_k^2, \bar{s})})$, $\lambda(a,b,c) = a^2 + b^2 + c^2 - 2ab - 2bc - 2ca$. The coefficients A_s, B_s, etc. are given in Tab. 3 – 6.

$\tilde{\chi}_i^0 \to$	$\tilde{\chi}_k^0 l^+ l^-$	$\tilde{\chi}_k^0 \nu \bar{\nu}$	$\tilde{\chi}_k^0 q \bar{q}$
$A_s = B_s = C_s$	$4\,\frac{(O''^L_{ki})^2 (L_l^2 + R_l^2)}{\cos^4\theta_W}$	$\frac{(O''^L_{ki})^2}{\cos^4\theta_W}$	$12\,\frac{(O''^L_{ki})^2 (L_q^2 + R_q^2)}{\cos^4\theta_W}$
$A_t^L = A_u^L = A_{tu}^L$	$(f^L_{lk} f^L_{li})^2$	$(f^L_{\nu k} f^L_{\nu i})^2$	$3\,(f^L_{qk} f^L_{qi})^2$
$A_t^R = A_u^R = A_{tu}^R$	$(f^R_{lk} f^R_{li})^2$	0	$3\,(f^R_{qk} f^R_{qi})^2$
$A_{st}^L = B_{st}^L = A_{su}^L = B_{su}^L$	$2\,\frac{f^L_{lk} f^L_{li} O''^L_{ki} L_l}{\cos^2\theta_W}$	$\frac{f^L_{\nu k} f^L_{\nu i} O''^L_{ki}}{\cos^2\theta_W}$	$6\,\frac{f^L_{qk} f^L_{qi} O''^L_{ki} L_q}{\cos^2\theta_W}$
$A_{st}^R = B_{st}^R = A_{su}^R = B_{su}^R$	$2\,\frac{f^R_{lk} f^R_{li} O''^R_{ki} R_l}{\cos^2\theta_W}$	0	$6\,\frac{f^R_{qk} f^R_{qi} O''^R_{ki} R_q}{\cos^2\theta_W}$

Tab. 5: Coefficients in eqs. (A.25 – A.30) for $\tilde{\chi}_i^0 \to \tilde{\chi}_k^0 f \bar{f}$. The couplings are given in the Appendix A.6.

$\tilde{\chi}_i^0 \to$	$\tilde{\chi}_k^+ l^- \bar{\nu}$	$\tilde{\chi}_k^+ \bar{u} d$
A_s	$2\,(O^L_{ik})^2$	$6\,(O^L_{ik})^2$
B_s	$2\,(O^R_{ik})^2$	$6\,(O^R_{ik})^2$
C_s	$-2\,O^L_{ik} O^R_{ik}$	$-6\,O^L_{ik} O^R_{ik}$
A_t^L	$(f^L_{\nu i} V_{k1})^2$	$3\,(f^L_{ui} V_{k1})^2$
A_u^L	$(f^L_{li} U_{k1})^2$	$3\,(f^L_{di} U_{k1})^2$
A_{tu}^L	$f^L_{\nu i} f^L_{li} V_{k1} U_{k1}$	$3\,f^L_{ui} f^L_{di} V_{k1} U_{k1}$
A_{st}^L	$-\sqrt{2} f^L_{\nu i} V_{k1} O^L_{ik}$	$-3\sqrt{2} f^L_{ui} V_{k1} O^L_{ik}$
B_{st}^L	$\sqrt{2} f^L_{\nu i} V_{k1} O^R_{ik}$	$3\sqrt{2} f^L_{ui} V_{k1} O^R_{ik}$
A_{su}^L	$\sqrt{2} f^L_{li} U_{k1} O^R_{ik}$	$3\sqrt{2} f^L_{di} U_{k1} O^R_{ik}$
B_{su}^L	$-\sqrt{2} f^L_{li} U_{k1} O^L_{ik}$	$-3\sqrt{2} f^L_{di} U_{k1} O^L_{ik}$

Tab. 6: Coefficients in eqs. (A.25 – A.30) for $\tilde{\chi}_i^0 \to \tilde{\chi}_k^+ f \bar{f}'$. The couplings are given in the Appendix A.6. Notice that $A_t^R = A_u^R = A_{ut}^R = A_{st}^R = A_{su}^R = B_{st}^R = B_{su}^R = 0$.

A.6 Couplings

$$
\begin{aligned}
f_{fk}^{L} &= -\sqrt{2}\left[\frac{1}{\cos\theta_W}\left(T_{3f}-e_f\sin^2\theta_W\right)N_{k2}+e_f\sin\theta_W N_{k1}\right] && \text{(A.31)}\\
f_{fk}^{R} &= -\sqrt{2}e_f\sin\theta_W\left(\tan\theta_W N_{k2}-N_{k1}\right) && \text{(A.32)}\\
L_f &= T_{3f}-e_f\sin^2\theta_W && \text{(A.33)}\\
R_f &= -e_f\sin^2\theta_W && \text{(A.34)}\\
O_{ij}^{L} &= -\left(N_{i4}\cos\beta-N_{i3}\sin\beta\right)\frac{V_{j2}}{\sqrt{2}}+\left(N_{i1}\sin\theta_W+N_{i2}\cos\theta_W\right)V_{j1} && \text{(A.35)}\\
O_{ij}^{R} &= \left(N_{i4}\sin\beta+N_{i3}\cos\beta\right)\frac{U_{j2}}{\sqrt{2}}+\left(N_{i1}\sin\theta_W+N_{i2}\cos\theta_W\right)U_{j1} && \text{(A.36)}\\
O_{ij}^{\prime L} &= -V_{i1}V_{j1}-\frac{1}{2}V_{i2}V_{j2}+\delta_{ij}\sin^2\theta_W && \text{(A.37)}\\
O_{ij}^{\prime R} &= -U_{i1}U_{j1}-\frac{1}{2}U_{i2}U_{j2}+\delta_{ij}\sin^2\theta_W && \text{(A.38)}\\
O_{ij}^{\prime\prime L} &= -\frac{1}{2}\left(N_{i3}\cos\beta+N_{i4}\sin\beta\right)\left(N_{j3}\cos\beta+N_{j4}\sin\beta\right)\\
&\quad+\frac{1}{2}\left(-N_{i3}\sin\beta+N_{i4}\cos\beta\right)\left(-N_{j3}\sin\beta+N_{j4}\cos\beta\right) && \text{(A.39)}\\
O_{ij}^{\prime\prime R} &= -O_{ij}^{\prime\prime L}. && \text{(A.40)}
\end{aligned}
$$

Here T_{3f} and e_f are the third component of weak isospin and the charge (in units of e) of the fermion f. We have used a convention where the matrices U, V, and N are real.

References

[1] For a general introduction to supersymmetry, see J. Wess, J. Bagger, Supersymmetry and Supergravity, Princeton University Press, 1983

[2] H.-P. Nilles, these Lecture Notes

[3] H.P. Nilles, Phys. Rep. 110 (1984) 1; B. Lahanas, D.V. Nanopoulos, Phys. Rep. 145 (1985) 1; R. Barbieri, Riv. Nuov. Cim. 11 (1988) 1

[4] H.E. Haber, G.L. Kane, Phys. Rep. 117 (1985)

[5] L.J. Hall, J. Polchinski, Phys. Lett. 152B (1985) 335

[6] J.F. Gunion, H.E. Haber, Nucl. Phys. B272 (1986) 1; J.F. Gunion, H.E. Haber, G.L. Kane, S. Dawson, "The Higgs Hunter's Guide", Addison-Wesley, 1990

[7] G. Ridolfi, Joint Int. Lepton-Photon Symposium and Europhysics Conference on High Energy Physics, Geneva, 1991, and references therein, to appear

[8] J. Kalinowski, these Lecture Notes, and references therein;
S. Pokorski, these Lecture Notes

[9] M. Davier, Proceedings of the Joint Int. Lepton-Photon Symp. and Europhysics Conf. on High Energy Physics, Geneva, 1991

[10] H. Baer, X. Tata, J. Woodside, Phys. Rev. D44 (1991) 207

[11] H. Baer et al., FSU-HE-910308, BU/TH/91-1, UCD-91-4, UH-511-922-91

[12] H. Baer, A. Bartl, D. Karatas, W. Majerotto, X. Tata, Int. J. of Mod. Phys. A4 (1989) 4111

[13] J.F. Gunion, H.E. Haber, Phys. Rev. D37 (1988) 2515

[14] M. Chen, C. Dionisi, M. Martinez, X. Tata, Phys. Rep. 159 (1988) 201; C. Dionisi et al., Proc. of the ECFA Workshop on LEP 200, Aachen, 1986, Vol. II, p. 380, CERN 87 - 08, ECFA 878/108

[15] A. Bartl, H. Fraas, W. Majerotto, B. Mößlacher, preprint UWThPh-1991-44, HEPHY-PUB-561-91

[16] Proceedings of the Workshop on Physics and Experiments with Linear Colliders, Lapland, Finland, 1991, to appear

[17] Proceedings of the Workshop on e^+e^- Collisions at 500 GeV: The Physics Potential, DESY, Hamburg, 1991, to appear

[18] A. Brignole et. al., in [17]; P. Janot, in [17]

[19] A. Bartl, W. Majerotto, B. Mößlacher, in [17]

[20] J.F. Grivaz, in [17]

[21] A. Bartl, H. Fraas, W. Majerotto, N. Oshimo, Phys. Rev. D40 (1989) 1594

[22] C. Vander Velde, in [17]

[23] R. Becker, R. Starosta, in [17]

[24] V. Angelopoulos et al., Proc. of the Workshop on Physics at Future Accelerators, La Thuile, 1987, Vol. I, p. 80, CERN 87–07; A. Savoy-Navarro, N. Zaganidis, ibidem, Vol. II, p. 82; R. Batley, ibidem, Vol. II, p.109; H. Baer, X. Tata and J. Woodside, Proc. of the Summer Study on High Energy Physics in the 1990's, Snowmass, 1988, p. 220; R.M. Barnett et al., ibidem, p.226

[25] F. Pauss, Proc. of the ECFA Large Hadron Collider Workshop, Aachen, 1990, Vol. I, p. 118, CERN90–10, ECFA90–133

[26] S. Dawson, E. Eichten and C. Quigg, Phys. Rev. D31 (1985) 1581 and references therein

[27] Report of the Supersymmetry Working Group (Conveners: G.G. Ross and F. Zwirner), Proc. of the ECFA Large Hadron Collider Workshop, Aachen, 1990, Vol. II, CERN90–10, ECFA90–133

[28] A. Bartl, W. Majerotto, B. Mößlacher, N. Oshimo, S. Stippel, Phys. Rev. D43 (1991) 2214

[29] H. Baer, V. Barger, D. Karatas and X. Tata, Phys. Rev. D36 (1987) 96; R.M. Barnett, J.F. Gunion and H.E. Haber, Phys. Rev. Lett. 60 (1988) 401; Phys. Rev. D37 (1988) 1892

[30] H. Baer, X. Tata and J. Woodside, Phys. Rev. D42 (1990) 1568

[31] A. Bartl, W. Majerotto, B. Mößlacher, N. Oshimo, Z. Phys. C52 (1991) 477

[32] S.K. Jones and C.H. Llewellyn Smith, Nucl. Phys. B217 (1983) 145; P.R. Harrison, Nucl. Phys. B249 (1985) 704; J.A. Bagger and M.E. Peskin, Phys. Rev. D31 (1985) 2211; J. Bartels and W. Hollik, Z. Phys. C39 (1988) 433; T. Kon, K. Nakamura, T. Kobayashi, Z. Phys. C45 (1990) 567

[33] H. Komatsu and R. Rückl, Nucl. Phys. B299 (1988) 407; A. Bartl, H. Fraas and W. Majerotto, Nulc. Phys. B297 (1988) 479

[34] A. Bartl, H. Fraas and W. Majerotto, Z. Phys. C41 (1988) 475

[35] Proceedings of the HERA Workshop, DESY, Hamburg, 1991, to appear

[36] R. Rückl, Proc. of the ECFA Large Hadron Collider Workshop, Aachen, 1990, Vol. I, p. 229, CERN 90–10, ECFA 90–133

[37] A. Bartl, W. Majerotto, B. Mößlacher, N. Oshimo and S. Stippel, Proc. of the ECFA Large Hadron Collider Workshop, Aachen, 1990, Vol. II, p. 1033, CERN 90–10, ECFA 90–133

[38] A. Bartl, W. Majerotto, B. Mößlacher, N. Oshimo Z. Phys. C52 (1991) 677

[39] M. Besançon, Proc. of the ECFA Large Hadron Collider Workshop, Aachen, 1990, Vol. II, p. 1040, CERN 90–10, ECFA 90–133

[40] A. Bartl, H. Fraas, W. Majerotto, Z. Phys. C30 (1986) 441; Nucl. Phys. B278 (1986) 1; Z. Phys. C34 (1987) 411

Supersymmetry Searches Using High Energy Photon Beams

Frank Cuypers[a], Geert Jan van Oldenborgh[a] and Reinhold Rückl[a,b]

[a] *Sektion Physik, Universität München, Theresienstraße 37, D-8000 München 2, Germany*
[b] *Max-Planck-Institut für Physik, Werner-Heisenberg-Institut, Föhringer Ring 6, D-8000 München 40, Germany*

1 Introduction

Bartl, Majerotto and Mößlacher [1] have presented a comprehensive study of supersymmetry searches in pp, e^+e^- and ep collisions in the range of LEP and Tevatron energies to LHC and SSC energies. For completeness, we briefly discuss the prospects of supersymmetry searches with 200-1000 GeV photon beams. Such beams are expected to be obtainable at linear $e^\pm e^-$ colliders through back-scattering of high intensity laser rays on high energy electron beams [2]. The energy spectrum of the resulting beam of real photons is very hard, and the conversion efficiency can reach almost 100%. Thus, in addition to $e^\pm e^-$ collisions, linear colliders may provide $e\gamma$ and $\gamma\gamma$ collisions with energies and luminosities similar to those obtained with $e^\pm e^-$ collisions. In the following, we concentrate on a few particularly promising reactions involving the production of selectrons and their decay into a single electron accompanied only by invisible particles. More specifically, we compare the event rates which can be expected in e^+e^-, e^-e^-, $e^-\gamma$ and $\gamma\gamma$ collisions at linear colliders with $e^\pm e^-$ centre of mass energies $\sqrt{s_{ee}} = 500 - 2000$ GeV and luminosities of the order of $\mathcal{L} = 10^{33}$ cm^{-2}s^{-1}. Furthermore, we consider the same range of supersymmetry parameters as in Ref. 1.

2 Photon Energy Spectrum

The production of energetic photon beams by back-scattering laser rays on high energy electron beams from linear colliders was proposed and is described in Ref. 2. The distribution $P(y)$ of the energy fraction y of an electron transferred to a photon, $y = E_\gamma/E_e$, is given by

$$P(y) = \frac{1}{N}\left(1 - y + \frac{1}{1-y} - \frac{4y}{x(1-y)} + \frac{4y^2}{x^2(1-y)^2}\right) , \qquad (1)$$

where

$$0 \le y \le \frac{x}{x+1} \qquad (2)$$

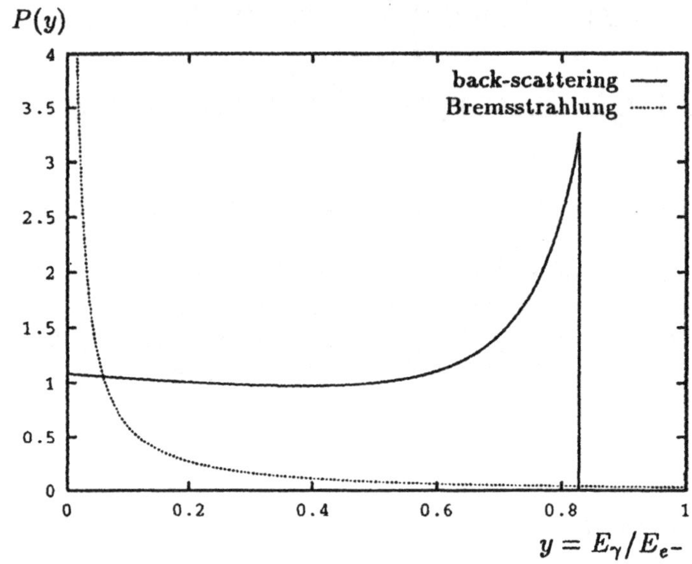

Figure 1: Distribution of photon energies for laser Compton back-scattering ($x = 2+2\sqrt{2}$) and Bremsstrahlung in the Weizsäcker-Williams approximation for a 250 GeV electron beam.

and

$$x = \frac{4E_e E_{\text{laser}}}{m_e^2} \; . \tag{3}$$

The factor N normalizes $\int dy\, P(y)$ to 1. The electron and laser beams are taken to be aligned and their respective energies are E_e and E_{laser}. In what follows, we assume a 100% conversion efficiency and neglect the angular dispersion of the back-scattered photons. When x reaches the value $2(1+\sqrt{2}) \approx 4.83$, the back-scattered and laser photons have enough relative energy to produce e^+e^- pairs. Since the conversion efficiency drops considerably for larger values of x, we assume the laser energy to be tuned in such a way as to obtain $x = 2(1+\sqrt{2})$. The corresponding energy spectrum (1) is displayed in Fig. 1. It sharply contrasts with the energy spectrum of Bremsstrahlung photons, which is much softer and, hence, less useful for heavy particle searches.

The cross sections of $e\gamma$ and $\gamma\gamma$ reactions involving such a back-scattered photon beam have to be folded with the energy distribution (1). The laboratory frame is thus not the centre of mass frame and the $e\gamma$ and $\gamma\gamma$ centre of mass energies, $\sqrt{s_{e\gamma}}$ and $\sqrt{s_{\gamma\gamma}}$ respectively, are given by

$$s_{e\gamma} = y s_{ee} \tag{4}$$

$$s_{\gamma\gamma} = y_1 y_2 s_{ee} \; , \tag{5}$$

where $\sqrt{s_{ee}} = 2E_e$ is the collider energy and the different y's are the photon energy fractions. The convoluted cross sections are obtained from

$$\sigma(s_{ee}) = \int_{y_{min}}^{y_{max}} dy\ P(y)\sigma(s_{e\gamma}) \tag{6}$$

$$\sigma(s_{ee}) = \int_{y_{min}}^{y_{max}} dy_1 \int_{y_{min}/y_1}^{y_{max}} dy_2\ P(y_1)P(y_2)\sigma(s_{\gamma\gamma})\ , \tag{7}$$

where the upper integration limits y_{max} are given by Eq. 2 whereas the lower limits are set by the kinematical threshold of the considered process, $y_{min} = (m_{\tilde{e}} + m_{\tilde{\chi}_1^0})^2/s_{ee}$ or $y_{min} = 4m_{\tilde{e}}^2/s_{ee}$.

3 Supersymmetric Processes

We concentrate here on the simplest possible signals: one or two high $p_\perp$ electrons and otherwise only invisible particles giving rise to missing $p_\perp$. This kind of signal is obtained in minimal supersymmetric extensions of the standard model [3] where the lightest supersymmetric particle (LSP) is the lightest neutralino $\tilde{\chi}_1^0$. A selectron pair $\tilde{e}^\pm\tilde{e}^-$ or selectron-neutralino pair $\tilde{e}^-\tilde{\chi}_1^0$ is produced with the selectron subsequently decaying into an electron-neutralino pair $e^\pm\tilde{\chi}_1^0$. This scenario assumes R-parity to be conserved, so that the LSP is stable and remains unobserved.

We examine the following processes in the narrow width approximation:

$$e^+e^- \rightarrow \tilde{e}^+\tilde{e}^- \rightarrow e^+\tilde{\chi}_1^0\, e^-\tilde{\chi}_1^0 \tag{8}$$

$$e^-e^- \rightarrow \tilde{e}^-\tilde{e}^- \rightarrow e^-\tilde{\chi}_1^0\, e^-\tilde{\chi}_1^0 \tag{9}$$

$$e^-\gamma \rightarrow \tilde{e}^-\tilde{\chi}_1^0 \rightarrow e^-\tilde{\chi}_1^0\, \tilde{\chi}_1^0 \tag{10}$$

$$\gamma\gamma \rightarrow \tilde{e}^+\tilde{e}^- \rightarrow e^+\tilde{\chi}_1^0\, e^-\tilde{\chi}_1^0\ . \tag{11}$$

The corresponding lowest order Feynman diagrams are shown in Figs 2 and the resulting first three differential cross sections[1] can be found in Refs 4,5,6. They depend on four supersymmetry parameters [3]:

- the soft supersymmetry breaking mass parameters M_2 and μ associated with the $SU(2)_L$ gauginos and higgsinos, respectively (for the $U(1)_Y$ gaugino mass parameter M_1 we assume $M_1 = M_2 5/3\tan^2\theta_w$, where θ_w is the weak mixing angle, in accordance with the renormalisation group evolution from a common value $M_1 = M_2$ at the GUT scale);
- the mass of the selectron $m_{\tilde{e}}$ (for simplicity we assume the supersymmetric partners of the left- and right-handed electrons to have equal masses: $m_{\tilde{e}_L} = m_{\tilde{e}_R}$);
- the ratio $\tan\theta_v = v_2/v_1$ of the Higgs vacuum expectation values (as long as $|\mu| \gtrsim M_2/2$, this is not an essential parameter).

[1]We have computed the cross sections to the processes (8) and (9) in the approximation where $m_{\tilde{\chi}_1^0} \ll m_{\tilde{\chi}_i^0}$ (i=2–4).

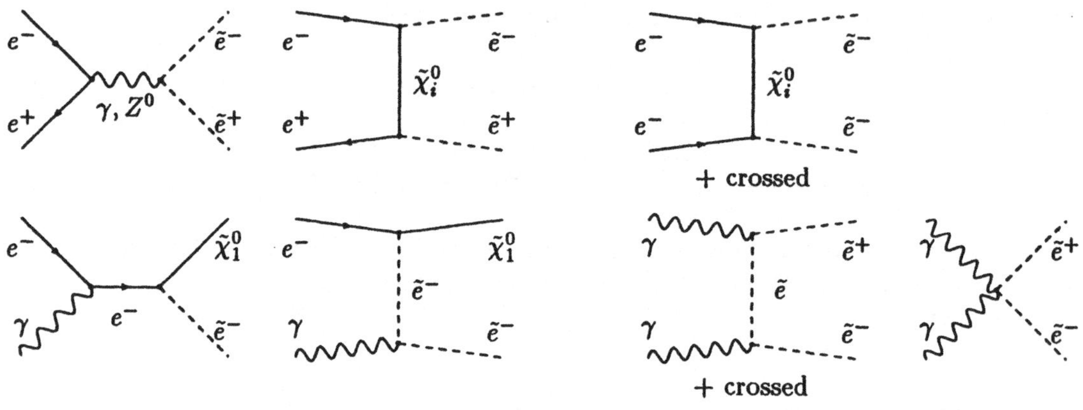

Figure 2: Feynman diagrams contributing to the supersymmetry signal in e^+e^- collisions (top left), e^-e^- (top right), $e^-\gamma$ (bottom left) and $\gamma\gamma$ (bottom right).

For arbitrary values of M_2 and μ the gauginos and higgsinos mix to form neutralino and chargino mass eigenstates [3]. Roughly, for $|\mu| \gtrsim M_2/2$ the higgsino admixture to the lightest neutralino is small [7]. For some of the processes studied here this is an essential condition to obtain measurable cross sections since the higgsino-electron Yukawa coupling is suppressed by the mass of the electron. In this region of parameter space the dependence of the gaugino masses and couplings on θ_v remains very small. If, in addition, $M_2 \gtrsim m_Z$, one finds roughly $m_{\tilde{\chi}_1^0} \approx M_2/2$ and $m_{\tilde{\chi}_2^0} \approx m_{\tilde{\chi}_1^\pm} \approx M_2$. For consistency with the assumption that $\tilde{\chi}_1^0$ is the LSP we must in this case require $M_2 \lesssim 2m_{\tilde{e}}$. If the selectron is lighter than the second lightest chargino or neutralino, the $\tilde{e}^- \to e^-\tilde{\chi}_1^0$ branching ratio [8] is 100%. If there are charginos or other neutralinos which are lighter than the selectron, the $\tilde{e}_L^- \to e_L^-\tilde{\chi}_1^0$ branching ratio can be considerably less than 100%. The $\tilde{e}_R^- \to e_R^-\tilde{\chi}_1^0$ branching ratio, however, remains close to 100% over most of the (μ, M_2) parameter space.

In what follows we have considered a typical scenario:

$$\begin{aligned} \mu &= 375 \text{ GeV} \\ M_2 &= 250 \text{ GeV} \\ m_{\tilde{e}} &= 250 \text{ GeV (when it is not allowed to vary)} \\ \tan\theta_v &= 4\,. \end{aligned}$$

As a result of this choice we have $m_{\tilde{\chi}_1^0} = 120$ GeV and the $\tilde{e}_L^- \to e_L^-\tilde{\chi}_1^0$ branching ratio is 53%.

4 Cross Sections

The total cross sections for the processes (8–11) are plotted in Figs 3 as functions of the selectron mass for two collider energies, $\sqrt{s_{ee}} = 500$ and 1000 GeV. It appears that a

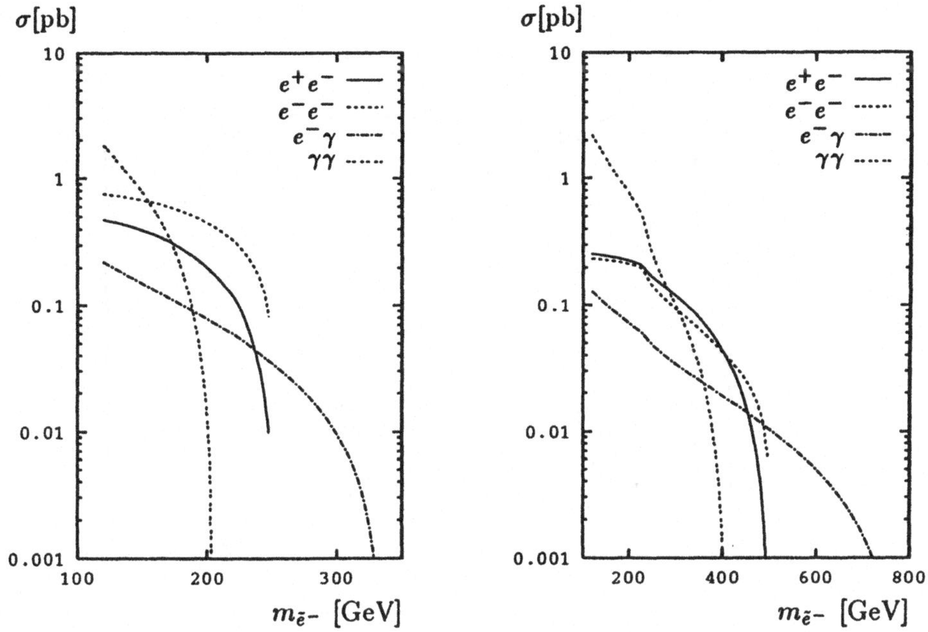

Figure 3: Total cross sections times branching ratios to electron for the four production mechanisms $e^+e^- \to \tilde{e}^+\tilde{e}^-$, $e^-e^- \to \tilde{e}^-\tilde{e}^-$, $e^-\gamma \to \tilde{e}^-\tilde{\chi}_1^0$ and $\gamma\gamma \to \tilde{e}^+\tilde{e}^-$ for collider energies $\sqrt{s_{ee}} = 500$ GeV (left) and 1000 GeV (right).

light selectron can be most efficiently produced in the $\gamma\gamma$ mode, an intermediate mass one in $e^\pm e^-$, whereas the $e^-\gamma$ channel allows one to probe heavier selectrons, beyond the kinematical limit of pair production. The $\gamma\gamma$ mode turns out to be a particularly interesting option at higher energy accelerators, because away from threshold its cross section depends very little on $s_{\gamma\gamma}$. This is due to the absence of a propagator in the diagram involving the four boson vertex. Note that the curves all start at 120 GeV, which is the mass of the LSP (here the lightest neutralino) for the chosen set of parameters. The kink in the curves at $m_{\tilde{e}} \approx 220$ GeV (which is particularly visible for $\sqrt{s_{ee}} = 1000$ GeV and for the reactions involving two selectrons) is due to the fact that for this choice of parameters the lightest chargino has a mass of 220 GeV. Heavier selectrons thus have more decay channels open than the one leading to the single or two electron signal.

The question remains whether this signal is large enough to be separated from the standard model background. Specializing on the $e^-\gamma$ process, Fig. 4 compares the behaviour of the total $e^-\gamma \to \tilde{e}^-\tilde{\chi}_1^0 \to e^-\tilde{\chi}_1^0\tilde{\chi}_1^0$ cross section to those of the standard model background processes $e^-\gamma \to e^-Z^0 \to e^-\bar{\nu}\nu$ and $e^-\gamma \to W^-\nu \to e^-\bar{\nu}\nu$, as functions of the collider energy. In spite of a reasonable number of signal events (400 for 10 fb^{-1} at $\sqrt{s_{ee}} = 500$ GeV in the scenario considered in Fig. 4), it appears that the supersymmetric signal is completely swamped. However, since most of the standard model cross sections is due to the u-channel exchange in $e^-\gamma \to e^-Z^0$ and the t-channel exchange in $e^-\gamma \to W^-\nu$ (since $m_e, m_W \ll \sqrt{s_{ee}}$), these cross sections can be drastically reduced by angular or rapidity cuts [9]. Moreover, the standard model background can be computed with great precision. The accuracy of these calculations can even be checked in the case of the W^- channel, by comparing with the $e^-\gamma \to \mu^-\bar{\nu}\nu$ signal. In principle, thus, any

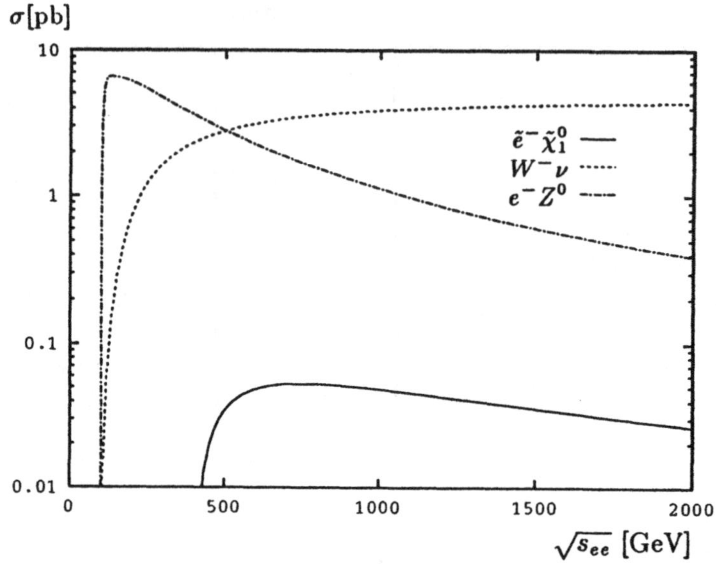

Figure 4: Total cross sections times branching ratios for obtaining a single high $p_\perp$ electron accompanied by invisible particles from $\tilde{e}^- \tilde{\chi}_1^0$ production (assuming the scenario described in the text with $m_{\tilde{e}^-} = 250$ GeV) as compared to the standard model $W^-\nu$ and $e^- Z^0$ production.

deviation from the predicted rates for the standard model processes can be a signal for supersymmetry. For $m_{\tilde{e}^-} = 250$ GeV, $M_2 \leq 300$ GeV and $|\mu| \gtrsim M_2/2$, it turns out that one needs about 20 fb^{-1} at a $\sqrt{s_{ee}} = 500$ GeV collider in order to establish a 3σ effect. This is well matched to the range projected for these colliders.

5 Conclusions

In summary, the $e^-\gamma$ and $\gamma\gamma$ modes at $e^\pm e^-$ colliders turn out to be competitive cptions to discover or study selectrons. The $e^-\gamma$ mode is particularly interesting for relatively heavy selectrons with masses above the pair production limit $m_{\tilde{e}} = \sqrt{s_{ee}}/2$. In contrast, the $\gamma\gamma$ mode is favourable for relatively light selectrons with masses $m_{\tilde{e}} \lesssim \sqrt{s_{ee}}/3$. In the mass range $\sqrt{s_{ee}}/3 \lesssim m_{\tilde{e}} < \sqrt{s_{ee}}/2$, the usual $e^\pm e^-$ modes are the most promising. In addition, the process $e^-\gamma \to \tilde{e}^- \tilde{\chi}_1^0$ can serve as probe of the lightest neutralino, an elusive particle which is difficult to detect. Finally, the standard model backgrounds are generally large, but can be handled in a sizable region of the supersymmetry parameter space.

References

1. A. Bartl, W. Majerotto and B. Mößlacher, these Lecture Notes.
2. I.F. Ginzburg, G.L. Kotkin, V.G. Serbo and V.I. Telnov, *Nucl. Instr. Meth.* **205**, 47 (1983).
3. H.E. Haber and G.L. Kane, *Phys. Rep.* **117**, 75 (1985).
4. A. Bartl, H. Fraas and W. Majerotto, *Z. Phys.* **C 34**, 411 (1987).
5. W.-Y. Keung and L. Littenberg, *Phys. Rev.* **D 28**, 1067 (1983).
6. G. Altarelli, G. Martinelli, B. Mele and R. Rückl, *Nucl. Phys.* **B 262**, 204 (1985).

   J.A. Grifols and R. Pascual, *Phys. Lett.* **B 135**, 319 (1984).
7. H. Komatsu and R. Rückl, *Nucl. Phys.* **B 299**, 407 (1988).
8. A. Bartl, H. Fraas and W. Majerotto, *Nucl. Phys.* **B 297**, 479 (1988).
9. F. Cuypers, G.J. van Oldenborgh and R. Rückl, Munich preprint MPI-Ph/92/14 (1992).

RARE DECAYS, HEAVY TOP AND MINIMAL SUPERSYMMETRY

Stefano Bertolini

Istituto Nazionale di Fisica Nucleare
Sez. di Trieste, c/o SISSA
Via Beirut 4, I-34013 Trieste

1. INTRODUCTION

Flavour changing neutral current (FCNC) transitions can proceed in the standard electroweak model (SM) only through loops induced by the exchange of the $W^{\pm}$ bosons and quarks. The loop suppression and the heaviness of the $W^{\pm}$ and Z^0 gauge bosons provide in first approximation a sensible explanation for the rarity of these processes. Even so, the SM would not have had such a spectacular success in confronting experiments without the presence of a further "process dependent" suppression mechanism related to the field theoretical structure of the quark mixings: the unitarity of the Cabibbo-Kobayashi-Maskawa (CKM) matrix [1]. This property of the quark mixing matrix, residue of the diagonalization of the up and down quark fields in the charged current, makes FCNC transitions vanish identically if all quarks of the same isospin are degenerate in mass. This effect, known as the Glashow-Iliopoulos-Maiani (GIM) [2] mechanism, is at the root of the potential relevance of a heavy top quark for this class of processes.

When extensions of the standard model are considered, other new sources of FCNC effects may arise. If neutrinos are massive, lepton flavour violating transitions appear as well, in analogy with the quark sector. If the Higgs sector is extended, physical charged scalars may replace the W boson in the loop and lead to additional or genuinely new effects. Neutral Higgs induced flavour changing transitions may also arise at the tree level and require these particles to be much heavier than the Fermi scale, in order to replace the missing loop and GIM suppressions.

It is therefore important to carefully carry on and perfect the study of this sector of the electroweak phenomenology because of its implications on the detailed structure of the standard electroweak model and on the quest for the missing pieces (top quark and Higgs sector). Also non-standard physics, which may still hyde from us in direct searches, has a chance of being indirectly probed in this world of "low energy" phenomena.

A plethora of FCNC processes is available to us for theoretical and experimental studies. How to efficiently spend our efforts in trying to disentangle a specific effect? How to look for processes that are most sensitive to the physical scenario that we want to test? In this lecture we shall see that, just with the help of simple and general considerations, we can recognize a systematic structure in FCNC transitions which provides us with simple but effective criteria for dedicated searches. We will first try to analyze the relevance of a heavy top quark for electroweak FCNC transitions and then study its impact on some extension of the SM. In particular our present discussion will focus on the minimal supersymmetric scenario.

2. RARE FLAVOUR VIOLATING TRANSITIONS

The world of loop-induced flavour changing transitions is populated by the so called "penguin" and "box" diagrams. Some examples are given in fig. 1, where the quark diagrams contributing to $K^+ \to \pi^+\nu\bar{\nu}$ are shown. Since, quite independently on how we name the various fermion lines, the structure of the Feynman diagrams is basically the same, it is sensible to expect that there exist properties of these transition amplitudes that are common to wide classes of processes. Our goal is a systematic classification of the many FC transitions that the theory allows, according to their specific sensitivity to the nature and heaviness of the particles running in the loops.

To begin with, let us consider the effective electroweak vertex $\bar{d}(p')\Gamma_\mu(q)A^\mu(q)s(p)$, where $A^\mu \equiv A^\mu_a T^a$ represents the gauge field (γ, Z^0 or gluons) with fermions in the representation T^a, and $q = p - p'$ is the momentum transferred in the transition. With the help of gauge and Lorentz invariance we can then write

$$\begin{aligned}\Gamma^\mu(q) &= A_1\,(q^2g^{\mu\nu} - q^\mu q^\nu)\,F_1(m_i^2,q^2)\,\gamma_\nu(1-\gamma_5)\\ &+ A_2\,m_s q_\nu\,F_2(m_i^2,q^2)\,i\sigma^{\mu\nu}(1+\gamma_5)\\ &+ A_2'\,m_d q_\nu\,F_2'(m_i^2,q^2)\,i\sigma^{\mu\nu}(1-\gamma_5)\\ &+ A_0\,m_W^2\,F_0(m_i^2,q^2)\,\gamma^\mu(1-\gamma_5)\,, \end{aligned} \qquad (1)$$

where the constant coefficients A_i depend on the interaction strengths. The form factors F_k depend generally on the masses and momenta of the real and virtual particle involved in the transition. If the gauge current is conserved (photon and gluon penguins) then $F_0 = 0$. The form factor F_0 is therefore present only for Z^0 induced verteces. We will refer to F_1 and F_2 as the charge and dipole form factors respectively. In the following we will also neglect m_d compared to m_s and therefore discard the third term in eq. (1).

Let us now consider for instance the semileptonic transition $s \to d\,e^+e^-$. The penguin diagram contribution to this process, once the photon and Z^0 propagators and their

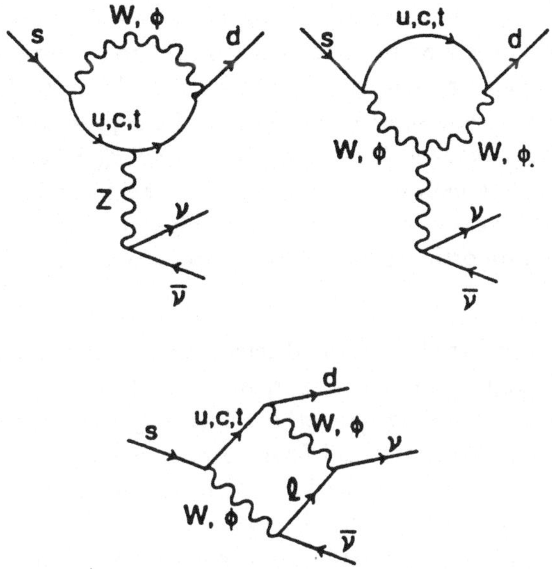

Figure 1: Standard Penguin and Box diagrams contributing to $K^+ \to \pi^+ \nu\bar{\nu}$.

couplings to to the lepton current are included, can be written as

$$
\begin{aligned}
\mathcal{A}_{peng}(s \to d\ e^+e^-) \ &\approx\ \frac{G_F}{\sqrt{2}}\frac{\alpha}{\pi}\sum_i \lambda_i \left\{ F_1'\left(\frac{m_i^2}{m_W^2}, \frac{q^2}{m_W^2}\right)\ J^{Q\ \mu}_{V-A} J^L_{V\ \mu} \right. \\
&+\ \frac{m_s q_\nu}{q^2}\, F_2\left(\frac{m_i^2}{m_W^2}\right)\ J^{Q\ \mu\nu}_{T+PT} J^L_{V\ \mu} \\
&+\ \left. \frac{1}{\sin^2\theta_W}\, F_0\left(\frac{m_i^2}{m_W^2}\right)\ J^{Q\ \mu}_{V-A} J^L_{V-A\ \mu} \right\}\ , \qquad (2)
\end{aligned}
$$

where α is the electromagnetic coupling constant. The Lorentz properties of the quark and lepton currents $J^{Q,L}$ are explicitly indicated. The factors $\lambda_i \equiv K_{is}K^*_{id}$ denote the product of the CKM angles involved in the transition and a sum over $i = u, c, t$ is understood. It is worth recalling that the form of eq. (2) follows from the conservation of the lepton current $\partial_\mu J^{L\mu}_V = 0$. Before discussing some properties of the three form factors, we have to consider also the contribution arising from box diagrams, as in fig. 1. One finds

$$
\mathcal{A}_{box}(s \to d\ e^+e^-) \approx \frac{G_F}{\sqrt{2}}\frac{\alpha}{\pi \sin^2\theta_W}\sum_i \lambda_i\ B\left(\frac{m_i^2}{m_W^2}\right)\ J^{Q\ \mu}_{V-A} J^L_{V-A\ \mu}\ . \qquad (3)
$$

From the form of eq. (3) it is evident that the box contribution can be included in eq. (2) together with the Z^0 penguin form factor to give $F_0' = F_0 + B$. In turn, the

Z^0 contribution to the transition amplitude has been already split in two parts. The component proportional to the electron weak isospin (T_3) contributes to F_0', whereas that proportional to the lepton electric charge ($g^2 \sin^2 \theta_W$) contributes to F_1', together with the photon penguin form factor. The short-distance part of the process $K \to \pi e^+ e^-$ can then be described by three *gauge invariant* functions F_1', F_2 and F_0', whose specific dependence on the quark masses and momenta we will discuss in the following. Although each of the form factors does not correspond to an observable, they must be separately gauge invariant since they are proportional to different interaction strenghts and/or different currents in the amplitude.

These considerations, inferred from the discussion of a specific transition, depend in fact very little on it. Changing the particles involved in the process may modify overall factors in the amplitudes, as coupling strengths, select only some of the components and, of course, modify their detailed functional form. However it does not affect, as we will see, some characteristic behaviours of the form factors, which are related to the structure of the currents involved. One may be therefore tempted to systematize the study of these loop-induced phenomena by constructing a basis of gauge invariant and, to some extent, process independent functions which are common blocks of apparently unrelated processes. Following this approach it may become easier to identify and isolate the effects related to the presence of heavy quarks, or more exotic physics, and identify those processes that are potentially promising for their search. The "Penguin-Box Expansion" of ref. [3] is in fact an attempt toward a systematic study of FCNC processes based on these considerations. We will not enter here the many technically subtle aspects of such an analysis (for a review we refer the reader to ref. [4]). We will however show that even the simple knowledge of some semiquantitative properties of the effective amplitudes, eqs. (2–3), gives us already enough information to justify such an investigation. The crucial point rests in a remarkable property of the form factor F_1 in the photon and gluon penguins. Studying the "infrared" (m_i^2/m_W^2, $q^2/m_W^2 \to 0$) behaviour of the various contributions to this term, one can in fact convince oneself that there exist a component which diverges logarithmically. This shows that F_1 exhibits a genuine logarithmic dependence on the internal quark masses, which is absent in the other form factors. One finds

$$F_1 \sim \text{const} \int_0^1 dx\, x(1-x) \ln \frac{m_i^2 - x(1-x)q^2}{m_W^2} + \widehat{F}_1\left(\frac{m_i^2}{m_W^2}, \frac{q^2}{m_W^2}\right) , \tag{4}$$

where the coefficient of the logarithm is gauge invariant since no other pure logarithmic terms arise in the amplitude. When the GIM mechanism is applied, the component $\widehat{F}_1$ vanishes quadratically in the limit $m_i \to 0$, analogously to F_2, F_0 and B. These contributions, unlike the first term in eq. (4), are dominated by the exchange of heavy quarks and, for light quark decays, we may neglect their q^2 dependence altogether (see eqs. (2–3)).

This feature shows two different realizations of the GIM mechanism: a power-like suppression ($\Delta m_i^2/m_W^2$) in the F_2, F_0 and B terms; a logarithmic "soft" suppression in F_1. As a consequence, in a world of light quarks ($m_i^2 \ll m_W^2$) the component F_1 dominates the amplitude, and a quite good estimate of the process is obtained by solely computing the logarithmic term in eq. (4). Life was somewhat easier in the past, when the top quark was still believed to be light! The presence of a heavy top makes the power terms in the amplitude a priori no longer negligible, and calls also for a reanalysis of the QCD effects in light quark transitions (for an extensive discussion see ref. [4]).

It is interesting to consider the question of the interplay between a heavy top and light quark decays with more detail. Does the knowledge of this "anomalous" behaviour of F_1 tell us still something about the sensitivity of FCNC processes to physics scales much heavier than the few GeV of the light quarks? In order to answer this question one has to consider separately $s \to d$ (a) and $b \to s$ (b) transitions. In fact, it is well known that in processes of the type (a) top exchange is disfavoured by CKM angles, namely

$$\frac{|\lambda_t|}{|\lambda_c|} \equiv \frac{|K_{ts}K_{td}^*|}{|K_{cs}K_{cd}^*|} \approx \sin^4\theta_C \approx 10^{-3} \tag{5}$$

In spite of that, for $m_t > 90\ GeV$ we have

$$|\lambda_t|\frac{m_t^2}{m_W^2} \gtrsim |\lambda_c|\frac{m_c^2}{m_W^2}\ln\frac{m_W^2}{m_c^2} \gg |\lambda_c|\frac{m_c^2}{m_W^2} \tag{6}$$

and therefore top and charm exchange in the form factors F_2, F_0 and B can be of comparable size. However, whenever F_1 is present we have

$$|\lambda_t|\ln\frac{m_t^2}{m_W^2} \ll |\lambda_c|\ln\frac{m_W^2}{m_c^2}, \tag{7}$$

and light quark exchange largely dominates.

In conclusion top exchange is negligible whenever a soft GIM suppression is present. This amounts to saying that a necessary condition (albeit not sufficient !) for sizeable effects of the top quark in FCNC $s \to d$ transitions is to consider processes *which cannot be induced by electromagnetic and/or gluon penguins.* Without much effort, we have reached indeed a valuable result. This simple criterium is in fact quite selective and allows us to focus our attention on a limited class of processes.

However, before proceeding to the discussion of two explicit examples, it is worthwhile to remark that this criterium does not apply to CP *violating* $s \to d$ transitions. In fact, at $O(\sin^3\theta_C)$ the physical phase in the standard parametrization of the CKM matrix appears only in the K_{ub} and K_{td} entries. It is then obvious that a loop induced CP violating transition involving virtual up-type quarks must include top exchange. In other words, although the real part of the amplitude is dominated by the exchange of light quarks (an example is $K_{L,S} \to \pi\pi$), the imaginary part is not, and we can roughly write

$$\frac{\mathrm{Im}\mathcal{A}}{\mathrm{Re}\mathcal{A}} < O\left(\frac{\lambda_t}{\lambda_c}\right) \approx 10^{-3}\ . \tag{8}$$

The smallness of CP violating effects in the SM is therefore simply justified. Recalling our discussion of the relevance of the top quark for the various form factors, it follows immediately that in the calculation of ϵ'/ϵ one cannot neglect anymore Z^0 penguins and box diagrams compared to the photon and gluon penguins, whose imaginary parts are CKM suppressed. A re-evaluation of ϵ'/ϵ in the presence of a heavy top has been performed by different authors in the last few years [4]. However, what is still missing to make the analysis complete (at least for what concerns the short-distance part) is the inclusion of the dipole components (form factor F_2) of the gluon and electromagnetic penguins.

Traditionally, only the component F_1 of the gluon penguin has been included in the evaluation of ϵ'/ϵ, because of the soft GIM suppression. As we have seen, this is a good approximation only when all the quarks are light on the weak scale. The same argument that compels us to consider Z penguin and box diagrams applies a priori to the dipole components of the photon and gluon penguins as well. Even more so for the latter which contributes to the leading gluon diagram. The size of the effect depends however on the complete analysis of the QCD renormalization and mixings of the new operators with those up to now considered, and, more crucially, on the evaluation of the hadronic matrix elements. At least two additional operators have to be included in the effective Lagrangian for $\mathcal{A}(K_{L,S} \to \pi\pi)$

$$\mathcal{O}_{ph} = \frac{\alpha}{\pi}\left(\frac{m_s q_\nu}{q^2}\right)(\bar{s}_R\, i\sigma^{\mu\nu}\, d_L) \sum_{f=u,d,s} e_f\, (\bar{f}\,\gamma_\mu\, f)\,, \tag{9}$$

$$\mathcal{O}_{glue} = \frac{\alpha_s}{\pi}\left(\frac{m_s q_\nu}{q^2}\right)(\bar{s}_R\, i\sigma^{\mu\nu}T^a\, d_L) \sum_{f=u,d,s} (\bar{f}\,\gamma_\mu T^a\, f)\,. \tag{10}$$

Their Wilson coefficients at the weak scale read respectively ($\mathcal{A} = -(G_F/\sqrt{2})\lambda_t\, C_i\, \mathcal{O}_i$)

$$C_{ph}(m_W) = \frac{x^2(2-3x)}{2(x-1)^4}\ln x + \frac{x(8x^3-3x^2-12x+7)}{12(x-1)^4}\,, \tag{11}$$

$$C_{glue}(m_W) = \frac{3x^2}{2(x-1)^4}\ln x + \frac{x(x^3-6x^2+3x+2)}{4(x-1)^4}\,, \tag{12}$$

where $x = m_t^2/m_W^2$.

The absence of these operators in the existing analysis of ϵ'/ϵ was pointed out by the present author about two years ago [5]. One attempt to estimate their effect has recently appeared. The authors of ref. [6] study the effect of the operator in eq. (10) and conclude that its contribution to ϵ'/ϵ is about one order of magnitude smaller than that induced by F_1 in the gluon penguin. The authors include the multiplicative QCD renormalization neglecting any mixing effect. The QCD induced mixing of the operators in eqs. (9–10) with four-quark operators of the type $(\bar{s}_L\gamma^\mu c_L)(\bar{c}_L\gamma_\mu d_L)$ occurs in fact at the two-loop level. However, if we are allowed to make an analogy with the case of the $b \to s\gamma$ decay, this effect can be large, since gluon renormalization may bring back a soft GIM suppression, absent in eqs. (11–12). Due to the presence of the charm quark, the two-loop mixing can be competitive in size with the one-loop operators in eqs. (9–10).

A study of the renormalization of the complete set of effective operators, together with a consistent estimate of the hadronic matrix elements $\langle\pi\pi|\, \mathcal{O}_{ph,\ glue}\, |K\rangle$ is required in order to ascertain the relevance of these additional contributions to ϵ'/ϵ in the SM.

Coming back to CP conserving $s \rightarrow d$ transitions, we have learned that in order to uncover possible effects due to heavy particles running in the loop (the importance of the top quark is strictly related to the potential relevance of other non-standard physics, as we will explicitly see for supersymmetric models) we have to search for processes which cannot be induced by electromagnetic and /or gluon penguin diagrams. Two examples of such processes are (i) $K_L \rightarrow \mu^+\mu^-$ and (ii) $K^+ \rightarrow \pi^+\nu\bar{\nu}$. Both decays can only occur through Z^0 penguins and box diagrams. While for the second process this conclusion is obvious, in the case (i) it follows from the conservation of the *vector* component of the lepton current: $\langle 0|\, \bar{s}\gamma^\mu\gamma_5 d\, |K(q)\rangle\, (\bar{\mu}\gamma_\mu\mu) \sim q^\mu(\bar{\mu}\gamma_\mu\mu) = 0$.

Experimentally, there exist two recent measurements of $BR(K_L^0 \rightarrow \mu^+\mu^-)$, namely $(8.4 \pm 1.1) \times 10^{-9}$ (KEK) [7] and $(7.0 \pm 0.5) \times 10^{-9}$ (BNL) [8]. Since this transition is measured, and since the short distance contributions are dominated by top exchange, we would expect this process to be a good probe for indirect top searches. Unfortunately, long-distance (LD) contributions turn out to be dominant. The transition can in fact occur via a two photon intermediate state. The absorptive part of this long-distance effect can be estimated by convoluting $\Gamma(K_L \rightarrow \gamma\gamma)$ and $\Gamma(\gamma\gamma \rightarrow \mu^+\mu^-)$. Using the measured value of $BR(K_L \rightarrow \gamma\gamma)$ [9] one obtains $BR(K_L \rightarrow \mu^+\mu^-)_{\rm abs} = (6.8 \pm 0.3) \times 10^{-9}$. Since the absorptive part of the short distance contributions is negligible (light quark exchange is subleading), the value above is in fact a lower bound for the total branching ratio.

In order to use this information to bound the (dispersive) short-distance component, we should get a handle on the dispersive part of the long-distance contribution (off-shell photon form factor). This task would be made easier if we had data on the invariant mass distributions for $K_L \rightarrow e^+e^-e^+e^-$, for which only recently the branching ratio has been measured at the level of 5×10^{-8} [10], in agreement with the SM expectation. Using the better experimental knowledge of $K_L \rightarrow e^+e^-\gamma$, one can nevertheless attempt an extrapolation to the 2γ off-shell form factor relevant for $K_L \rightarrow \mu^+\mu^-$. With the help of vector meson dominance the authors of ref. [11] estimate the dispersive part of the long distance component to be at most a few percent of the absorptive part. Using this result they study, as a function of the top quark mass, the range of values of $BR(K_L \rightarrow \mu^+\mu^-)$ allowed by our knowledge of the relevant electroweak parameters. By comparing the results of KEK and BNL with fig. 2, taken from ref. [11], it is apparent that no definite statement on the top quark mass can be presently made. In addition to the uncertainty in the evaluation of the LD contribution, another source of uncertainty is related to our poor knowledge of $\mathrm{Re}\{K_{td}^*K_{ts}\}$. A more recent analysis of the short distance contributions to $K_L \rightarrow \mu^+\mu^-$, which takes into account the combined constraints coming from other measured FCNC processes and CP violating transitions has been performed in ref. [3]. For $m_t \approx 100\ GeV$ the branching ratio induced only by short distance contributions can be as small as 3×10^{-10}, while for $m_t \approx 200\ GeV$, for which the preferred values of the

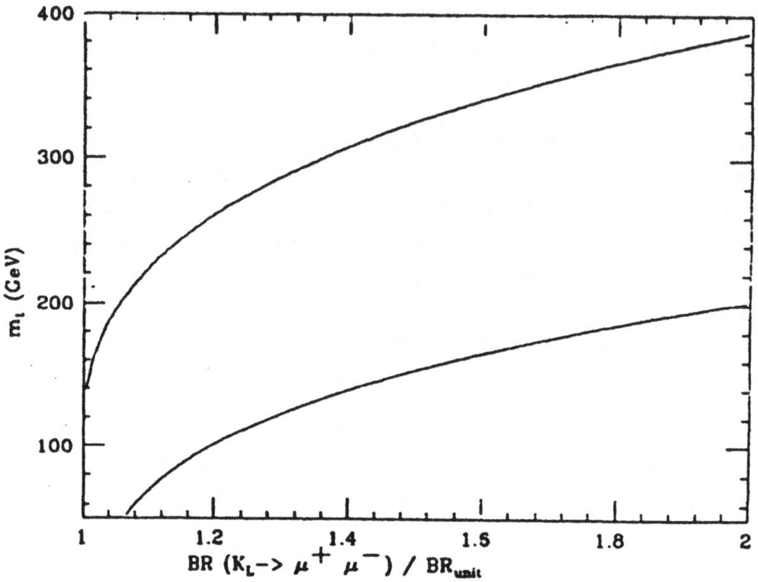

Figure 2: The allowed range for $BR(K_L \to \mu^+\mu^-)$ is shown as a function of the top mass, for $BR_{\text{unit}} = 6.8 \times 10^{-9}$. The figure is taken from ref. [11].

CKM phase lie in the first quadrant, it does not exceed 2×10^{-9}. It is clear that until we obtain a precise evaluation of the LD contributions, this decay cannot help us in the study of the subleading short-distance effects.

At variance with $K_L \to \mu^+\mu^-$, the decay $K^+ \to \pi^+\nu\bar{\nu}$ can be considered from the point of view of the theory a "clean" process. In fact, a property of the latter decay which is crucial for our short-distance analysis is that long-distance contributions have been shown to be negligible [12]. Short-distance contributions exhibit also some nice features. In addition to the hard GIM suppression, which makes top exchange competitive with charm exchange, it turns out that QCD corrections suppress the leading logarithmic terms ($\sim x_c^n \ln x_c$), thus further decreasing the charm contribution by about 30% [13]. Finally, a partial cancellation of the subleading ($\sim x_c^n$) terms, occurring between the Z^0 penguin and the box diagram, makes this process less sensitive to next-to-leading QCD corrections. Unfortunately, we do not have yet experimental evidence for this decay. When summing over three light neutrinos, the expected branching ratio varies between 1.5 and 8×10^{-10}, for $m_t < 200\ GeV$ (for a detailed discussion see ref. [3]). The highest values are reached for the CKM phase in the second quadrant and $m_t \approx 150\ GeV$. The present experimental bound is 4.4×10^{-9} [14], about one order of magnitude above the needed sensitivity. In view of the interesting properties of this process, an experimental effort to cover the present gap between measurement and theoretical expectations is more than justified.

With the previous two examples we have seen that the Kaon system may offer the possibility of testing physics at the Fermi scale, provided we choose processes in which light quark exchange is strongly GIM suppressed. This amounts to disregarding FCNC transition which can be induced by electromagnetic and/or gluon penguins. This criterium is for the Kaon system quite selective. Is the distinction between softly and strongly GIM suppressed transitions as effective when considering heavier meson systems ?

The preferred sector for investigating indirect effects of weak scale physics is that of FCNC transitions involving the B mesons. If we consider loop-induced $b \rightarrow s$ transitions, charm exchange is no longer favoured by CKM mixings. In connection to that and because of the heaviness of the B meson long-distance contributions may be expected to play a less relevant role than in the Kaon system. A further consequence of the heaviness of the b-quark is that inclusive B-decays can be reliably modelled by quark decays (non-spectator diagrams are wave-function suppressed: $f_B^2/m_b^2 \approx 1/400$). Finally, when the top quark exchange dominates the amplitude, then the SM prediction for the branching ratio depends very little on the uncertainties related to our poor knowledge of the relevant CKM angles. In fact, by normalizing the various transition to $b \rightarrow ce\bar{\nu}$, from the unitarity of the three-generation CKM matrix one obtains $|K_{ts}^* K_{tb}| \,/\, |K_{cb}| = 1 + O(\sin^4 \theta_C)$. Therefore, the short distance components of the transitions depend, including QCD corrections, only on two parameters: m_t and Λ_{QCD}.

On the other hand, if a soft GIM suppression is present (once again via electromagnetic and gluon penguins), the sensitivity to heavy scales may be hindered by the relevance of the charm (or external momenta) contribution : $|K_{cs}^* K_{cb} \ln x_c| \gtrsim |K_{ts}^* K_{tb}| \, x_t$. We expect therefore that , even in the B-sector, the processes that are most promising for indirect searches of exotic heavy physics are those which cannot be induced by the form factor F_1 in the effective $\bar{b}\, \Gamma_\mu A^\mu\, s$ vertex. To this class of processes belong $B_s^0 \rightarrow \tau^+\tau^-$ and $B \rightarrow X_s\, \nu\bar{\nu}$, in analogy with the Kaon case. However, because of the combined effects of low branching ratios and low detection efficiencies these two processes do not offer at present concrete possibilities of experimental test. We would like instead to focus our attention on the radiative $B \rightarrow X_s + \gamma$ transition, where a real photon is emitted together with a strange hadronic "jet". At the quark level, this decay is induced by the effective $b \rightarrow s + \gamma$ transition, which, because of the on-shell photon, can occur only through the dipole form factor F_2 (non-spectator tree-level contributions, which would be dominant for the analogous transition in the Kaon sector, are here negligible). As a consequence, the genuine electroweak amplitude is strongly GIM suppressed, and therefore sensitive to virtual top effects.

Surprisingly, QCD corrections turn out to play an important role. A soft GIM suppression in fact appears when an additional gluon loop is present, mixing the renormalization of the original dipole operator with effective four-quark operators of the type $(\bar{b}_L\gamma^\mu c_L)(\bar{c}_L\gamma_\mu s_L) \equiv \mathcal{O}_1$. The presence of large logarithms involving the charm quark mass enhances the two-loop contribution which turns out to be comparable to the one-loop electroweak amplitude even for a heavy top [15]. This effect, confirmed by detailed renor-

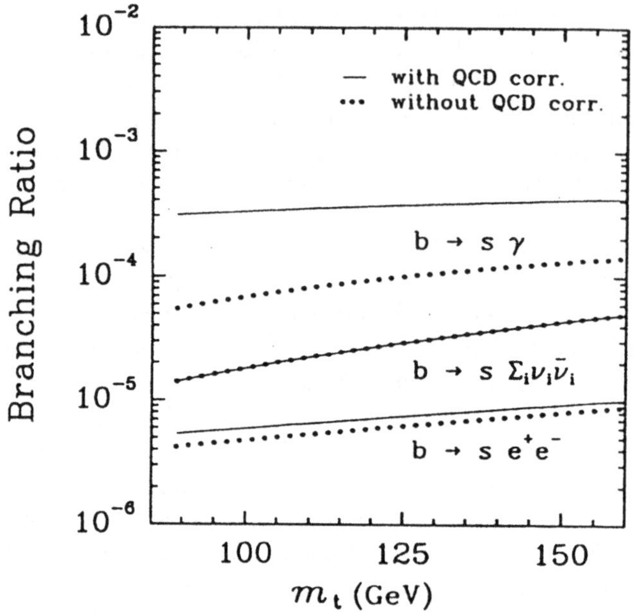

Figure 3: Standard model expectations for some $b \rightarrow s$ transitions as a function of the top mass. Dotted lines do not include QCD corrections. The figure is taken from ref. [21].

malization group analysis [16], enhances the branching ratio for the *inclusive* $b \rightarrow s + \gamma$ transition above 3×10^{-4} for $m_t > 90$ GeV, although it smooths the original top dependence (see fig. 3).

Although the photon signature is a welcome feature of this decay, comparison with experiment is not straighforward. Assuming the dominance of two-body intermediate state production via a Kaon resonance and a photon, the lightest exclusive final state allowed is $K^*(892) + \gamma$, followed by $K^* \rightarrow K\pi$. The theoretical estimates of the branching fraction $R_{K^*} \equiv \Gamma(B \rightarrow K^*\gamma)/\Gamma(B \rightarrow X_s\gamma)$ vary between 5% and 40% [17,18]. The difficulty in obtaining a reliable result rests on the fact that non-relativistic quark models are unsuitable for describing the overlapping between the B meson and the K^* wavefunctions since the s-quark is relativistic. At the inclusive level, one may try to detect the most energetic photons, having care to cut at photon energies $E_\gamma \lesssim 2.25$ GeV, in such a way to exclude contamination from the background decay $B \rightarrow D^*(2010) + \gamma$. This would nevertheless include the first few Kaon resonances and may amount to measuring a substantial fraction of the inclusive transition [17] (compare also with the analysis of ref. [18], based on the study of $b \rightarrow s\gamma + gluon$). As for the lightest exclusive mode, assuming a phenomenological model for the B meson wavefunction Ali and Greub [18] estimate $0.1 < R_{K^*} < 0.2$, for any value of m_t in the $100 - 200$ GeV range.

These expectations have to be compared with the already close experimental bounds. The best experimental limits on the $b \to s + \gamma$ process come from the recent results of CLEO II [19]. At the inclusive level, based on the analysis of ref. [18], they obtain $BR(b \to s + \gamma) < 8.4 \times 10^{-4}$, whereas for the lightest exclusive mode they have reached the limit $BR(B \to K^* + \gamma) < 0.92 \times 10^{-4}$. If, somewhat conservatively, we take $R_{K^*} = 10\%$ we see that the two bounds above are roughly equivalent and are only a factor 2-3 away from the SM expectations, depending on m_t. This can be seen from fig. 3 where we report, as a function of the top quark mass, the SM expectations for various $b \to s$ transitions. One can recognize the different sensitivities to the top mass and to the leading QCD corrections, according to the discussion presented in this section. A more detailed description of the processes shown in fig. 3 can be found in refs. [20] and [21].

After these introductory considerations, in the next section we will examine the impact that these FCNC processes have on minimal supersymmetric extensions of the standard scenario. In order to include the leading logarithmic QCD corrections in extended frameworks it is convenient to observe that, if (i) no *new* effective operators are added to the basis relevant for the process considered (this is, for instance, the case if all the *additional* particles exchanged in the loop are "heavy" with respect to the characteristic scale of the process), (ii) possible additional tree level contributions to effective four-fermion operators, like $\mathcal{O}_1$, are suppressed (a typical case is when Yukawa couplings are involved), and (iii) the newly exchanged particles can be integrated away together with the W gauge boson (we will assume the same range of applicability as for the case of the top quark), then, it follows that (a) the new contributions to the amplitude are only multiplicatively renormalized (no new induced mixing effects), and (b) the QCD correction factors are the same as for the SM amplitudes. Conditions (i) and (ii) are generally satisfied in the models considered here. In addition, we are presently interested in mass ranges for the new particles which satisfy (iii). In this case, the extension of QCD corrections to non-standard contributions is straightforward and does not require further work (for a more detailed discussion see ref. [20]).

3. THE MINIMAL SUPERSYMMETRIC STANDARD MODEL

In this section we examine a particularly simple supersymmetric (SUSY) extension of the SM, derived from spontaneously broken minimal N=1 supergravity (for a review see ref. [22]). Spontaneously broken N=1 supergravity theories with a residual N=1 global supersymmetry (SUSY) preserved down to the Fermi scale represent the most attractive possibility of solving the naturalness problem in the scalar sector of the SM, while incorporating gravity in a unified picture of all elementary interactions. If, unquestionably, the direct searches for supersymmetric particles at LEP, CDF, LHC and SSC constitute the primary tool in this effort of testing low-energy SUSY, nonetheless its manifestation through virtual effects in FCNC rare processes is still of interest and, at any rate, complementary to the other direction. We will show that some bounds on rare B decays reachable by CLEO II conceivably in the near future can be translated into limitations on the available SUSY parameter space which are comparable to those obtainable at LEP.

Even limiting ourselves to low energy theories derived from spontaneously broken N=1 supergravity, we are faced by a host of models which substantially differ in the nature and number of their free parameters and, in particular, on their predictions for the topic of interest for us here, i.e. heavy flavour physics. The first major distinction concerns the way one avoids a fatally short proton decay lifetime. It is known that the supersymmetrized version of the SM allows for the presence of baryon and lepton violating terms in the superpotential. Here we take the most "conservative" attitude of forbidding all these dangerous terms by imposing a discrete symmetry, known as R-parity, which discriminates the ordinary particles from their SUSY partners. Even restricting ourselves to SUSY models with conserved R-parity, we are left with a very large number of free parameters in the sector which softly breaks the residual N=1 global SUSY. To further reduce this arbitrariness we consider the simplest possible SUSY version of the SM where one introduces the minimal number of superfields necessary to supersymmetrize the SM and chooses a flat Kähler metric [22]. In this minimal SUSY version of the SM only four additional new parameters appear. Moreover, it can be easily seen that the Higgs sector of such a minimal SUSY model with two Higgs doublets, needed to provide a mass to both the up- and down-quark sectors, is not large enough to allow for the required electroweak symmetry breaking at the tree level. If one insists on the minimal prescription of only two doublets in the Higgs sector, then the correct spontaneous breaking of $SU(2) \times U(1)$ can be achieved radiatively, i.e. making use of the renormalization effects on the SUSY parameters (in particular on the Higgs masses) when evolving them from the superlarge scale of supergravity breaking down to $\sim m_W$. Imposing such a radiative breaking creates a functional relation between the four SUSY parameters, on one side, and the weak scale on the other side. Hence, the number of independent free SUSY parameters in the MSSM with electroweak radiative breaking (RMSSM) further reduces to *three*. Our analysis will be performed in the context of the RMSSM. This is the SUSY model that offers the highest degree of predictivity and correspondingly the scenario where we can put the sharpest constraints on the parameter space.

Let us briefly survey the structure of the model. As previously mentioned, minimality strongly constrains the structure of the Lagrangian which describes spontaneously broken N=1 supergravity models. The requirements that only the fields due to the supersymmetrization of the SM are present and that matter parity is preserved, immediately dictate the expression of the $SU(3)_C \times SU(2)_L \times U(1)_Y$ invariant superpotential:

$$W = h_U^{ij} Q_i U_j^c H_2 + h_D^{ij} Q_i D_j^c H_1 + h_E^{ij} L_i E_j^c H_1 + \mu H_1 H_2 \, , \tag{13}$$

where the chiral matter superfields Q, U^c, D^c, L, E^c, H_1 and H_2 transform as follows under

$SU(3)_C \times SU(2)_L \times U(1)_Y$:

$$\begin{array}{llll} Q & \equiv (3,2,1/6); & U^c \equiv (\bar{3},1,-2/3); & D^c \equiv (\bar{3},1,1/3); \\ L & \equiv (1,2,-1/2); & E^c \equiv (1,1,1); & \\ H_1 & \equiv (1,2,-1/2); & H_2 \equiv (1,2,1/2) \; . & \end{array} \tag{14}$$

Isospin and colour indices are contracted in the usual way. The couplings h_U, h_D and h_L are 3×3 matrices in the generation space $(i,j = 1,2,3)$.

The expression of the soft supersymmetry breaking operators turns out to be quite simplified by the assumption of having a flat Kähler metric. At the grand unified scale M_X they appear as:

i) a cubic gauge-invariant polynomial in the complex scalar fields:

$$S = mA\left[h_U\tilde{Q}\tilde{U}^cH_2 + h_D\tilde{Q}\tilde{D}^cH_1 + h_E\tilde{L}\tilde{E}^cH_1\right] + Bm\mu H_1H_2 + h.c. \tag{15}$$

The tilde denotes here the *scalar* component of the chiral matter superfields Q, U^c, D^c, L and E^c, while, for simplicity, we denote by H_1 and H_2 also the scalar components of the Higgs superfields H_1 and H_2. A and B are c-numbers and, in the presence of a flat Kähler metric, the equality $B = A - 1$ holds;

ii) a universal mass term for the scalar components y_i of the chiral superfields:

$$\mathcal{M}^2 \equiv m^2\Sigma_i|y_i|^2 \; ; \tag{16}$$

iii) gaugino Majorana mass terms:

$$\hat{M} \equiv \frac{M}{2}(\lambda_1\lambda_1 + \lambda_2\lambda_2 + \lambda_3\lambda_3) + h.c. \; , \tag{17}$$

where λ_1, λ_2 and λ_3 denote the two-component gaugino fields of $U(1)_Y$, $SU(2)_L$ and $SU(3)_C$, respectively. In eq. (17), the grand unification constraint of equal gaugino masses at M_X has been added to the usual requirement of minimality.

Hence, the class of minimal theories which we consider here is described at the grand unification scale by the Lagrangian:

$$L_{MSSM} = W + S + \mathcal{M}^2 + \hat{M} + \text{kinetic terms} \tag{18}$$

This Lagrangian contains four new parameters in addition to those which are already present in the SM, i.e. A, μ, m and M (for the present discussion we neglect CP violation effects and consider all parameters to be real). The correct electroweak breaking is achieved on the vacuum:

$$\langle H_1\rangle = \begin{pmatrix} v_1 \\ 0 \end{pmatrix}; \quad \langle H_2\rangle = \begin{pmatrix} 0 \\ v_2 \end{pmatrix}; \quad \langle\tilde{q}\rangle = \langle\tilde{\ell}\rangle = 0 \; , \tag{19}$$

where $\tilde{q}$ and $\tilde{\ell}$ denote the scalar quark and lepton fields, respectively. The tree-level scalar potential does not produce the correct vacuum. On the other hand, in order to discuss the physical implications of L_{MSSM} at low energy (i.e. at the electroweak scale), we need to renormalize the relevant parameters in eq. (18) from M_X down to $\sim m_W$. Imposing that the renormalization group improved potential leads to the vacuum in eq. (19) at m_W further reduces the number of genuinely new parameters introduced by SUSY to three.

Obviously, in addition to these parameters, our results will depend also on the mass of the top quark, m_t, as was already the case in the SM. The constraints we derive from rare B physics will be reported on a two-dimensional SUSY parameter space for fixed values of m_t and of the third remaining SUSY free parameter that we choose to be the ratio of the vacuum expectation values (VEV) of the two doublets H_1 and H_2, i.e. v_2/v_1.

The whole SUSY mass spectrum can be expressed in terms of the three SUSY parameters and m_t. We focus here only on the structure of the squark mass matrices, while referring the interested reader to reviews on the subject for a study of the chargino and neutralino mass matrices [23].

The 6×6 matrix of the $Q = 2/3$ sector is formally written in terms of the 3×3 submatrices M_{U_L}, M_{U_R} and $M^2_{U_{LR}}$ as follows:

$$\widetilde{M}^2_U = \begin{pmatrix} M_{U_L}M^\dagger_{U_L} & M^2_{U_{LR}} \\ M^{2\dagger}_{U_{LR}} & M^\dagger_{U_R}M_{U_R} \end{pmatrix} . \tag{20}$$

The expression of the above 3×3 blocks is very simple at the superlarge scale M_X where, roughly, the breaking of $N = 1$ supergravity takes place and where we start the running of our parameters. At that scale (unbroken $SU(2)_L \times U(1)_Y$):

$$\begin{aligned} M_{U_L}M^\dagger_{U_L} &= m^2\,\mathbf{1} \\ M^\dagger_{U_R}M_{U_R} &= m^2\,\mathbf{1} \\ M^2_{U_{LR}} &= 0\,. \end{aligned} \tag{21}$$

The renormalization of $\widetilde{M}^2_U$ from M_X down to the electroweak scale produces relevant effects. At the low-energy scale, the 3×3 blocks can be written as:

$$\begin{aligned} (M_{U_L}M^\dagger_{U_L})_{ij} &= a_{ij}M^2 + b_{ij}m^2 + c_{ij}m^2A^2 + d_{ij}mAM \\ &\quad + (M_U M^\dagger_U)_{ij} + M_Z^2\cos 2\beta\left(\frac{1}{2} - \frac{2}{3}\sin^2\theta_W\right)\delta_{ij} \end{aligned} \tag{22}$$

$$(M^2_{U_{LR}})_{ij} = (e_{ij}M + f_{ij}mA + \mu/\tan\beta)(M_U)_{ij} \tag{23}$$

$$\begin{aligned} (M^\dagger_{U_R}M_{U_R})_{ij} &= p_{ij}M^2 + q_{ij}m^2 + r_{ij}m^2A^2 + s_{ij}mAM \\ &\quad + (M^\dagger_U M_U)_{ij} + M_Z^2\cos 2\beta\left(\frac{2}{3}\sin^2\theta_W\right)\delta_{ij} \end{aligned} \tag{24}$$

where i, j are generation indices, M_U denotes the up-quark mass matrix, $\tan\beta = v_2/v_1$ and the dimensionless matrix coefficients $a, ..., s$ depend only on the gauge and Yukawa couplings and are determined by numerical solution of the RG equations driving the renormalization of the SUSY parameters from M_X down to M_W. Taking a look at the expression of these equations (see, for instance, App. A of ref. [21]), one can see that the various coefficients in eq. (22) receive contributions from both the up- and down-quark Yukawa couplings, at variance with the coefficients in eq. (24) which are affected by contributions only from the up-quark Yukawa couplings. Hence, when considering the diagonal up-quark basis (i.e. the up-quark basis which diagonalizes $M_U M_U^\dagger$) one simultaneously obtains the diagonalization of the $M_{U_R}^\dagger\, M_{U_R}$ block in eq. (24), while $M_{U_L} M_{U_L}^\dagger$, having contributions proportional to $M_D M_D^\dagger$, remains off-diagonal. The amount of this "lack of diagonality" traces back to the difference in the rotation for the diagonalization of the up- and down-quark mass matrices, i.e. the effect is roughly proportional to the elements of the Cabibbo-Kobayashi-Maskawa mixing matrix. This is the key element to understand a peculiarity of SUSY theories in the FCNC sector. If one considers a vertex gluino-quark-squark ($\tilde{g} - q - \tilde{q}$), for instance $\tilde{g} - u - \tilde{u}$, the absence of simultaneous diagonalization in the u and $\tilde{u}$ mass matrices leads to the possibility of a $\tilde{g}$ converting a u–quark into a $\tilde{c}$-squark at its vertex. Gluinos can mediate FCNC [24]. The same applies to neutralino vertices in general. For the purpose of our analysis, there is a second relevant observation on the $\tilde{q}$ mass matrices to be made. The down squark ($\tilde{d}$) mass matrix has a structure similar to that of $\widetilde{M}_U$ replacing M_U by M_D, with some new values of the $e, ..., s$ dimensionless coefficients established by the RG equations (the coefficients $a, ..., d$ remain unchanged) and with a different sign in front of the terms proportional to M_Z^2 in eqs. (22–24), namely:

$$\begin{aligned}(M_{D_L} M_{D_L}^\dagger)_{ij} &= a_{ij} M^2 + b_{ij} m^2 + c_{ij} m^2 A^2 + d_{ij} mAM \\ &+ (M_D M_D^\dagger)_{ij} - M_Z^2 \cos 2\beta \left(\frac{1}{2} - \frac{1}{3} \sin^2 \theta_W\right) \delta_{ij} \end{aligned} \qquad (25)$$

$$(M_{D_{LR}}^2)_{ij} = (e'_{ij} M + f'_{ij} mA + \mu \tan\beta)(M_D)_{ij} \qquad (26)$$

$$\begin{aligned}(M_{D_R}^\dagger M_{D_R})_{ij} &= p'_{ij} M^2 + q'_{ij} m^2 + r'_{ij} m^2 A^2 + s'_{ij} mAM \\ &+ (M_D^\dagger M_D)_{ij} - M_Z^2 \cos 2\beta \left(\frac{1}{3} \sin^2 \theta_W\right) \delta_{ij} \end{aligned} \qquad (27)$$

There is a major difference in the structure of $\widetilde{M}_U^2$ and $\widetilde{M}_D^2$: the presence of terms proportional to m_t in the off-diagonal block ($M_{U_{LR}}^2$) of $\widetilde{M}_U^2$ (eq. (23)). Given the large value of m_t, such $\tilde{t}_L \tilde{t}_R$ mass entry tends to produce one light (resp. one heavy) mass eigenstate $\tilde{u}_1$ ($\tilde{u}_6$) in the up squark sector, which is mainly a combination of $\tilde{t}_L$ and $\tilde{t}_R$. Just requiring $m_{\tilde{u}_1}$ to be positive forces the elements along the diagonal, and, hence, m and/or M to be rather large. Consequently also the diagonal elements of $\widetilde{M}_D^2$ tend to become correspondingly large, thus preventing some eigenvalue to be light. In addition to this effect, the presence of the D-term contributions, proportional to M_Z^2 (which also become relevant for large m_t since $v_1 \neq v_2$ is favoured) increases the average values of $m_{\tilde{d}}^2$ while decreasing $m_{\tilde{u}}^2$.

Since the d-squark masses cluster about an average value determined by the size of the elements along the diagonal (with the possible exception of one eigenstate, lighter, due the off-diagonal renormalization effects induced by the top quark in the M_{D_L} block), they turn out to be substantially heavy for most of the parameter space. These considerations are important, since the d-squarks eigenstates appear together with the gluinos, in potentially relevant contributions to the loop B processes here discussed. This feature, together with the rapidly growing bounds on gluino and squark masses from collider data, is at the root of the general loss of relevance of FCNC gluino mediated amplitudes in the RMSSM for rare B decays. As a consequence, a complete study of all the possible SUSY contributions to a given process is required.

In the SM the $b \to s$ and $b \to d$ transitions are dominated by one-loop contributions with the exchange of a virtual W and the top quark. When supersymmetry is considered several competing sources of FCNC are present. To begin with, in SUSY models the Higgs sector is richer than in the SM, since at least two Higgs doublets must be present. Consequently, there exists at least one physical charged scalar H^- which can be exchanged in the one-loop contributions to $b \to s$ or $b \to d$, together with an up-quark. The second obvious source of FCNC comes from the supersymmetrization of the W and the charged Higgs contribution, where the up quark is replaced by an up-squark $\tilde{u}$ and W^- and H^- are respectively replaced by their SUSY partners, w-ino ($\widetilde{W}^-$) and higgsino ($\widetilde{H}^-$). To be more precise, since $\widetilde{W}^-$ and $\widetilde{H}^-$ are only current eigenstate, the eigenstates of the 2×2 charged fermion mass matrix, the so-called charginos ($\widetilde{\chi}^-$) have actually to be considered. A less obvious source of FCNC, typical of SUSY theories, comes from the FC vertices $q - \tilde{q}' - \tilde{\chi}^0$ or $q - \tilde{q}' - \tilde{g}$, where $\tilde{\chi}^0$ is a neutralino and $\tilde{g}$ is the gluino. This, as we have previously mentioned, is due to a characteristic renormalization effect of the quark and squark mass matrices which arises when the effective low-energy Lagrangian is derived.

Thus, in SUSY there are five classes of one-loop diagrams which contribute to FCNC $b \to s$ and $b \to d$ transitions. They can be distinguished according to the virtual particles running in the loop:

1) W^- and up-quarks;

2) H^- and up-quarks;

3) $\tilde{\chi}^-$ and up-squarks;

4) $\tilde{\chi}^0$ and down-squarks;

5) $\tilde{g}$ and down-squarks.

As we said, the actual novelty of SUSY in the context of FCNC is the possibility for gluinos and neutralinos to realize flavour changes at their vertices with quarks and squarks. Gluino exchange in particular looks very promising for a possible SUSY enhancement in rare B physics. Indeed it was shown [25] that in the radiative decays $b \rightarrow s+\gamma$ and $b \rightarrow s+g$ the SUSY-GIM mechanism realizes a suppression comparable to that present in the SM, so that, taking into account the presence of the strong couplings in the case of $\tilde{g}$ exchange, one could correctly argue that such FCNC radiative b decays should be SUSY enhanced. This may remain true in some low energy SUSY realizations, however this hope is not fulfilled in the RMSSM [21]. The reason is twofold: on one hand, the search for $\tilde{g}$ and $\tilde{q}$ at CDF has pushed up the lower bounds on their masses at a level of 100 GeV and, on the other hand, in the RMSSM there is a general tendency for the down squarks to be heavier than the up squarks, as we have previously discussed. Notice that for the other class of superpenguins with virtual γ or g exchange, producing $b \rightarrow s\ell^+\ell^-$ or $b \rightarrow sq\bar{q}$, the standard GIM mechanism is soft whereas the corresponding SUSY-GIM suppression produces effects comparable to those which are present in $b \rightarrow s\gamma$ and $b \rightarrow sg$. Hence, in this case even the a priori hope of a substantial gluino enhancement is absent. In conclusion, in the RMSSM the contributions with $\tilde{g}$ and $\tilde{d}$ exchanged in the loop do not represent the leading contribution in FCNC b decays like $b \rightarrow s\gamma$, $b \rightarrow s\ell\bar{\ell}$ and $b \rightarrow sq\bar{q}$. Clearly the role of neutralinos is even more negligible given the smallness of the their couplings in comparison with the strong coupling of the gluinos.

The chargino and charged Higgs exchanges constitute the dominant SUSY contributions to rare B decays [21]. The best lower bounds on their masses come from LEP and amount roughly to 40 GeV. Notice that the lower bound of CDF on squark masses of $O(100\ GeV)$ refers to the case of degenerate squarks, but does not rule out the possibility of a single squark much lighter than all the remaining squarks. In view of our previous considerations on the large off-diagonal element $\tilde{t}_L - \tilde{t}_R$ in $\widetilde{M}_U^2$ it is indeed conceivable that one light stop (of, say, 25-30 GeV) may exist. Obviously, its exchange in a loop with a light chargino represents a potentially important contribution in FCNC rare B decays.

We come now to a quantitative analysis of our results. It is useful to recall that, apart from the implementation of the bounds on the SUSY particle masses, what makes a strict selection of the points at disposal in the SUSY parameter space is the requirement of the electroweak radiative breaking [26]. We shall show how much of the SUSY parameter space is already ruled out by this request and which further reduction is obtained by imposing the experimental bounds on rare B decays.

In our view $b \rightarrow s+\gamma$ represents one of the most interesting FCNC rare B decays. The presence of large QCD corrections in the SM has been previosuly discussed. In this section we shall point out that this process offers the best chance in rare B physics to constrain the SUSY parameter space in a way complementary to the results achievable in direct accelerator searches. Most of the material presented in this section is taken from ref. [21] and we refer the interested reader to that work for a more detailed discussion.

In fig. 4 we give a quantitative content to our previous considerations on the relative importance of the various SUSY contributions. The individual components of the amplitude for $b \rightarrow s+\gamma$, corresponding to the exchange of the charged Higgs, the charginos,

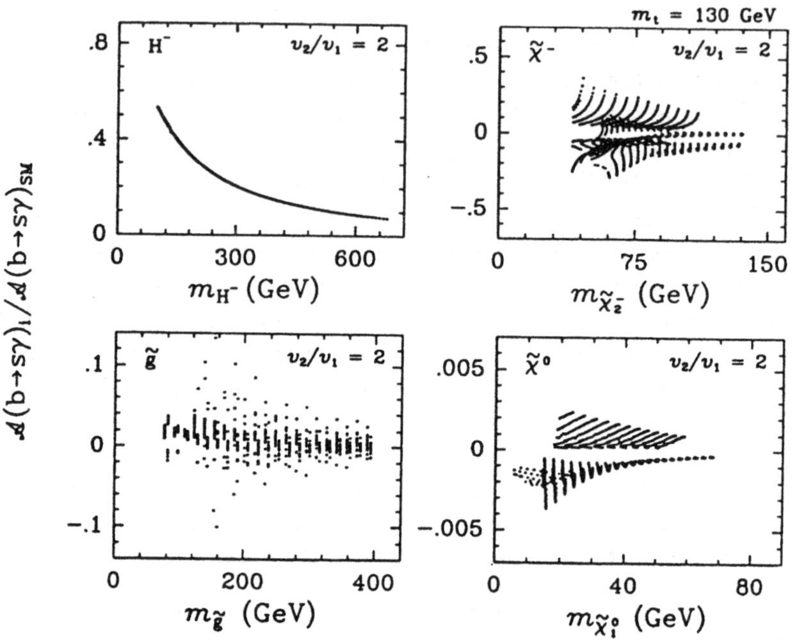

Figure 4: Ratios of SUSY to SM amplitudes contributing to $b \to s + \gamma$ (leading QCD corrections are included). The figure is taken from ref. [21].

the gluinos and the neutralinos, are compared to the value of the SM amplitude. QCD corrections are included in these ratios. It is apparent from this figure that charged Higgs and chargino contributions dominate over the gluino exchange contributions, at least for charged Higgs masses below 300 GeV. A second remark concerns the size of the separate SUSY amplitudes in comparison with the SM amplitude. Fig. 4 clearly shows that the exchange of SUSY particles never prevails over the standard W exchange in $\mathcal{A}(b \to s+\gamma)$. At this point it becomes crucial to determine whether there exists a definite pattern of interference between the SUSY and the SM contributions. The answer is exhibited in fig. 5. The figure shows the size of the various processes as a function of the soft breaking mass m. The other parameter, needed to determine completely the structure of the model, is taken to be the gaugino mass M, which is varied between ± 200 GeV. The solid lines refer to the SM expectations for $m_t = 130$ GeV, with the inclusion of the QCD corrections. The dotted areas denote the RMSSM predictions as a function of the SUSY breaking parameter m for two different values of v_2/v_1, 2 and 8 (fig. 5a and 5b, respectively). Remember that, having fixed v_2/v_1, two SUSY parameters remain free. This explains why to a certain value of m, more values of $BR(b \to s + \gamma)$ correspond. The interesting feature of fig. 5 concerning $BR(b \to s + \gamma)$ is that the RMSSM dotted band lies almost entirely above the SM solid line. The SUSY contributions to $b \to s + \gamma$ interfere positively with the SM contribution (Higgs exchange generally dominates) and

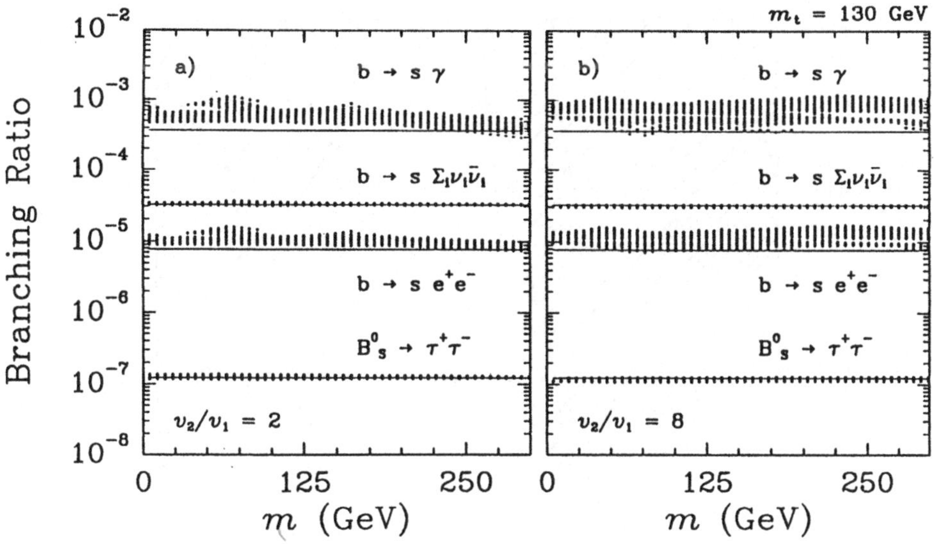

Figure 5: Expected branching ratios of various loop-induced $b \to s$ transitions in the RMSSM for $m_t = 130\ GeV$ and $\tan\beta = 2,\ 8$, as a function of the soft breaking scalar mass m. The horizontal solid lines represent the corresponding SM predictions ($f_{B_s} = 150\ MeV$). The figure is taken from ref. [27].

the conclusion is that in the RMSSM the $BR(b \to s + \gamma)$ is expected to be larger than the SM prediction, reaching a maximal enhancement of about a factor three.

As discussed in sect. 2, the SM prediction for $BR(b \to s + \gamma)$ varies between 3×10^{-4} and 4×10^{-4} for $90\ GeV < m_t < 160\ GeV$ when the large QCD corrections are included. Hence the CLEO II bounds are only roughly a factor three larger than the *lowest* predicted value in the SM (a factor six when considering the very conservative value $R^* = 5\%$ for the lightest exclusive channel). Given that $BR(b \to s + \gamma)$ is predicted to be even larger in the RMSSM, we conclude that an improvement of the present CLEO II bound by a factor 2–3 is of utmost importance in testing the SUSY contribution and the presence of the large QCD corrections in SM.

To help clarifying the message concerning the RMSSM which should be drawn from the results depicted in fig. 5, consider the case of the top mass being measured at 130 GeV with an uncertainty of 10 GeV. The *inclusive* rate for $b \to s + \gamma$ is then known in the SM with an accuracy of $10 - 15\%$ (the accuracy of the leading-log estimate of the QCD corrections). In figs. 15-17 of ref. [21], we showed, for different combinations of the SUSY parameters, the regions of the parameter space allowed when the radiative breaking of $SU(2) \times U(1)$ is required, compared to the tiny regions which are left after imposing a bound on $BR(b \to s + \gamma)$ just above the SM prediction. In fig. 6 we extend further this

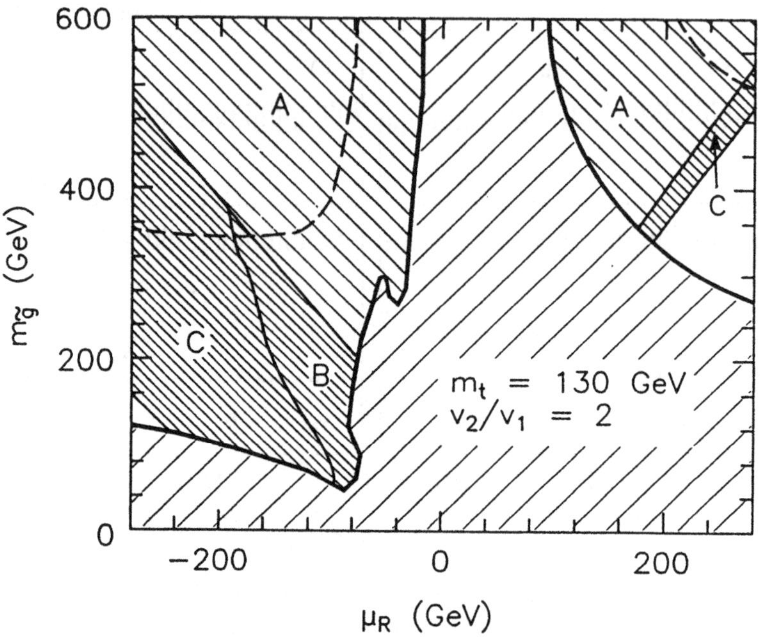

Figure 6: The regions of the plane $m_{\tilde{g}} - \mu$ excluded by SUSY searches at LEP I (area enclosed by the thick solid line), and the projected limits obtainable at LEP II (area below the dashed line), as reported in ref. [28], are compared with the regions excluded by: i) the sole requirement of radiative electroweak breaking (A); ii) a bound on $BR(b \to s+\gamma)$ 50% above the SM prediction ($A+B$); iii) 15% above the SM prediction ($A+B+C$). The latter is equivalent to a measurement consistent with the SM expectation for $m_t = 130\ GeV$. The figure is taken from ref. [27].

analysis by directly comparing the potentiality of the radiative $b \to s+\gamma$ decay with that of SUSY searches at LEP I and II. As an indicative example, we consider the area of parameter space which is excluded by a bound on $BR(b \to s+\gamma)$ at the level of 1.5 times the SM prediction and that excluded by a measurement consistent with the SM estimate (within 15%). For $m_t = 130\ GeV$ the two cases correspond to *inclusive* branching ratios of 5.6 and 4.3 times 10^{-4} respectively ($BR_{SM} \approx 3.7 \times 10^{-4}$). In fig. 6 we show the regions of parameter space excluded in the two aforementioned cases (dark shaded areas $A+B$ and $A+B+C$) confronted with those excluded by present SUSY searches at LEP I (light shaded area) and by the projected searches at LEP II (area below the dashed lines). It is apparent that an improvement of $BR(b \to s+\gamma)$ by at least a factor of two may offer a test of the SUSY parameter space which is comparable with what LEP II can produce.

We finally come to what was considered in the past a kind of benchmark for rare FCNC b decays, i.e. b semileptonic decays with strangeness in the final products. In fig. 5 we have reported the results concerning $b \to s\nu\bar{\nu}$ $b \to s\ell^+\ell^-$, and $B_s^0 \to \tau^+\tau^-$ in the RMSSM and, for comparison, in the SM. Although the last two decays exhibit a quite clear experimental signature, their rates are too small for the present machines and we must wait for the advent of B factories to exploit the information coming from these rare processes. For $b \to s\ell^+\ell^-$ we can repeat what we said for $b \to s + \gamma$. The SUSY contributions interfere constructively with the SM W-exchange contribution for almost the entire available SUSY parameter space. This is due to the dominance of Higgs exchange in the dipole component of the amplitude. In fact, all the Higgs-induced contributions in which a helicity flip between b and s is not needed are suppressed by $\tan^{-2}\beta$. This is clearly shown by the snallness of the corrections to $b \to s\nu\bar{\nu}$, where the electromagnetic penguin, and therefore the dipole form factor, is missing. If the bound on the inclusive $BR(b \to s\ell^+\ell^-)$ is pushed down to $O(10^{-5})$ (which corresponds to bounds on exclusive modes, as $B \to K\ \ell^+\ell^-$, an order of magnitude lower [29]) a large area of the SUSY parameter space (comparable to that excluded by $b \to s\gamma$) could be ruled out. This means that a significant result for SUSY can be achieved only by an improvement of at least two orders of magnitude with respect to the present experimental bounds.

We conclude our analysis of SUSY contributions to FCNC processes in B physics with the $B^0 - \bar{B}^0$ oscillations. We address the following two questions: can the exchange of SUSY particles in the box diagrams responsible for ΔM in the $B_d^0 - \bar{B}_d^0$ system produce a significant enhancement over the value of $\Delta M(B_d)$ predicted by SM ? Is the ratio $\Delta M(B_d)/\Delta M(B_s)$ in the RMSSM the same as in the SM ? A clear answer is provided by fig. 7. The inclusion of SUSY particles in the ΔM box diagrams yields a quite moderate enhancement of no more than 30% over the SM prediction. This result is merely a consequence of the suppression of the charged Higgs induced diagrams for $\tan\beta > 1$. As for the second abovementioned question, the answer comes from a comparison of $\Delta M_{SUSY}/\Delta M_{SM}$ for the $B_d^0 - \bar{B}_d^0$ and $B_s^0 - \bar{B}_s^0$ cases. Their close resemblance indicates that also in the RMSSM the usual SM scaling of ΔM by $(K_{ts}/K_{td})^2$ remains true. This result, which is obvious for the Higgs induced component, holds with large accuracy also for the chargino, gluino and neutralino contributions. This shows, as one naively expects, that the standard CKM elements K_{td} and K_{ts} play also a relevant role in the rotation of the squark mass matrices.

It is important to recall that these results are obtained in the minimal framework, and therefore do not exclude in general the possibility of measurable supersymmetric effects in other contexts. At any rate, they can provide a useful guidance for further analysis.

An updated SUSY study of FCNC effects and CP violation in the Kaon sector is missing. The authors of ref. [30] study the impact of supersymmetry on rare B decays, mixings and related CP asymmetries. However, they do not perform a consistent numerical study of the renormalization of the relevant low-energy parameters. In particular, the splitting of the down-squark mass eigenstates is assumed to be simply of the form $c\ m_{\tilde{t}}^2$, with $c = 0.5$. As we have seen in our present discussion, this amounts in most cases to overestimating the real effect. In addition, it does not account for the crucial

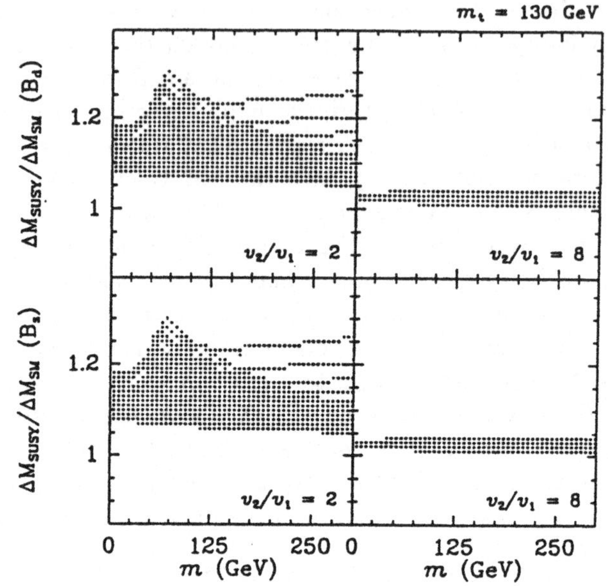

Figure 7: The ratio of the *total* RMSSM contribution to Δm_{B_q} (q = d, s) over the corresponding SM prediction is shown for $m_t = 130\ GeV$ and $\tan\beta = 2,\ 8$, as a function of the soft breaking mass m [21].

interplay between the renormalization of the up-squark and down-squark mass matrices, which turns out to preclude any relevant role of the gluino induced contributions.

4. CONCLUSIONS

In this lecture we have discussed some aspects related to FCNC transitions in the K and B meson systems in the standard electroweak scenario and beyond. We have tried to individuate criteria for a systematic understanding of the sensitivity of these low energy phenomena to a heavy top quark or exotic physics at the electroweak scale. We have focussed our analysis on the minimal supersymmetric extension of the standard model.

The lesson we have learned is that, while the combined study of rare K and B processes is quite effective in limiting extensions of the standard scenario, theoretical and experimental improvements are needed for a detailed search of heavy top effects. There is certainly no competition with the precision tests of the standard model performed at the high-energy colliders.

The evaluation of the hadronic matrix elements and long-distance contributions to the various processes represent from the point of view of the theory the major source of uncertainty. The analysis of the short-distance components and of the leading QCD corrections is instead, for $m_t < 200\ GeV$, well under control. In the B meson system, short-distance contributions are known, for a given top mass, with a precision of 10%–20%. Therefore, if the top quark is found (at CDF?) then the study of rare B phenomena may provide very stringent tests on non-standard physics, and, what is perhaps more important, give us precise information ($O(10\%)$) on the long-distance components of the transitions. Among else, this is a valuable input for the study of CP violation in the B system, which is the preferred way of performing a redundant test of the structure of the CKM matrix, and, ultimately, the motivation for building a B factory.

REFERENCES

[1] N. Cabibbo, Phys. Rev. Lett. 10 (1963) 531; M. Kobayashi and M. Maskawa, Prog. Theor. Phys. 49 (1973) 652.

[2] S.L. Glashow, J. Iliopoulos and L. Maiani, Phys. Rev. D 2 (1970) 1285.

[3] G. Buchalla, A.J. Buras and M.K. Harlander, Nucl. Phys. B 349 (1991) 1.

[4] A.J. Buras and M.K. Harlander, in "Heavy Flavours", eds. A.J. Buras and M. Lindner, World Scientific, Singapore, 1992.

[5] S. Bertolini, Plenary session report at the DESY Theory Workshop, Hamburg, 1990 (unpublished).

[6] A.A. Penin and A.A. Pivovarov, Moscow Institute for Nuclear Research, Preprint 725 (1991).

[7] T. Inagaki *et al.* , Phys. Rev. D 40 (1989) 1712.

[8] BNL AGS E791 Exp., A.P. Heinson *et al.* , Phys. Rev. D 44 (1991) R1.

[9] Particle Data Group, Phys. Lett. B 239 (1990) 1.

[10] NA31 Coll., G.D. Barr *et al.* , Phys. Lett. B 259 (1991) 389; AGS E845 Coll., W.M. Morse, Preprint BNL-46282 (1991).

[11] L. Bergström, E. Masso and P. Singer, Phys. Lett. B 249 (1990) 141.

[12] D. Rein and L.M. Sehgal, Phys. Rev. D 39 (1989) 3325.

[13] C. Dib, I. Dunietz and F.J. Gilman, Mod. Phys. Lett. A 6 (1991) 3573.

[14] BNL E787 Exp., D.S. Akerib, Ph.D. Thesis, Preprint DOE/ER-3072-64 (1991).

[15] S. Bertolini, F. Borzumati and A. Masiero, Phys. Rev. Lett. 59 (1987) 180; N.G. Deshpande, P. Lo, J. Trampetic, G. Eilam and P. Singer, Phys. Rev. Lett. 59 (1987) 183.

[16] B. Grinstein, R. Springer and M. Wise, Phys. Lett. B 202 (1988) 138; Nucl. Phys. B 339 (1990) 269; G. Cella, G. Curci, G. Ricciardi and A. Viceré, Phys. Lett. B 248 (1990) 181.

[17] T. Altomari, Phys. Rev. D 37 (1988) 677; C.A. Dominguez, N. Paver and Riazuddin, Phys. Lett. B 214 (1988) 459; P.J. O'Donnel, in "Quark Gluons and Hadronic Matter", Eds. R. Viollier and N. Warner, World Scientific, Singapore (1987).

[18] A. Ali and C. Greub, Zeit. für Physik C 49 (1991) 431; Phys. Lett. B 259 (1991) 182.

[19] M. Danilov, Plenary session report at the LP-HEP/91 International Conference, 29 July - 2 August 1991, Geneva, CH.

[20] S. Bertolini, in "Higgs Particles", ed. A. Ali, Plenum Press, New York, 1990.

[21] S. Bertolini, F. Borzumati, A. Masiero and G. Ridolfi, Nucl. Phys. B 353 (1991) 591.

[22] H.P. Nilles, Phys. Rep. 110 (1984) 1.

[23] H.E. Haber and G.L. Kane, Phys. Rep. 117 (1985) 75;
J.F. Gunion and H.E. Haber, Nucl. Phys. B272 (1986) 1.

[24] M.J. Duncan, Nucl. Phys. B221 (1983) 285;
J.F. Donoghue, H.P. Nilles and D. Wyler, Phys. Lett. B128 (1983) 55;
M.J. Duncan and J. Trampetic, Phys. Lett. B134 (1984) 439;
E. Franco and M. Mangano, Phys. Lett. B135 (1984) 445;
J.M. Gerard, W. Grimus, A. Raychauduri and G. Zoupanos, Phys. Lett. B140 349;
J.M. Gerard, W. Grimus, A. Masiero, D.V. Nanopoulos and A. Raychaudhuri, Phys. Lett. B141 (1985) 93;
P. Langacker and R. Sathiapalan, Phys. Lett. B144 (1984) 401;
A. Bouquet, J. Kaplan and C.A. Savoy, Phys. Lett. B148 (1984) 69; Nucl. Phys. B262; J.-P. Deredinger and C.A. Savoy, Nucl. Phys. B237 (1984) 307;
N. Dugan, B. Grinstein and L. Hall, Nucl. Phys. B255 (1985) 413.

[25] S. Bertolini, F. Borzumati and A. Masiero, Phys. Lett. B192 (1987) 437; Nucl. Phys. B294 (1987) 321; Phys. Lett. B194 (1987) 545; (E) ibid. 198 (1987) 590.

[26] K. Inoue, A. Kakuto, H. Komatsu and S. Takeshita, Progr. Theor. Phys. 68 (1982) 927; ibid. 71 (1984) 413;
L.E. Ibanez and G.G. Ross, Phys. Lett. B110 (1982) 214;
H.P. Nilles, Phys. Lett. B118 (1982) 193;
L. Alvarez-Gaumé, M. Claudson and M.B. Wise, Nucl. Phys. B207 (1982) 96;
L. Alvarez-Gaumé, J. Polchinski and M.B. Wise, Nucl. Phys. B221 (1983) 495;
J. Ellis, J.S. Hagelin, D.V. Nanopoulos and K. Tamvakis, Phys. Lett. B125 (1983) 275; L.E. Ibanez and C. Lopez, Phys. Lett. B126 (1983) 54; Nucl. Phys. B233 (1984) 511; C. Kounnas, A.B. Lahanas, D.V. Nanopoulos and M. Quiros, Nucl. Phys. B236

(1984) 438; U. Ellwanger, Nucl. Phys. B238 (1984) 665;
B. Gato, J. Leon, J. Perez-Mercader and M. Quiros, Nucl. Phys. B253 (1985) 285;
L.E. Ibanez, C. Lopez and C. Muñoz, Nucl. Phys. B256 (1985) 218;
B. Gato, Zeit. für Physik C37 (1988) 407;
G.F. Giudice and G. Ridolfi, Z. Phys. C41 (1988) 447;
M. Olechowski and S. Pokorski, Phys. Lett. B214 (1988) 239;
P. Li and H. Navelet, Nucl. Phys. B319 (1989) 239;
W. Majerotto and B. Mösslacher, Z. Phys. C48 (1990) 273.

[27] S. Bertolini, F. Borzumati and A. Masiero, in " B Decays", ed. S. Stone, World Scientific, Singapore, 1991.

[28] H. Baer *et al.* , Proceedings of the 1990 DPF Summer Study on High Energy Physics, Snowmass, CO.

[29] N.G. Deshpande and J. Trampetic, Phys. Rev. Lett. 60 (1988) 2583;
C.A. Dominguez, N. Paver and Riazzudin, Zeit. für Physik C 48 (1990) 55.

[30] I.I. Bigi and F. Gabbiani, Nucl. Phys. B 352 (1991) 309.

SUPERSYMMETRIC QUANTUM EFFECTS ON ELECTROWEAK PRECISION OBSERVABLES

J. SOLÀ

Grup de Física Teòrica
Universitat Autònoma de Barcelona
08193 Bellaterra (Barcelona), Catalonia, Spain

ABSTRACT

I review the analysis of one-loop SUSY radiative corrections to the weak gauge boson masses, W and Z, and to the low energy ρ parameter in the context of a specific complete renormalization framework for electroweak interactions based on neutral currents processes. I also discuss the prospects for detecting SUSY radiative effects from the combined use of high energy and low energy data to be obtained in the next generation of high precision experiments. Details are provided of the electroweak renormalization framework used throughout.

1. Introduction

The present day "burning" activity going on at the various fronts of experimental endeavour in Particle Physics, whether it be in electron-positron, proton-antiproton, neutrino-nucleon or in neutrino-electron scattering experiments, is pushing day by day the precision at which the Standard Model (SM) is being tested [1] to the very frontiers reached by the most sophisticated theoretical calculations [2]-[9]. These calculations are able to account for the 1-loop (and in some respects 2-loop [10]) quantum effects on the amplitudes of the physical processes, evaluated in the field theory context of the SM with only one Higgs doublet. Several schemes have been devised to face the complexity of these calculations [2]-[7], but apart from minor numerical differences among them, due to unavoidable effects from truncation in the loop expansion, the convergence of the various approaches is a most valuable piece of information to trust in practice the correcteness of the results as well as the healthiness of the whole theoretical construct underlying the SM.

In view of these facts, one can hardly strengthen, for obvious, the conclusion that the "golden age" of the radiative corrections is born in electroweak physics. The "old" days of the first real production of the intermediate gauge bosons, with very rough determination of their masses, are now superseded by impressively good accurate measurements of the mass and width of the Z and fairly good, though still far not comparable, measurements of the corresponding W parameters. We are presently living the LEP I era, the era of Z physics, but the LEP II era, that of W physics, is not far away from us.

Above all, the ultimate endeavour for a physicist is to assess to which extension theory matches experiment. Indeed we are now closer than ever to the very crucial confrontation of theory with experiment at the level of the fine renormalization effects predicted by our favourite quantum field theories. On the theoretical side, calculations of the radiative shifts induced on the W and Z boson masses in the framework of the SM have been performed by several authors and in different schemes [2]-[9]. On the experimental side, recent measurements of the W and Z boson masses and precise determinations of $sin^2\theta_W$ and of the ρ-parameter patently show fairly good agreement between theory and experiment at the level of quantum corrections [1].

The successfulness of this program reinforces to the fullest the necessity of testing the predictions of the SM to the limit where hints of potential new physics may be disentangled, or at least be severely constrained, once the remainder of ignorance on the parameters of the SM (e.g., the top quark mass and the Higgs mass) will be pinned down. There are still important gaps to fill in our descriptive knowledge of the pure SM electroweak physics, such as the finding of the Higgs scalar and of the top quark [11], not to mention our complete ignorance on deep theoretically rooted questions, such as the whys and wherefores of having such a complicated structure for the strong and electroweak interactions, with so many uncorrelated parameters, flavours, mixings, families etc., whose precise values, or even their mere existence, looks fully divorced from the presence of the feeblest, though ubiquitous and uttermost important- cosmic featuring!- force in nature: gravity.

Since the old days of supersymmetry (SUSY) [12], the idea of using the new Fermi-Bose paradigm to reconcile the gravitational force with the $SU(3)_c \times SU(2)_L \times U(1)_Y$ interactions has been cherished by theoretical physicists [13]. In modern days one expects that the final unification may come from the realm of supersymmetric string theory, i.e. Superstring Theory [14], although it is by no means clear to everybody that such a theoretical creature will ever survive at all. Be as it may, the idea that supersymmetry could play a significant role in the important issue of building a consistent grand unified theory (GUT) of all interactions is worth exploring [15] and should not be abandoned, despite all previous unsuccessfull efforts to tag a genuine SUSY signal, and still less in a moment that highly precision data are being obtained to test the correcteness of the SM to its ultimate consequences [1]. Perhaps the most fashionable motivation that can be adduced at present to support SUSY is that the extrapolation of the LEP high precision data, with the inclusion of supersymmetric matter contributions in the renormalization group equations that govern the running of the three gauge coupling constants, displays an spectacular coupling constant unification at $M_X \simeq 10^{16}\ GeV$ (Cf. ref. [16]), provided the sparticle masses lie in the range $100\ GeV \leq M_{SUSY} \leq 10\ TeV$. Of course, this is not a proof that SUSY exists, but it is difficult not to be impressed by this result. It could be a virtual manifestation of SUSY through quantum effects from the Fermi scale to the GUT scale.

In these lectures I would like to elaborate on the complementary issue of the radiative corrections on the basic electroweak parameters from SUSY, with special emphasis on the comparison between (neutrino) low energy and (LEP) high energy data. There exist in the literature various partial analyses of SUSY radiative corrections to the ρ parameter and to the weak gauge boson masses, $M_{W,Z}$, and in different renormalization schemes [17]-[21]. Here I will develop a more detailed exposition of the analysis of the results presented schematically in references [20] and [21]. The main point raised in these references was to show that for a large, phenomenologically relevant, region of SUSY parameter space (involving sparticle masses of order 100 GeV) only a fully-fledged calculation (i.e. not just a two-point function estimate) can shed light on the answer to the final SUSY corrections predicted in a given renormalization framework. This is in contradistinction to the SM case, where the bulk of the radiative corrections just comes from the universal two-point functions, i.e. from the so-called indirect or "oblique" [22] fermionic contributions (quarks and leptons) to the self energies of the electroweak gauge bosons [3]-[5]. Whereas the "direct" (process-dependent) corrections (namely: boxes, vertices and external wave function renormalization) corrections are comparatively negligible in the SM, this is no longer the case in SUSY. All sources of SUSY quantum corrections may be equally important in the aforementioned parameter range and for a given scheme.

The plan is the following. Section 2 develops the specific renormalization framework used throughout. Section 3 introduces the necessary SUSY formalism. Section 4 presents the analytical and numerical analysis of the SUSY radiative corrections. Finally, Section 5 is devoted to the conclusions and some speculations on 2-loop SUSY quantum effects on the basic electroweak parameters. An appendix at the very end provides some useful formulas.

2. Renormalization Framework

Our renormalization framework is based on the on-shell scheme developed in ref.[2], which was extensively used for 1-loop SM calculations in refs. [3] and [5], and in 2-loop calculations in ref.[10]. The following three processes are used to define the basic electroweak parameters: $\nu_\mu e$ and $\bar{\nu}_\mu e$ elastic scattering is used to define the electroweak mixing angle $s_\theta^2 \equiv \sin^2\theta_W$; muon decay to define the Fermi constant, G_F, and finally Coulomb scattering to define the fine structure constant $\alpha(k^2 = 0)$. In many applications, it is convenient to take the on-shell renormalization framework of ref.[4], where M_Z is taken as an input datum. In view of the present high precision measurements of this quantity, it proves to be an extremely convenient choice [7]-[8]. But certainly not so interesting if one is addressing the issue of the radiative corrections to M_Z itself!.

Clearly, the source of largest error in our set-up comes from the first parameter mentioned. We shall comment later on about the uncertainty induced in our calculations from the inherent error in the parameter s_θ^2, which is inputed from low energy neutrino-electron scattering, and specially on the prospects to improve that error in the future (also in deep inelastic scattering). But for the moment we take the point of view that such a scheme can be useful to illustrate the kind of effects to be expected in the context of a fully-fledged analysis involving the complete structure of the 1-loop SUSY corrections. Needless to say, the calculations in this framework are purely electroweak (no uncertainty from strong interactions), and although they are anyway quite cumbersome, they are at least clean and straightforward. From the results obtained one can assess whether facing a similar, but still more complex analysis, namely an analysis in which $\nu_\mu e$-scattering is replaced by deep νN-scattering (with a corresponding error on s_θ^2 likely to be improved substantially in the future), could be useful or not to unravel potential SUSY quantum effects.

Specifically, in our framework s_θ^2 is fixed through the ratio

$$R = \frac{\sigma(\bar{\nu}_\mu e \to \bar{\nu}_\mu e)}{\sigma(\nu_\mu e \to \nu_\mu e)} = \frac{\xi^2 + \xi + 1}{\xi^2 - \xi + 1} , \tag{1}$$

where the parameter ξ is given by

$$\xi = -a/b, \tag{2}$$

with a and b the vector and axial coefficients in the neutral current interaction

$$\mathcal{L}_Z = \frac{g}{4c_\theta} \bar{l}\gamma^\mu (a - b\gamma_5) l \; Z_\mu \, . \tag{3}$$

Here $c_\theta \equiv \cos\theta$, and the values of a and b for the leptons are

$$a = -1 + 4s_\theta^2, \; b = -1 \;\text{ for } l = e, \mu, \text{ etc}, \tag{4}$$

and

$$a = b = 1 \;\text{ for }\; l = \nu_e, \nu_\mu, \text{ etc},. \tag{5}$$

Therefore, the value of s_θ^2 obtained from the measured parameter ξ is given by

$$\frac{1}{4}(1 + \xi) = s_\theta^2 \, . \tag{6}$$

This is our (to all orders) definition of the (physical) electroweak mixing angle.

The expressions for the cross sections used to compute the ratio R are:

$$\sigma(\nu_\mu e^- \to \nu_\mu e^-) = G_F^2 \frac{m_e E_\nu}{8\pi} [(a + b)^2 + \frac{1}{3}(a - b)^2] \tag{7}$$

and

$$\sigma(\bar{\nu}_\mu e^- \to \bar{\nu}_\mu e^-) = G_F^2 \frac{m_e E_{\bar{\nu}}}{8\pi} [(a - b)^2 + \frac{1}{3}(a + b)^2] \, , \tag{8}$$

where E_ν ($E_{\bar{\nu}}$) is the incident neutrino (antineutrino) energy in the laboratory. In these formulas, a and b are those corresponding to the charged leptons (eq.(4)).

At the tree level, the bare values of the parameters are finite and given by $a^{(0)} = a$ and $b^{(0)} = b$. Therefore, in lowest order of perturbation theory

$$\xi = -\frac{a^{(0)}}{b^{(0)}} \, , \tag{9}$$

and so at this order the physical relation (6) remains true

$$\frac{1}{4}(1 + \xi) = s_\theta^{(0)2} \, . \tag{10}$$

But at higher orders the coefficients a and b in equations (3)-(8) are shifted from the bare values by the radiative corrections. For example, let the 1-loop bare values be $a^{(1)}$ and $b^{(1)}$. The coefficients a and b are now given by

$$a = a^{(0)} = a^{(1)} + \delta a \;\; , \;\; b = b^{(0)} = b^{(1)} + \delta b \tag{11}$$

and the 1-loop bare value of s_θ^2 by

$$s_\theta^{(1)2} = \frac{1}{4}(1 - \frac{a^{(1)}}{b^{(1)}}) , \tag{12}$$

while the parameter ξ reads

$$\xi = \frac{a^{(1)} + \delta a}{b^{(1)} + \delta b} . \tag{13}$$

Eq.(12) is the model Lagrangian definition of the bare mixing angle and remains formally true order by order in perturbation theory.

From eqs.(10-12) it is easy to see that

$$s_\theta^{(1)2} = \frac{1}{4}(1 - \frac{a^{(0)} - \delta a}{b^{(0)} - \delta b}) = s_\theta^{(0)2} + \frac{a^{(0)}}{4b^{(0)}}(\frac{\delta a}{a^{(0)}} - \frac{\delta b}{b^{(0)}}). \tag{14}$$

This equation shows that, in contradistinction to the tree level result (10), the radiative corrections modify the relation between ξ and the bare vector and axial coefficients; it is no longer of the form (6). In fact, at 1-loop that relation gets shifted in the following way:

$$\frac{1}{4}(1 + \xi) = s_\theta^{(1)2}(1 - \delta_3) , \tag{15}$$

where

$$\delta_3 = \frac{a^{(0)}}{4s_\theta^{(0)2} b^{(0)}}(\frac{\delta a}{a^{(0)}} - \frac{\delta b}{b^{(0)}}), \tag{16}$$

as it follows from eqs.(10), (14) and (15) [23].

Clearly, δ_3 acts as a counterterm in the definition of the electroweak mixing angle. Its value follows after computing all possible (1-loop) radiative corrections to $\nu_\mu e$-scattering. Thus it is in general an ultraviolet (UV) divergent quantity, as it also is the bare value $s_\theta^{(1)2}$. The UV-divergences must cancel on the RHS of eq.(15) in order that the LHS is finite. In this way δ_3 takes care that eq.(6) be a consistent definition of the physical mixing angle.

Concerning the definition of the Fermi constant, one has

$$\Gamma(\mu^- \to e^- \nu_\mu \bar{\nu}_e) = \frac{G_F^2 m_\mu^5}{192\pi^3}(1 - 8\frac{m_e^2}{m_\mu^2}) , \tag{17}$$

where it is understood that we have already subtracted from the experimental value, Γ_{exp}, the pure (i.e. the so-called "conventional") electromagnetic radiative corrections, which are well known to be finite:

$$\Gamma(\mu^- \to e^- \nu_\mu \bar{\nu}_e) \equiv \Gamma_{exp} - \frac{G_F^2 m_\mu^5}{192\pi^3}\frac{\alpha}{2\pi}(\frac{25}{4} - \pi^2) . \tag{18}$$

At the tree level, the relation between G_F and the parameters of the charged-current Lagrangian

$$\mathcal{L}_W = \frac{g}{2\sqrt{2}}[\bar{l}\gamma^\mu(1 - \gamma_5)\nu_l W_\mu^- + h.c.] \tag{19}$$

reads as follows:

$$\frac{G_F}{\sqrt{2}} = \frac{g^{(0)2}}{8M_W^{(0)2}} . \tag{20}$$

We call $\mathcal{L}^{(0)} = \mathcal{L}(s_\theta^{(0)2}, g^{(0)}, M_W^{(0)})$ the bare Lagrangian at the tree-level, i.e. the classical Lagrangian $\mathcal{L}$ expressed in terms of the bare parameters in lowest order of perturbation theory (or tree-level parameters). These are finite, but not physical.

As in eq.(15), the previous tree-level relation gets shifted when written in terms of the 1-loop bare parameters $g^{(1)}$ and $M_W^{(1)}$:

$$\frac{G_F}{\sqrt{2}} = \frac{g^{(1)2}}{8M_W^{(1)2}}(1+\delta_1) . \tag{21}$$

Here δ_1 is the corresponding counterterm taking care that G_F is consistently defined through the relation (17). Its value is to be obtained by explicitly computing all possible radiative corrections to μ-decay.

Finally, the fine structure constant, $\alpha \equiv \alpha(k^2 = 0)$, is defined as the residue of the one-photon exchange pole at zero momentum transfer in a Coulomb scattering process, say $\mu e \to \mu e$. At lowest order, we obviously have

$$\alpha = \frac{g^{(0)2} s_\theta^{(0)2}}{4\pi}, \tag{22}$$

whilst at the 1-loop level

$$\alpha = \frac{g^{(1)2} s_\theta^{(1)2}}{4\pi}(1+\delta_2) , \tag{23}$$

where δ_2 is the counterterm associated to our definition of the fine structure constant. It follows upon computing all possible (1-loop) radiative corrections to Coulomb scattering.

By inverting the simultaneous set of fundamental relations (15), (21) and (23), we obtain the bare 1-loop values of the Lagrangian fundamental parameters in terms of the various counterterms and the physical parameters:

$$s_\theta^{(1)2} = \frac{1}{4}(1+\xi)(1+\delta_3) = s_\theta^{(0)2}(1+\delta_3) , \tag{24}$$

$$g^{(1)2} = \frac{16\pi\alpha}{1+\xi}(1-\delta_2-\delta_3) = g^{(0)2}(1-\delta_2-\delta_3) , \tag{25}$$

$$M_W^{(1)2} = \frac{4\pi\alpha}{\sqrt{2}G_F(1+\xi)}(1+\delta_1-\delta_2-\delta_3) = M_W^{(0)2}(1+\delta_1-\delta_2-\delta_3) . \tag{26}$$

On the second RHS's of these equations, I have replaced just for notational convenience the physical parameters by the Lagrangian tree-level parameters. The latters are, of course, of no physical significance, since their values depend on the particular renormalization framework that one adheres to. However, I repeat, I am using them here just as a bookkeeping device.

The bare 1-loop Lagrangian, i.e. the classical Lagrangian $\mathcal{L}$ expressed in terms of the bare 1-loop parameters (24)-(26), will be called $\mathcal{L}^{(1)} = \mathcal{L}(s_\theta^{(1)2}, g^{(1)}, M_W^{(1)})$. To get the

1-loop finite value of the W mas from the 1-loop bare value $M_W^{(1)}$, we just add the free propagator obtained from $\mathcal{L}^{(1)}$ plus a 1-loop correction (self-energy insertion) to it, and require the ensuing expression (call it the unrenormalized 1-loop propagator) to have a pole at the physical mass. This is nothing else than the on-shell renormalization prescription, well known from QED. If we call $\Pi_W(k^2)$ the vacuum polarization function associated to the W-boson self-energy insertion, the 1-loop propagator reads

$$\frac{-ig_{\mu\nu}}{k^2 - [M_W^{(1)2} + \Pi_W(k^2)]} \,. \tag{27}$$

I have disregarded the contribution from the $k_\mu k_\nu$ terms in the W propagator. In practice they would afford negligible contributions in our framework, since they would be proportional to the external fermion masses squared. However, for our processes $m_l^2/M_W^2 << 1\,; l = e, \mu, \nu$.

The physical (up to 1-loop) value of the W mass therefore is

$$M_W^2 = M_W^{(1)2} + \Pi_W(k^2 = M_W^2) = M_W^{(0)2}(1 + \delta_1 - \delta_2 - \delta_3)_{k^2=0} + \Pi_W(M_W^2) \,, \tag{28}$$

where in the second equality I used eq.(26). The label $k^2 = 0$ means that the various contributions to the δ_i $(i = 1, 2, 3)$ are to be evaluated at low energies $(k^2 << M_W^2)$ (more precisely, at low momentum transfers), in accordance to the processes I have chosen.

Although the contributions to the δ_i 's and $\Pi_W(M_W^2)$ are individually UV-divergent, eq.(28) must be finite in a renormalizable theory, e.g. in the SM or in any (softly broken) global SUSY version of it.

As for the bare Z-mass parameter, our model Lagrangian definition of it is $M_Z = M_W/c_\theta$. This relation must be preserved by the bare parameters order by order in perturbation theory, in particular

$$M_Z^{(0)} = M_W^{(0)}/c_\theta^{(0)} \quad \text{and} \quad M_Z^{(1)} = M_W^{(1)}/c_\theta^{(1)} \,. \tag{29}$$

Imposing on-shell renormalization as before, it is easy to see that the physical (up to 1-loop) value of the Z mass is

$$M_Z^2 = M_Z^{(0)2}(1 + \delta_1 - \delta_2 - \delta_3 + \frac{s_\theta^2}{c_\theta^2}\delta_3)_{k^2=0} + \Pi_Z(M_Z^2) \,, \tag{30}$$

where the last term is the on-shell value of the vacuum polarization function associated to the Z self-energy insertion. Eq.(30) must be finite in a renormalizable theory. Notice that relation (29) does not hold for c_θ- as obtained from (6)- and the physical masses M_W and M_Z- given by (28) and (30) -in contrast to the scheme of ref.[4], where c_θ is defined that way. Of course, that c_θ could be computed in our scheme by simply performing the quotient between the physical weak boson masses inferred from (28) and (30).

Let us specialize formulas (28) and (30) to include only the 1-loop self-energy (SE) contributions to the δ_i 's (see Fig. 3, Secc. 4) from our basic three independent processes. We can easily check that

$$\delta_1^{SE} = -\frac{\Pi_W(k^2)}{M_W^2} \;, \quad \delta_2^{SE} = \frac{\Pi_\gamma(k^2)}{k^2} \,. \tag{31}$$

As for the radiative shifts to the vector and axial coefficients, we have

$$\delta a^{SE} = -4 s_\theta c_\theta \frac{\Pi_{\gamma Z}(k^2)}{k^2} \quad , \quad \delta b^{SE} = 0 \, , \tag{32}$$

so that from eq.(16)

$$\delta_3^{SE} = \frac{c_\theta}{s_\theta} \frac{\Pi_{\gamma Z}(k^2)}{k^2} . \tag{33}$$

Formulas (28) and (30) therefore boil down to

$$(\delta M_W^2)^{SE} = \left(-\Pi_W(k^2) - \frac{c_\theta}{s_\theta} M_W^2 \frac{\Pi_{\gamma Z}(k^2)}{k^2} - M_W^2 \frac{\Pi_\gamma(k^2)}{k^2}\right)_{k^2=0} + \Pi_W(M_W^2) \tag{34}$$

and

$$(\delta M_Z^2)^{SE} = \left(-\frac{\Pi_W(k^2)}{c_\theta^2} + \frac{s_\theta^2 - c_\theta^2}{s_\theta c_\theta} M_Z^2 \frac{\Pi_{\gamma Z}(k^2)}{k^2} - M_Z^2 \frac{\Pi_\gamma(k^2)}{k^2}\right)_{k^2=0} + \Pi_Z(M_Z^2) \, , \tag{35}$$

where Π_γ is the vacuum polarization function of the photon (from Coulomb scattering) and $\Pi_{\gamma Z}$ is the mixed self-energy function (both from Coulomb and from $\nu_\mu e$ scattering).

In general, the full contribution to formulas (34)-(35) from a renormalizable theory is not finite. For example, in the SM is not finite. However, if the various vacuum polarization functions include only the contributions from fermions, then, it is finite and as a matter of fact it yields the bulk of the radiative corrections to the weak gauge boson masses in the pure SM context [3]-[5].

Let us now turn our attention to the ρ-parameter and its renormalization by higher order quantum effects. The low energy ρ-parameter is defined as the ratio between the weak neutral and charged-current Fermi constants at zero momentum transfer [24]. To be more precise, let J^μ_{CC} and J^μ_{NC} be the standard charged and neutral weak currents. The low energy effective Lagrangian accounting for our weak processes may be written

$$\mathcal{L}_{eff} = \frac{G_F}{\sqrt{2}} (J^\mu_{CC}(\mu) J^\dagger_{CC\mu}(e) + \rho J^\mu_{NC}(e) J_{NC\mu}(\nu) + \ h.c.) \, . \tag{36}$$

Specifically, $J^\mu_{CC}(\mu \text{ or } e)$ is the charged weak current for the muon or electron vertices in μ-decay, and $J^\mu_{NC}(e \text{ or } \nu)$ is the neutral weak current for the electron or neutrino vertices in $\nu_\mu e$-scattering .

Within the context of the SM, ρ can be easily identified. At the tree-level, i.e. using $\mathcal{L}^{(0)}$ to compute the currents, it simply reads

$$\rho^{(0)} = \frac{M_W^{(0)2}}{M_Z^{(0)2} c_\theta^{(0)2}} = 1 \, , \tag{37}$$

where the second equality presumes a weak isodoublet structure in the Higgs sector of the SM or of the SUSY SM, in accordance with eq.(29). Therefore, any departure of the experimentally measured ρ-parameter from 1 has to be attributed, within this context, to higher order quantum effects induced by SM, and potential non-SM, radiative corrections.

To see that ρ is no longer 1 at higher orders, let us identify it from eq.(36) at 1-loop order by using $\mathcal{L}^{(1)}$ to compute the currents. Of course, the first term of eq.(36)–the charged current interaction– remains unmodified, for that term is used to define G_F through the counterterm δ_1 (see eq.(21)). It is the second term of (36), the one defining the neutral-current Fermi constant $G_F^0 \equiv \rho G_F$, that gets renormalized with respect to the first. One source of renormalization just comes from the aforementioned condition that the charged-current Fermi constant remains untouched by radiative corrections. Therefore, from eq.(21) we set $G_F \rightarrow G_F(1-\delta_1)$ only on the second term of (36). On the other hand, the coefficient ρ in eq.(36) will undergo on its own an over-all rescaling $\rho \rightarrow \rho(1+\delta_4)$, where δ_4 is the shift induced on that parameter by those radiative corrections to $\nu_\mu e$-scattering for which $\delta_3 = 0$ identically. These are the subclass of radiative corrections to $\nu_\mu e$-scattering that verify $\delta a/a - \delta b/b = 0$ (see eq.(16)). The net effect on the structure of the Lagrangian (36), therefore, is

$$\mathcal{L}_{eff}^{(1)} = \frac{G_F}{\sqrt{2}}(J_{CC}^{\mu}(\mu)J_{CC\mu}^{\dagger}(e) + (1-\delta_1)\rho^{(0)}(1+\delta_4)J_{NC}^{(1)\mu}(e)J_{NC\mu}(\nu) + h.c.) \,. \tag{38}$$

Here $J_{NC}^{(1)\mu}(e) \equiv J_{NC}^{\mu}(a^{(1)}+\delta a, b^{(1)}+\delta b)$ is the weak neutral current associated to the electron. It absorbs the subclass of renormalization effects on $\nu_\mu e$-scattering that go into the definition of s_θ^2. The other type of renormalization effects, those that go into $J_{NC}^{(1)\mu}(\nu) \equiv (1+\delta_4)J_{NC}^{\mu}(\nu)$ have been factored out explicitly and are responsible for $\rho \rightarrow \rho(1+\delta_4)$.

Hence the total renormalization of the ρ parameter is

$$\rho^{(1)} = (1-\delta_1)(1+\delta_4)\rho^{(0)} \equiv 1 + \delta\rho \,, \tag{39}$$

where the full 1-loop shift is

$$\delta\rho = -\delta_1 + \delta_4 \,. \tag{40}$$

If we only consider the universal effects from the self-energies (i.e. the total "oblique" contribution), then δ_1 is given in eq.(31) while δ_4 gets a similar contribution from $\nu_\mu e$-scattering. Thereupon eq.(40) boils down to [25]

$$\delta\rho^{SE} = \left(\frac{\Pi_W(k^2)}{M_W^2} - \frac{\Pi_Z(k^2)}{M_Z^2}\right)_{k^2=0} \,, \tag{41}$$

which must be finite for $k^2 = 0$. In general, however, the complete formula (eq.(40)) will involve a lot more of calculations than the simple eq.(41); namely, it will require the computation of both the indirect ("oblique") as well as the direct (process-dependent) effects. In SUSY all of them turn out to be of comparable size, at least in our framework (Cf. Secc. 4).

To finish this section, let me remind the reader that deviations from eq.(37), induced by nonvanishing $\delta\rho$ within the context of the SM with a Higgs doublet sector, can be viewed as an intrinsec breaking of an accidental global isospin symmetry, the so-called "custodial" $SU(2)_V$ symmetry [25]-[26]. It would be a remnant tree-level symmetry of the spontaneously broken gauge theory, in the absence of weak hypercharge interactions and of unequal Yukawa couplings for the $T_3 = 1/2$ and $T_3 = -1/2$ weak-isospin components in each fermion isodoublet. In lowest order of perturbation theory, the $SU(2)_V$-breaking terms do not affect ρ at all ($\delta\rho^{(0)} = 0$), but at higher orders they induce $\delta\rho \neq 0$ contributions. In general, beyond the tree level (i.e. for $n \geq 1$) all the parameters of the effective n-loop Lagrangian become sensitive to the breaking of $SU(2)_V$. An example is afforded by eqs. (34) and (35). Notice that, for $s_\theta^2 \to 0$, the difference in sensitivity of M_W and M_Z to hypercharge interactions vanishes (the expressions inside the parentheses on the RHS's of these equations become identical) and only the mass splittings within isodoublets (caused by the unequal Yukawas) remain: $(\Pi_W - \Pi_Z)|_{s_\theta^2 \to 0} \neq 0$. At the 1-loop level ($n = 1$), however, non-vanishing contributions to $\delta\rho$ arise only from mass splittings within isodoublets but not from hypercharge interactions. This situation changes substantially in the SUSY SM, where the presence of certain soft SUSY breaking terms (like sfermion and gaugino masses, and trilinear couplings in the superpotential) and of asymmetric vacuum expectation values in the multidoublet Higgs sector, may introduce additional sources of breakdown of custodial symmetry, both explicitly and spontaneously (Cf. Sections 3-4) [27].

3. SUSY Formalism

The renormalization framework developed in the previous section is completely general and can be used in a pure SM context or in any renormalizable extension of it. As stated in the introduction, I want to concentrate on the kind of potential new quantum effects afforded by supersymmetric extensions of the SM. Let me not to insist on the theoretical motivations for supersymmetry (SUSY), see ref. [15]-[16], and go directly to a quick review of the necessary formalism for our calculations [18].

We shall perform our calculations in a mass-eigenstate basis. One goes from the weak-eigenstate basis to the mass-eigenstate basis via appropriate unitary transformations. Two classes of SUSY particles enter our calculations: the fermionic partners of gauge bosons and Higgs bosons (called gauginos and higgsinos, respectively) and on the other hand the scalar partners of quarks and leptons (called squarks and sleptons, respectively, or sfermions generically). In the minimal SUSY extension of the SM, hereafter called MSSM, we need two Higgs doublets with weak hypercharges $Y_{1,2} = \mp 1$:

$$\Phi_1 = \begin{pmatrix} H_1^0 \\ H_1^- \end{pmatrix} \quad \text{and} \quad \Phi_2 = \begin{pmatrix} H_2^+ \\ H_2^0 \end{pmatrix} . \tag{1}$$

This is due to the fact that the Yukawa couplings responsible for the masses of the up- and down-like quarks are generated from the superpotential, a SUSY structure that cannot support LH and RH fields at the same time. The Higgs doublet Φ_1 (Φ_2) gives mass to the down (up) -like quarks. The corresponding higgsino doublets are denoted by $\tilde{\Phi}_1$ and $\tilde{\Phi}_2$. Their components are Weyl spinors. From the higgsinos and the various gauginos we form the following three sets of two-component Weyl spinors:

$$\Gamma_i^+ = \{\tilde{W}^+, \tilde{H}_2^+\} , \quad \Gamma_j^- = \{\tilde{W}^-, \tilde{H}_1^-\} , \tag{2}$$

$$\Gamma_\alpha^0 = \{\tilde{B}^0, \tilde{W}_3^0, \tilde{H}_2^0, \tilde{H}_1^0\} . \tag{3}$$

These states get mixed up when the neutral Higgs fields acquire nonvanishing v.e.v's, $< H_1^0 >= v_1$ and $< H_2^0 >= v_2$, giving mass to the gauge bosons: $M_W^2 = (1/2)(v_1^2 + v_2^2), M_Z = M_W/c_\theta$. (In general $v_1 \neq v_2$, so that we note, in passing, that custodial symmetry becomes spontaneously broken). Arranging the fields (2)-(3) in column vectors $|\Gamma^\pm >$ and $|\Gamma^0 >$, the undiagonal mass Lagrangian for the gaugino-higgsino (GH) sector reads

$$\mathcal{L}_M = -(< \Gamma^+|\mathcal{M}|\Gamma^- > + h.c.) - \frac{1}{2} < \Gamma^0|\mathcal{M}^0|\Gamma^0 > , \tag{4}$$

where the charged and neutral GH mass matrices, after $SU(2)_L \times U(1)_Y$ breaking, are the following:

$$\mathcal{M} = \begin{pmatrix} M & M_W\sqrt{2}\, s_\beta \\ M_W\sqrt{2}\, c_\beta & \mu \end{pmatrix} \tag{5}$$

and

$$\mathcal{M}^0 = \begin{pmatrix} M' & 0 & -M_Z\, c_\beta\, s_\theta & M_Z\, s_\beta\, s_\theta \\ 0 & M & M_Z\, c_\beta\, c_\theta & -M_Z\, s_\beta\, c_\theta \\ -M_Z\, c_\beta\, s_\theta & M_Z\, c_\beta\, c_\theta & 0 & -\mu \\ M_Z\, s_\beta\, s_\theta & -M_Z\, s_\beta\, c_\theta & -\mu & 0 \end{pmatrix} , \tag{6}$$

with the following notation:

$$s_\beta \equiv \sin\beta \quad , \quad c_\beta \equiv \cos\beta \quad , \quad \tan\beta = \frac{v_2}{v_1} . \tag{7}$$

The mass parameters M and M' come from $SU(2)_L \times U(1)_Y$-invariant gaugino mass terms that "softly" break global SUSY (i.e. that do not induce quadratical divergences [15]) while μ comes from the SUSY mass term $\mu \hat{H}_1{}^\dagger \hat{H}_2$ built out of the two chiral Higgs superfields $\hat{H}_{1,2}$. The origin of the softly SUSY breaking gaugino mass terms can be traced back to a remnant of the flat space-time limit in local SUSY, i.e. of Supergravity (SUGRA) [13], which on the other hand is thought to be the point limit of Superstrings [14]. We shall assume that the MSSM can be embedded in a GUT, in which case the parameters M' and M are related as follows [15]:

$$\frac{M'}{M} = \frac{5}{3}\tan^2\theta_W \simeq 0.5 . \tag{8}$$

The 2×2 mass matrix $\mathcal{M}$ is in general non-symmetrical and its diagonalization is accomplished by two unitary matrices U and V, while the symmetrical 4×4 mass matrix $\mathcal{M}^0$ can be diagonalized by a single unitary matrix N. The entries of the matrices U,V and N are complex in general in order to obtain positive-definite mass eigenvalues:

$$U^*\mathcal{M}V^\dagger = diag\{M_1, M_2\} \text{ and } N^*\mathcal{M}^0 N^\dagger = diag\{M_1^0, ...M_4^0\} . \tag{9}$$

We are now ready to construct the charged mass-eigenstate 4-spinors (charginos) associated to the mass eigenvalues M_i $(i = 1, 2)$. Call them Ψ_i^+, and let Ψ_i^- be the corresponding charge conjugate states. We have

$$\Psi_i^+ = \begin{pmatrix} U_{ij}\Gamma_j^+ \\ V_{ij}^*\bar{\Gamma}_j^- \end{pmatrix} \text{ and } \Psi_i^- = C\bar{\Psi}_i{}^{-T} = \begin{pmatrix} V_{ij}\Gamma_j^- \\ U_{ij}^*\bar{\Gamma}_j^+ \end{pmatrix} . \tag{10}$$

As for the neutral mass-eigenstate 4-spinors (neutralinos) associated to the mass eigenvalues M^0_α $(\alpha = 1, ..., 4)$, they are the following Majorana spinors

$$\Psi^0_\alpha = \left(\begin{array}{c} N_{\alpha,\beta}\Gamma^0_\beta \\ N^*_{\alpha,\beta}\bar{\Gamma}^0_\beta \end{array} \right) = C\bar{\Psi}^{0T}_\alpha \ . \tag{11}$$

The gauge interactions with charginos and neutralinos (referred to generically as "inos") in the mass-eigenstate basis are the following:

$$\mathcal{L}^{GH}_W = J^\mu_0 Z_\mu + J^\mu_{em} A_\mu + (J^\mu_+ W^-_\mu + h.c.) \ , \tag{12}$$

where

$$\begin{aligned} J^\mu_0 &= \frac{g}{c_\theta}\{\sum_{i,j} \bar{\Psi}^+_i \gamma^\mu (U^L_{ij} P_L + U^R_{ij} P_R)\Psi^+_j \\ &\quad + \frac{1}{2}\sum_{\alpha\beta} \bar{\Psi}^0_\alpha \gamma^\mu (O^L_{\alpha\beta} P_L + O^R_{\alpha\beta} P_R)\Psi^0_\beta\} && (13) \\ J^\mu_{em} &= g s_\theta \sum_i \bar{\Psi}^+_i \gamma^\mu \Psi^+_i && (14) \\ J^\mu_+ &= g\sum_\alpha \sum_i \bar{\Psi}^0_\alpha \gamma^\mu (C^L_{\alpha i} P_L + C^R_{\alpha i} P_R)\Psi^+_i \ , && (15) \end{aligned}$$

with $P_{L,R} = (1/2)(1 \pm \gamma_5)$. Notice that the neutral weak current is not diagonal, as could be expected from the fact that the mixed weak-eigenstate gauginos and higgsinos belong to different weak-isospin representations. The various coupling matrices are the following:

$$\begin{aligned} U^L_{ij} &= U_{i1}U^*_{j1} + \frac{1}{2}U_{i2}U^*_{j2} - s^2_\theta \delta_{ij} \\ U^R_{ij} &= V^*_{i1}V_{j1} + \frac{1}{2}V^*_{i2}V_{j2} - s^2_\theta \delta_{ij} \\ O^L_{\alpha\beta} &= -O^{R*}_{\alpha\beta} = \frac{-1}{2}N_{\alpha 3}N^*_{\beta 3} + \frac{1}{2}N_{\alpha 4}N^*_{\beta 4} \\ C^L_{\alpha i} &= \frac{1}{\sqrt{2}}N_{\alpha 3}U^*_{i2} - N_{\alpha 2}U^*_{i1} \\ C^R_{\alpha i} &= \frac{-1}{\sqrt{2}}N^*_{\alpha 4}V_{i2} - N^*_{\alpha 2}V_{i1} \ . && (16) \end{aligned}$$

I turn now to the scalar sector and their interactions. The mass matrices for sfermions mix left-and right-handed type fields, i.e. those SUSY scalar fields associated respectively to the two chiral components of each SM fermion field. Consequently, they are not-in general-mass eigenstates. For each sfermion flavour, f ,we have an angle, φ_f, which defines the rotation that diagonalizes the scalar mass matrix. I shall not dwell into the actual structure of these matrices in special cases but rather work with general unspecified matrix elements. Let us illustrate the case of sleptons $(\tilde{l} = \tilde{e}, \tilde{\mu}, ...)$. Squarks $(\tilde{q} = \tilde{u}, \tilde{d}, ...)$ are to be treated analogously, of course, but I will spare the reader the ordeal of reading up all the details. As before, we shall work in the mass-eigenstate basis (the physical basis). Thus we introduce a rotation matrix for each slepton flavour

$$R^{(l)} = \left(\begin{array}{cc} c_l & s_l \\ -s_l & c_l \end{array} \right) \ ; \ s_l \equiv \sin\varphi_l \ , \ c_l \equiv \cos\varphi_l \ , \tag{17}$$

relating the weak-eigenstate basis $\tilde{l}'_a = \{\tilde{l}_L, \tilde{l}_R\}$ to the mass-eigenstate basis $\tilde{l}_a = \{\tilde{l}_1, \tilde{l}_2\}$ through

$$\tilde{l}_a = \sum_b R^{(l)}_{ab} \tilde{l}'_b \; . \tag{18}$$

Let us call $m_1(\tilde{l})$ and $m_2(\tilde{l})$ the mass eigenvalues associated to the two scalar fields (18). Denoting by $\tilde{\nu}_l$ the sneutrino associated to $\tilde{l}_L$, they together form an $SU(2)_L$ doublet while $\tilde{l}_R$ forms an $SU(2)_L$ singlet:

$$\begin{pmatrix} \tilde{\nu}_l \\ \tilde{l}_L \end{pmatrix} \; , \; \tilde{l}_R \tag{19}$$

i.e. they follow the same pattern as the corresponding charged lepton $l = l_L + l_R$ and neutrino ν_l. Of course, in the case of squarks one needs a rotation matrix for both the up and down components in each squark doublet, and they are in general different, but as I mentioned before I will omit details. As for the the sneutrino masses $m_{\tilde{\nu}_l}$, there obviously are no non-trivial mass matrices in the MSSM. The interaction Lagrangian of a slepton generation with the gauge bosons reads, in the physical basis (1 am suppressing flavour-lower and upper-indices here):

$$\begin{aligned}
\mathcal{L}^{\tilde{l}}_W \;=\; & \frac{ig}{\sqrt{2}} \sum_a \{ R^*_{a1} \tilde{\nu}^* \overleftrightarrow{\partial^\mu} \tilde{l}_a W^+_\mu + h.c. \} - igs_\theta \sum_a \tilde{l}^*_a \overleftrightarrow{\partial^\mu} \tilde{l}_a A_\mu \\
& + \frac{ig}{2c_\theta} \{ \tilde{\nu}^* \overleftrightarrow{\partial^\mu} \tilde{\nu} + \sum_{a,b} S_{ab} \tilde{l}^*_a \overleftrightarrow{\partial^\mu} \tilde{l}_b \} Z_\mu \\
& + \frac{1}{2} g^2 \{ \sum_{ab} R^*_{a1} R_{b1} \tilde{l}^*_b \tilde{l}_a + \tilde{\nu}^* \tilde{\nu} \} W^+_\mu W^{-\mu} + g^2 s_\theta \sum_a \tilde{l}^*_a \tilde{l}_a A_\mu A^\mu \\
& + \frac{g^2}{4c^2_\theta} \{ \tilde{\nu}^* \tilde{\nu} + \sum_{ab} \hat{S}_{ab} \tilde{l}^*_a \tilde{l}_b \} Z_\mu Z^\mu - g^2 \frac{s_\theta}{c_\theta} \sum_{ab} S_{ab} \tilde{l}^*_a \tilde{l}_b A_\mu Z^\mu \; ,
\end{aligned} \tag{20}$$

where the following notation has been introduced:

$$S_{ab} = \sum_c \lambda_c R_{ac} R^*_{bc} \quad \text{and} \quad \hat{S}_{ab} = \sum_c {\lambda_c}^2 R_{ac} R^*_{bc} \; , \tag{21}$$

with (see eq.(1.4))

$$\lambda_1 = \frac{a+b}{2} \quad , \quad \lambda_2 = \frac{a-b}{2} \; . \tag{22}$$

Finally, both sectors-fermionic and scalar- are subject to mutual interaction through Yukawa couplings of the form chargino-fermion-sfermion or neutralino- fermion-sfermion. In building up the physical Yukawa couplings from the weak-eigenstate Yukawa coupligs, we may ignore the higgsino components, coupled to leptons and sleptons, since their strength is negligible for the electrons and muons involved in the external legs of our low-energy processes. In general, the only relevant Yukawa coupling of this type is the "higgsino-top-stop" component, but it never enters our framework at 1-loop (see, however, an intriguing remark on this particular in Secc.5). Therefore, the relevant pieces are the following:

$$\mathcal{L}_Y \;=\; -ig \sum_i U^*_{i1} \tilde{\nu}^* \bar{\Psi}^-_i P_L l^- - ig \sum_i \sum_a R_{a1} V^*_{i1} \tilde{l}^*_a \bar{\Psi}^+_i P_L \nu$$

$$+\frac{ig}{\sqrt{2}c_\theta}\sum_\alpha(s_\theta N^*_{\alpha 1}-c_\theta N^*_{\alpha 2})\tilde{\nu}^*\bar{\Psi}^0_\alpha P_L\nu$$
$$+\frac{ig}{2\sqrt{2}c_\theta}\sum_a\sum_\alpha \tilde{l}^*_a\bar{\Psi}^0_\alpha(A_{a\alpha}-B_{a\alpha}\gamma_5)l^- + h.c. \ , \tag{23}$$

where

$$\begin{aligned} A_{a\alpha} &= R_{a1}(s_\theta N^*_{\alpha 1}+c_\theta N^*_{\alpha 2})-2s_\theta R_{a2}N_{\alpha 1} \\ A_{a\alpha} &= R_{a1}(s_\theta N^*_{\alpha 1}+c_\theta N^*_{\alpha 2})+2s_\theta R_{a2}N_{\alpha 1} \ . \end{aligned} \tag{24}$$

To summarize, the SUSY interaction Lagrangian needed to perform our calculations is obtained from eqs.(12), (20) and (23):

$$\mathcal{L}^{SUSY}_{int.} = \mathcal{L}^{GH}_W + \mathcal{L}^{\tilde{l}}_W + \mathcal{L}^{\tilde{q}}_W + \mathcal{L}_Y \ , \tag{25}$$

where the third piece is the squark analogue of the second one. The corresponding Feynman rules are trivially read off this Lagrangian, with perhaps only two exceptions worth mentioning. They refer to two aspects of the neutralino interactions with the Z boson connected with the Majorana nature of these fields; namely, on one hand the Feynman rule associated to the $Z\bar{\Psi}^0_\alpha\Psi^0_\beta$-vertex gets an "unexpected" additional factor of 2 that cancels the 1/2 standing out in the corresponding Lagrangian term (see eq.(13)), and on the other hand the neutralino contribution to the vacuum polarization of the Z-boson demands an over-all exchange symmetry factor of 1/2.

4. Analysis of the SUSY Radiative Corrections

Using the formalism and Lagrangian interaction terms presented in Seccs.2-3, we are now ready to compute all possible 1-loop SUSY radiative corrections to our basic three low-energy processes: $\nu_\mu e$-scattering, μ-decay and Coulomb scattering. By isolating the full 1-loop structure of the three counterterms δ_i $(i = 1,2,3)$, and of the radiative shift δ_4, we will infer (from eqs.(2.28), (2.30) and (2.40)) the radiative corrections to the basic electroweak parameters themselves: M_W, M_Z and ρ. In this section I will provide some details of the analytical formulae resulting from the 1-loop calculations. However, since the complete analytical results [18] are rather long and repetitive, I will concentrate myself on just the analytical computation of SUSY radiative corrections to the first process, namely $\nu_\mu e$-scattering. Nonetheless the numerical analysis that I present at the end of this section contains, of course, complete information from the three processes.

As explained in Secc. 2, there are two general classes of radiative corrections to $\nu_\mu e$-scattering: 1) those that contribute to δ_3 and are used to define s^2_θ through eq.(2.15), and 2) those that are absorbed into δ_4 -identically giving $\delta_3 = 0$-and so contribute (in part) to the renormalization of the ρ parameter. Let us start with the first class of radiative corrections. The 10 diagrams of Fig. 1 correspond to SUSY vertex corrections and wave function renormalization of the external fermions. Summing the contribution from the first 4 diagrams (a.1)-(a.4), a straightforward calculation shows that the UV divergences cancel and no logarithm of dimensionful quantities is left:

$$\delta_3(a.1+a.2+a.3+a.4) = \frac{-g^2}{16\pi^2}\{\sum_{ij}[U_{i1}U^L_{ji}U^*_{j1}\hat{E}_0(M_i, M_j, m(\tilde{\nu}_e))$$

$$-\frac{2M_iM_j}{M_i^2-M_j^2}U_{i1}U^R_{ji}U^*_{j1}F(M_i, M_j, m(\tilde{\nu}_e))] + c_\theta^2[1-\sum_i |U_{i1}|^2\hat{E}_0(M_i, m(\tilde{\nu}_e), m(\tilde{\nu}_e))]\}\ ,(1)$$

where the finite (i.e. UV-convergent) functions F and $\hat{E}_0$ are related to standard renormalized 3-point functions (Cf. Appendix). The next two diagrams (b.1)-(b.2) in Fig. 1 are of the neutrino-charge-radius type. By expanding their amplitudes in powers of the photon momentum, one easily verifies that they render no net contribution to the neutrino charge and that the first non-vanishing contribution is indeed of the charge-radius type. Another straightforward calculation yieds

$$\delta_3(b.1+b.2) = \frac{g^2}{16\pi^2}\sum_a\sum_i |R^{(\mu)}_{a1}|^2|V_{i1}|^2\left(\frac{M_W}{M_i}\right)^2\left[\frac{f_1(x_{ai})}{(1-x_{ai})^4}+\frac{x_{ia}f_2(x_{ia})}{(1-x_{ia})^4}\right], \qquad (2)$$

where

$$\begin{aligned} f_1(x) &= -\frac{7}{18}+2x-\frac{5}{2}x^2+\frac{8}{9}x^3+x^2(1-\frac{2}{3}x)\log x\ , \\ f_2(x) &= \frac{1}{9}-\frac{1}{2}x+x^2-\frac{11}{18}x^3+\frac{1}{3}x^3\log x\ , \\ x_{ai} &= \frac{m_a^2(\tilde{\mu})}{M_i^2}\quad , \quad x_{ia} = x_{ai}^{-1}\ . \end{aligned} \qquad (3)$$

The contribution from diagrams (c.1)-(c.3) is also (globally) finite but it turns out to be rather involved:

$$\delta_3(c.1+c.2+c.3) = \frac{-g^2}{16\pi^2}\sum_a\sum_\alpha[\frac{1}{4}|N_{\alpha 2}+t_\theta N_{\alpha 1}|^2\{\lambda_1|R^{(e)}_{a1}|^2[\log\frac{m_a^2(\tilde{e})}{M_\alpha^{02}}$$

$$+\hat{E}_0(M^0_\alpha, m_a(\tilde{e}), m_a(\tilde{e}))] - \sum_b S^*_{ab}R^{(e)}_{a1}R^{*(e)}_{b1}\hat{E}_0(m_a(\tilde{e}), m_b(\tilde{e}, M^0_\alpha)\}$$

$$+\frac{g^2}{16\pi^2}\sum_a\sum_\alpha\frac{a+b}{4c_\theta^2}|N_{\alpha 1}|^2\{\lambda_2|R^{(e)}_{a2}|^2[\log\frac{m_a^2(\tilde{e})}{M_\alpha^{02}}+\hat{E}_0(M^0_\alpha, m_a(\tilde{e}), m_a(\tilde{e}))]$$

$$-\sum_b S^*_{ab}R^{(e)}_{a2}R^{*(e)}_{b2}\hat{E}_0(m_a(\tilde{e}), m_b(\tilde{e}), M^0_\alpha)\}\ . \qquad (4)$$

Here $t_\theta = s_\theta/c_\theta$ and the various coupling matrices and remaining notation have been defined in Secc. 3. Notice that the last diagram of Fig. 1 (diagram (c.4)) has no counterpart in the SM case, for there is no triple Z-boson vertex in the SM. Here the non-diagonal coupling of the Z to the Majorana neutralinos furnishes the following finite contribution:

$$\delta_3(c.4) = \frac{g^2}{16\pi^2}\sum_a\sum_{\alpha\beta}\{|R^{(e)}_{a1}|^2(N_{\alpha 2}+t_\theta N_{\alpha 1})(N_{\beta 2}+t_\theta N_{\beta 1})^*$$

$$\times[O^R_{\alpha\beta}\frac{1}{2}\hat{E}_0(M^0_\alpha, M^0_\beta, m_a(\tilde{e})) - O^L_{\alpha\beta}M^0_\alpha M^0_\beta C_0(M^0_\alpha, M^0_\beta, m_a(\tilde{e}))] + \frac{a+b}{c_\theta^2}|R^{(e)}_{a2}|^2N^*_{\alpha 1}N_{\beta 1}$$

$$\times[O^L_{\alpha\beta}\frac{1}{2}\hat{E}_0(M^0_\alpha, M^0_\beta, m_a(\tilde{e})) - O^R_{\alpha\beta}M^0_\alpha M^0_\beta C_0(M^0_\alpha, M^0_\beta, m_a(\tilde{e}))]\}\ , \qquad (5)$$

where the renormalized 3-point function C_0 is defined in the Appendix.

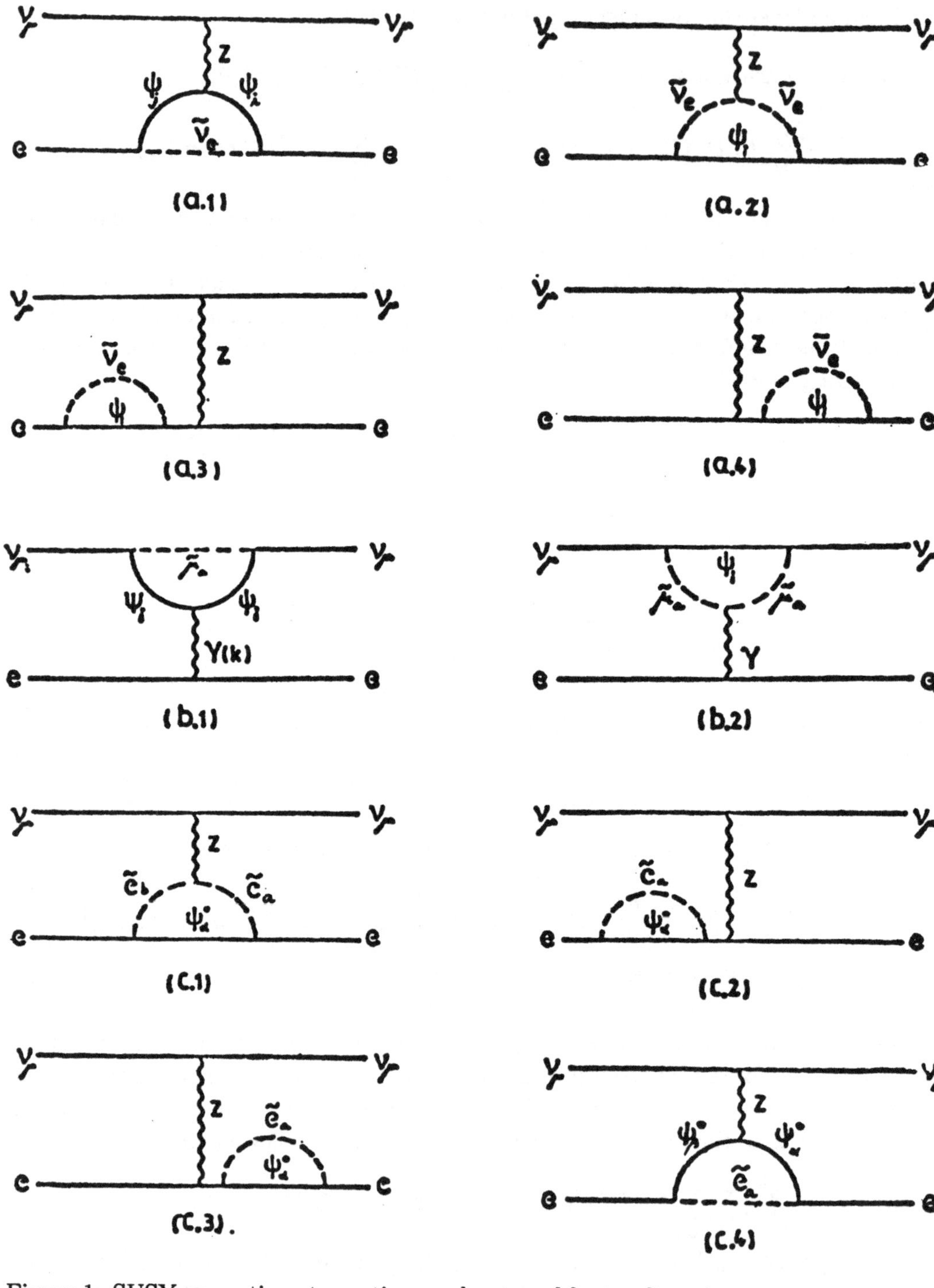

Figure 1: SUSY corrections to vertices and external lepton lines in $\nu_\mu e$-scattering.

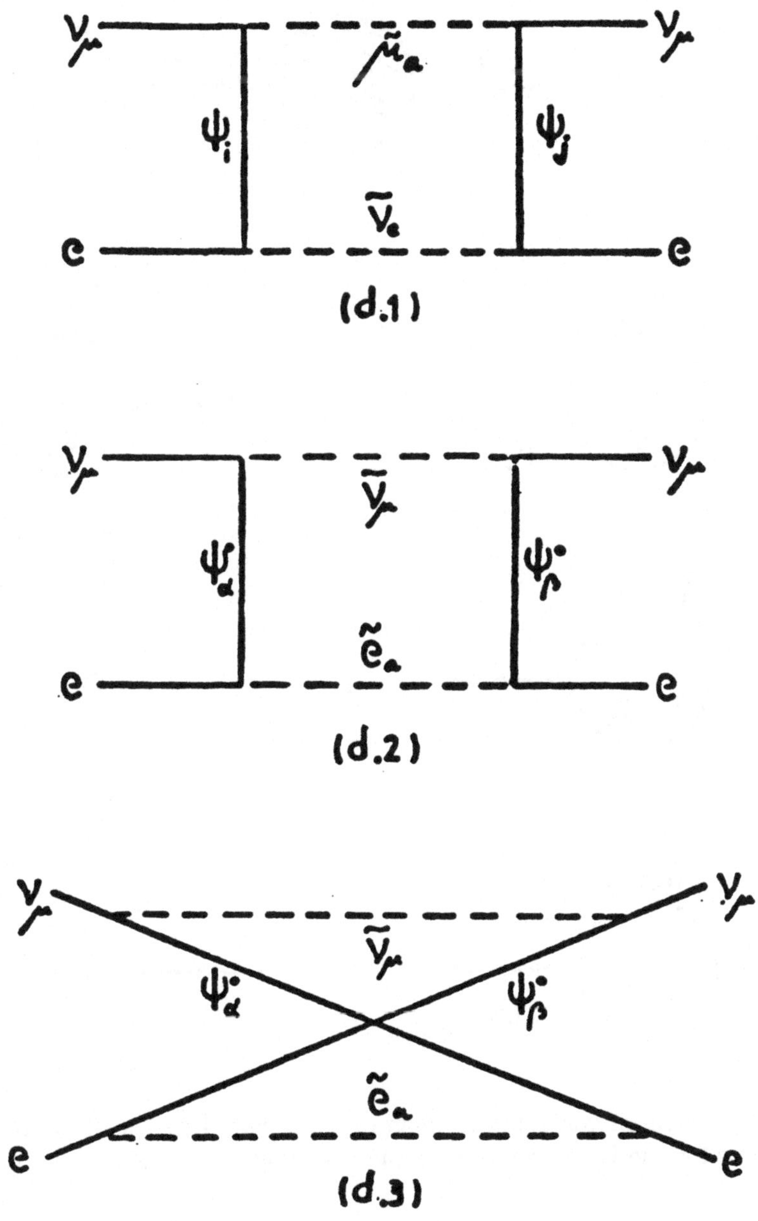

Figure 2: SUSY box diagrams in $\nu_\mu e$-scattering.

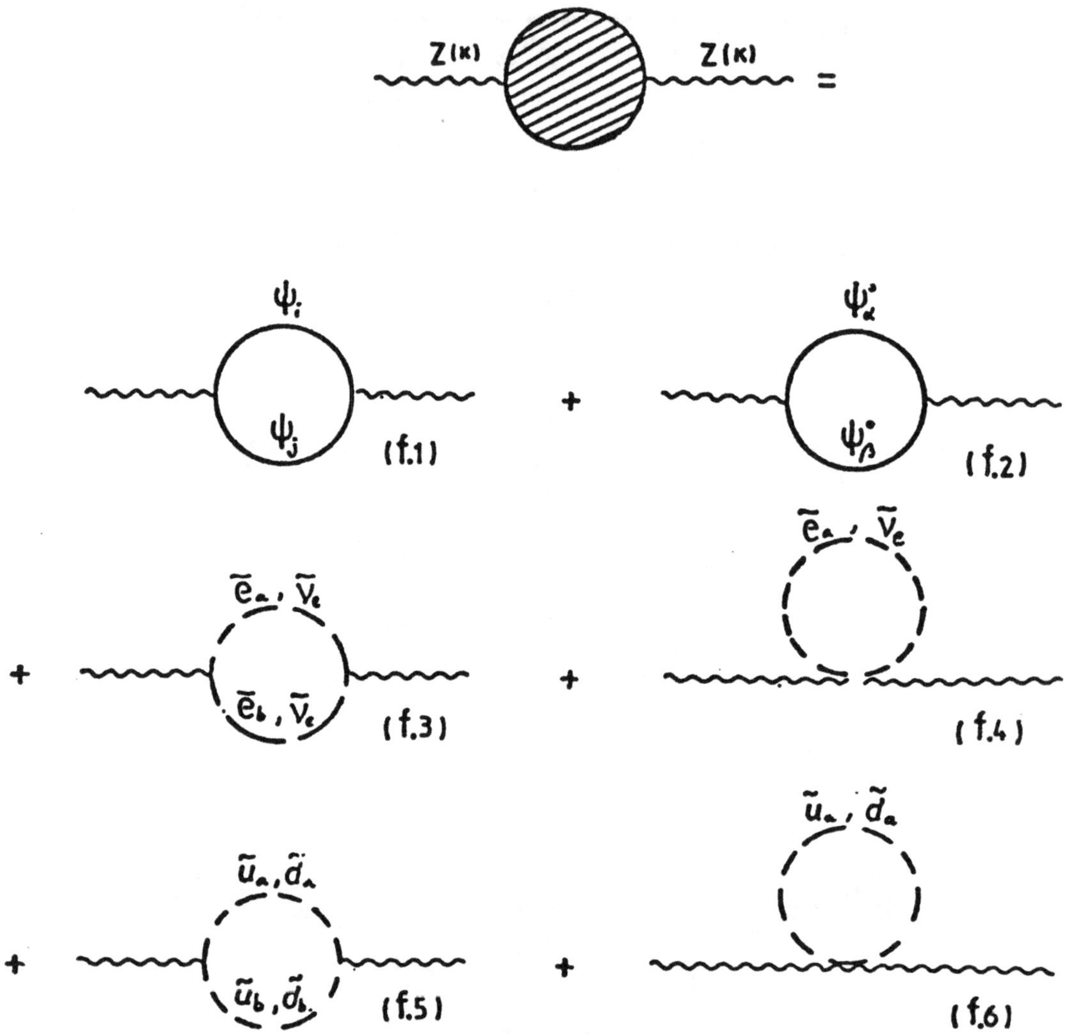

Figure 3: SUSY vacuum polarization diagrams associated to the Z boson. There are similar diagrams for the W boson and for the γ and mixed $\gamma - Z$ self-energies.

Turning now to Fig. 2, we see that we have contribution from several box diagrams. They require some Fierz reordering before proper identification of the radiative shifts on the vector and axial coefficients. The total output from this sector is a rather cumbersome expression:

$$\delta_3(Fig.2) = -\frac{g^2}{16\pi^2}\sum_{ij}\sum_{a}|R_{a1}^{(\mu)}|^2(U_{i1}^*U_{j1})(V_{i1}^*V_{j1})\frac{2M_W^2}{M_i^2-M_j^2}\{\frac{M_iM_j}{M_i^2-m_a^2(\tilde{\mu})}$$
$$\times F(M_i,m_a(\tilde{\mu}),m(\tilde{\nu}_e)) - \frac{M_iM_j}{M_j^2-m_a^2(\tilde{\mu})}F(M_j,m_a(\tilde{\mu}),m(\tilde{\nu}_e))\} - \frac{g^2}{16\pi^2}$$
$$\times\sum_{\alpha\beta}\sum_{a}\frac{M_W^2}{4(M_\alpha^{02}-M_\beta^{02})}\{\frac{a+b}{c_\theta^2}|R_{a2}^{(e)}|^2(N_{\alpha 1}N_{\alpha 2}^*-t_\theta|N_{\alpha 1}|^2)(N_{\beta 1}^*N_{\beta 2}-t_\theta|N_{\beta 1}|^2)$$
$$\times[\frac{M_\alpha^{02}}{M_\alpha^{02}-m_a^2(\tilde{e})}F(M_\alpha^0,m_a(\tilde{e}),m(\tilde{\nu}_\mu)) - \frac{M_\beta^{02}}{M_\beta^{02}-m_a^2(\tilde{e})}F(M_\beta^0,m_a(\tilde{e}),m(\tilde{\nu}_\mu))]$$
$$+|R_{a1}^{(e)}|^2(N_{\alpha 2}^{*2}-t_\theta^2N_{\alpha 1}^{*2})(N_{\beta 2}^2-t_\theta^2N_{\beta 1}^2)$$
$$\times[\frac{2M_\alpha^0M_\beta^0}{M_\alpha^{02}-m_a^2(\tilde{e})}F(M_\alpha^0,m_a(\tilde{e}),m(\tilde{\nu}_\mu)) - \frac{2M_\alpha^0M_\beta^0}{M_\beta^{02}-m_a^2(\tilde{e})}F(M_\beta^0,m_a(\tilde{e}),m(\tilde{\nu}_\mu))]$$
$$+\frac{a+b}{c_\theta^2}|R_{a2}^{(e)}|^2(N_{\alpha 1}^*N_{\alpha 2}^*-t_\theta N_{\alpha 1}^{*2})(N_{\beta 1}N_{\beta 2}-t_\theta N_{\beta 1}^2)$$
$$\times[\frac{2M_\alpha^0M_\beta^0}{M_\alpha^{02}-m_a^2(\tilde{e})}F(M_\alpha^0,m_a(\tilde{e}),m(\tilde{\nu}_\mu)) - \frac{2M_\alpha^0M_\beta^0}{M_\beta^{02}-m_a^2(\tilde{e})}F(M_\beta^0,m_a(\tilde{e}),m(\tilde{\nu}_\mu))]$$
$$+|R_{a1}^{(e)}|^2(|N_{\alpha 2}|^2-t_\theta^2|N_{\alpha 1}|^2)(|N_{\beta 2}|^2-t_\theta^2|N_{\beta 1}|^2)$$
$$\times[\frac{M_\alpha^{02}}{M_\alpha^{02}-m_a^2(\tilde{e})}F(M_\alpha^0,m_a(\tilde{e}),m(\tilde{\nu}_\mu)) - \frac{M_\beta^{02}}{M_\beta^{02}-m_a^2(\tilde{e})}F(M_\beta^0,m_a(\tilde{e}),m(\tilde{\nu}_\mu))]\} \; . \quad (6)$$

Finally, we have to compute the SUSY 1-loop vacuum polarization functions associated to the Z and W bosons. We need their expressions for $k^2 = 0$ and $k^2 = M_{Z,W}^2$ (Cf. eqs.(2.34), (2.35) and (2.41)). We also need the mixed $\gamma - Z$ self-energy function involved in the $\nu_\mu e$-scattering process at $k^2 = 0$. For completeness I shall provide general formulas for arbitrary k^2 and shall also quote the SUSY photon self-energy function from Coulomb scattering. All of them receive contributions from "inos" and sfermions. Fig. 3 illustrates the case of the Z vacuum polarization function. Similar diagrams hold for the other cases.

The analytical results from the "ino" sector read as follow (see Appendix for the various 1-and 2-point functions used below)

$$i\Pi_Z(k^2)(inos) = \frac{g^2}{c_\theta^2}\{2\sum_{ij}[X_{ij}\{2k^2[B_{21}(k^2,M_i,M_j) - B_1(k^2,M_i,M_j)]$$
$$+(M_j^2-M_i^2)B_1(k^2,M_i,M_j)\} + M_i(M_iX_{ij}-M_jY_{ij})B_0(k^2,M_i,M_j)]$$
$$+\sum_{\alpha\beta}[X_{\alpha\beta}^0\{2k^2[B_{21}(k^2,M_\alpha^0,M_\beta^0) - B_1(k^2,M_\alpha^0,M_\beta^0)]$$
$$+(M_\beta^{02}-M_\alpha^{02})B_1(k^2,M_\alpha^0,M_\beta^0)\} + M_\alpha^0(M_\alpha^0X_{\alpha\beta}^0 - M_\beta^0Y_{\alpha\beta}^0)B_0(k^2,M_\alpha^0,M_\beta^0)]\} \; ,$$

$$i\Pi_W(k^2)(inos) = 2g^2 \sum_\alpha \sum_i \{P_{\alpha i}\{2k^2[B_{21}(k^2, M^0_\alpha, M_i) - B_1(k^2, M^0_\alpha, M_i)] + (M_i^2 - M_\alpha^{02})$$
$$\times B_1(k^2, M^0_\alpha, M_i)\} + M^0_\alpha(M^0_\alpha P_{\alpha i} - M_i Q_{\alpha i}) B_0(k^2, M^0_\alpha, M_i)\} \ ,$$

$$i\Pi_{\gamma Z}(k^2)(inos) = 4g^2k^2s_\theta^2 \sum_i [B_{21}(k^2, M_i, M_i) - B_1(k^2, M_i, M_i)] \ ,$$

$$i\Pi_\gamma(k^2)(inos) = g^2k^2s_\theta^2 \sum_i (U^L + U^R)_{ii}[2B_{21}(k^2, M_i, M_i) - B_0(k^2, M_i, M_i)] \ , \tag{7}$$

with

$$\begin{aligned} X_{ij} &= U^L_{ij}U^{L*}_{ij} + U^R_{ij}U^{R*}_{ij} \ , \ Y_{ij} = U^L_{ij}U^{R*}_{ij} + U^{L*}_{ij}U^R_{ij} \ , \\ X^0_{\alpha\beta} &= O^L_{\alpha\beta}O^L_{\beta\alpha} + O^R_{\alpha\beta}O^R_{\beta\alpha} \ , \ Y^0_{\alpha\beta} = O^L_{\alpha\beta}O^R_{\beta\alpha} + O^L_{\beta\alpha}O^R_{\alpha\beta} \ , \end{aligned} \tag{8}$$

whereas sleptons yield

$$i\Pi_Z(k^2)(sleptons) = \frac{g^2}{4c_\theta^2}\{k^2[(2s_\theta^2 - c_l^2)^2[B_0(k^2, m_1, m_1) - 4B_{21}(k^2, m_1, m_1)]$$
$$+(2s_\theta^2 - s_l^2)^2 B_0(k^2, m_2, m_2) - 4B_{21}(k^2, m_2, m_2)] + 4s_l^2c_l^2[B_1(k^2, m_1, m_2)$$
$$-2B_{21}(k^2, m_1, m_2)] + B_0(k^2, m_{\tilde\nu}, m_{\tilde\nu}) - 4B_{21}(k^2, m_{\tilde\nu}, m_{\tilde\nu})]$$
$$+2s_l^2c_l^2(m_1^2 - m_2^2)[2B_1(k^2, m_1, m_2) - B_0(0, m_1, m_2)]\} \ ,$$

$$i\Pi_W(k^2)(sleptons) = g^2\{c_l^2[k^2[B_1(k^2, m_1, m_{\tilde\nu}) - 2B_{21}(k^2, m_1, m_{\tilde\nu})]$$
$$+(m_1^2 - m_{\tilde\nu}^2)B_1(k^2, m_1, m_{\tilde\nu})]$$
$$+s_l^2[k^2[B_1(k^2, m_2, m_{\tilde\nu}) - 2B_{21}(k^2, m_2, m_{\tilde\nu})]$$
$$+(m_2^2 - m_{\tilde\nu}^2)B_1(k^2, m_2, m_{\tilde\nu})]$$
$$-\frac{1}{2}[c_e^2A(m_1) + s_e^2A(m_2) - A(m_{\tilde\nu})]\} \ ,$$

$$i\Pi_{\gamma Z}(k^2)(sleptons) = -\frac{1}{2}g^2k^2t_\theta\{(2s_\theta^2 - c_l^2)[B_0(k^2, m_1, m_1) - 4B_{21}(k^2, m_1, m_1)]$$
$$+(2s_\theta^2 - s_l^2)[B_0(k^2, m_2, m_2) - 4B_{21}(k^2, m_2, m_2)]\} \ ,$$

$$i\Pi_\gamma(k^2)(sleptons) = g^2k^2s_\theta^2 \sum_a [B_0(k^2, m_a, m_a) - 4B_{21}(k^2, m_a, m_a)]\} \ . \tag{9}$$

Here $m_a (a = 1, 2)$ and $m_{\tilde\nu}$ are the mass eigenvalues corresponding to one slepton generation. One must, of course, sum over generations $(\tilde e, \tilde\mu, \tilde\tau)$ and add up similar contributions from squarks.

With regard to the second class $(\delta_3 = 0, \delta_4 \neq 0)$ of radiative corrections to $\nu_\mu e$-scattering, we get contributions from Figs. 4 and 5. The net δ_4 output from diagrams (c.1)-(c.3) adds up to zero, due to a Ward identity. Diagram (c.4), however, is non-

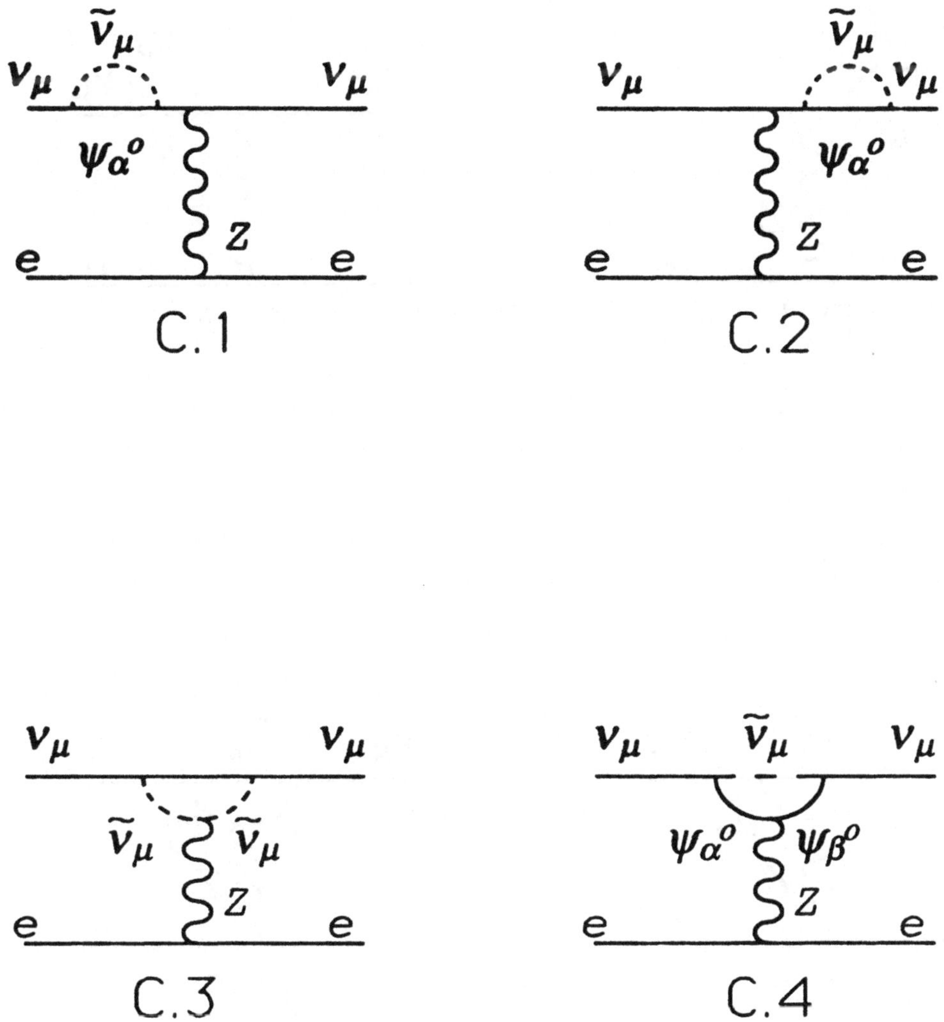

Figure 4: SUSY corrections to the ρ-parameter from neutralinos and muonic sneutrino.

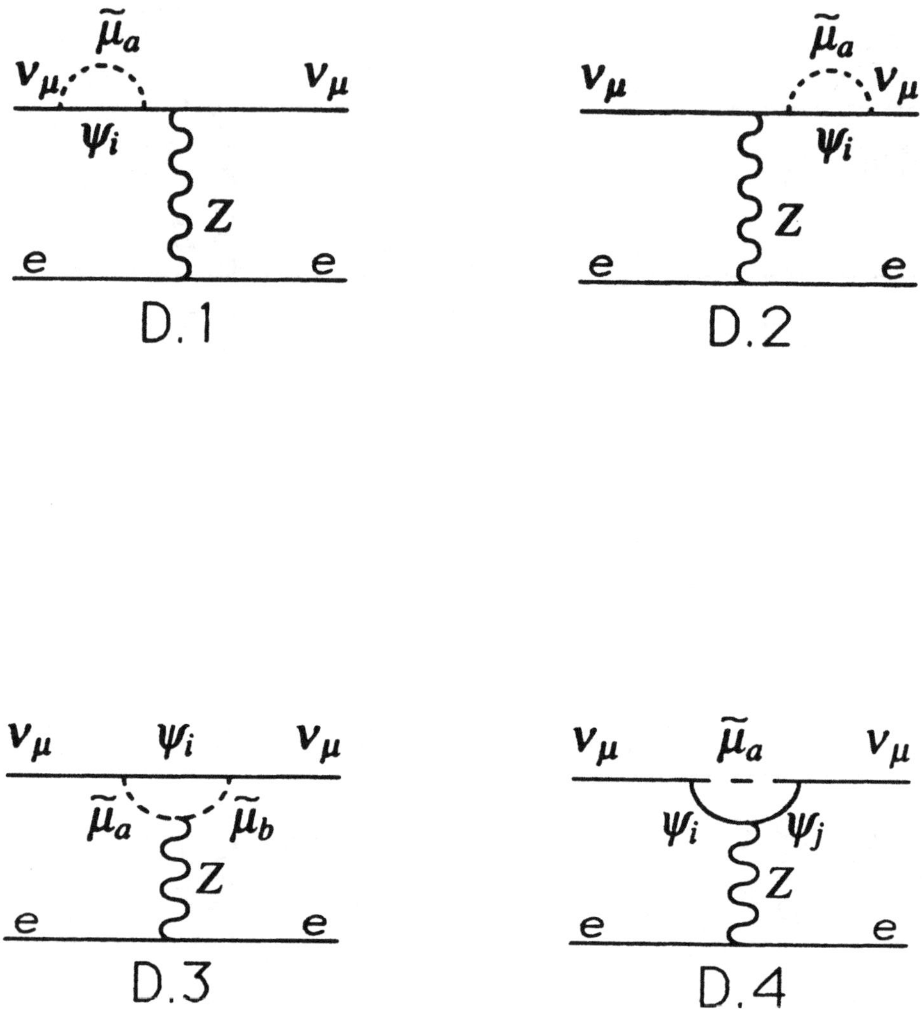

Figure 5: SUSY corrections to the ρ-parameter from charginos and smuons.

vanishing and UV-finite. Therefore

$$\delta_4(Fig.4) = -\frac{g^2}{16\pi^2}\sum_{\alpha\beta}(N^*_{\alpha 2} - t_\theta N^*_{\alpha 1})(N_{\beta 2} - t_\theta N_{\beta 1})$$
$$\times[\frac{1}{2}O^R_{\beta\alpha}\hat{E}_0(m_{\tilde{\nu}_\mu}, M^0_\alpha, M^0_\beta) - O^L_{\beta\alpha}\frac{M_{\alpha^0}M^0_\beta}{m^2_{\tilde{\nu}_\mu} - M^{02}_\alpha}F(m_{\tilde{\nu}_\mu}, M^0_\alpha, M^0_\beta)] \ . \tag{10}$$

Finally, the pay-off from the diagrams of Fig. 5 is also (globally) finite and reads

$$\delta_4(Fig.5) = -\frac{g^2}{16\pi^2}\{c^2_\theta + \sum_a\sum_{ij}|R^{(\mu)}_{a1}|^2(V^*_{i1}V_{j1})$$
$$\times[U^R_{ji}\hat{E}_0(M_i, M_j, m_a(\tilde{\mu})) - U^L_{ji}\frac{2M_iM_j}{M^2_i - M^2_j}F(M_i, M_j, m_a(\tilde{\mu}))]$$
$$+\frac{1}{2}\sum_{ab}\sum_i |V_{i1}|^2(R^{(\mu)}_{a1}R^{(\mu)*}_{b1})S_{ba}\hat{E}_0(M_i, m_a(\tilde{\mu}), m_b(\tilde{\mu}))$$
$$-\frac{1}{2}\sum_a\sum_i |V_{i1}|^2|R^{(\mu)}_{a1}|^2\hat{E}_0(M_i, m_a(\tilde{\mu}), m_a(\tilde{\mu}))\} \ . \tag{11}$$

From these eqs. it is clear that, in contradistinction to the SM, one gets nonvanishing hypercharge contributions to $\delta\rho$ at 1-loop in the MSSM.

Concerning the full 1-loop structure of δ_1, associated to the definition (2.21) of the Fermi constant, it gets multiple contributions from Figs. 6 and 7. Its full analytical form is a rather cumbersome expression, similar to the ones we have just derived for $\delta_{3,4}$, and will not be quoted here [18]. There is, in addition, the universal two-point function contribution from $\Pi_W(inos + sfermions)$ already given in eqs.(7) and (9).

The last step in our renormalization program is to compute δ_2, which is associated to the definition (2.23) of the fine structure constant. By explicit computation one can easily check that, in the static limit $k^2 = 0$, the various SUSY vertex corrections involved in Coulomb scattering (of muons on electrons, for example) are exactly cancelled by wave function renormalization of external legs (as expected from the SUSY analogue of standard Ward identites). Furthermore, the $\gamma - Z$ mixed vacuum polarization function, although it develops a pole at $k^2 = 0$, its contribution is depressed by an extra factor of $\Pi_{\gamma-Z}(k^2)/M^2_Z \propto k^2/M^2_Z$ with respect to the tree-level amplitude. To finish, notice that SUSY box diagrams in Coulomb scattering obviously contribute $\delta_2 = 0$ in the static limit, so we are left with $\Pi_\gamma(inos + sfermions)$ (Cf. eqs.(7) and (9)) as the only non-vanishing type of SUSY contribution to δ_2 (Cf. eq.(2.31)).

Gathering up the complete analytical formulae for δ_i $(i = 1, 2, 3, 4)$, I am now ready to present the numerical analysis of SUSY 1-loop corrections to ρ, M_W and M_Z. This I do in Figs. 8-10, to be next discussed in more detail. As mentioned earlier, I have not committed myself to any particular model, but just respected the current phenomenological (more or less stringent) bounds on sparticle masses [28]. For charged sleptons:

$$m_a(\tilde{l}) \geq 45\ GeV \ , \tag{12}$$

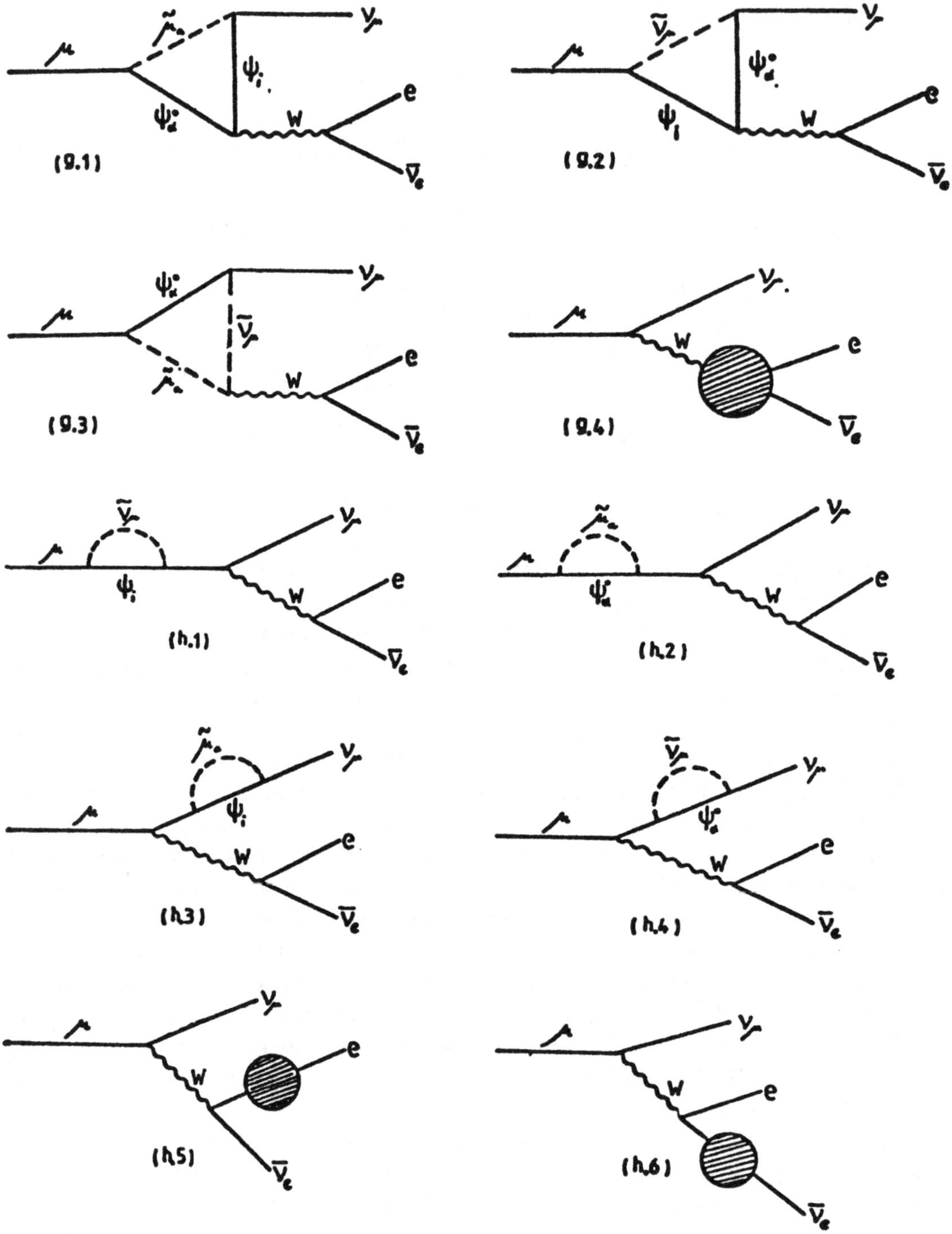

Figure 6: SUSY vertex and external line corrections to μ-decay.

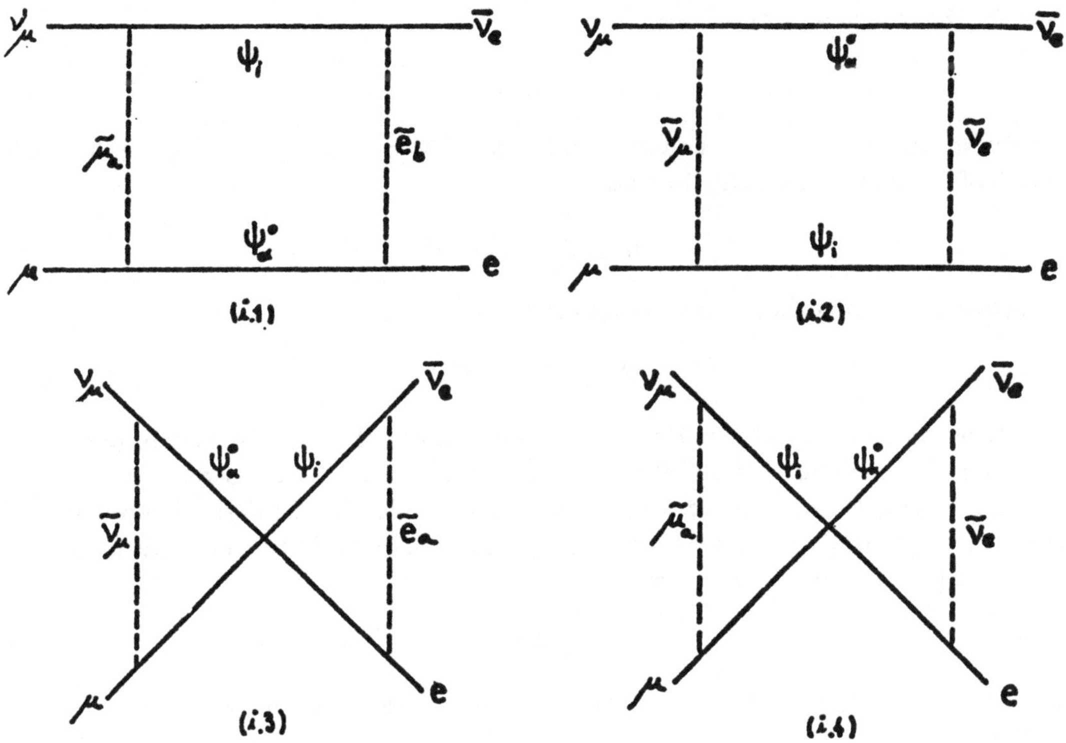

Figure 7: SUSY box diagrams contributing to μ-decay.

and for the chargino (Ψ^+) masses

$$M_i \geq 45 \; GeV \; . \tag{13}$$

As for sneutrinos, there is a bound from the non-SM contribution to the non-hadronic width of the Z, which roughly implies

$$m_{\tilde{\nu}_l} \geq 30 - 40 \; GeV \; . \tag{14}$$

Neutralino (Ψ^0_α) bounds are not so much stringent

$$M^0_\alpha \geq 20 - 30 \; GeV \tag{15}$$

and they are automatically satisfied by the eigenvalues of the 4×4 matrix (3.6) after imposing the chargino bound (13) on the eigenvalues of the 2×2 matrix (2.5). Finally, concerning squarks the situation is less clear. For example, the fact that the top-flavoured squark ("stop") could be lighter than its SM partner, changes the usual bounds from CDF and UA2 [29]. In any case a bound

$$m_{\tilde{q}} \geq 70 \; GeV \tag{16}$$

for all the generations looks reasonable to estimate the kind of effects to to be expected in our analysis.

The window in the (M, μ) parameter space, that I have selected in Figs. 8-10, covers a wide spectrum of mass eigenvalues of the chargino and neutralino mass matrices. The lightest eigenvalue in both cases ranges from the lower experimental bounds mentioned before up to about 200 GeV -around the upper corners of the plots. The blank regions on these plots are phenomenologically exluded by the bound (13). Although some new experimental information could restrict these regions even more severely [28], I feel that they will suffice to illustrate my point. I have selected the two extreme values $\tan\beta = 2, 8$ for the quotient between the Higgs v.e.v.'s. Let me point out that there is nothing special in the selected window. By explicit numerical analysis, it has been check that the complementary window, namely the one obtained by interchanging the ranges of M and μ chosen in Figs. 8-10, give very similar results.

Let us now focus our attention specifically on Fig. 8, where isolines of $\delta\rho^{SUSY}$ (in %) are exhibited [30]. They are full 1-loop corrections, as given by eq.(2.40) (not just eq.(2.41)!), except that I have not included corrections from the two-doublet Higgs sector of the MSSM (see below). Moreover I have decoupled the "oblique" correction (2.41) from squarks, the reason being that squarks enter the game only as universal type corrections and I want, for the purposes of illustration, to maximize the direct process-dependent effects. The point is that whereas the universal contributions (2.41) from squarks and sleptons are positive-definite (as can be easily checked analytically and numerically [17]-[19]), the process-dependent corrections may be of either sign and of the same order of magnitude. Therefore, in order to illustrate consistently the possibility that $\delta\rho^{SUSY}$ could be driven to negative values by the process-dependent corrections, I have to include the

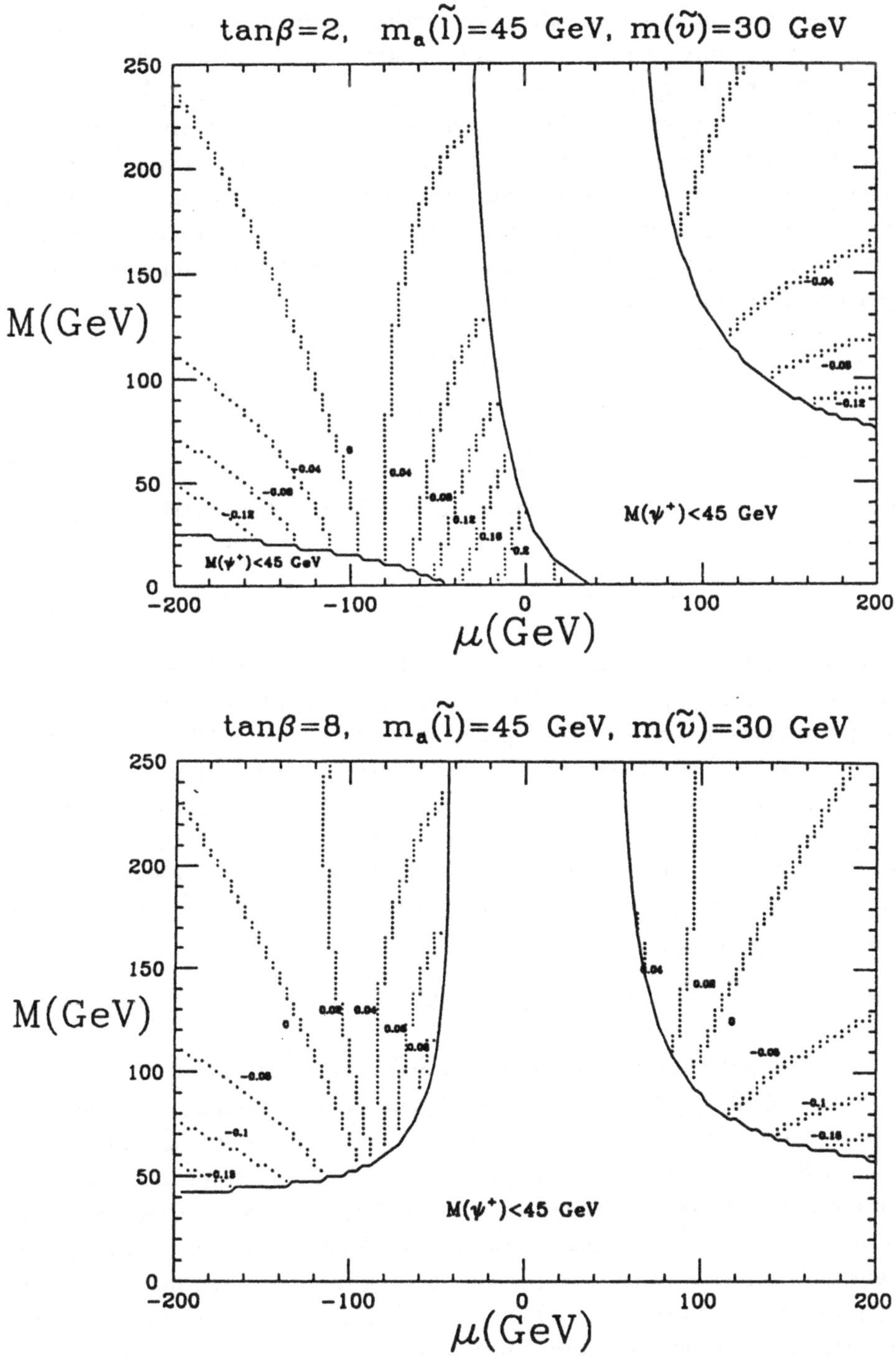

Figure 8: Contour plots of constant $\delta\rho^{SUSY}$ (in %) for indicated values of $\tan\beta$. The blank regions are excluded by the sparticle mass bounds given on the text.

full 1-loop (universal plus process-dependent) corrections from sleptons, charginos and neutralinos, and omit the universal contribution from squarks. The results shown in Fig. 8 indeed prove that there are regions in parameter space where the global $\delta\rho^{SUSY} \leq 0$. In the pure SM case, $\delta\rho$ can also get individual negative contributions from the Higgs sector being of the same order of magnitude than the genuine SUSY ones. However, in the MSSM, the two-doublet Higgs sector gives positive or negative individual contributions to $\delta\rho$ that are one order of magnitude less than those shown in Fig. 8 and so, as stated above, have not been included in the plots. Experimentally we know from global fits to the most important neutral current data, that [31]

$$\rho = 1.006 \pm 0.004 \ , \tag{17}$$

whence

$$0.002 \leq \delta\rho \leq 0.01 \ . \tag{18}$$

If we saturate this bound with the SM contribution from the top quark, we get the following allowed window for its mass

$$80 \ GeV \leq m_t \leq 180 \ GeV \tag{19}$$

-the lowest limit would be already surpassed by the CDF bound $m_t \geq 89 \ GeV$ if one assumes canonical (SM) decays [32]. From the analysis of Fig. 8 it follows that SUSY can change the prospects of the upper limit on m_t. See also Fig. 5 of ref. [20], where another (complementary) window of parameter space is explored [30]. If the sign of the SUSY corrections keeps pace with that of the top quark, namely if $\delta\rho^{SUSY} \geq 0$, then the upper limit on m_t may decrease about 30 GeV. If it is the other way around, then, m_t could be as much as 30 GeV heavier than expected from the SM alone. In this respect let us remember that the upshot of various recent SM analysis of all the precision electroweak experiments roughly agree that [1]

$$m_t = 125 \pm 30 \ GeV \ . \tag{20}$$

We have, however, exemplified an scenario in which SUSY quantum effects could change the prospects on m_t expected from the pure SM physics. And viveversa, knowledge on m_t (by direct production at the colliders) could result on hints of SUSY quantum effects, if the predicted deviations from the SM prospects are eventually confirmed. Let me emphasize, however, that the kind of effects from SUSY that I have focused my attention on could be different if $\delta\rho^{SUSY}$ becomes dominated by "oblique" squark contributions (2.41). This alternative scenario, characterized by a large stop-sbottom mass splitting, has already been widely discussed earlier [17]-[18], and also recently [19], in the literature, whereas the present complementary picture was largely ignored and should also be taken into account [20].

Turning now to Figs. 9-10, we consider the SUSY radiative corrections to the W and Z masses [21]. Here we restore the squark contributions and consider the full 1-loop yield of the complete analytical calculation. The maximum (negative) contributions

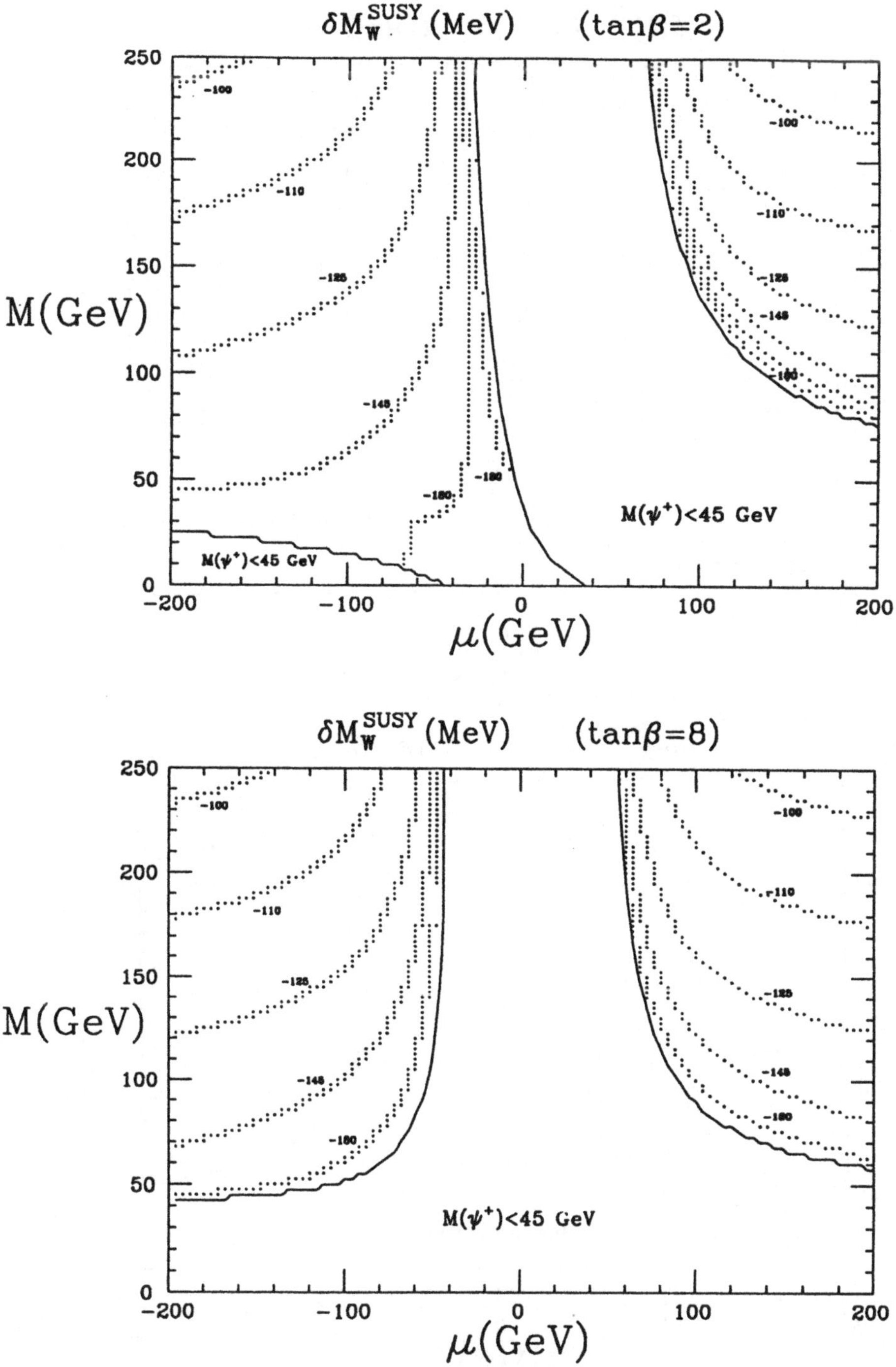

Figure 9: Contour plots of constant δM_W^{SUSY} (in MeV) for the indicated values of $\tan\beta$. The blank regions are excluded phenomenologically as in Fig. 8.

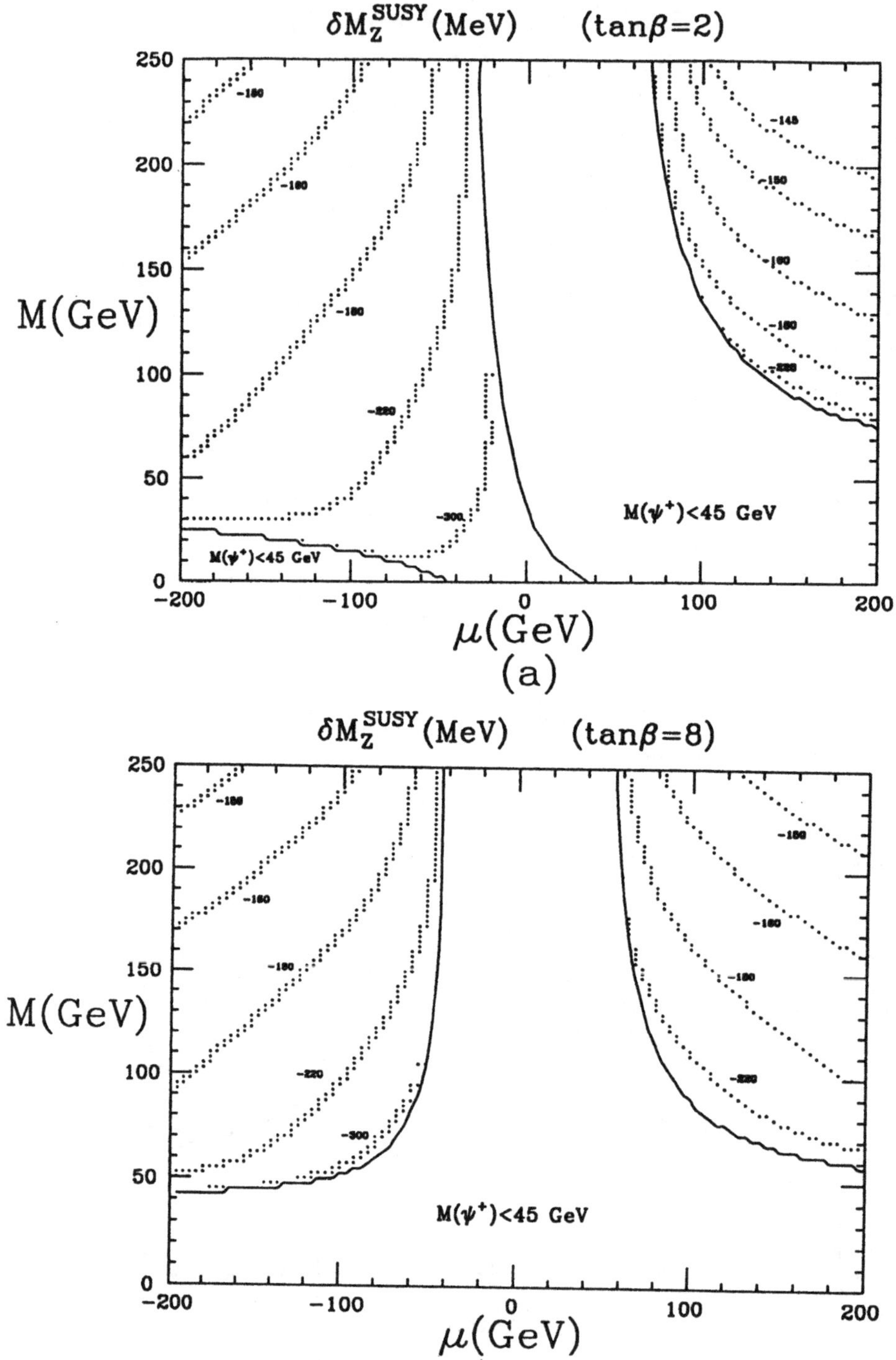

Figure 10: Contour plots of constant δM_Z^{SUSY} (in MeV) similar to those in Fig. 9.

corresponding to the situation of sfermion masses equal to the lower bounds stated above are computed to be

$$\delta M_W^{SUSY} = -230\ MeV \quad , \quad \delta M_Z^{SUSY} = -380\ MeV \quad \text{for} \quad \tan\beta = 2 \tag{21}$$

and

$$\delta M_W^{SUSY} = -215\ MeV \quad , \quad \delta M_Z^{SUSY} = -340\ MeV \quad \text{for} \quad \tan\beta = 8\ , \tag{22}$$

and come from the region of the (M, μ) parameter space almost over the rim of the physical region. Our emphasis, however, is placed upon the radiative corrections obtained in the main body of the allowed area. The total SUSY corrections to the W and Z masses, in the case under consideration, stabilize around $100\ MeV \leq |\delta M_W^{SUSY}| \leq 150\ MeV$ and $150\ MeV \leq |\delta M_Z^{SUSY}| \leq 250\ MeV$, respectively. A relevant feature of this stabilization is that even in remote corners of our windows, where the charginos and neutralinos are rather heavy $(100-200\ GeV)$, the radiative shifts to M_W and M_Z persist to be above 100 MeV and 150 MeV, respectively. I have checked that very similar results hold when changing in several ways the mixing angles in the sfermionic sector. Furthermore, although I have chosen to display detailed numerical plots only for the case corresponding to the present lower limits on the sparticle masses, I have also explored regions of a heavier spectrum. For instance, consider the case of $\tan\beta = 2$, and assume that all squarks and charged sleptons have a mass around 100 GeV while sneutrinos (much more difficult to detect or to bound) remain as before. The ensuing SUSY radiative corrections to M_Z corresponding to this situation can be as big as $-500\ MeV$. Finally, I want to point out that the results obtained from the present complete analysis constitute a revision of the partial results presented earlier on this subject in ref. [18]. Although the general conclusion- namely, that the maximum SUSY radiative corrections to $M_{W,Z}$ are of at most a few hundred MeV- remains qualitatively unmodified, the sign of these corrections has been fixed to be negative. This is in contrast to the partial numerical results exhibited in ref.[18], which were mainly based on vacuum polarization ("oblique") estimates, and explicitly demonstrates-once again- the relevance of the process-dependent corrections. We shall see in the next section that the sign of the corrections may be of crucial phenomenological (hence, practical) relevance.

5. Concluding Remarks

The present experimental precision achieved in the measurement of the Z mass at LEP can be considered highly satisfactory [1]: $M_Z = 91.175 \pm 0.021\ GeV$. In practice, it is a quantity "without" error. As for W-mass measurements, the situation is not bad but still far not comparable. Direct determinations from the colliders give a value for M_W, centered at essentially 80 GeV, with errors near 400 MeV [33]. On the other hand, the error on the W mass from the SM analysis of precision electroweak data, given by $M_W = 80 \pm 0.15\ GeV$ [1], is nearly twice better than the collider error. Clearly, we have to wait for LEP II before improving the W-mass measurements substantially and model-independently. There is no way at present to disentangle our SUSY radiative corrections

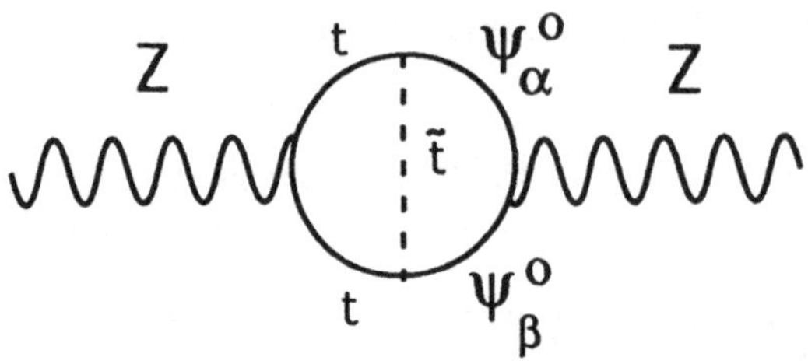

Figure 11: Potentially relevant SUSY 2-loop "oblique" contribution.

to M_W from the large experimental errors. Fortunately (perhaps optimistically), the expected precision at which M_W is to be measured in future experiments with hadron colliders and LEP II is about $80-100\ MeV$ [34], hence at the level of our SUSY radiative corrections. Nevertheless, this does not yet mean that these corrections can be measured, even if really there and with the predicted maximum size, unless the precision in which our inputed s_θ^2 from the ratio R in eq.(2.1) achieves the level of 1% or better, which is certainly not the case at present. The ratio R suffers from systematic uncertainties connected with the relative normalization of the neutrino and antineutrino fluxes. It is currently measured with an error of $\delta R = 0.1$, which translates into an error on the mixing angle of $\delta s_\theta^2/s_\theta^2 = 4\%$. In future experiments [35], one expects to lessen this error at the level of 2%, though it will be difficult enough [36]. However, using the special ratio

$$R' = \frac{\sigma(\nu_\mu e \to \nu_\mu e)}{\sigma(\bar{\nu}_\mu e \to \bar{\nu}_\mu e) + \sigma(\nu_e e \to \nu_e e)}\ , \tag{1}$$

which has the advantatge [37] of highly reducing the systematic uncertainties, s_θ^2 will eventually be pinned down at the desired level of 1% or even better. In the meanwhile the current 4% error on s_θ^2 is the real stumbling block to unraveling potential SUSY radiative corrections from the very accurate ("zero error") present measurement of the Z mass. In spite of the planned improvements on this issue, the computation of M_Z in the MSSM, within our framework, inherits an error at the tree level (namely, $\delta M_Z \approx 300\ MeV$ for $\delta s_\theta^2/s_\theta^2 = 1\%$) which could still be too big to hope for seeing any clear footprint of the new physics. There is, however, a circumstance that may help here. From the analytical results of ref.[3], one can check that the radiative corrections to M_Z from the top-bottom family are, for relatively large values of m_t $(150-200\ GeV)$, negative and of the same order of magnitude than our genuine SUSY radiative corrections. Therefore, an intriguing possibility is left open for the future: the sum of both contributions could produce an enhanced signal (say, $\delta M_Z = -(500-800)\ MeV$, depending on whether we consider the case of Figs. 9-10 or the more favorable situations mentioned at the end of Secc. 4), well above the expected errors, that should be visible and could be interpreted as evidence of both of a (relatively heavy) top quark and of SUSY.

I would like to point out, in passing, the possibility that a class of 2-loop "oblique"

SUSY corrections could play a significant role in the previous picture. A typical case (among many others) is shown in Fig. 11, where a scalar top, $\tilde{t}$, is exchanged across the fermion loop. An estimate of the contribution from it (after subtraction of UV-divergences) yields:

$$|\delta M_Z^2| \approx \frac{3g^2}{c_\theta^2}\Big(\frac{m_t}{16\pi^2}\Big)^2\Big(\frac{m_t}{M_W}\Big)^2 \log\frac{m_{\tilde{t}}^2}{m_{\tilde{t}}^2} . \tag{2}$$

To maximize this effect, take $m_t = 200\ GeV$ and $m_{\tilde{t}} = 2000\ GeV$. It follows that $|\delta M_Z| = 400\ MeV$, i.e. of the order of the 1-loop corrections!. Thus, depending on the sign of these corrections, they could enhance in a remarkable way the final SUSY radiative signal, or on the contrary they could even cancel it to a large extent. Although we would, of course, prefer the first possibility, both alternatives are interesting from the theoretical point of view: 2-loop corrections being of the same order of magnitude than 1-loop corrections looks an unusual situation and certainly deserves further study [38].

To conclude, a practical observation. As recognized previously, despite the theoretical cleanness of the calculations in this framework (which is free from strong interaction effects), $\nu_\mu e$-scattering experiments are affected with experimental uncertainties greater than those in deep inelastic scattering. Therefore, an analysis of the latters would be more realistic and useful. This situation becomes more dramatic and noticeable at the prospect of lessening the error on δs_θ^2 down to 0.6% in future deep inelastic neutrino experiments, as has already been emphasized by members of the CCFR Collaboration [39]. With such a precision one could even reverse the direction followed by our calculations and try to predict s_θ^2 obtained from these experiments using the very precise value of M_Z as an experimental input [40]. Thus we may also view the present work as an exemplification of the corresponding effects that can be expected in the more precise framework of present and future neutrino deep inelastic experiments. We have not yet carried out such an analysis, but we expect that it should exhibite the same kind of behaviour [41]. May be in the meantime we shall be able to produce a relatively light ($\approx 50\ GeV$) physical supersymmetric particle before achieving the required precision, thus obviating SUSY. If not, at least we may be confident that the potential SUSY quantum effects on $M_{W,Z}$ and ρ could be sizeable and stable, even for a SUSY spectrum involving masses within a range ($\approx 100\ GeV$) which certainly cannot be reached directly from clean e^+e^- collisions in the foreseeable future.

Acknowledgements: I would like to thank Prof. Hollik for inviting me to participate in the Workshop on Phenomenological Aspects of Supersymmetry at the Max Planck Institute in Munich, for his warm hospitality and for various technical comments and useful suggestions concerning the present and future planned calculations in the field of SUSY radiative corrections to the electroweak parameters.

5. Appendix

I display in this appendix, very briefly, the essentials on 1-loop integrals associated to 1-, 2-, 3- and 4-point Green's functions used throughout the text. I adopt in part the notation of ref.[42] (where many more details are given), but I use the metric $g_{00} = 1, g_{ii} = -1$.

The basic 1-point loop integral is given by

$$A(m) = \int \frac{d^n q}{(2\pi)^n} \frac{1}{q^2 - m^2} = \frac{-i}{16\pi^2} m^2 (\Delta + \log m^2 - 1) \ , \tag{1}$$

where the ultraviolete (UV), dimensionally regularized, pole is denoted by

$$\Delta \equiv \frac{2}{\epsilon} + \gamma_E - \log 4\pi \ , \tag{2}$$

with $\epsilon \equiv n - 4 \to 0$ and γ_E being Euler's constant. The basic set of functions B_0, B_1, B_{21} and B_{22} associated to two-point loop integrals is defined by

$$\begin{aligned}
\int \frac{d^n q}{(2\pi)^n} \frac{1}{(q^2 - m_1^2)[(q-k)^2 - m_2^2]} &= B_0(k^2, m_1, m_2) \ , \\
\int \frac{d^n q}{(2\pi)^n} \frac{q_\mu}{(q^2 - m_1^2)[(q-k)^2 - m_2^2]} &= k_\mu B_1(k^2, m_1, m_2) \ , \\
\int \frac{d^n q}{(2\pi)^n} \frac{q_\mu q_\nu}{(q^2 - m_1^2)[(q-k)^2 - m_2^2]} &= k_\mu k_\nu B_{21}(k^2, m_1, m_2) + g_{\mu\nu} B_{22}(k^2, m_1, m_2) \ . \qquad (3)
\end{aligned}$$

I need the explicit form of the B-functions for $k^2 = 0$ and $k^2 = M^2_{W,Z}$, so I compute them for arbitrary k^2:

$$\begin{aligned}
B_0(k^2, m_1, m_2) &= \frac{-i}{16\pi^2}[\Delta + I_0(k^2, m_1, m_2)] \ , \\
B_1(k^2, m_1, m_2) &= \frac{-i}{16\pi^2}[\frac{1}{2}\Delta + I_1(k^2, m_1, m_2)] \ , \\
B_{21}(k^2, m_1, m_2) &= \frac{-i}{16\pi^2}[\frac{1}{3}\Delta + I_2(k^2, m_1, m_2)] \ ,
\end{aligned} \tag{4}$$

where

$$\begin{aligned}
I_n(k^2, m_1, m_2) &= \frac{1}{n+1}\{\log k^2 + \sum_{\Gamma=\pm} [\log(1 - x_\Gamma) + F_{n+1}(x_\Gamma)]\} \ , \\
F_n(x) &= -x^n \log \frac{x-1}{x} - \sum_{\alpha=1}^{n} \frac{x^{n-\alpha}}{\alpha} \ , \\
x_\pm &= \frac{1}{2k^2}[k^2 + m_1^2 - m_2^2 \pm \lambda^{1/2}(k^2, m_1, m_2)] \ , \\
\lambda(k^2, m_1, m_2) &= [k^2 - (m_1 + m_2)^2][k^2 - (m_1 - m_2)^2] \ .
\end{aligned} \tag{5}$$

As far as 3-and 4-point loop integrals are concerned, it suffices to evaluate them at $k^2 = 0$, since the low momentum transfer in the three processes considered is negligible as compared to the average mass of the SUSY particles. Thus the basic 3-point loop integrals needed are C_0, which is finite, and $C^0_{\mu\nu}$, which is UV-divergent:

$$\begin{aligned}
\frac{-i}{16\pi^2} C_0(m_1, m_2, m_3) &= \int \frac{d^4 q}{(2\pi)^4} \frac{1}{(q^2 - m_1^2)(q^2 - m_2^2)(q^2 - m_3^2)} \\
&= \frac{-i}{16\pi^2 (m_1^2 - m_2^2)} F(m_1, m_2, m_3) \ , \\
F(m_1, m_2, m_3) &= \log \frac{m_1^2}{m_2^2} + \frac{m_3^2}{m_3^2 - m_1^2} \log \frac{m_3^2}{m_1^2} - \frac{m_3^2}{m_3^2 - m_2^2} \log \frac{m_3^2}{m_2^2}
\end{aligned} \tag{6}$$

and

$$C^0_{\mu\nu}(m_1,m_2,m_3) = \int \frac{d^nq}{(2\pi)^n}\frac{q_\mu q_\nu}{(q^2-m_1^2)(q^2-m_2^2)(q^2-m_3^2)} ,$$
$$= \frac{-i}{16\pi^2} g_{\mu\nu}[\frac{1}{4}\Delta + \hat{C}_0(m_1,m_2,m_3)] ,$$
$$\hat{C}_0(m_1,m_2,m_3) = \frac{1}{4}[\frac{-3}{2} + \log m_3^2 + \hat{E}_0(m_1,m_2,m_3)] ,$$
$$\hat{E}_0(m_1,m_2,m_3) = \frac{m_2^2}{m_3^2-m_2^2}\log\frac{m_3^2}{m_2^2} + \frac{m_1^2}{m_1^2-m_2^2}F(m_1,m_2,m_3) . \quad (7)$$

The two necessary 4-point loop integrals are UV-convergent:

$$\frac{-i}{16\pi^2}D_0(m_1,m_2,m_3,m_4) = \int \frac{d^4q}{(2\pi)^4}\frac{1}{(q^2-m_1^2)(q^2-m_2^2)(q^2-m_3^2)(q^2-m_4^2)}$$
$$= \frac{-i}{16\pi^2(m_1^2-m_2^2)}[\frac{F(m_1,m_3,m_4)}{m_1^2-m_3^2} - \frac{F(m_2,m_3,m_4)}{m_2^2-m_3^2}] \quad (8)$$

and

$$D^0_{\mu\nu}(m_1,m_2,m_3,m_4) = \int \frac{d^4q}{(2\pi)^4}\frac{q_\mu q_\nu}{(q^2-m_1^2)(q^2-m_2^2)(q^2-m_3^2)(q^2-m_4^2)}$$
$$= \frac{-i}{64\pi^2}\frac{g_{\mu\nu}}{(m_1^2-m_2^2)}[\frac{m_1^2}{m_1^2-m_3^2}F(m_1,m_3,m_4) - \frac{m_2^2}{m_2^2-m_3^2}F(m_2,m_3,m_4)] . \quad (9)$$

References

[1] For a recent review of the precision tests of the SM, see for example, R. D. Peccei, "Status of the Standard Model in 1991", UCLA preprint UCLA/91/TEP/40 and Proc. of the 1991 Meeting of the Division of Particles and Fields of the American Physical Society, Vancouver, B.C. Canada, August 1991.

[2] M. Green and M. Veltman, Nucl. Phys. B169 (1980) 137.

[3] M. Veltman, Phys. Lett. B91 (1980) 95.

[4] A. Sirlin, Phys. Rev. D22 (1980) 971; W. Marciano and A. Sirlin, Phys. Rev. D22 (1980) 2695; D29 (1984) 75 and 945.

[5] F. Antonelli, M. Consoli and G. Corbò, Phys. Lett. B91 (1980) 90; F. Antonelli, G. Corbò, M. Consoli and O. Pellegrino, Nucl. Phys. B183 (1981) 195.

[6] B. Lynn, M. E. Peskin and R. Stuart, in: Physics at LEP, eds. J. Ellis and R. Peccei, CERN Yellow Report CERN 86-02 (1986); D. Kennedy and B. Lynn, Nucl.

Phys. B232 (1989) 1; M. Peskin and T. Takeuchi, Phys. Rev. Lett. 65 (1990) 964; D. Kennedy and P. Langacker, Phys. Rev. Lett. 65 (1990) 2967; G. Altarelli and R. Barbieri, Phys. Lett. B253 (1991) 161.

[7] M. Consoli, W. Hollik and F. Jegerlehner, in: Z physics at LEP 1, eds. G. Altarelli, R. Kleiss and C. Verzegnassi, CERN Yellow Report CERN 89-08 (1989); see also the detailed reports by W. Hollik, Fortschr. Phys. 34 (1986) 687 and 38 (1990) 165.

[8] W. Hollik, R. Kleiss and C. Verzegnassi, in: Precision Tests of the Standard Model at High Energy Colliders, eds. F. del Águila, A. Méndez and A. Ruiz, (World Scientific, Singapore, 1991).

[9] P. Langacker and M. Luo, Phys. Rev. D44 (1991) 817.

[10] J. van der Bij and F. Hoogeveen, Nucl. Phys. B283 (1987) 477.

[11] J. Gunion, H. Haber, G. Kane and S. Dawson, "The Higgs Hunter's Guide" (Addison-Wesley , 1990).

[12] Y. Golfand and E. Lichtman, JETP Lett. 13 (1971) 323; D. Volkov, V. Akulov, Phys. Lett. B46 (1973) 109; J. Wess and B. Zumino, Phys. Lett. B49 (1974) 52.

[13] P. Fayet and S. Ferrara, Phys. Rep. 32 (1977) 249; P. van Nieuwenhuizen, Phys. Rep. 117 (1985) 75.

[14] M. Green, J. Schwarz and E. Witten, "Superstring Theory" (Cambridge Univ. Press, 1987); M. Kaku, "Introduction to Superstrings" (Springer Verlag, 1990).

[15] H. Nilles, Phys. Rep. 110 (1984) 1; H. Haber and G. Kane, Phys. Rep. 117 (1985) 75; A. Lahanas and D. Nanopoulos, Phys. Rep. 145 (1987) 1; see also the exhaustive reprint collection "Supersymmetry" (2 vols.), ed. S. Ferrara (North Holland/World Scientific, Singapore, 1987).

[16] U. Amaldi, W. de Boer, H. Fürstenau, Phys. Lett. B260 (1991) 447; J. Ellis, S. Kelly and D. Nanopoulos, Phys. Lett. B249 (1990) 441; 260B (1991) 131; see also P. Langacker and M. Luo in [9].

[17] L. Alvarez-Gaumé, J. Polchinski and M. Wise, Nucl. Phys. B221 (1983) 495; R. Barbieri and L. Maiani, Nucl. Phys. B224 (1983) 32; C. Lim, T. Inami and N. Sakai, Phys. Rev. D29 (1984) 1488; E. Eliasson, Phys. Lett. B147 (1984) 65; Z. Hioki, Prog. Theor. Phys. 73 (1985) 1283; K. Schwarzer, Oxford Univ. preprint 40/84 (1984), unpublished; B. Lynn, SLAC-PUB-3358 (1984), unpublished; B. Lynn and M. Peskin, SLAC-PUB-3724 (1985), unpublished.

[18] J. Grifols and J. Solà, Phys. Lett. B137 (1984) 257; Nucl. Phys. B253 (1985) 47; J. Solà, **Thesis**, Univ. Autònoma de Barcelona (1985).

[19] R. Barbieri, M. Frigeni, F. Giuliani and H. Haber, Nucl. Phys. B341 (1990) 309; M. Dress and K. Hagiwara, Phys. Rev. D42 (1990) 1709; A. Bilal, J. Ellis and G. Fogli, Phys. Lett. B246 (1990) 459; M. Dress, K. Hagiwara and A. Yamada, DTP/91/34 (1991).

[20] P. Gosdzinsky and J. Solà, Phys. Lett. B254 (1991) 139.

[21] P. Gosdzinsky and J. Solà, Mod. Phys. Lett. A6 (1991) 1943; J. Solà, "Radiative Corrections to the Electroweak Parameters from Supersymmetry", to appear in the Proc. of the XXVI Rencontres de Moriond, Electroweak Interactions and Unified Theories (Frontières, 1992).

[22] B. Lynn, M. Peskin and R. Stuart, in [6].

[23] The sign of Eq. (3.2) of the second ref.[18] must be changed to conform with that of eq. (16) in the text. Only the latter is correct.

[24] M. Veltman, Nucl. Phys. B123 (1977) 89.

[25] M. Einhorn, D. Jones and M. Veltman, Nucl. Phys. B191 (1981) 146.

[26] S. Weinberg, Phys. Rev. D19 (1979) 1277; L. Susskind, Phys. Rev. D20 (1979) 2619; R. Lytel, Phys. Rev. D22 (1980) 505.

[27] For a detailed study of $SU(2)_V$-breaking by SUSY oblique corrections, see M. Drees and K. Hagiwara in [19].

[28] D. Decamp et al., ALEPH collab., Phys. Lett. B244 (1990) 541; P. Abreu et al., DELPHI collab., Phys. Lett. B247 (1990) 157; M. Z. Akrawy et al., OPAL collab., Phys. Lett. B248 (1990) 211.

[29] F. Abe et al., CDF collab., Phys. Rev. Lett. 62 (1989) 1825; J. Alitti et al., UA2 collab., Phys. Lett. B235 ((1990) 363.

[30] The plots in Figs. 5 and 6 of Ref.[20] contain a typographical error: μ and M have to be interchanged, i.e. the μ- axis is in reality the M-axis, and conversely. However, as mentioned in the text, there is nothing special in the selected window, hence the interchange of ranges of μ and M does not affect significantly the results.

[31] P. Langacker, Review of Particle Properties, Phys. Lett. B239 (1990) III.56; "Precision tests of the Standard Model", Univ. of Pennsylvania preprint UPR-0435T (1990).

[32] P. Sinervo, Theoretical Advanced Study Institute, Boulder (1990).

[33] F. Abe et al., CDF collab., Phys. Rev. Lett. 65 (1990) 2243; J. Alitti, UA2 collab., Phys. Lett. B241 (1990) 150.

[34] M. L. Swartz, "Precision experiments in electroweak interactions", SLAC preprint SLAC-PUB-5219 (1990).

[35] The future error on R from CHARM II is projected to be 0.05, which leads to $\delta s_\theta^2 = 0.005$, see D. Geiregat et. al., Phys. Lett. B232 (1989).

[36] I want to thank Prof. F. Dydak for a private conversation on the status of $\nu_\mu e$-scattering experiments.

[37] P. Langacker, M. Luo and A. Man, " High precision electroweak experiments". Univ. of Pennsylvania preprint UPR-0458T (1991).

[38] J. Solà, in preparation.

[39] S. Mishra, CCFR collab., "A next generation of high energy neutrino experiments", to appear in the Proc. of the XXVI Rencontres de Moriond, Electroweak Interactions and Unified Theories (Frontières, 1992). I wish to thank Dr. Mishra for informing me on the details of this experiment.

[40] I am grateful to Prof. W. Hollik for this suggestion and for an illuminating discussion on this particular.

[41] R. Jiménez and J. Solà, in preparation.

[42] G. 't Hooft and M. Veltman, Nucl. Phys. B153 (1979) 365.

Radiative Corrections in the Supersymmetric Higgs Sector

Stefan Pokorski

Institute for Theoretial Physics
Warsaw University
Hoza 69, 00-681 Warsaw, Poland

Abstract
A non-technical review is presented of the subject of radiative corrections in the Higgs sector of the Minimal Supersymmetric Standard Model and their implications for phenomenology.

Content:

1. Intoduction
2. More motivation for studying the MSSM Higgs sector
3. Closer look at the MSSM Higgs sector and the origin of large radiative corrections
4. Survey of the existing 1-loop calculations
5. Choice of parameters and renormalization scheme dependence
6. The MSSM Higgs sector after radiative corrections and LEP phenomenology
7. Summary: where is the lightest supersymmetric Higgs boson?

To be published in "Phenomenological Aspects of Supersymmetry" in the series Lecture Notes in Physics, by Springer-Verlag.

Supported in part by the Committee for Scientific Research under the grant 2 0165 91 01.

1 Introduction

The electroweak theory based on the spontaneously broken $SU(2) \times U(1)$ gauge symmetry is consistent with all available experimental data, including high precision LEP experiments. However, the ***actual mechanism*** of the spontaneous gauge symmetry breaking remains unknown. It is very likely that its understanding, necessary to complete the electroweak theory, will also be a link to new physics beyond the Standard Model.

Since long we know two important constraints on the mechanism of the $SU(2)\times U(1)$ symmetry breaking. These are: the experimental value of the ρ parameter, $\rho \cong 1$, and strong suppression of flavour changing neutral current (FCNC) reactions. The former translates into custodial $SU(2)_V$ symmetry of the electroweak vacuum and the latter has to be reconciled with the fermion mass generation in a model considered. The simplest possibility consistent with those constraints is the minimal version of the Higgs mechanism, with only one Higgs doublet responsible for the symmetry breaking and for the fermion mass generation. Extensions to models with several Higgs doublets are also acceptable provided their couplings to fermions are consistent with the FCNC constraint. The easiest way to achieve this is to impose such symmetries that only one scalar doublet couples to right-handed fermions of the same charge.

However, theories with elementary scalars suffer from the well known naturalness problem and it is strongly felt on the theoretical side that the actual mechanism of the gauge symmetry breaking goes beyond the minimal version of the Standard Model. The three main ideas proposed so far to remedy the naturalness problem are technicolour models, strongly interacting Higgs sector and supersymmetry. Recent high precision LEP data seem to go against generic (extended) technicolour models [1], adding to the known difficulties with such models in reconciling the suppression of FCNC with the quark mass generation. The remaining two alternatives wait for experimental judgement.

Supersymmetry, effectively broken at some scale $M > M_W$, is at present the most interesting theoretical framework which naturally accommodates elementary scalars. The simplest and most economical supersymmetric model is the Minimal Supersymmetric Standard Model (MSSM). It is defined by a) gauge group $SU(3)\times SU(2)\times U(1)$; b) minimal matter content: three generations of quarks and leptons and two Higgs doublets, plus their superpartners; c) an exact discrete symmetry: R-parity; d) explicit "soft" breaking of supersymmetry by gaugino and scalar mass terms and trilinear scalar couplings.

The MSSM can be embedded into grand unification schemes with the correct value of $\sin^2 \Theta_W(M_Z)$ [2] and the lightest supersymmetric particle is a natural candidate for dark matter. Various non-minimal supersymmetric extensions of the Standard Model, which have typically more free parameters, often give the MSSM as the low energy effective theory [3]. Thus, the MSSM is an interesting laboratory, both for theoretical and experimental study (and to discover effects) of supersymmetry.

The Higgs sectors of supersymmetric extensions of the Standard Model require at least two Higgs doublets (for the supersymmetric structure of Yukawa couplings and for the triangle anomaly cancellation). The minimal such an extension (the MSSM) is *naturally* a two-Higgs doublet model and, moreover, with only one doublet coupled to right-handed fermions of a given charge (i.e. with no conflict with the suppression of the FCNC). The two-Higgs doublet structure combined with supersymmetry offers one more attractive, though speculative, feature namely the possibility of partly understanding the pattern of fermion masses. The full solution to this problem most likely requires really new ideas and may be closely related to our understanding of the fermion generation puzzle. However, it is still conceivable that at least part of the mass problem has more conventional character. Here I mean the question: how is the isotopic spin symmetry of the quark masses broken? Supposing that in the first approximation only the masses of the third family are generated the question actually is why $m_t \gg m_b$. It is an old idea that the up - and down - quark masses are driven by two different Higgs doublets with *vevs* such that $m_t/m_b \cong v_2/v_1 \equiv \tan\beta$ (where $v_2(v_1)$ is the *vev* of the Higgs field coupled to the up (down) quarks) whereas the Yukawa couplings for the top and bottom quarks are approximately equal, $h_t \approx h_b$. This mechanism of the isotopic spin symmetry breaking looks particularly attractive in supersymmetric models which not only must, for consistency, have at least two Higgs doublets but also offer the possibility of the $SU(2) \times U(1)$ symmetry breaking by radiative corrections, with $\tan\beta \gg 1$. Indeed, it has been found [4] that for $h_t \cong h_b$ the $SU(2) \times U(1)$ symmetry breaking with $\tan\beta \gg 1$ is driven by the right-handed squark masses and indirectly by gaugino mass (M_1): su_R and sd_R masses evolve differently simply because of different $U(1)$ charge assignment.*So, large* $\tan\beta$ *values belong to the very interesting region of the parameter space in the MSSM.*

Given good reasons for supersymmetry, in this lecture we concentrate on the Higgs sector of the MSSM and in particular on the recently discovered large loop effects in this sector. Our purpose is to give a not too technical overview of the problem. There seems to be little point in repeating technical details which are available in the original papers. Usually, loop effects are associated with high precision data and here there is no data whatsoever. Therefore it seems appropriate first to recall the motivation for this effort.

2 More motivation for studying the MSSM Higgs sector

The Higgs sector in the MSSM has several very distinct features. In particular, in the tree approximation there is in this model always at least one light neutral scalar ($M_h < M_Z$), no matter how high is the scale of soft supersymmetry breaking. This looks very fortunate as the scale of supersymmetry breaking is likely to be relatively high, say, $0(1\ TeV)$ and therefore the Higgs boson may well be the first experimental trace of supersymmetry (or one may hope to rule out the MSSM by non-observation of the scalar in the predicted mass range). However, even if softly broken supersymmetry is indeed realized in nature, it is quite likely that the other Higgs particles are much heavier, with masses of the order of the supersymmetry breaking scale. This scenario is suggested by the naturalness arguments, as it will be discussed later.

Clearly, we are facing two important questions:
a) what are the potential new signatures of the supersymmetric Higgs particles and, in particular, of the lightest one as compared to the minimal Standard Model Higgs scalar search?
b) suppose a Higgs particle is observed: shall we be able to identify it as the minimal Higgs scalar or the lightest supersymmetric Higgs scalar or still something else?

Those questions can easily be answered in the tree approximation [see J. Kalinowski, this Volume, for details]. Let us recall only the main facts. The only free, unknown, parameter of the minimal SM Higgs sector is the scalar mass whereas all its couplings are known in terms of the gauge couplings and the fermion masses. The dominant production mechanisms are then obvious: this is the Bjorken process in the e^+e^- collisions and the gluon-gluon fusion in hadronic collisions. In the MSSM the Higgs sector depends at the tree level on two unknown parameters. In terms of only two free parameters one predicts masses of three physical scalars and one pseudoscalar as well as their couplings to gauge boson, to fermions and their self-couplings. The two free parameters can be conveniently taken as e.g. M_A (the mass of the pseudoscalar) and $\tan\beta$. Among important tree level results are the well known relations:

$$\begin{aligned} M_h &< \min(M_A, M_Z) \\ M_H &> \max(M_A, M_Z) \\ M^+ &> \max(M_W, M_A) \end{aligned} \tag{1}$$

where M_h, M_H and M^+ are the lighter, the heavier and the charged scalar masses, respectively. The couplings of the MSSM Higgs particles are summarized below (in units of the minimal SM Higgs scalar couplings)

	ZZ	ZA	$U\overline{U}$	$D\overline{D}$
h	$\sin(\beta-\alpha)$	$\cos(\alpha-\beta)$	$\frac{\cos\alpha}{\sin\beta}$	$\frac{-\sin\alpha}{\cos\beta}$
H	$\cos(\beta-\alpha)$	$\sin(\alpha-\beta)$	$\frac{\sin\alpha}{\sin\beta}$	$\frac{\cos\alpha}{\cos\beta}$
A	—	—	$\cot\beta$	$\tan\beta$

where the angle α is the mixing angle which diagonalizes the tree level mass matrix for neutral scalars. We observe that there are now three potentially important production mechanisms (for sake of definiteness we discuss only the e^+e^- collisions at energies below the threshold for the $\bar{t}t$ production):

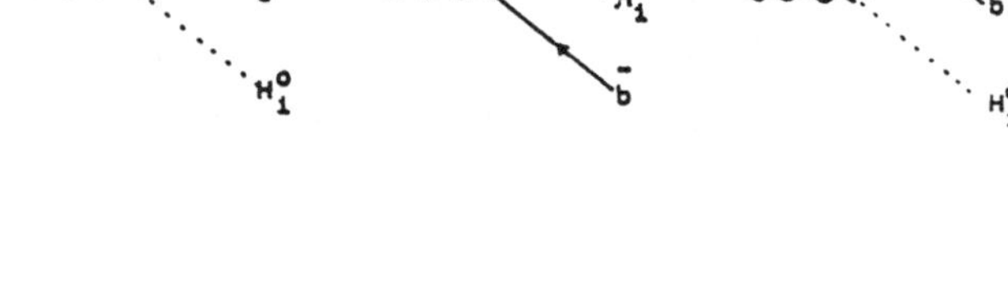

They are complementary in the sense that different mechanisms dominate the production process in different regions of the $(\tan\beta, M_A)$ parameter space. [5,6] The width and decay modes of the Higgs particles for different values of $(\tan\beta, M_A)$ can, of course, be easily discussed, too.

We also note that for certain values of $(\tan\beta, M_A)$ the lightest supersymmetric Higgs particle h resembles closely the Higgs scalar of the minimal SM. This is shown in Fig. 9: the cross sections for $e^+e^- \rightarrow Zh$ are very similar to the analogous cross sections for the SM Higgs scalar for $M_A > M_Z$ and for any value of $\tan\beta$. It is clear from the Table that simultaneously the cross sections for $e^+e^- \rightarrow hA$ are strongly suppressed. The widths and branching ratios of the h are somewhat more sensitive to its actual nature but, in general, the tree level results suggest that it may be rather difficult to distinguish experimentally between the lightest supersymmetric Higgs scalar and the minimal SM one. Clearly, it is very important to known how much modified are all those tree level results by loop corrections.

The interest in the Higgs sector of the MSSM follows also from the fact that its strong dependence on $\tan\beta$ links it to several other very interesting investigations. First of all, Yukawa couplings of the charged Higgs bosons in two-doublet models depend on $\tan\beta$, too. Therefore, potential departures from the SM predictions for the charged current mediated processes are sensitive to constraints on $\tan\beta$ derived from studying the Higgs sector or, conversly, present experimental data for those processes give some limits on $\tan\beta$ and M^+. Upper limits on $\tan\beta$ for different values of M^+ can be obtained from τ and b decays [7] and lower limits - mainly from $K-\overline{K}$ and $B-\overline{B}$ mixing [8]. Those results refer to $\tan\beta$ defined as the tree level parameter and give:$1/2 < \tan\beta < M^+/1GeV$. Another physical quantities dependent on $\tan\beta$ are sparticle masses. Again, there is then a link between them and the MSSM Higgs sector.

Those considerations become particularly interesting and constrained in the MSSM defined in terms of very few (four of five) free parameters specified at the Planck scale (with the low energy parameters obtained by the renormalization group evolution via supersymmetric renormalization group equations) and supplemented by the requirement of the radiative $SU(2)\times U(1)$ symmetry breaking.*Remaining at the tree level of the low energy softly broken supersymmetric lagrangian*, one then obtains strong interconnection between:

radiative $SU(2)\times U(1)$ breaking	Higgs sector
neutralinos as dark matter	sparticle mass spectrum

The programme which has been followed by many authors is to study the parameter space of the model which is allowed after imposing presently available experimental constraints from all the three sector and from the requirement of the radiative $SU\times U(1)$ breaking. In particular the latter one results in the limits $1 < \tan\beta < m_t/m_b$ [9,4], i.e. much stronger than the experimental limits quoted earlier.

Let us stress again: so far we have been talking about tree level results. Some of them are striking enough to ask the question: are they stable against radiative corrections? For instance, if this were true for the tree level structure of the Higgs sector in the MSSM it would enormously facilitate the experimental task of seeing there the signal of supersymmetry or of ruling out the model. However, we envisage two potential sources of large 1-loop corrections: reminiscence of soft supersymmetry breaking and large top quark Yukawa coupling.

The rest of this article is devoted mainly to 1-loop corrections to the Higgs sector but we also try to keep in mind the question of a consistent approach simultaneously to all the problems interconnected at the tree level. It has been pointed out already some time ago that 1-loop corrections to the effective potential are important for the radiative $SU(2) \times U(1)$ symmetry breaking [10].

3 Closer look at the MSSM Higgs sector and the origin of large radiative corrections

The Higgs potential in the MSSM reads:

$$\begin{aligned} V &= m_1^2 \overline{H}_1 H_1 + m_2^2 \overline{H}_2 H_2 - m_{12}^2 (\epsilon_{ab} H_1^a H_2^b + c.c) \\ &\quad + \frac{1}{8}(g_1^2 + g_2^2)(\overline{H}_1 H_1 - \overline{H}_2 H_2)^2 - \frac{g_2^2}{2} \mid \overline{H}_1 H_2 \mid^2 \end{aligned} \tag{2}$$

where $\epsilon_{12} = -1, H_1$ and H_2 are the two Higgs doublet fields and $m_i^2 = \hat{m}_i^2 + \mu^2; \mu$ is the coefficient of the $H_1 H_2$ mixing term in the superpotential, $\hat{m}_1^2, \hat{m}_2^2$ and m_{12}^2 are soft supersymmetry breaking Higgs boson mass parameters with m_{12}^2 defined to be negative. The crucial point about the potential (2) is that, due to supersymmetry, its quartic couplings are $SU(2)$ and $U(1)$ gauge couplings. The only free parameters are the three mass parameters. It is convenient to introduce the tree level mass eigenstate basis:

$$\begin{aligned} H_1 &= \begin{pmatrix} \frac{1}{\sqrt{2}}(v_1 + \cos\alpha H - \sin\alpha h + i \sin\beta A - i\cos\beta G) \\ \sin\theta H^- - \cos\theta G^- \end{pmatrix} \\ H_2 &= \begin{pmatrix} \cos\theta H^+ + \sin\theta G^+ \\ \frac{1}{\sqrt{2}}(v_2 + \sin\alpha H + \cos\alpha h + i\cos\beta A + i\sin\beta G) \end{pmatrix} \end{aligned} \tag{3}$$

where h, H, $H^\pm$ and A are physical particles intoduced earlier and G and $G^\pm$ are neutral and charged Goldstone bosons. For v_1 and v_2 which minimalize the tree level potential one has $\theta = \beta$ with $\tan\beta = v_2/v_1$. The potential has its minimum for

$$v_1 = v\cos\beta \qquad v_2 = v\sin\beta \tag{4}$$

where

$$\sin 2\beta = \frac{2 \mid m_{12} \mid^2}{m_1^2 + m_2^2} \tag{5}$$

$$v^2 = \frac{8(m_1^2 - m_2^2 \tan^2\beta)}{(g_1^2 + g_2^2)(\tan^2\beta - 1)} \tag{6}$$

Since $\sqrt{v_1^2+v_2^2} \equiv v$ is fixed by the value of the Fermi constant, all the physical Higgs boson masses and their couplings are expressed in terms of only two unknown parameters. They can be conveniently taken e.g. as $\tan\beta$ and M_A where:

$$M_A = m_1^2 + m_2^2 \tag{7}$$

The other two Higgs scalar masses read:

$$M_{h,H}^2 = \frac{1}{2}(M_A^2 + M_Z^2 \mp \sqrt{(M_A^2+M_Z^2)^2 - 4M_A^2 M_Z^2 \cos^2 2\beta}\ ) \tag{8}$$

We see that the tree level mass spectrum is symmetric with respect to $\tan\beta = 1$. Tan β is not an observable but is a convenient parametrization. The real content of the supersymmetric tree level mass relations is in "natural" relations relating physical observables to physical observables, such as Eq.(1) and

$$\begin{aligned} \Delta &= M_h^2 + M_H^2 - M_A^2 - M_Z^2 \equiv 0 \\ M^{+2} &= M_A^2 + M_W^2 \end{aligned} \tag{9}$$

(for more discussion on best choice of parametrization see Section 5)

It is easy to understand now the origin and magnitude of the radiative corrections to the Higgs boson masses. Let M be the scale of the soft supersymmetry breaking sfermion masses (responsible for fermion-sfermion mass splitting). In the static approximation, i.e. approximating the effective action by the 1-loop effective potential, and neglecting terms suppressed by inverse powers of M in the expansion of the effective potential all the loop corrections induced by the soft supersymmetry breaking can be absorbed into renormalization of the parameters in the Higgs potentials. One gets

$$\begin{aligned} \tilde{V} \ = \ & \tilde{m}_1^2 \overline{H}_1 H_1 + \tilde{m}_2^2 \overline{H}_2 H_2 - \tilde{m}_{12}^2(\varepsilon_{ab} H_1^a H_2^b + c.c) \\ & +\lambda_1 \mid H_1 \mid^4 +\lambda_2 \mid H_2 \mid^4 +\lambda_3 \mid H_1 \mid^2 \mid H_2 \mid^2 +\lambda_4 \mid \overline{H}_1 H_2 \mid^2 \end{aligned} \tag{10}$$

The appearence of other quartic couplings is protected by the symmetries of the model. It is clear on dimensional grounds and from no-renormalization theorem for theories with unbroken supersymmetry that the loop corrections to the parameters are as follows:

$$\delta m_i^2 = \tilde{m}_i^2 - m_i^2 \sim 0(M^2) \tag{11}$$

and logarithmically divergent;

$$\delta\lambda_1 = \lambda_1 - \frac{g_1^2+g_2^2}{4}, \quad \delta\lambda_2 = \lambda_2 - \frac{g_1^2+g_2^2}{4} \tag{12}$$

$$\delta\lambda_3 = \lambda_3 + \frac{g_2^2+g_1^2}{4}, \quad \delta\lambda_4 = \lambda_4 + \frac{g_2^2}{2} \tag{13}$$

all finite and $0(\ln M)$.

We see that the largest ($0(M^2)$) and divergent loop corrections can be absorbed into the free parameters of the model, to be fixed from the data (Appelquist-Carrazone decoupling). The supersymmetric mass relations get finite, calculable corrections determined by $\delta\lambda_i$. The dominant 1-loop contribution to the effective potential is given

by the following diagrams

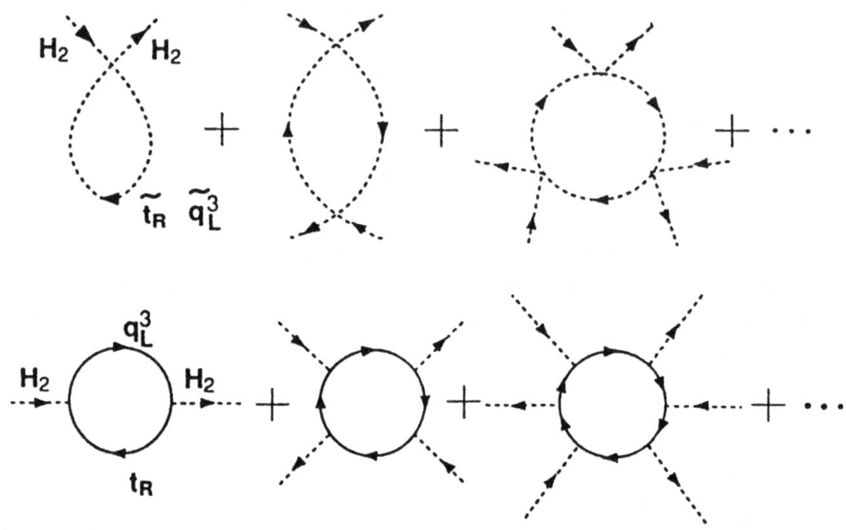

Therefore one gets (schematically):

$$\delta\lambda_i \sim Y_t^4 \ln \frac{m_t}{m_{stop}} \Rightarrow \Delta \sim 0(g_2^2 \frac{m_t^4}{m_W^2} \ln \frac{m_t}{m_{stop}}) \tag{14}$$

At this level of approximation the physical Higgs boson couplings remain the tree level couplings with new mixing angles α and β. There is one comment to be made here. The $0(M^2)$ corrections to the m_i^2 parameters suggest, in line with the standard naturalness arguments, that "natural" values for the $m_i's$, and therefore also for the pseudoscalar mass M_A, are $0(M)$.

4 Survey of the existing 1-loop calculations

Radiative corrections to the Higgs sector in the MSSM have recently drawn a lot of attention [11-17]. Those calculations differ in their method and the level of completeness. One can single out three different approaches. Each approach has its virtues and limitations when judged according to the two criteria:
a) precision in the Higgs sector
b) universality (simultaneous analysis of the four problems from the previous Section).

The first one is the effective potential approach, introduced in the previous Section. The most extensive analysis along this line is presented in ref.[13]. The full 1-loop effective potential reads:

$$V_1(Q) = V_o(Q) + \Delta V_1(Q) \tag{15}$$

where

$$\Delta V_1(Q) = \frac{1}{64\pi^2} Str \mathcal{M}^4 (\ln \frac{\mathcal{M}^2}{Q^2} - \frac{3}{2}) \tag{16}$$

Here, $V_o(Q)$ is the tree-level potential evaluated with couplings renormalized at some scale Q and Str denotes the supertrace. The Higgs mass-squared matrix is taken to be the matrix of the second derivatives of V_1 with respect to the Higgs fields, i.e. all the p^2 dependent effects present in the exact formula:

$$m^2_{phys} = \frac{d^2V}{d\varphi^2} |_{\varphi=v} + \Sigma(0) - \Sigma(m^2_{phys}) \tag{17}$$

are neglected. The method can include, however, all the details of the sparticle mass spectrum like e.g. $\tilde{t}_R - \tilde{t}_L$ mass splitting and, at least in principle, also the contribution from the gauge sector (the latter is technically not easy and has been neglected in the actual calculations). In this approach the physical Higgs boson couplings remain the tree level couplings with the mixing angles β and α determined by the minimalization of the V_1 and by the diagonalization of the new mass matrix, respectively.

The method offers a relatively simple connection with the question of the $SU(2) \times U(1)$ symmetry breaking by radiative corrections: this question and the corrections to the Higgs sector can be discussed together for the same set of parameters chosen at the grand unification scale. Consistent calculation of the 1-loop corrected other sparticle masses and couplings is, however, impossible.

The second approach is based on the renormalization group evolution of the parameters of the effective low energy theory down to the energy scale $0(M_W)$, according to the renormalization group equations obtained after decoupling of the heavy particles at their thresholds. We begin with the softly broken supersymmetric lagrangian, with the scalar potential (2), valid at the energy scale M where M is the scale of the soft supersymmetry breaking. If we make the simplifying assumption that all the sparticles aquire masses close to $M (M \gg M_W$ say, $M \sim 0(1\ TeV))$ and can be decoupled at the scale M, then the parameters of scalar potential evolve down with the scale according to the RGE of the SM with two Higgs doublets. Our assumption is rather crude as we know that generally some of the sparticles remain light (e.g. some charginos and neutralinos). The method can be, however, improved by using the recently derived [20] RGEs for all the parameters of the MSSM after decoupling of only squarks and gluinos at the common scale M.

The RG approach has its obvious limitation: it is impossible in practice to take care of all the details of the sparticle mass spectrum, $\tilde{t}_R - \tilde{t}_L$ mass splitting in particular. Also are present such shortcomings of the effective potential approach as no p^2 dependence and no 1-loop corrections to the ZZh and ZhA vertices. However, the RG approach has at least two important virtues. It summs up all the leading logarithms $\ln(M/m_t)$ and can, therefore, be trusted even for relatively high values of M. Secondly, it provides a very simple and consistent framework for studying at the same time all four constraints of the previous section on the parameter space of the minimal supergravity model [18].

Here we present some results obtained in the RG approach under the assumption that all sparticles can be decoupled at the common scale M. The RG equations of the SM with two Higgs doublets are used to evolve the parameters of the scalar potential from the scale M (where they satisfy the supersymmetric boundary conditions) to the scale Q_{top} where the Higgs boson masses are calculated. The scale Q_{top} is chosen so that for a given $\tan\beta$ (which in this approach is always the tree level $\tan\beta$ and for which we want to study the scalar masses) the 1-loop corrections to our low energy potential are minimal:

$$\left.\frac{\partial \Delta V_1}{\partial v_i}\right|_{v_i^{(0)}} \sim 0 \tag{18}$$

The main contribution to ΔV_1 comes from the top quark. Therefore we fix Q_{top} by the condition:

$$\frac{\partial}{\partial v_2}\left(Y_t^4 v_2^4\left(-\frac{3}{2} + \ln\frac{Y_t^2 v_2^2}{Q_{top}^2}\right)\right) = 0 \tag{19}$$

which gives $Q_{top} \sim m_t/\sqrt{e}$.

In Fig.10-15 we show some results obtained in this approach. Their comparison with the results of ref. [13] (effective potential approach) and with the complete 1-loop analysis of ref. [16] will be briefly discussed in Section 6.

The third approach to the radiative corrections in the Higgs sector is based on systematic perturbative calculation of all the relevant Green's functions. Potentially, this is the most complete approach to 1-loop corrections in the Higgs sector and indeed such a complete calculation has been reported in ref. [16]. On-shell renormalization programme and t'Hooft-Feyman gauge are the best choices for this calculation. Of course, this analysis includes all the p^2 dependent effects, loop corrections to all the vertices and can accommodate any values of the soft supersymmetry breaking parameters.

This approach can be systematically extended to all other sectors of the MSSM, providing a general renormalization scheme for calculating loop corrections in this model. There are, however, two limitations. Firstly, as any 1-loop calculation, it becomes less accurate with increasing value of the $\ln M/m_t$. Secondly, it is now less straightforward to implement the constraints from the requirement of the radiative $SU(2)\times U(1)$ symmetry breaking (see the next Section).

The details of this third approach are given in ref. [17]. Here we would like to concentrate on the problem of the renormalization scheme dependence and the most convenient choice of parameters for the Higgs sector. Those topics are discussed in the next Section.

5 Choice of parameters and renormalization scheme dependence

In the previous Section we have been advocating the mass M_A and $\tan\beta$ as a good set for parametrizing the Higgs sector in the MSSM. The reasons are basically two: they parametrize the Higgs sector in an unambigous way and $\tan\beta$ is convenient also for other purposes. However, as with any other parameter in the lagrangian, the definition of $\tan\beta$ in terms of physical observables such as e.g. masses is renormalization scheme dependent. Other parameters of the Standard Model, like e.g. $v = 240\ GeV$ or $\sin^2\Theta_W$, also face the same problem but in the high precision tests they can be easily traded for measured physical observables such as the Fermi constant, the electric charge and M_Z (and predictions for observables expressed in terms of observables show only residual higher order renormalization scheme dependence). Analogous strategy in our case is to use two observables as the two unknown parameters and the natural choice is M_A and M_h. Now, however, there is a two-fold ambiguity at the tree level: due to the symmetry of the mass spectrum with respect to $\tan\beta = 1$ each pair of values (M_A, M_h) gives two sets of values for other observables in the Higgs sector. The symmetry is broken by radiative corrections (for fixed values of m_t and m_b Yukawa couplings take different values in the two regions) but in certain regions of the parameter space the ambiquity remains. Still, this is a very natural approach: calculate cross sections for given values of M_A and M_h (using $\tan\beta$ only as an intermediate step) and present the experimental limits as the excluded region in those variables.

On the other hand, presenting theoretical predictions for masses and cross sections (and experimental limits) in terms of $\tan\beta$ has the two virtues mentioned above.In particular it is motivated by theoretical considerations on the radiative $SU(2)\times U(1)$ symmetry breaking which constrains $\tan\beta$ to the region $1 < \tan\beta < m_t/m_b$. One should remember, however, that those results have been always obtained in some specific renormalization scheme (and with some at least implicit choice of the gauge fixing conditions) and the separation between $\tan\beta < 1$ and $\tan\beta > 1$ regions *is* renormalization scheme dependent. Of course, it is quite legitimate to use $\tan\beta$ as a useful bookkeeping for constraints from different sectors provided the same renormalization scheme has been used in those sectors. This is actually the case in the effective potential (and also in the renormalization group) calculation of the loop corrections to the Higgs sector and of the radiative $SU(2)\times U(1)$ symmetry breaking . The renormalization scheme of ref. [16] has been also designed to give $\tan\beta$ effectively very close to that of ref. [13].

The renormalization scheme dependence of $\tan\beta$ can be demonstrated most dramatically when we compare calculations in the Landau and in the t'Hooft-Feynman gauges . Let us first recall that there are two commonly used approaches to perturbative calculations in theories with spontaneously broken (gauge) symmetries [19]. The first one is to expand around the tree level vacuum v_o: for the scalar field we have then $<\phi>= 0$ in the tree approximation but not beyond. We then calculate 2, 3, 4...-point Green's functions and the renormalization programme is set up for them. This approach is not convenient for our programme as large tadpole contributions appear not only in the Higgs boson but also in the fermion 2-point functions.

Let us concentrate on the second approach which is to expand around the "true vacuum" i.e. around the *vev* v adjusted to give $<\phi>= 0$ order by order in perturbation theory. Here we make contact with the effective potential approach. At the 1-loop level we get then the following equation(s) for the scalar field(s):

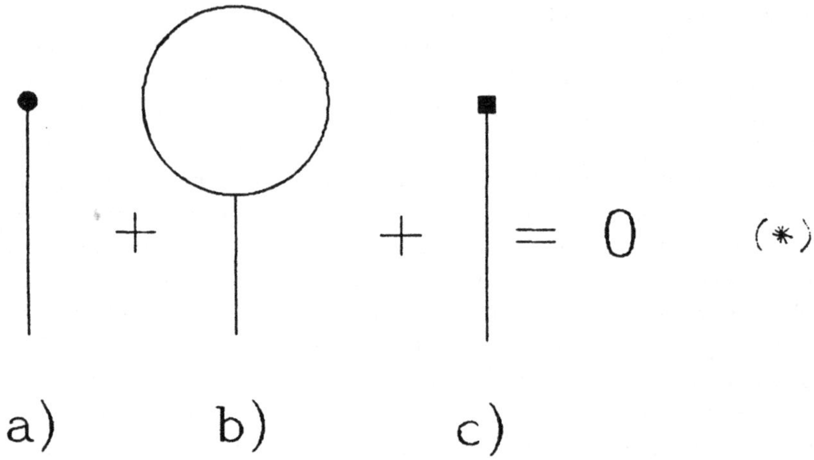

where a), b), and c) represent the tree level tadpole, 1-loop tadpole and tadpole counterterm, respectively. Among the tadpole graphs are Goldstone boson loops which are explicitly gauge dependent (gauge dependence of the gauge boson tadpole cancells out with the ghost tadpoles). It turns out that only in Lorentz gauges Eq.(*) can be solved with the counterterms of the symmetric (unbroken) phase (i.e. with mass, δm, and coupling constant, $\delta\lambda$, counterterms). Hence, e.g. in the Landau gauge $v \equiv v_R$ is fixed in terms of the renormalized parameters m_R, λ_R solely by imposing renormalization conditions on δm and $\delta\lambda$. This is the basis of the effective potential approach. In the R_ξ class of gauges an additional counterterm δv is needed to solve Eq.(*),to cancell the ξ-dependence of the Goldstone boson tadpoles[19]. In these gauges the renormalization scheme dependence of the *vevs* is particularly clear: the definition of v_R depends on one more renormalization condition imposed on δv.

One particular way of solving Eq.(*) is to impose the conditions

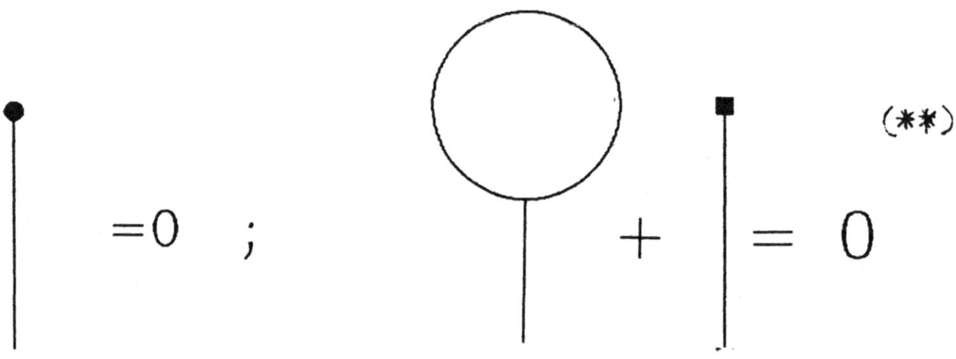

Then, the tree level relation between the renormalized parameters and the *vevs* is preserved at the 1-loop level. In the Landau gauge the tadpole counterterm is expressed in terms of δm and $\delta\lambda$, so imposing Eq. (**) means fixing one of the renormalization conditions for those parameters (in the minimal substraction scheme Eq. (**) can be satisfied by properly adjusting the renormalization scale). In the R_ξ gauges we have additional freedom of fixing δv. In ref. [16] we have chosen the $\delta v_i's$ to cancell only this contribution to tadpoles which appear when we change from the Lorentz to R_ξ gauges. More precisely we do it for infinite parts of tadpoles only (to avoid the spurious infrared divergences which such procedure introduces for the finite parts). The remaining parts of tadpoles are cancelled, as in the Landau gauge, by properly adjusting $\delta m_i s$. Our renormalized $\tan\beta$ is effectively very close to the $\tan\beta$ defined in ref. [13]. The renormalization schemes of ref. [13] and [16] can be contrasted with that of ref. [15] where δv_i are used to cancell the full tadpole contributions. Our choice of renormalization conditions has the virtue of absorbing large finite tadpole contributions into Higgs boson renormalization constants without affecting other sectors such as fermion, chargino and neutralino mass renormalization with large, finite $\delta v_i s$.

6 The MSSM Higgs sector after radiative corrections and LEP phenomenology

In this Section we present the main results obtained in ref. [12-17] and their implications for the LEP phenomenology. Qualitatively, all the papers reach similar conclusions. Here we first follow the most complete analysis of ref. [16] for the Higgs boson masses and then show some results for the cross sections obtained in the effective potential and RG approaches. Both sets of parameters ($\tan\beta, M_A$) and (M_h, M_A) are used complementarily to present the results. Finally, we address the question: are the minimal SM Higgs boson and the lightest MSSM Higgs boson distinguishable experimentally?

As said earlier, there is the "theoretically suggested" plausible region of $\tan\beta$ which is of particular interest for the MSSM: $1/2 < \tan\beta < m_t/m_b$. The lower limit comes from constraints on the tree level Yukawa couplings [8], the upper limit - from the mechanism of the radiative $SU(2) \times U(1)$ symmetry breaking. It was stressed in the previous Section that at the 1-loop level $\tan\beta$ is renormalization scheme dependent. However, the particular choice of the scheme in ref. [16] is such that a) large corrections are absorbed into Higgs boson renormalization constants and the Yukawa vertices receive only generic electroweak corrections - hence the tree level lower limit for $\tan\beta$ can be used as a sensible constraint for the renormalized $\tan\beta$ of ref. [16]; b) our $\tan\beta$ is effectively very close to the one used in the effective potential approach to the radiative $SU(2) \times U(1)$ symmetry breaking - hence we can use the upper limit as well.

Let us address, therefore, the question: how does the Higgs sector look like for the plausible values of our renormalized $\tan\beta$: $1/2 < \tan\beta < m_t/m_b$? The first important result is the upper limit for the lightest scalar h (i.e. violation of the tree level result(1)) as a function of the top quark mass. This is shown in Fig. 7. We see that the lightest Higgs boson may be beyond the reach of LEP II. Let us also note that the lower limit for $\tan\beta$ is crucial for obtaining those results: because of the growing

Yukawa top quark coupling, for $\tan\beta \to 0$ the 1-loop corrected $M_h \to \infty$ (eventually, perturbation theory breaks down).

For a more detailed presentation, it is useful to discuss our results for the neutral Higgs boson masses separately for $\tan\beta < 1$, intermediate $\tan\beta$, say, $1 < \tan\beta < 5$ and large $\tan\beta$. For, say, $\tan\beta > 5$ the tree level pattern of masses is basically preserved with positive corrections of similar magnitude to M_H for $M_A < 100\ GeV$ and to M_h for $M_A > 100\ GeV$ (Fig.1). They are typically of the order of $0(10\ GeV)$ for $m_t = 120\ GeV$ and $0(20-30\ GeV)$ for $m_t = 160\ GeV$. They grow like fourth power of m_t, logarithmically with M_{sq} and A_{sq} and are almost independent of μ (in this region of $\tan\beta$ the mass splitting in the stop sector is sensitive to A_{sq} and not to μ; $M_{sq}, \mu, A_{sq(sl)}$ are the soft squark mass parameter (taken the same for all squarks), the supersymmetric Higgs boson mixing parameter and the trilinear coupling for squarks (sleptons), respectively).The structure of corrections is easily understood by the fact that the top quark couples only to H_2 and H_2 is almost pure $H(h)$ for M_A $<$ $(>)$ $100\ GeV$. We notice in particular that, as at the tree level, both masses are almost independent of $\tan\beta$ (for $\tan\beta > 5$) and in addition they almost do not depend on M_A: M_H for $M_A < 100\ GeV$ and M_h for $M_A > 100\ GeV$. Nice thing about the corrections is that for this region of $\tan\beta$ they lift the mass degeneracy of the Higgs boson with the Z boson. The region $\tan\beta > 5$ and $M_A < 100\ GeV$ can be accessed experimentally as before: there $M_h \simeq M_A$ and the coupling ZhA is large.

In the intermediate $\tan\beta$ region the corrections modify the tree level structure quite drastically, in particular in the (M_h, M_A) sector. In general, they are bigger than for large $\tan\beta$ both for M_H and M_h (proportional to $m_t^4 \sin\alpha/\sin^2\beta$ and $m_t^4 \cos\alpha/\sin^2\beta$ respectively, with $\alpha = 0(\frac{\pi}{4})$). For $M_A < 40-60\ GeV$ the mass M_h is pushed well above its tree level limit $M_h < M_A$ and increases with decreasing $\tan\beta$. For instance, for $\tan\beta = 1$ and $m_t = 120(160)\ GeV$ (with $M_{sq} = 1\ TeV, A_{sq} = A_{sl} = 0, \mu = 70\ GeV$) we have $M_h \simeq 36(52)\ GeV$, respectively, at $M_A = 0$. An interesting feature is that for $\tan\beta$ close to 1 the 1-loop corrections to M_h are almost independent of M_A for arbitrary values of the pseudoscalar mass. Thus, the tree level value $M_h \simeq 0$ (also independent of M_A) is just pushed up remaining, however, for $M_A > 100-120\ GeV$ the lower limit for M_h. For very large squark mixing (A_{sq} and/or μ larger than M_{sq}) the 1-loop corrections to M_h begin to decrease for light M_A and eventually become negative.

For $\tan\beta < 1$ corrections to M_h and M_H are very large even for m_t as small as 100 GeV (as expected because of the growing Yukawa coupling). The mass M_H grows by 0(100 GeV) for $M_A < M_Z$ and the corrections to M_h depend on M_A in such a way that the final result (tree level plus 1-loop correction) is (as for $\tan\beta \simeq 1$) almost independent of M_A for any fixed $\tan\beta$ in the whole range $\frac{1}{2} < \tan\beta < 1$ (the M_A dependence grows with $m_t > 200\ GeV$) and the symmetry with the region $1 < \tan\beta < 2$ is clearly broken. Thus, even for $M_A = 0$ the mass M_h may be easily as large as 80-100 GeV and therefore unreachable at LEP.

For a representative choice of parameters, the bounds on M_H and on M_h as functions of M_A with contour lines of fixed $\tan\beta$ and on M_h as function of $\tan\beta$ with contour lines of fixed M_A are shown in Fig. 2,3 and 4, respectively. In particular Fig.4 shows clearly the discussed above very weak dependence of M_h on M_A for $\tan\beta < 1$ (M_A has been varied from 0 to 1 TeV) and, in consequence, the breakdown of the symmetry with the region $\tan\beta > 1$. Note, however, that the symmetry is largly restored if restricting M_A to $M_A > 100\ GeV$, i.e. for $\tan\beta > 1$ clear dependence of M_h on M_A is present only for $M_A < 100\ GeV$. Note also that the $\tan\beta < 1$ region is quite sensitive to large values of μ (Fig.4b).

Finally, let us discuss the charged Higgs boson mass. Here, the corrections to the tree level result are small for $1 < \tan\beta < 10$ and rapidly become sizably negative (positive) for $\tan\beta < 1 (\tan\beta > 10)$. This is shown in Fig.5. Particularly interesting feature is the strong dependence of M^+ on the μ parameter for $\tan\beta < 1$ (with very similar results for $\mu < 0$). It can be traced back to the quadratic (and not just logarithmic) dependence of ΔM^+ on the stop-sbottom mass splitting which in turn is for $\tan\beta < 1$ very sensitive to the value of μ. Large negative correction to M^+ for large μ and $\tan\beta < 1$ offer an interesting possibility to rule out this region by experimental limits on M^+, at least for light ($< 50\ GeV$)M_A. The region of $\tan\beta < 1$ and small μ can be on the other hand ruled out by limits on the light chargino mass. The combination of those two effects is illustrated in Fig.6 where we plot the M^+ and the chargino mass as functions of μ, for $\tan\beta = .5$. The chargino mass is the tree level chargino mass but within our renormalization scheme (tadpoles absorbed by Higgs boson counterterms) the radiative corrections to the chargino mass are expected to be small. We also recall that for $\tan\beta$ fixed in the region $\tan\beta < 1$ (we always refer to $\tan\beta$ in our renormalization scheme) and for small μ the theoretical limits on M_h are narrow enough (Fig.4) to make the experimental search at least well defined.

The results for the Higgs boson masses presented here are consistent with similar results obtained in the effective potential [13] and RG [17] approaches. This is illustrated in Fig.8 where some of the results obtained by the RG evolution [17] are shown. In general, the RG approach is very good for $A_{sq} \approx \mu \approx 0$ (small mass splitting in the squark sector) and for $M_{gau} \sim M_{sq}$ (all sparticles are decoupled at the same scale M). On the other hand, the gaugino contribution has been neglected in the effective potential calculation [13] (this corresponds to setting the $SU(2) \times U(1)$ gaugino masses to zero). The detailed comparison of the results obtained by the three methods shows how the complete calculation [16] interpolates between the two other methods for different sets of parameters.

Let us turn now to the cross sections for the Higgs boson production at LEP. The complete 1-loop calculation will be soon ready [17] and at present we can discuss only the effective potential [13,14] and the RG [17] results. Still, encouraged by the results for the masses one may expect them to be quite accurate. In both methods the cross sections are calculated in the tree approximation, with angles and masses obtained in the effective theories. So, the relations between the Higgs boson masses and the angles α and β which appear in the Higgs boson couplings to the Z boson and to fermions are different from the tree approximation in the MSSM. In addition there are obvious (and large) changes in kinematics of the reactions (induced by the large corrections to masses). In Fig.10-15 we show cross sections for the reactions $e^+e^- \to Zh$ and $e^+e^- \to hA$ at $\sqrt{s} = 180\ GeV$. They are presented as functions of the $(\tan\beta, M_A)$, and (M_h, M_A) parameters, to illustrate complementarity of those choices. It is, in fact, interesting to see how different projections of the same results fit each other and reflect the pattern of the corrected mass spectrum.

Let us concentrate on Fig.12-14 where the cross sections are plotted as functions of M_h with contour lines of constant M_A. First of all, we note the discussed earlier two-fold ambiguity of this parametrization: in some cases the same (M_h, M_A) give two different values for the cross sections. Of course, they correspond to two different values of $\tan\beta$ or, in terms of physical observables, to two different Higgs boson widths and branching ratios (Fig.16). The cross sections are such that for a not too large M_A (<50 - 60 GeV) the two reactions are complementary: for light h (i.e. large $\tan\beta$) the $e^+e^- \rightarrow hA$ dominates, for heavier h(i.e. smaller $\tan\beta$) the $e^+e^- \rightarrow Zh$ dominates. For heavier A, the first reaction is kinematically forbidden and the second one has reasonable cross sections only for light h (i.e. small $\tan\beta$).

It is very interesting to compare the cross sections $\sigma(e^+e^- \rightarrow Zh)$ with cross sections for the minimal SM Higgs boson production (dashed-dotted lines in Fig.12,13). We see that, as in the tree approximation (Fig.9), they are very close to each other in most of the parameter space and, in addition, the process $\sigma^+e^- \rightarrow Ah$ is then unrealistically weak. The only exception is the "large $\tan\beta$" branch for $M_A < M_Z$. This region is, however, already excluded in its large part by data analysis (combined search for $e^+e^- \rightarrow Zh$ and $e^+e^- \rightarrow Ah$). In some parts of the "similar" regions one may still hope to distinguish the MSSM and the SM scalars by their branching ratios into various channels. For small M_A the decay $h \rightarrow AA$ may be sizable (Fig.15) and for larger M_A the branching ratios into fermions remain different (Fig.16). In general, however, the heavier the pseudoscalar A the harder is the task of distinguishing the supersymmetric h from the minimal SM Higgs boson.

Finally, the RG calculation is compared in Fig.12-14 with the effective potential approach. It turns out that the difference is mainly due to the gaugino contribution which is neglected in ref.[13] and effectively included in the RG calculation (with $M_{gau} \sim M$). Indeed, the difference is sizably reduced by using the RG equations of ref.[20]. (Those equations can be used to decouple gauginos at any chosen M_{gau}.).

To summarize this Section: radiative corrections are very important for phenomenology of the supersymmetric Higgs boson search. They introduce quite large a dependence on the number of unknown parameters such as the top quark mass, squark masses etc. By far the most important for phenomenology are corrections to Higgs boson masses. And more optimistically: systematic search in the plane (M_h, M_A) is still possible. However, once a scalar is discovered it may be difficult to distinguish between the supersymmetric h and the minimal SM Higgs boson.

7 Summary: where is the lightest supersymmetric Higgs boson ?

In this lecture an overview has been given on the subject of the MSSM Higgs sector after radiative corrections and on its links to several other crucial questions. For technical details we purposely refer the reader to the original papers.

Let us now have some fun and make a "plausible" guess in what mass range is actually placed the lightest supersymmetric Higgs scalar. It is likely that

1. the scale of the soft supersymmetry breaking is, say, between 0.5 TeV - 1.5 TeV;
2. the top quark mass is in the range 120 $GeV - 160$ GeV;
3. the naturalness arguments of Section 3 are correct and the M_A is of the order of the soft supersymmetry breaking scale i.e. $M_A \gg M_Z$
4. renormalized $\tan\beta$ in "sensible" renormalization schemes is relatively large, say, $\tan\beta > 5$.

By inspection of Fig.1 and of similar ones for $m_t = 120$ GeV and for other values of the parameters M_{sq}, A_{sq}, μ one concludes that 100 $GeV < M_h < 120$ GeV.
Of course, such speculations do not free us from the duty of excluding experimentally the lower mass region. BUT, LET'S BE PATIENT.

REFERENCES

1. Peskin M E, Takeuchi T (1991) SLAC-PUB-5618
2. Amaldi U, de Boer W, Fürstenau H (1991) preprint CERN-PPE/91-44
3. Haber H, Nir Y (1991) Nucl Phys B335 363
4. Pokorski S (1989) Proceedings of the XIth Warsaw Conference on Elementary Particle Physics ed by Ajduk Z, Pokorski S, Wroblewski A K (World Scientific)
5. Giudice G F (1988) Phys Lett B208 315
6. Kalinowski J, Pokorski S (1989) Phys Lett B219 116
   Kalinowski J, Grzadkowski B, Pokorski S (1990) Phys Lett B241 534
   Kalinowski J, Nilles H P (1991) Phys Lett B255 134
7. Krawczyk P, Pokorski S (1988) Phys Rev Lett 60 182
8. Gunion J F, Grzadkowski B (1990) Phys Lett B243 301
   Buras J, Krawczyk P, Lautenbacher M E, Salazar C (1990) Nucl Phys B337 284
   Barger V, Hewett J L, Phillips R J N (1990) Phys Rev D41 3421
   Cocolicchio D, Cudell J R (1990) Phys Lett B245 591
9. See e.g. Lahanas A B, Nanopoulos D V (1987) Phys Rep 145 1
10. Gamberini G, Ridolfi G, Zwirner F (1990) Nucl Phys B331 331
    Chankowski P H (1990) Phys Rev D41 2877
11. Li S P, Sher M (1984) Phys Lett B140 339
    Gunion J, Turski A (1989) Phys Rev D40 2325 2333 Phys Rev D39 2701
    Berger M (1990) Phys Rev D41 225
12. Okada Y, Yamaguchi M, Yanagida T (1991) Prog Theor Phys Lett 85 1; Phys Lett B262 55
    Haber H, Hempfling R (1991) Phys Rev Lett 66 1815
    Barbieri R, Frigeni M, Caravagolis M (1991) Phys Lett B258 167
13. Ellis J, Ridolfi G, Zwirner F (1991) Phys Lett B257 83
    Ellis J, Ridolfi G, Zwirner F (1991) Phys Lett B262 477
    Brignole A, Ellis J, Ridolfi G, Zwirner F (1991) Phys Lett B271
    Brignole A preprint DFPD?91/TH/28 - October 1991
14. Barberi R, Frigeni M (1991) Phys Lett B258 395
    Kalinowski (1991) preprint TUM-T31-14/91
15. Yamada A (1991) B263 233
16. Chankowski P , Pokorski S, Rosiek J (1992) Phys Lett B274 191

17. Chankowski P H, Pokorski S, Rosiek J (1992) to be published
Chankowski P H (1991) Thesis unpublished

18. Olechowski M, Pokorski S, to be published

19. Appelquist T, Carazzone J, Goldman T, Quinn H R (1973) Phys Rev D8 1747

20. Chankowski P H (1990) Phys Rev D41 2877

FIGURE CAPTIONS

1. Masses M_h and M_H as functions of M_A for $\tan\beta = 5$ and $\tan\beta = 25$. Parameters not specified in the pannel are $M_{sl} = 300$ GeV and $M_{gau} = 400$ GeV.

2. Bounds on M_H as function of M_A with contour lines of fixed $\tan\beta$. M_{sl} and M_{gau} as in Fig. 1. $A = M_{sf}$ means $A_{sq} = M_{sq}, A_{sl} = M_{sl}$.

3. Bounds on M_h as function of M_A with contour lines of fixed $\tan\beta$. Parameters not specified in the pannel are as in Fig. 2.

4. Bounds on M_h as function of $\tan\beta$ with contour lines of fixed M_A. Parameters not specified in the pannel are: $M_{sl} = 150$ GeV, $M_{gau} = 200$ GeV.

5. Bounds on the charged Higgs boson mass as function of M_A with contour lines of fixed $\tan\beta$. Dashed lines correspond (from below) to $\tan\beta = .75,1,1.4,5,10,20$, respectively. Parameters not specified in the pannel are as in Fig. 2. Dotted line is the tree level result.

6. M^+ as function of μ for several values of M_A. Slepton and gaugino masses are 150 GeV and 200 GeV, respectively, for $M_{sq} = 500$ GeV; and 300 GeV and 400 GeV for $M_{sq} = 1000$ GeV. Dotted lines show the mass of the lighter chargino.

7. Upper limit for M_h as function of the top quark mass and for $1/2 < \tan\beta < m_t/m_b$.

8. Bounds on M_h as function of M_A with contour lines of fixed $\tan\beta$ obtained in the RG approach of ref.[17]. Comparison with Fig.3 illustrates the accuracy of the method. The agreement with the complete 1-loop calculation is even better when A$\approx \mu \approx 0$ and $M_{gau} \approx M_{SUSY}$.

9. Tree level results for the cross section $\sigma(e^+e^- \to Zh)$ as function of M_h with contour lines of fixed M_A. Each line represents two branches: $\tan\beta < 1$ and $\tan\beta > 1$ corresponding to the same values of (M_h, M_A) and degenerate at the tree level. The arrow indicates the direction of departure from $\tan\beta = 1$. The dashed lines represent the cross section for the SM Higgs boson production.

10. Cross sections for the production of the light scalar h in the e^+e^- collisions at $\sqrt{s} = 180$ GeV as functions of M_A with contour lines of fixed $\tan\beta$, obtained in the *RG* approach [17] described in Section 4.

11. The same as in Fig. 10 but in the effective potential approach of ref.[13]. In this case $M_{SUSY} \equiv M_{sq}$.The parameters A and μ are taken to be zero.

12. Cross sections for $e^+e^- \to Zh$ as functions of M_h with contour lines of fixed M_A. The arrow indicates the direction of growing $\tan\beta$ (from 1/2 to m_t/m_b). Solid lines: RG results of ref.[17]. Dashed lines: effective potential results of ref.[13] with parameters as in Fig.11. Dashed-dotted line: cross section for the SM Higgs scalar production. The curves for $M_A > 110$ GeV (not plotted) are inside the respective curves for $M_A = 110$ GeV.

13. Cross sections for $e^+e^- \to Zh$ as functions of M_h with contour lines of fixed M_A, obtained in the RG approach of ref.[17]. The arrow indicates the direction of growing $\tan\beta$.Dashed line: cross section for the SM Higgs scalar production.

14. Cross sections for $e^+e^- \to Ah$ plotted as in Fig. 12.

15. Results for the decay $h \to AA$ obtained in the RG approach of ref.[17]. Contour lines are for fixed $\tan\beta$.

16. Decay widths of the scalar h after radiative corrections.

Fig. 1

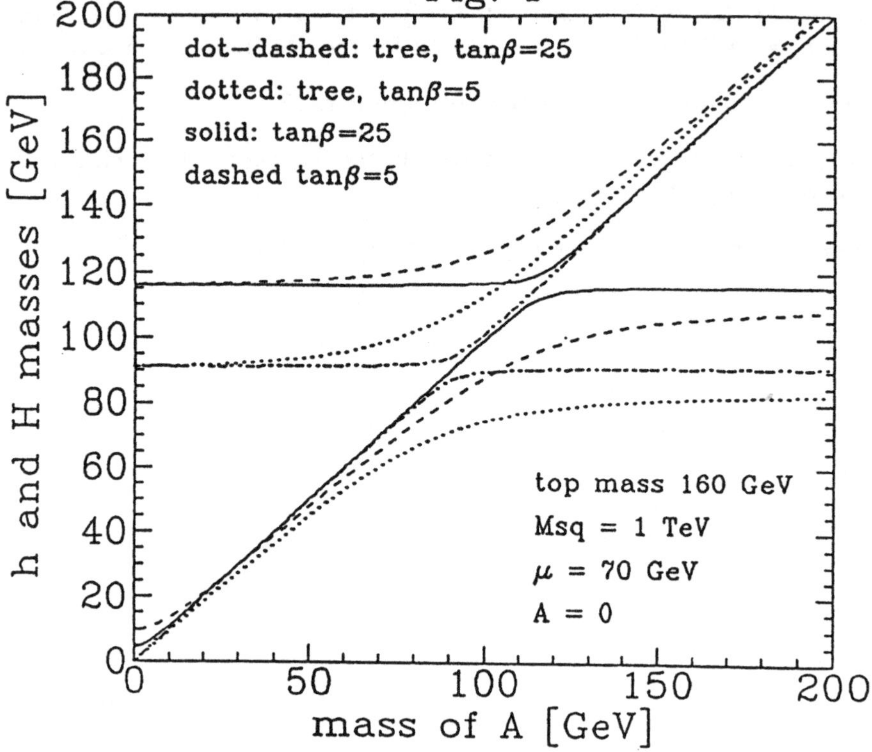
dot–dashed: tree, tanβ=25
dotted: tree, tanβ=5
solid: tanβ=25
dashed tanβ=5
top mass 160 GeV
Msq = 1 TeV
μ = 70 GeV
A = 0
h and H masses [GeV]
mass of A [GeV]
0
20
40
60
80
100
120
140
160
180
200
0
50
100
150
200

Fig. 2

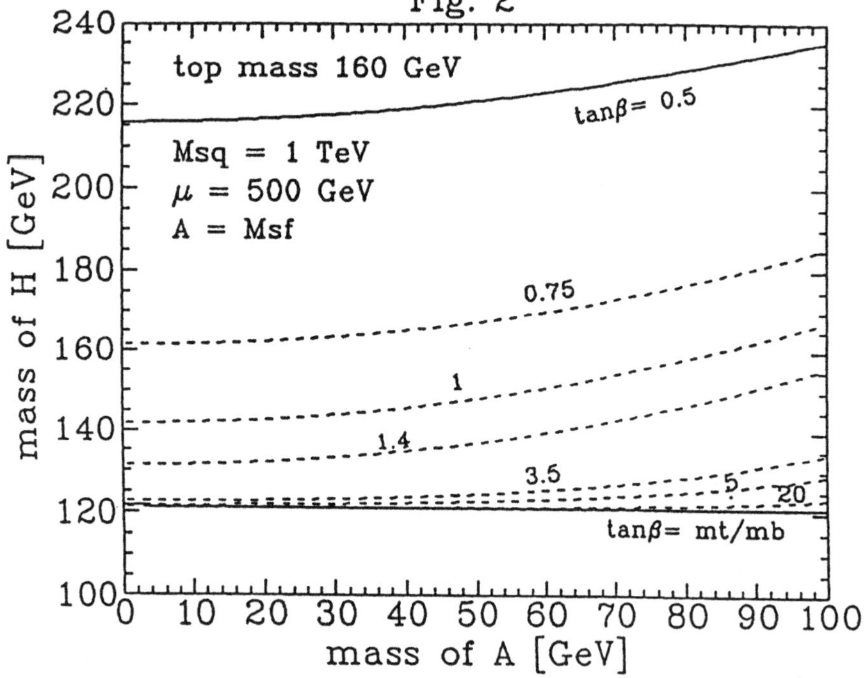
top mass 160 GeV
tanβ= 0.5
Msq = 1 TeV
μ = 500 GeV
A = Msf
0.75
1
1.4
3.5
5
20
tanβ= mt/mb
mass of H [GeV]
mass of A [GeV]
100
120
140
160
180
200
220
240
0
10
20
30
40
50
60
70
80
90
100

Fig. 3

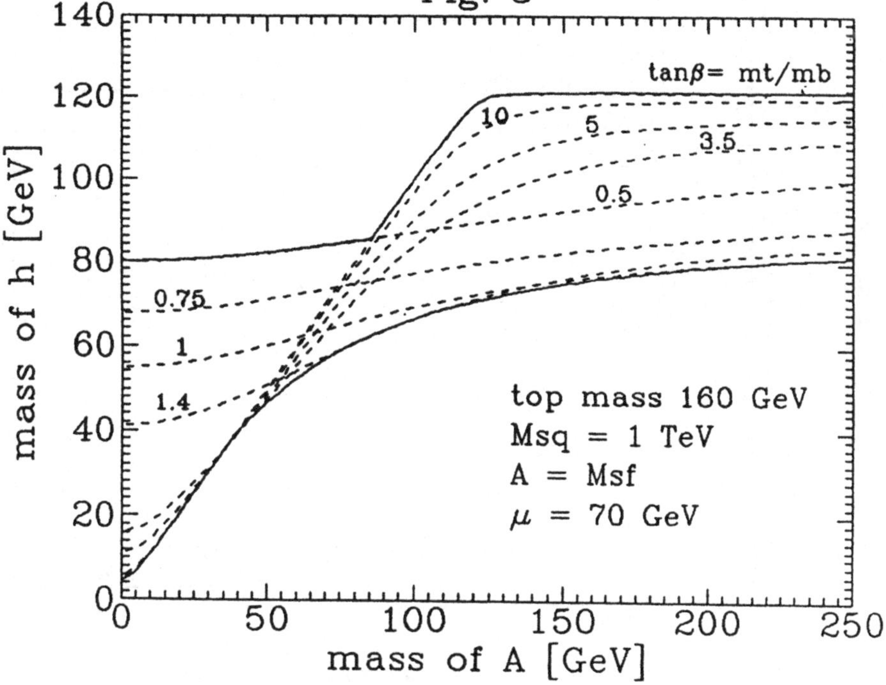
140
120
100
80
60
40
20
0
0
50
100
150
200
250
mass of h [GeV]
mass of A [GeV]
tanβ= mt/mb
10
5
3.5
0.5
0.75
1
1.4
top mass 160 GeV
Msq = 1 TeV
A = Msf
μ = 70 GeV

Fig.4a

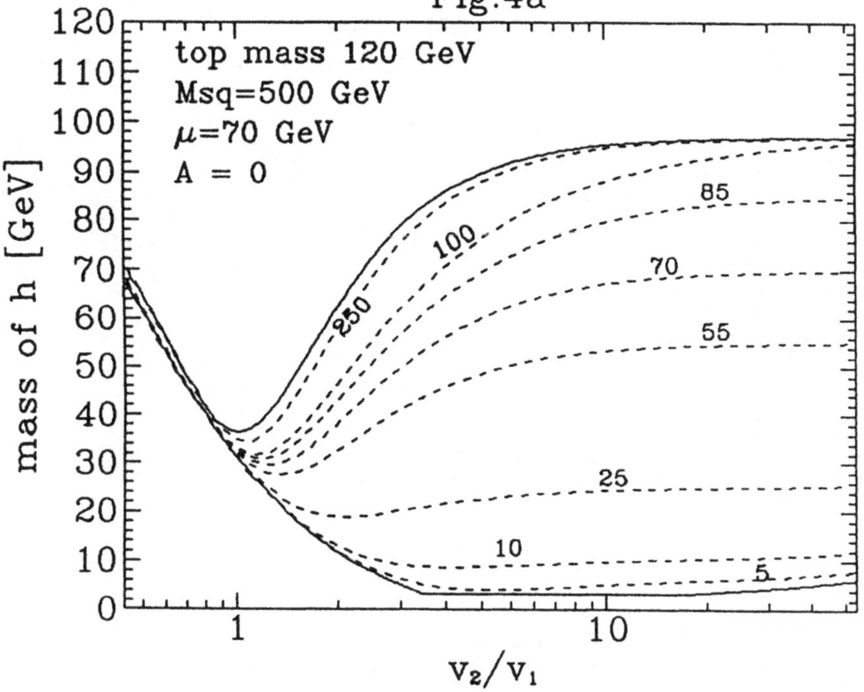
120
110
100
90
80
70
60
50
40
30
20
10
0
1
10
mass of h [GeV]
V2/V1
top mass 120 GeV
Msq=500 GeV
μ=70 GeV
A = 0
85
100
70
250
55
25
10
5

Fig. 4b

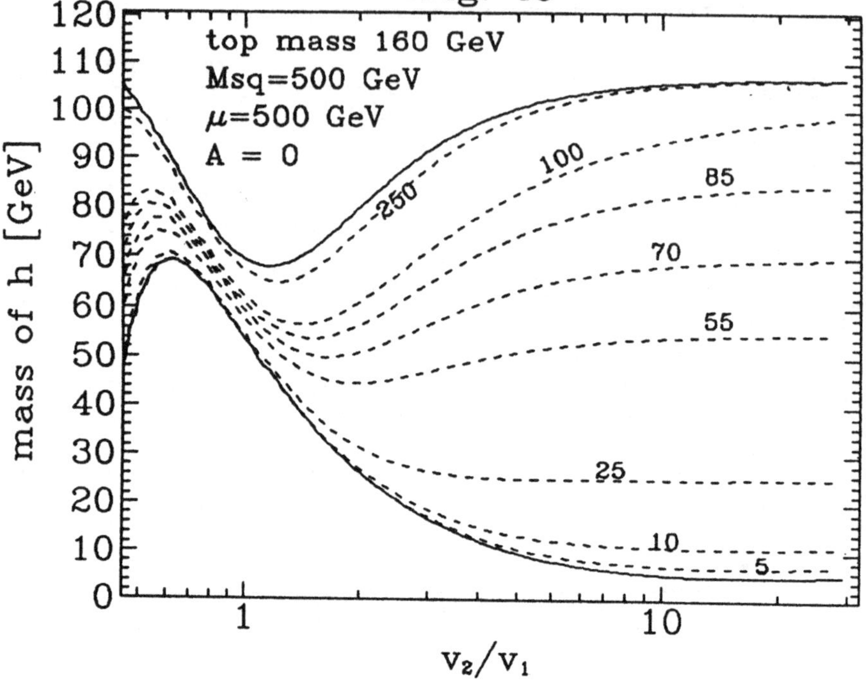
top mass 160 GeV
Msq=500 GeV
μ=500 GeV
A = 0
mass of h [GeV]
v2/v1
250
100
85
70
55
25
10
5

Fig. 5a

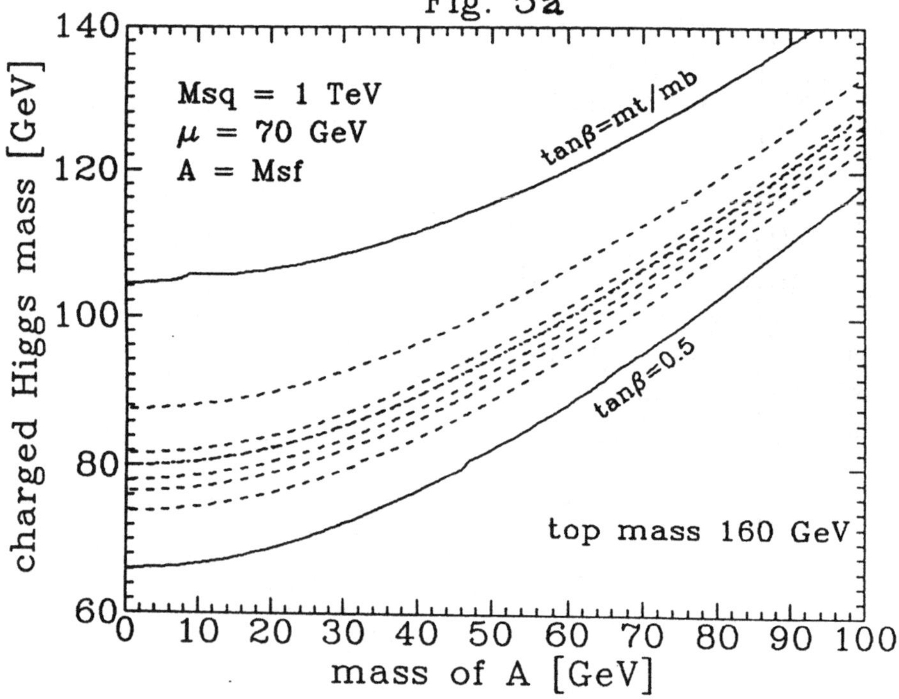
Msq = 1 TeV
μ = 70 GeV
A = Msf
tanβ=mt/mb
tanβ=0.5
top mass 160 GeV
charged Higgs mass [GeV]
mass of A [GeV]

Fig. 5b

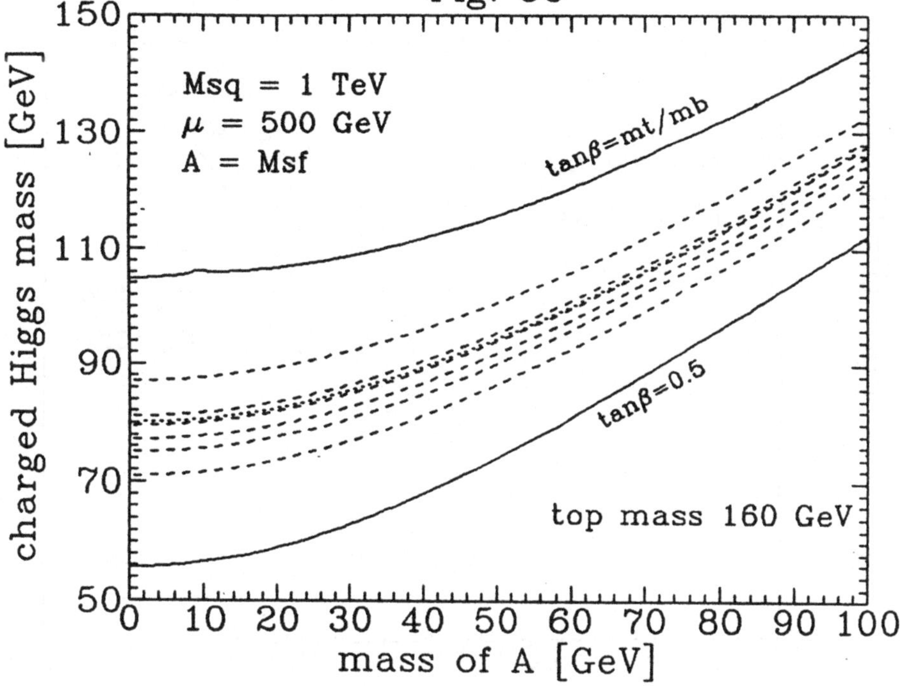
Msq = 1 TeV
μ = 500 GeV
A = Msf
tanβ=mt/mb
tanβ=0.5
top mass 160 GeV
charged Higgs mass [GeV]
mass of A [GeV]

Fig. 6

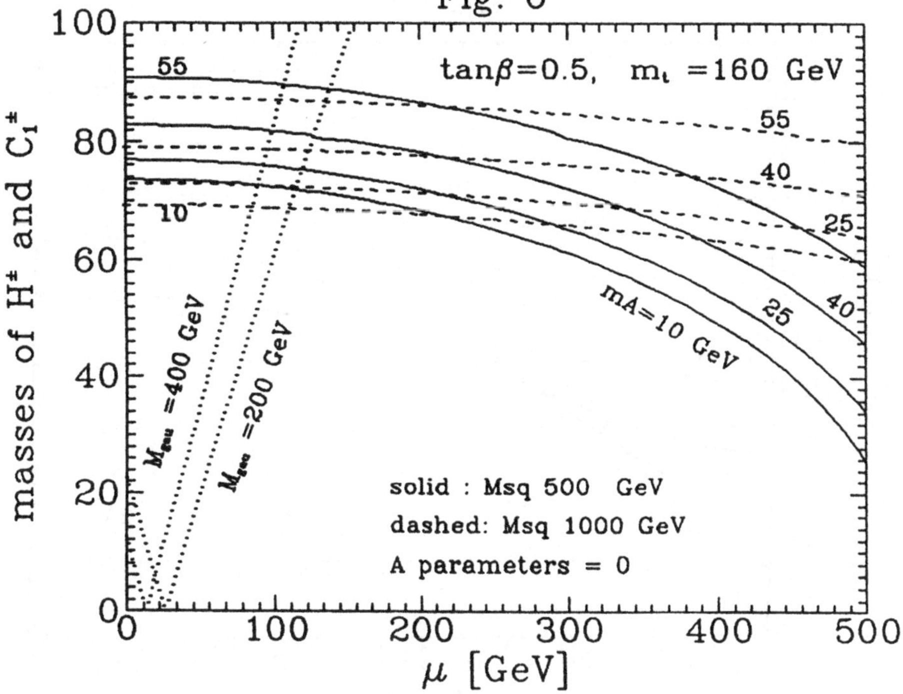
tanβ=0.5, m_t =160 GeV
55
40
25
10
mA=10 GeV
M_gau =400 GeV
M_gau =200 GeV
solid : Msq 500 GeV
dashed: Msq 1000 GeV
A parameters = 0
masses of H± and C1±
μ [GeV]

Fig. 7

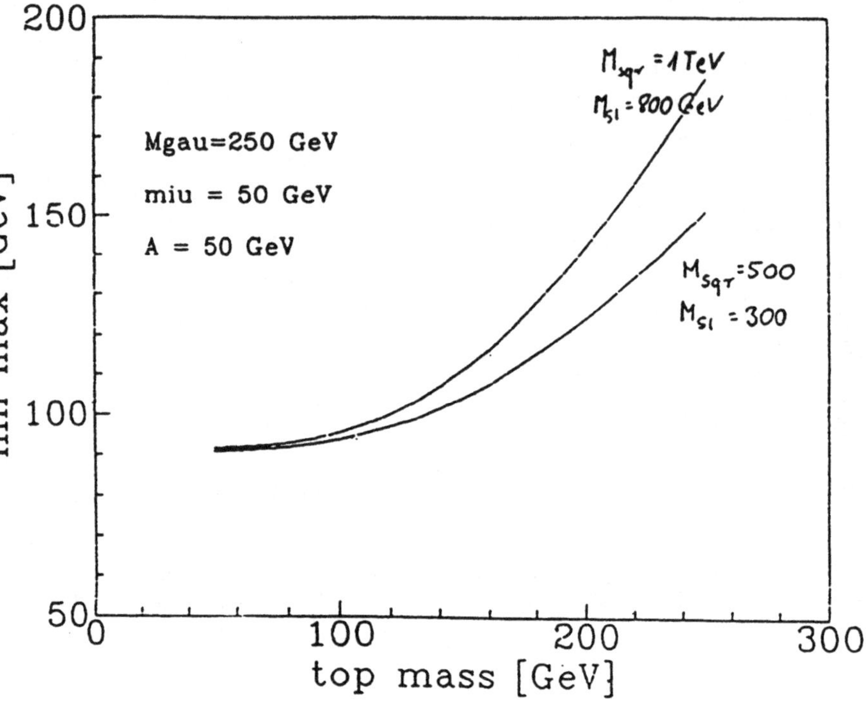
Mgau=250 GeV
miu = 50 GeV
A = 50 GeV
M_{sqr} = 1 TeV
M_{sl} = 800 GeV
M_{sqr} = 500
M_{sl} = 300
mh max [GeV]
top mass [GeV]
200
150
100
50
0
100
200
300

Fig. 8

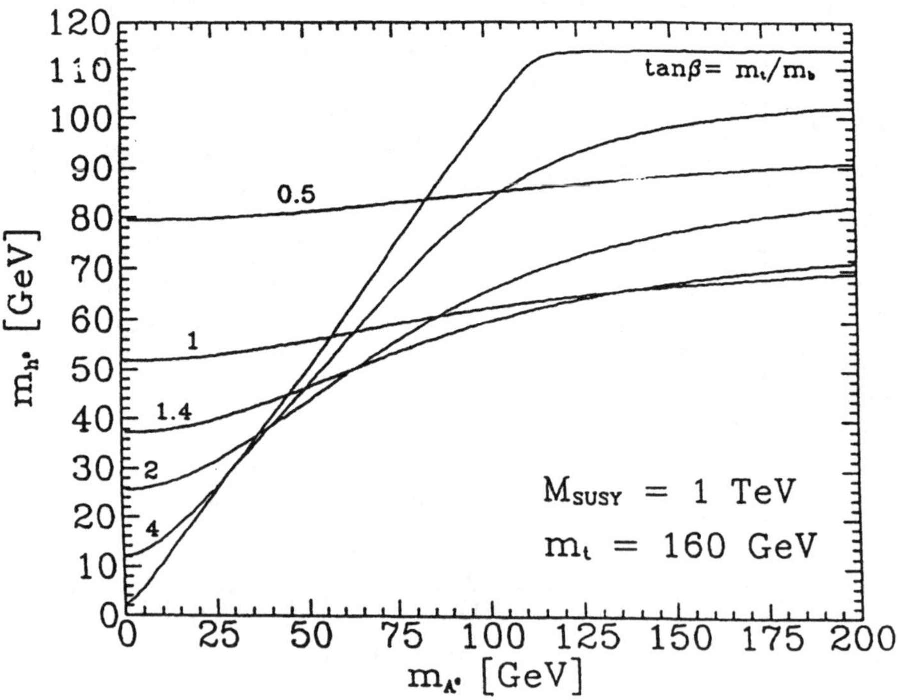
tanβ = m_t/m_b
0.5
1
1.4
2
4
M_{SUSY} = 1 TeV
m_t = 160 GeV
m_{h^0} [GeV]
m_{A^0} [GeV]
120
110
100
90
80
70
60
50
40
30
20
10
0
0
25
50
75
100
125
150
175
200

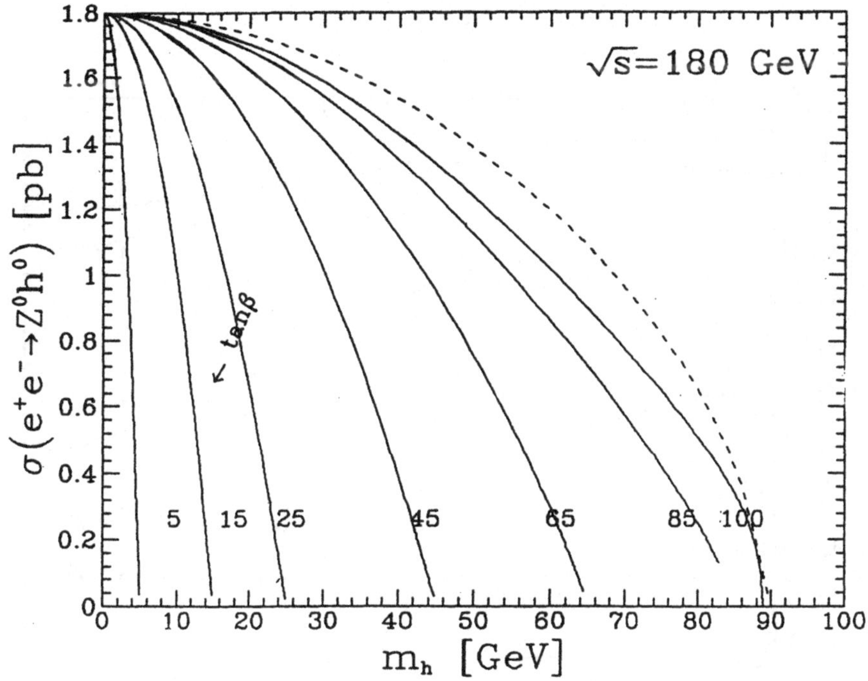

$\sqrt{s}$=180 GeV
σ(e⁺e⁻→Z⁰h⁰) [pb]
m_h [GeV]
tanβ
5
15
25
45
65
85
100

Fig. 9

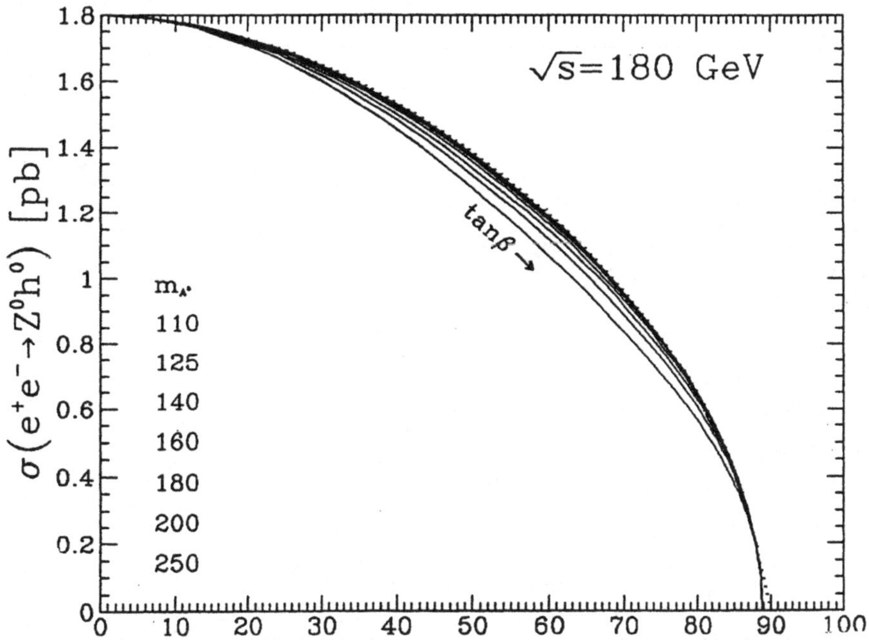

$\sqrt{s}$=180 GeV
σ(e⁺e⁻→Z⁰h⁰) [pb]
tanβ
m_A⁰
110
125
140
160
180
200
250

m_h [GeV]

Fig. 10

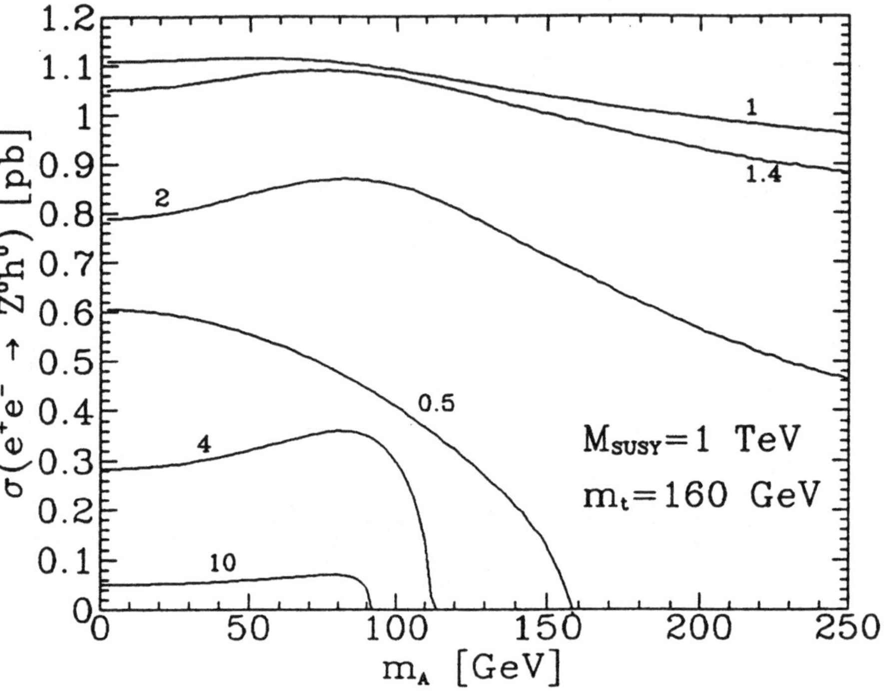
$\sigma(e^+e^- \to Z^0h^0)$ [pb]
m_A [GeV]
M_{SUSY}=1 TeV
m_t=160 GeV
1
1.4
2
0.5
4
10
0
50
100
150
200
250
0
0.1
0.2
0.3
0.4
0.5
0.6
0.7
0.8
0.9
1
1.1
1.2

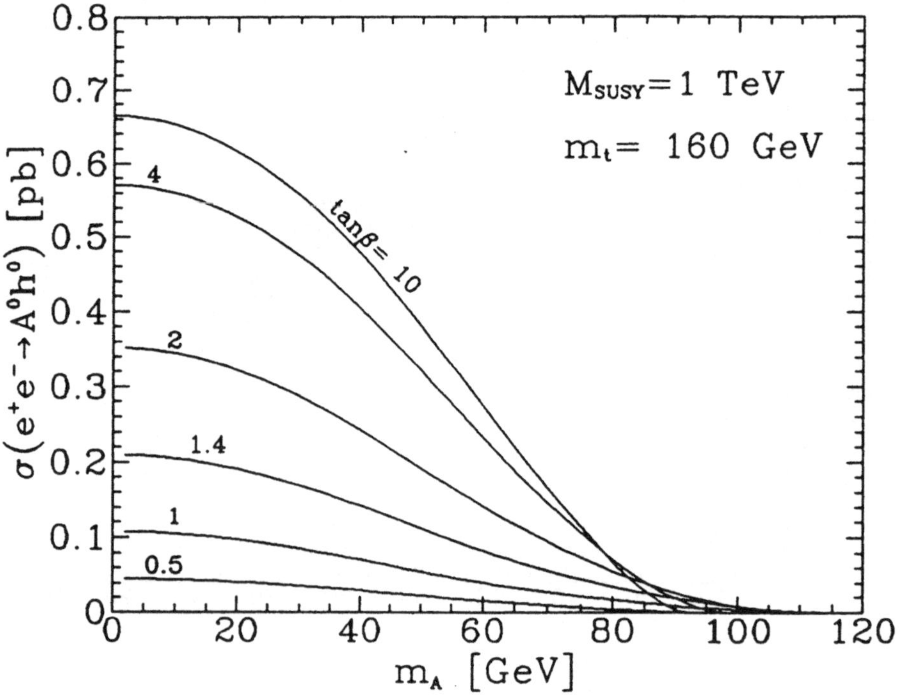
$\sigma(e^+e^- \to A^0h^0)$ [pb]
m_A [GeV]
M_{SUSY}=1 TeV
m_t= 160 GeV
$\tan\beta$= 10
4
2
1.4
1
0.5
0
20
40
60
80
100
120
0
0.1
0.2
0.3
0.4
0.5
0.6
0.7
0.8

Fig. 11

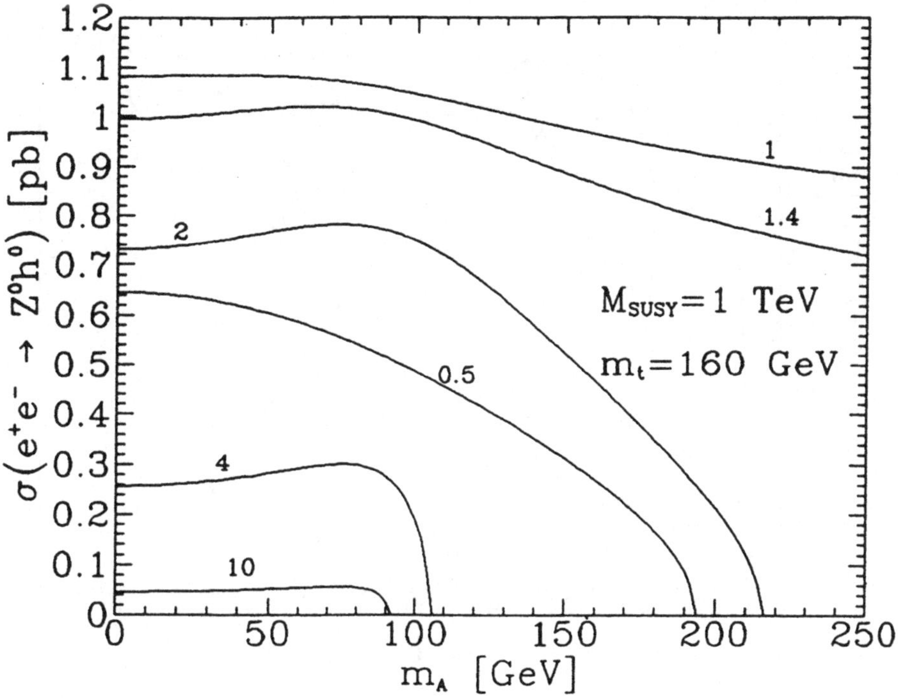
σ(e⁺e⁻ → Z⁰h⁰) [pb]
m_A [GeV]
M_SUSY=1 TeV
m_t=160 GeV
1
1.4
2
0.5
4
10

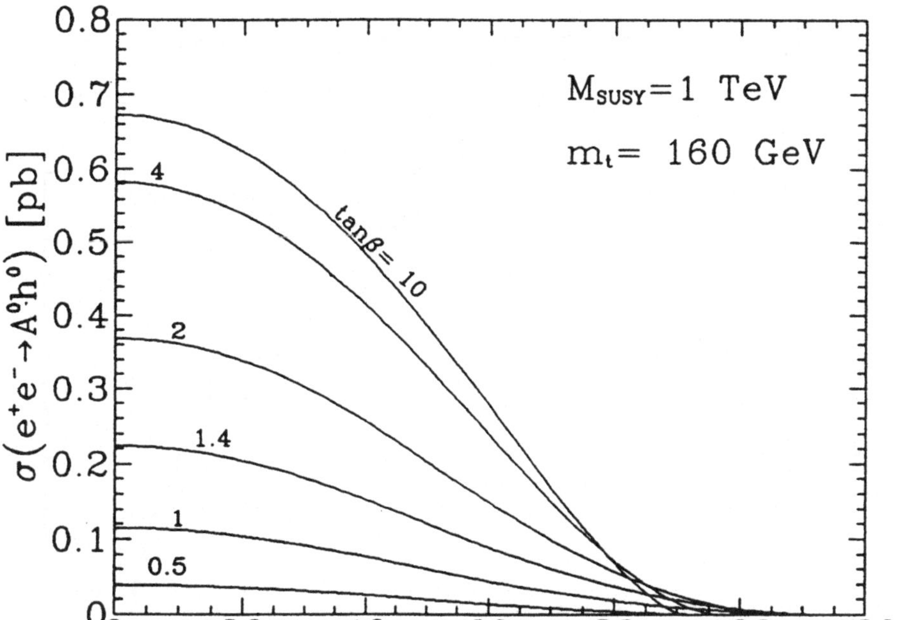
σ(e⁺e⁻→A⁰h⁰) [pb]
m_A [GeV]
M_SUSY=1 TeV
m_t= 160 GeV
tanβ= 10
4
2
1.4
1
0.5

Fig. 12

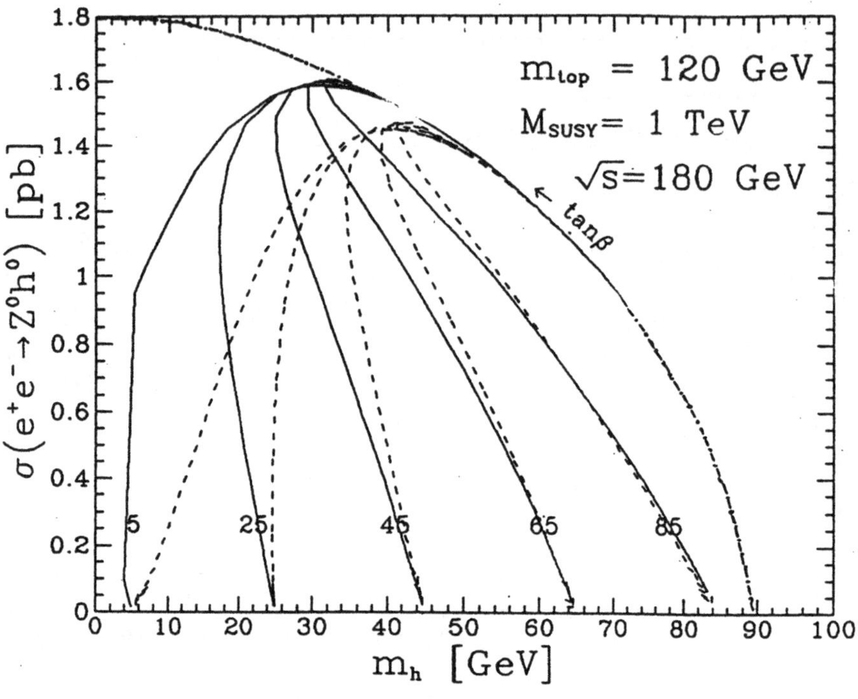

m_{top} = 120 GeV
M_{SUSY} = 1 TeV
$\sqrt{s}$=180 GeV
tanβ
5
25
45
65
85
σ(e⁺e⁻→Z⁰h⁰) [pb]
m_h [GeV]

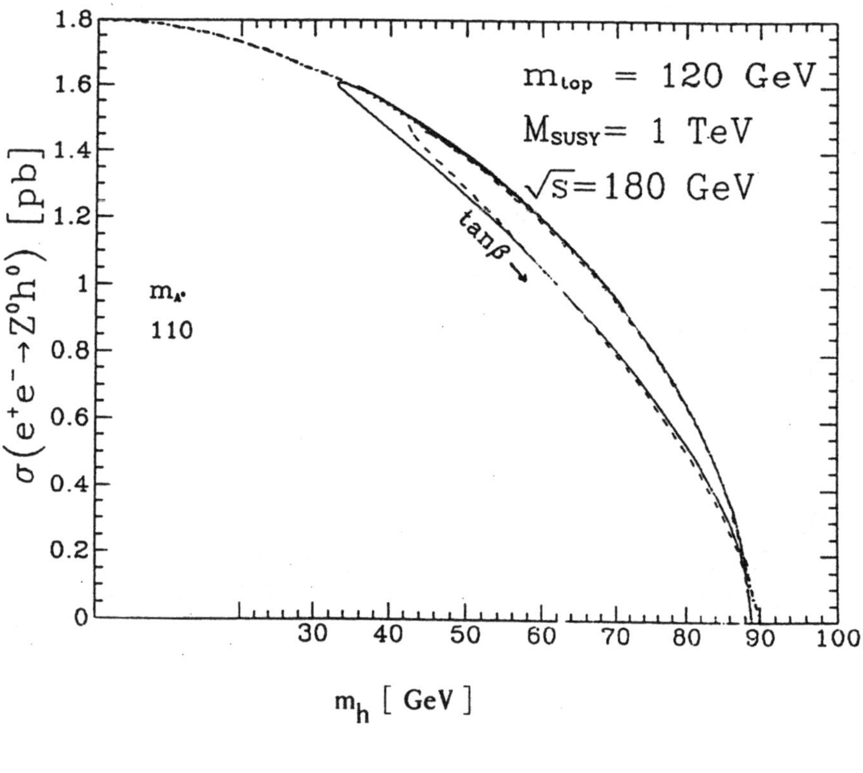

m_{top} = 120 GeV
M_{SUSY} = 1 TeV
$\sqrt{s}$=180 GeV
tanβ
m_{A^0}
110
σ(e⁺e⁻→Z⁰h⁰) [pb]
m_h [GeV]

Fig. 13

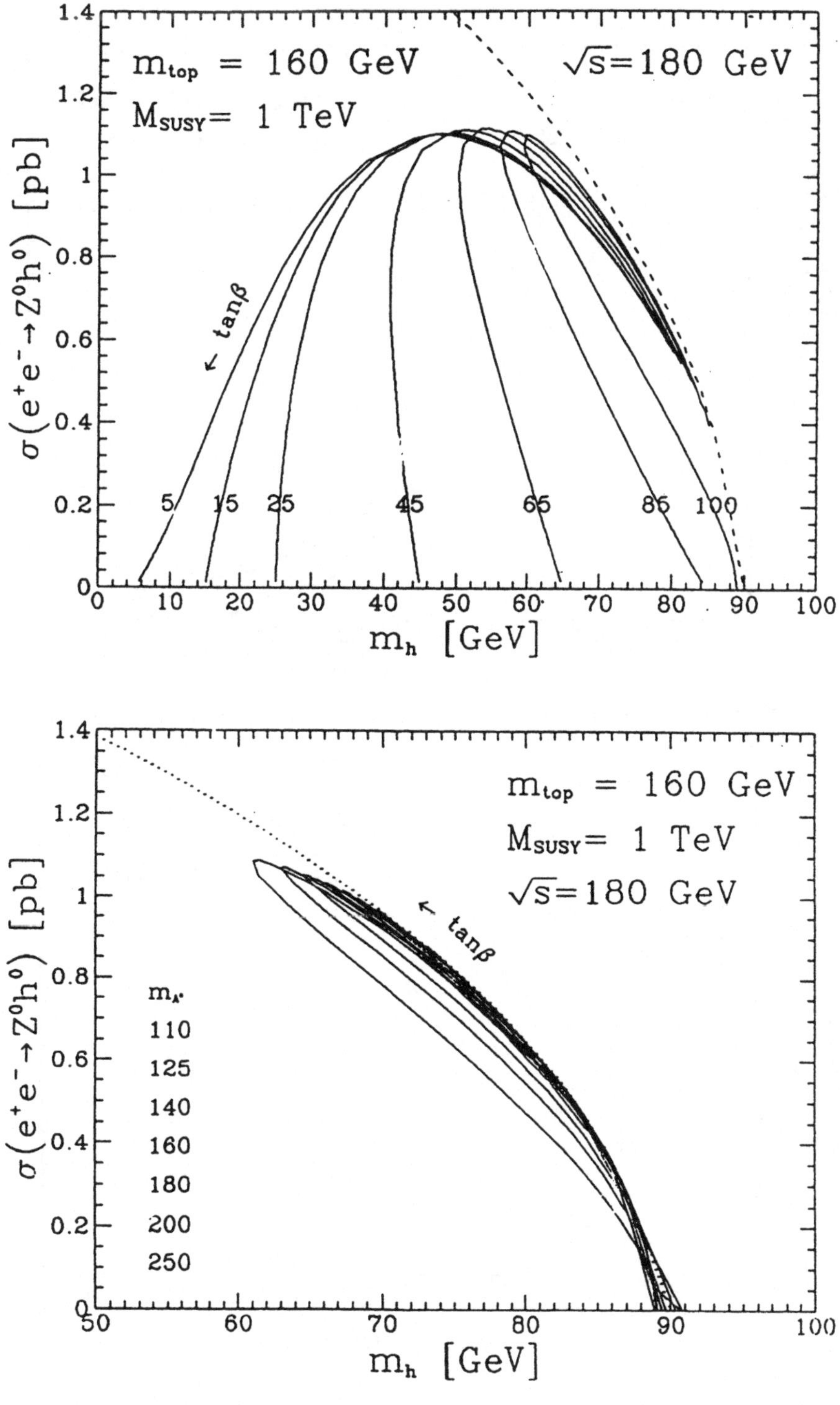
1.4
1.2
1
0.8
0.6
0.4
0.2
0
m_{top} = 160 GeV
M_{SUSY}= 1 TeV
$\sqrt{s}$=180 GeV
$\sigma(e^+e^- \to Z^0h^0)$ [pb]
← tanβ
5
15
25
45
65
85
100
10
20
30
40
50
60
70
80
90
100
m_h [GeV]
m_{top} = 160 GeV
M_{SUSY}= 1 TeV
$\sqrt{s}$=180 GeV
← tanβ
m_A
110
125
140
160
180
200
250
m_h [GeV]

Fig. 14

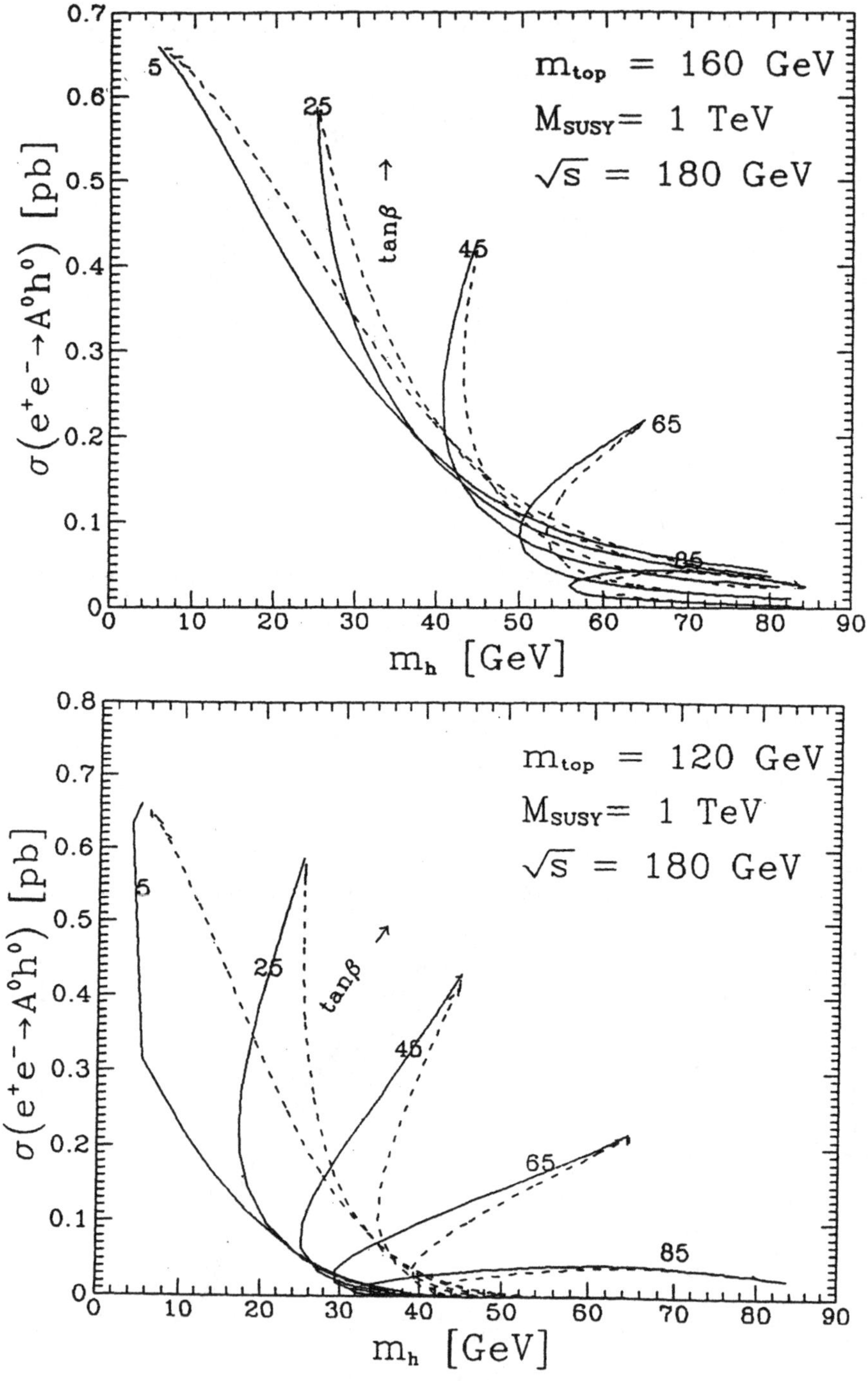
m_top = 160 GeV
M_SUSY = 1 TeV
√s = 180 GeV
σ(e⁺e⁻→A⁰h⁰) [pb]
m_h [GeV]
tanβ
5
25
45
65
85
m_top = 120 GeV
M_SUSY = 1 TeV
√s = 180 GeV
σ(e⁺e⁻→A⁰h⁰) [pb]
m_h [GeV]
tanβ
5
25
45
65
85

Fig. 15

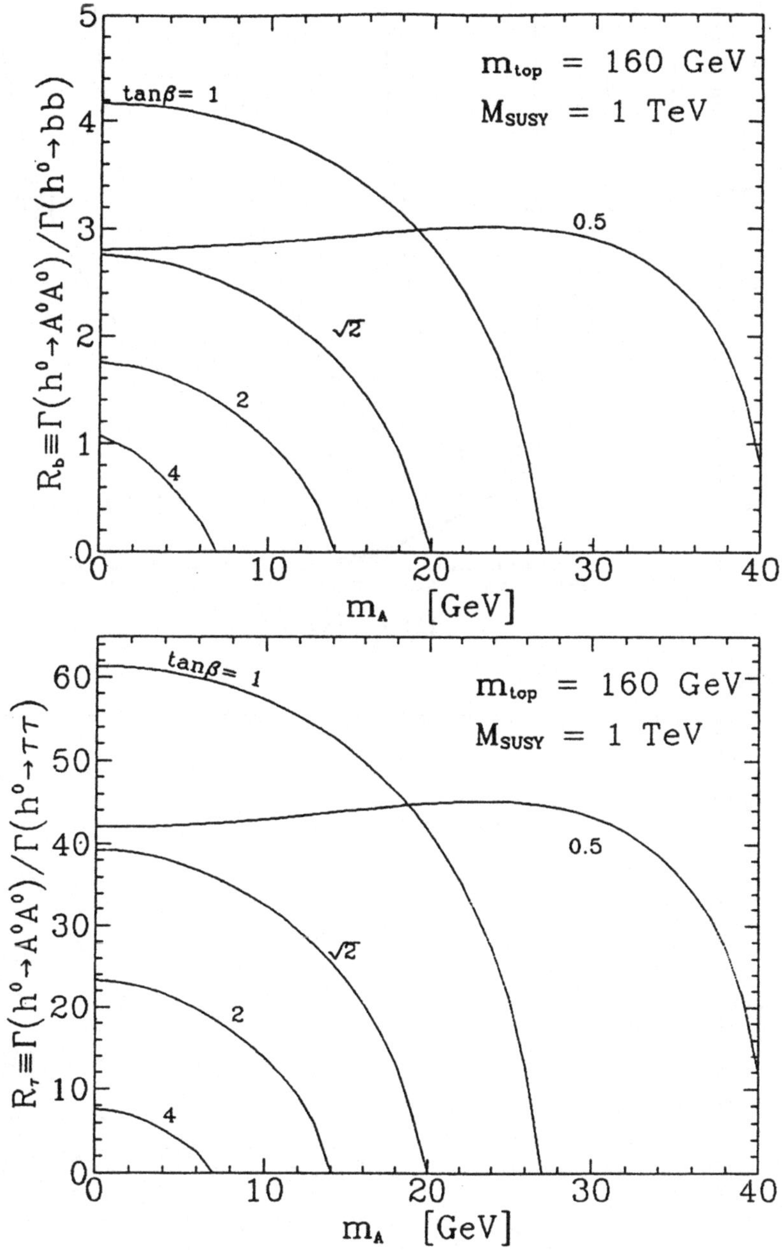

m_{top} = 160 GeV
M_{SUSY} = 1 TeV
tanβ= 1
0.5
√2
2
4
$R_b \equiv \Gamma(h^0 \to A^0A^0)/\Gamma(h^0 \to b\bar{b})$
m_A [GeV]
$R_\tau \equiv \Gamma(h^0 \to A^0A^0)/\Gamma(h^0 \to \tau\tau)$

Fig. 16a

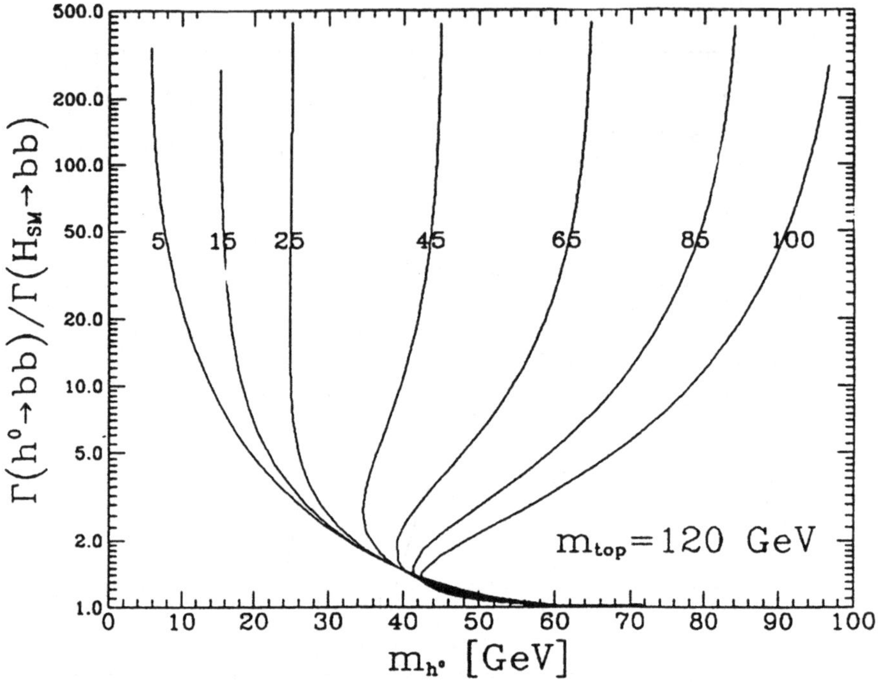
500.0
200.0
100.0
50.0
20.0
10.0
5.0
2.0
1.0
0
10
20
30
40
50
60
70
80
90
100
5
15
25
45
65
85
100
$\Gamma(h^0 \rightarrow bb)/\Gamma(H_{SM} \rightarrow bb)$
m_{top}=120 GeV
m_{h^0} [GeV]

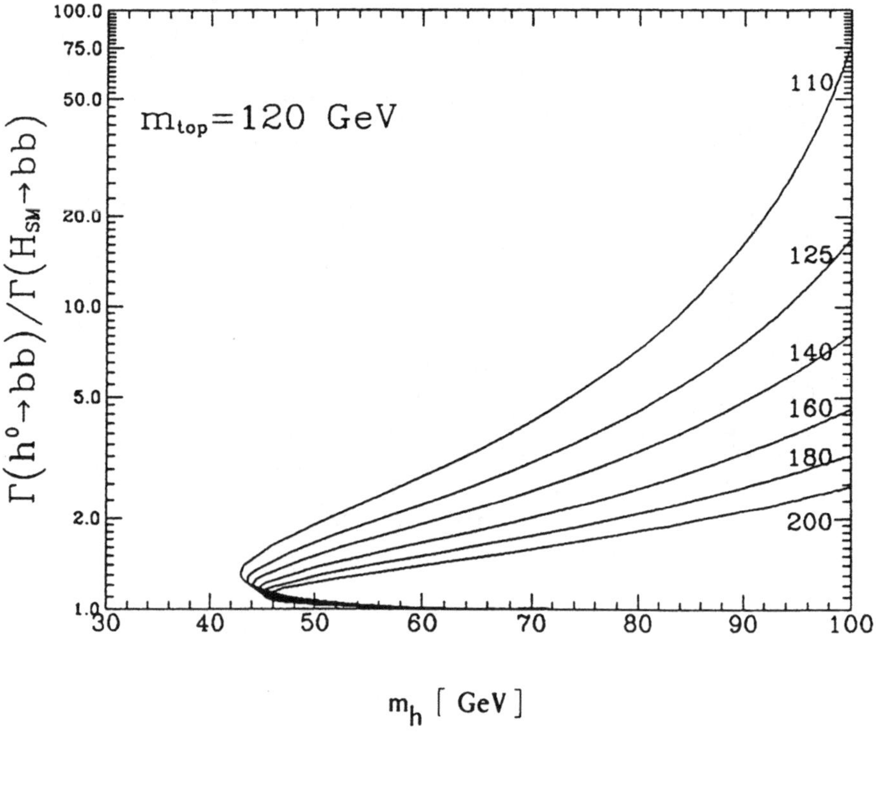
100.0
75.0
50.0
20.0
10.0
5.0
2.0
1.0
30
40
50
60
70
80
90
100
110
125
140
160
180
200
$\Gamma(h^0 \rightarrow bb)/\Gamma(H_{SM} \rightarrow bb)$
m_{top}=120 GeV
m_h [GeV]

Fig. 16b

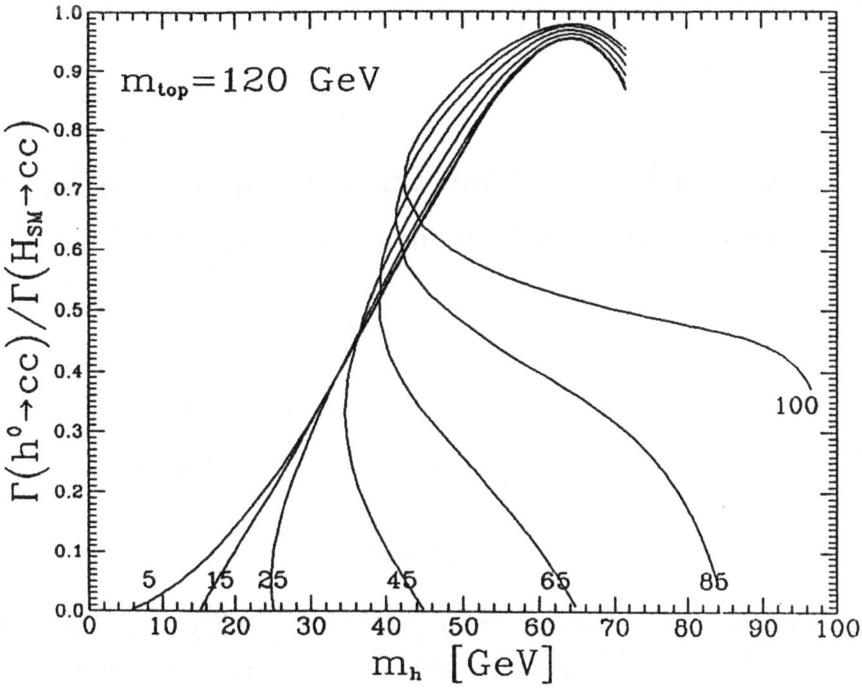
mtop=120 GeV
Γ(h⁰→cc)/Γ(H_SM→cc)
m_h [GeV]
5
15
25
45
65
85
100
1.0
0.9
0.8
0.7
0.6
0.5
0.4
0.3
0.2
0.1
0.0
0
10
20
30
40
50
60
70
80
90
100

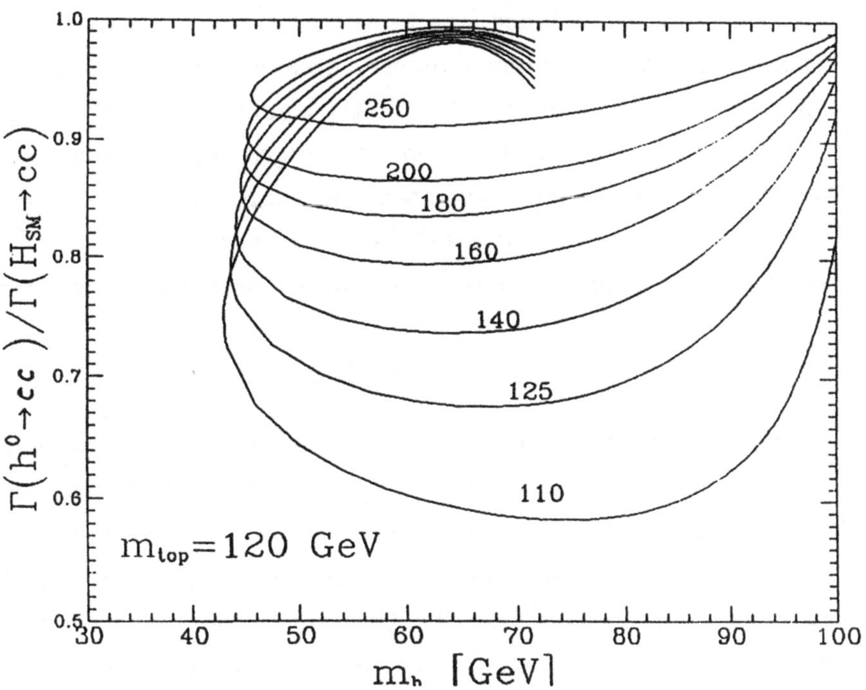
mtop=120 GeV
Γ(h⁰→cc)/Γ(H_SM→cc)
m_h [GeV]
250
200
180
160
140
125
110
1.0
0.9
0.8
0.7
0.6
0.5
30
40
50
60
70
80
90
100

The Upper Limit of the Light Higgs Boson Mass in the Minimal Supersymmetric Model

RALF HEMPFLING

Santa Cruz Institute for Particle Physics

University of California, Santa Cruz, CA 95064

Abstract

In the minimal supersymmetric model (MSSM), the Higgs self-interactions are related to the $SU_L(2) \times U_Y(1)$ gauge couplings. These relations impose an upper limit on the mass of the lightest Higgs boson of the model. When one-loop radiative corrections to this limit are included, one finds that the corrections are large and grow logarithmically with the supersymmetry breaking scale. Renormalization group techniques are employed in order to sum the leading log radiative corrections to all orders, and the result is that $m_{h^0} - m_Z$ converges to an infrared fixed point. However, due to the m_t^4 behavior of the dominant term, the radiative corrections can still lift the upper bound on the lightest Higgs mass above the reach of LEP-200. The radiatively corrected upper bound for m_{h^0} as a function of m_t is presented.

1. Introduction

In recent years supersymmetric theories have received a lot of attention. In these theories, quadratic divergences in unrenormalized Green functions are automatically cancelled by adding a bosonic (fermionic) partner to every fermion (boson) with the same quantum numbers and imposing supersymmetry (SUSY) on the resulting Lagrangian. This might hold the key to the solution of the *hierarchy problem* [1] one of the most fundamental puzzles of the Standard Model (SM). In building a supersymmetric extension of the SM, we also need at least two Higgs doublets H_n^i [$n = 1,2$; i is the $SU_L(2)$-index] in order to give masses to up- and down-type fermions. Following these steps we arrive at the minimal supersymmetric model (MSSM) [2]. In this model neither the gauge symmetry nor SUSY is broken spontaneously. To construct a realistic theory, one must assume that the MSSM is only the low energy part of a more complete theory and that SUSY is broken in a different sector. This will generate the following soft SUSY-breaking mass terms for the effective low energy theory [$\widetilde{q}_L \equiv (\widetilde{u}_L, \widetilde{d}_L)$, generation labels will be suppressed]:

$$\begin{aligned}
\mathcal{V}^{H}_{soft} &= m_1^2 H_1^{i*} H_1^i + m_2^2 (H_2^{j*} H_2^j) - (m_{12}^2 \epsilon_{ij} H_1^i H_2^j + \text{H.c.}) \\
\mathcal{V}^{\widetilde{Q}}_{soft} &= M_{\widetilde{Q}}^2 \widetilde{q}_L^* \widetilde{q}_L + M_{\widetilde{U}}^2 \widetilde{u}_R^* \widetilde{u}_R + M_{\widetilde{D}}^2 \widetilde{d}_R^* \widetilde{d}_R \\
\mathcal{V}^{\widetilde{G}}_{soft} &= \tfrac{1}{2} M_3 \widetilde{g}^a \widetilde{g}^a + \tfrac{1}{2} M_2 \widetilde{W}^a \widetilde{W}^a + \tfrac{1}{2} M_1 \widetilde{B}\widetilde{B} + \text{H.c.}
\end{aligned} \tag{1.1}$$

($\widetilde{g}$, $\widetilde{W}$ and $\widetilde{B}$ are the gauginos for the $SU_c(3)$, $SU_L(2)$ and $U_Y(1)$ symmetry groups, respectively). With the soft terms the $SU(2)_L \times U(1)_Y$ symmetry can be broken spontaneously and the masses of the superpartners can be pushed above the current experimental bounds. In addition the following soft interaction terms for the squarks are generated:

$$\mathcal{V}^{\widetilde{Q}}_{int} = h_U A_U \epsilon_{ij} H_2^i \widetilde{q}_L^j \widetilde{u}_R + h_D A_D \epsilon_{ij} H_1^i \widetilde{q}_L^j \widetilde{d}_R + \text{H.c.}. \tag{1.2}$$

The subscript U (D) corresponds to the up- (down-) type squarks. For the sleptons, the definitions are similar, except that there is no $\widetilde{\nu}_R$. The A-parameters and the squark mass parameters are in general arbitrary 3×3 matrices. However, in this talk I will confine myself to the special case where they are proportional to the unit matrix. This leaves only five free parameters: A_U, A_D, $M_{\widetilde{U}}$, $M_{\widetilde{D}}$ and $M_{\widetilde{Q}}$.

After electroweak symmetry breaking there remain five physical Higgs particles: two CP-even scalars (H^0 and h^0, with $m_{h^0} \leq m_{H^0}$), one CP-odd scalar (A^0) and a charged Higgs pair ($H^{\pm}$). The vacuum expectation values (VEV) of the two neutral Higgs fields are denoted by v_1 and v_2. Due to its minimal particle content the Higgs sector of the MSSM is well-constrained and thus allows some phenomenologically interesting predictions [3]. For example, the mass of the lightest Higgs scalar is

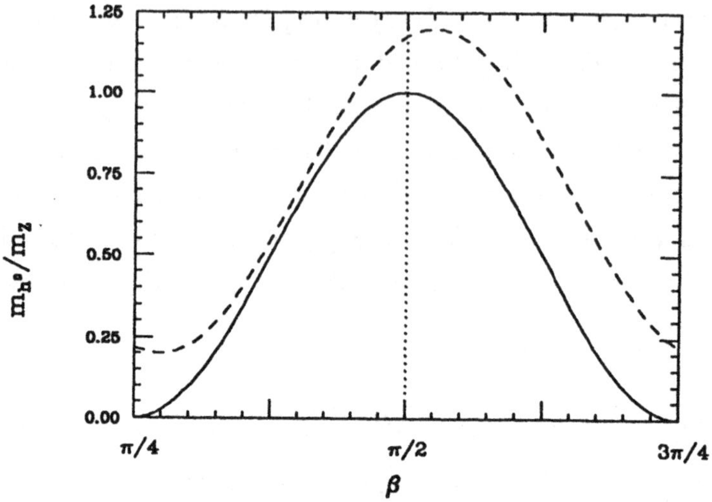

Fig. 1. The Higgs mass as a function of the angle β, in the limit of $m_{A^0} \to \infty$ (solid curve). The dashed line is the schematic plot of the one-loop corrected result. This curve depends on the renormalization scheme of β as well as the other parameters of the theory.

given by the tree-level relation:

$$m_{h^0}^2 = \frac{1}{2}\left[m_{A^0}^2 + m_Z^2 - \sqrt{(m_{A^0}^2 - m_Z^2)^2 + 4m_{A^0}^2 m_Z^2 \sin^2 2\beta}\right] \tag{1.3}$$

where $\tan\beta \equiv v_2/v_1$. In particular, h^0 must be lighter than the Z-boson. This prediction, however, is based on SUSY and need not be satisfied below the SUSY-breaking scale (M_{SUSY}) set by the parameters in eq. (1.1). It has recently been shown that the upper bound of m_{h^0} increases considerably when radiative corrections are included [4-6]. The goal of this talk is to present the one-loop corrections to the light Higgs mass bound and to improve the result by using renormalization group equation (RGE) techniques. This talk is based on work done in collaboration with Howard Haber [4,7].

2. Higgs Mass Shift at One-Loop

We see from eq. (1.3) that a prediction of m_{h^0} requires at least three experimental inputs at tree-level. However, the maximum value of m_{h^0} depends only on

Fig. 2. Possible radiative mechanisms for generating down-type quark masses.

m_Z. Therefore, it is clear that the one-loop correction to the upper bound of m_{h^0} will only require the renormalization of m_Z and not of any other quantity such as β or m_{A^0}. In fig. 1, we have plotted the tree-level prediction of m_{h^0} vs. β (solid line). The function has a maximum at β_0^{max}. The value of the maximum mass and β_0^{max} are independent of m_{A^0} (actually, in fig. 1, m_{A^0} is taken to infinity). The dashed line is a schematic plot of the one-loop radiatively corrected Higgs mass. We anticipate that this graph will have a maximum at β_1^{max} close to the tree-level maximum [that means $\beta_0^{max} - \beta_1^{max} = \mathcal{O}(g_2^2)$, where g_2 is the weak coupling constant]. In these plots m_{h^0} is defined as the pole of the propagator and is thus a renormalized physical quantity. The angle β is renormalized but it is not directly an observable. Rather it is defined through some scheme as an intermediate quantity which has no direct physical meaning. This means that the place of the maximum will in general depend upon the scheme. However, the value of m_{h^0} at the maximum is scheme independent. This is the quantity we are interested in. Now we would like to argue that within a first order approximation it is not necessary to find β_1^{max} in order to find the maximum Higgs mass. This is because the derivative at the maximum vanishes and thus $m_{h^0}(\beta_0^{max}) = m_{h^0}(\beta_1^{max})$ to first order in perturbation theory. We thus need to consider the case of the tree-level maximum. To achieve this, we must take either $\beta = \pi/2$ or $\beta = 0$ (that is either $v_1 = 0$ or $v_2 = 0$, respectively). Notice that the latter case cannot be even approximately correct because of the heavy top quark. On the other hand, $\beta = \pi/2$ is plausible since all the down-type quarks are light enough to be generated radiatively (see fig. 2). In fact the large value of m_t/m_b suggests that this scenario might be realized in nature. In ref. 4 we imposed an extra $U(1)$ symmetry that guarantied that $\beta_0^{max} = \beta_1^{max}$. However, this symmetry required implicitly that $\mu = 0$ and is thus not completely general.

We will now investigate the case of $\beta = \pi/2$. It is obtained by setting m_{12}, as defined in eq. (1.1), to zero. In this model, the tree-level Higgs mass spectrum consists of $m_{h^0} = m_Z$, $m_{H^0} = m_{A^0} \geq m_Z$, and $m_{H^\pm} = (m_W^2 + m_{A^0}^2)^{1/2}$. [Had we chosen $m_{A^0} < m_Z$, then we would have found $m_{h^0} = m_{A^0}$, and $m_{H^0} = m_Z$ (at tree-level), which is not of interest to us here.] Now we proceed to compute corrections to the value of m_{h^0}. We will derive an expression for:

$$\Delta m_{h^0}^2 \equiv m_{h^0}^2 - m_Z^2 . \tag{2.1}$$

There are two types of corrections that we will compute here. The first consists of the one-loop radiative corrections to the model specified above. This will be denoted by $(\Delta m_{h^0}^2)_{\beta=\pi/2}$. The second correction is due to RGE improvement. This shift is calculated numerically and will be denoted by $(\Delta m_{h^0}^2)_{RGIMP}$. Thus, the final result for the squared mass shift is:

$$\Delta m_{h^0}^2 = (\Delta m_{h^0}^2)_{\beta=\pi/2} + (\Delta m_{h^0}^2)_{RGIMP} . \tag{2.2}$$

We now turn to the computation of $(\Delta m_{h^0}^2)_{\beta=\pi/2}$. The full potential of the model is:

$$\begin{aligned}\mathcal{V} =&\lambda_1(H_1^{i*}H_1^i)^2 + \lambda_2(H_2^{i*}H_2^i)^2 + \lambda_3(H_1^{i*}H_1^i)(H_2^{j*}H_2^j) + \lambda_4 \mid H_1^{i*}H_2^i \mid^2 + \\ &\mid \mu \mid^2 (H_1^{i*}H_1^i + H_2^{j*}H_2^j) + \mathcal{V}_{soft}^H ,\end{aligned} \tag{2.3}$$

where

$$\begin{aligned}\lambda_1 &= \lambda_2 = -\tfrac{1}{2}\lambda_3 = \tfrac{1}{8}[g_2^2 + g_1^2] \\ \lambda_4 &= \tfrac{1}{2}g_2^2 .\end{aligned} \tag{2.4}$$

The relevant terms in our case will be the ones involving only the shifted neutral component of the second doublet $H_2 \to h^0/\sqrt{2} + v$:

$$\begin{aligned}\mathcal{V}_{h^0} &= th^0 + \tfrac{1}{2}m_{h^0}^2(h^0)^2 + \mathcal{O}[(h^0)^3] \\ \mathcal{V}_Z &= \tfrac{1}{2}m_Z^2 Z_\mu Z^\mu ,\end{aligned} \tag{2.5}$$

where

$$m_Z^2 = \tfrac{1}{2}(g_1^2 + g_2^2)v^2 , \tag{2.6}$$

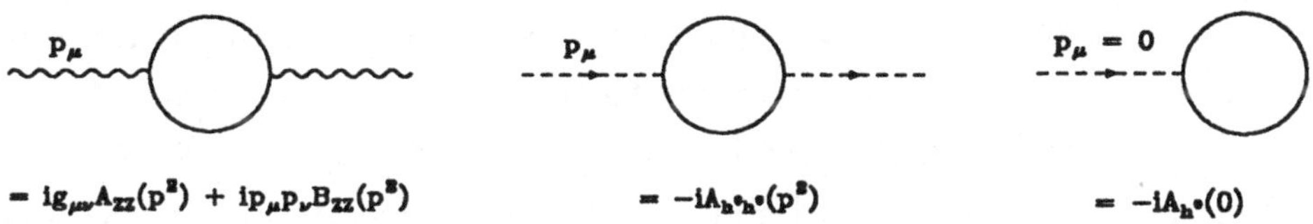

Fig. 3. The definitions of the self-energies for Higgs bosons and gauge bosons.

$$t = \sqrt{2}v\left[\tfrac{1}{4}v^2(g_1^2+g_2^2)+m^2\right], \tag{2.7}$$

$$m_{h^0}^2 = 6\lambda_2 v^2 + m^2 . \tag{2.8}$$

The mass parameter is defined as $m^2 \equiv m_2^2 + \mu^2$. If we eliminate $(g_1^2+g_2^2)v^2$ and m in favor of m_Z^2 and t, respectively, and use the boundary condition [eq. (2.4)] we end up with the following relation:

$$m_{h^0}^2 = m_Z^2 + \frac{g_2}{2m_W}t, \tag{2.9}$$

where we have used $m_W = g_2 v/\sqrt{2}$. Remember that m_{h^0} and m_Z are still unrenormalized parameters. The standard procedure at this point is to rewrite the bare parameters as renormalized parameters plus counterterms. The only requirements for these separations are that the renormalized parameters are finite and the counterterms are of $\mathcal{O}(g_2^2)$. Otherwise they are completely arbitrary. We avoid this step by expressing the bare quantities directly in terms of physical observables. These are the masses of h^0 and Z (indicated below with a subscript P) which are identified in the usual way as the poles in the corresponding propagators:

$$m_{ZP}^2 = m_Z^2 + \mathrm{Re}\ A_{ZZ}(m_Z^2), \tag{2.10}$$

$$m_{h^0P}^2 = m_{h^0}^2 + \mathrm{Re}\ A_{h^0h^0}(m_{h^0}^2). \tag{2.11}$$

Here the self-energies are defined in the standard fashion (see fig. 3). We now demand that v is the true VEV at one-loop. This means that the unphysical one-point functions (tadpoles) corresponding to a Higgs field disappearing into the vacuum are absent. This reads $t + A_{h^0}(0) = 0$, where $-iA_{h^0}(0)$ is the sum of all one-loop Feynman graphs contributing to the h^0 one-point function (see fig. 3). We find this choice convenient, since there will be no tadpole contributions to the calculation of A_{ZZ} and $A_{h^0h^0}$. Using eq. (2.9) we arrive at our first main result:

$$(\Delta m_{h^0}^2)_{\beta=\pi/2} \equiv m_{h^0P}^2 - m_{ZP}^2 = \text{Re}\left[A_{h^0h^0}(m_Z^2) - A_{ZZ}(m_Z^2)\right] - \frac{g_2}{2m_W} A_{h^0}(0) \,. \tag{2.12}$$

This is the mass shift due to one-loop radiative corrections. The error we make by evaluating the Higgs self-energy at m_Z instead of m_{h^0} is of second order in the perturbation series. Although our result has been derived using a specific convention for the shift v, it is easy to see that we could have expanded about any other point within $\mathcal{O}(g_2^2)$ of the tree-level minimum. For example, another possible convention would be to simply define $t = 0$. (In this case v would not be the true VEV, but this does not matter.) Then one would obtain: $(\Delta m_{h^0}^2)_{\beta=\pi/2} = \text{Re}\left[A_{h^0h^0}(m_Z^2) - A_{ZZ}(m_Z^2)\right]$. However, one would have to include the tadpole contributions to both $A_{h^0h^0}$ and A_{ZZ}. It is a simple exercise to check that these additional terms simply reproduce the term $-(g_2/2m_W)A_{h^0}(0)$ in eq. (2.12).

3. Numerical Results

We now turn to the numerical evaluation of the h^0 mass shift. Note that the contribution of each particle appearing in the self-energy loops to eq. (2.12) is separately divergent and depends on an arbitrary scale. The divergences and scale dependence will cancel only when one sums over a complete supersymmetric multiplet. This is true because m_{h^0} is calculable only in the supersymmetric model. In the SM the self-coupling constant λ is arbitrary and therefore the Higgs mass is an infinitely renormalized free parameter of the theory.

The one-loop correction can be split into a gauge-gaugino-Higgs-Higgsino sector and a quark-squark (slepton-lepton) sector which are separately finite. We shall first discuss the latter one since it gives the dominant contribution to the Higgs mass shift. As we argued above, there will be a cancellation of divergences generated by the quarks and their superpartners. Terms proportional to $\ln(M_{SUSY}^2/m_Z^2)$ are remnants of the cancellation. Here, we simply quote the leading log approximation

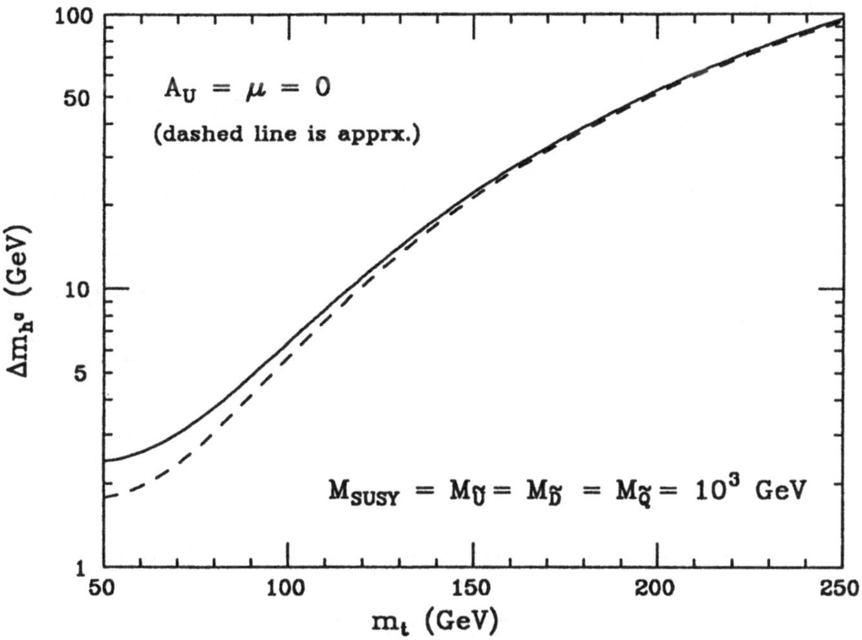

Fig. 4. Higgs mass shift $\Delta m_{h^0} \equiv m_{h^0} - m_Z$ due to one-loop radiative corrections. The solid line denotes the full one-loop result due to three generations of quarks, leptons and their superpartners. The dashed line corresponds to the leading log approximation given in eq. (3.1). All the mass parameters are set equal to the common SUSY scale, M_{SUSY}.

to our exact formulas, where we summed over six flavors of quarks/squarks and leptons/sleptons:

$$\left(\Delta m^2_{h^0}\right)_{\beta=\pi/2} = \frac{3g_2^2 m_Z^4}{16\pi^2 m_W^2}\left\{\ln\left(\frac{M^2_{SUSY}}{m_t^2}\right)\left[\frac{2m_t^4 - m_t^2 m_Z^2}{m_Z^4} + \tfrac{1}{6}\left(1 - \tfrac{8}{3}s_W^2 + \tfrac{32}{9}s_W^4\right)\right]\right.$$
$$+ \ln\left(\frac{M^2_{SUSY}}{m_Z^2}\right)\left[\tfrac{1}{3}\left(1 - \tfrac{8}{3}s_W^2 + \tfrac{32}{9}s_W^4\right) + \tfrac{1}{2}\left(1 - \tfrac{4}{3}s_W^2 + \tfrac{8}{9}s_W^4\right)\right.$$
$$\left.\left. + \tfrac{1}{3}\left(1 - 2s_W^2 + 4s_W^4\right)\right]\right\}, \tag{3.1}$$

where $s_W \equiv \sin\theta_W$. Here, we have assumed that all the soft squark mass parameters in eq. (1.1) are of order M_{SUSY}. The most dramatic aspect of this result is the m_t^4 growth, which arises due to the top quark/squark loops. In fig. 4 we have plotted the mass shift $\Delta m_{h^0} \equiv m_{h^0} - m_Z$ as function of the top-quark mass. The

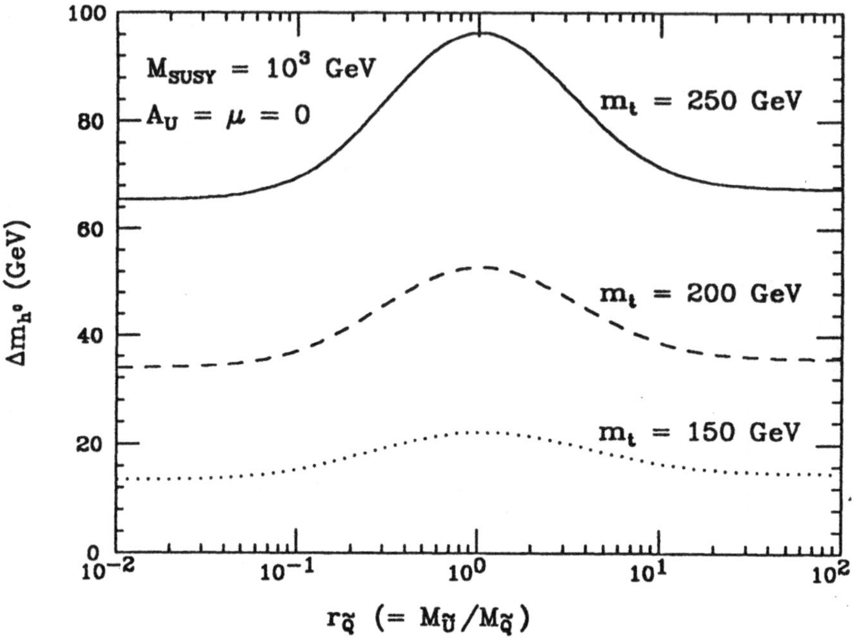

Fig. 5. Higgs mass shift Δm_{h^0} due to one-loop radiative corrections. The solid, dashed and dotted lines denote the full corrections due to three generations of quarks, leptons and their superpartners for a top quark mass of $m_t = 150$, 200 and 250 GeV, respectively. The common SUSY scale M_{SUSY} is 1 TeV.

solid (dashed) line is the exact one-loop result (leading log approximation). All the squark mass parameters are set to 1 TeV. The A-parameters and μ are set to zero (that is there is no $\widetilde{q}_L$–$\widetilde{q}_R$ -mixing). In fig. 5 we plot the mass shift as a function of the ratio $r_{\widetilde{Q}} \equiv M_{\widetilde{U}}/M_{\widetilde{Q}}$ while taking $(M^2_{\widetilde{Q}} + M^2_{\widetilde{U}}) = 2M^2_{SUSY}$ and $M_{\widetilde{D}} = M_{SUSY}$. The graphs show a maximum at $r_{\widetilde{Q}} = 1$ for all top quark masses considered.

Another important contribution to our exact result derives from the soft interaction terms $\mathcal{V}^{\widetilde{Q}}_{int}$ [eq. (1.2)] and the following F-terms:

$$\mathcal{V}^{\widetilde{Q}}_F = h_D \mu H_1^{i*} \widetilde{q}^{\,i}_L \widetilde{u}_R - h_U \mu H_2^{i*} \widetilde{q}^{\,i}_L \widetilde{d}_R + \text{H.c.}. \tag{3.2}$$

It is clear that contributions from these terms are not enhanced by a logarithm since they only come from the squark-sector. This becomes obvious in the down-squark sector since the down-type quarks do not contribute to the Higgs self-energy at all when $v_1 = 0$. After electroweak symmetry breaking these trilinear terms will generate mass terms that mix the left- and right-handed squarks in addition to the interaction terms. It is convenient to expand the result in powers of m_Z^2/M_{SUSY}^2. To leading order the mass shift due to squark mixing is:

$$\begin{aligned}(\Delta m_{h^0}^2)_U &= \frac{v^2}{32\pi^2}\sum_U N_c\left(\frac{a_U A_U^4}{M_{SUSY}^4}+\frac{b_U A_U^2}{M_{SUSY}^2}\right),\\ (\Delta m_{h^0}^2)_D &= \frac{v^2}{32\pi^2}\sum_D N_c\left(\frac{a_D \mu^4}{M_{SUSY}^4}+\frac{b_D \mu^2}{M_{SUSY}^2}\right),\end{aligned} \tag{3.3}$$

where we have dropped terms of order (m_Z^2/M_{SUSY}^2) since they are completely negligable for $M_{SUSY} \geq 1$ TeV. The number of color degrees of freedom is $N_c = 3$ (1) for squarks (sleptons). In the special case of $M_{SUSY} \equiv M_{\tilde{U}} = M_{\tilde{Q}} = M_{\tilde{D}}$ we find:

$$\begin{aligned} a_U &= -\tfrac{2}{3}h_U^4, \qquad & b_U &= \tfrac{4}{3}h_U^2(6h_U^2 - g_1^2 - g_2^2),\\ a_D &= -\tfrac{2}{3}h_D^4, \qquad & b_U &= \tfrac{2}{3}h_D^2(g_1^2+g_2^2)\,. \end{aligned} \tag{3.4}$$

In our model the up-type Yukawa coupling is $h_U = g_2 m_U/(\sqrt{2}m_W)$. The down-type Yukawa coupling h_D is a free parameter since we have argued that the down-type quark masses is generated radiatively. Clearly the contributions of the first two generations are negligible. In the case of the top-quark, the squared mass shift increases initially with A_U. It reaches its maximum for $A_U^2/M_{\tilde{Q}}^2 = -b_U/(2a_U) = 6-(g_1^2+g_2^2)/h_U^2$. Analogously for the bottom quark the maximum is at $\mu^2/M_{\tilde{Q}}^2 = (g_1^2+g_2^2)/(2h_D^2)$. The maximum shifts $(\Delta m_{h^0}^2)_Q^{max}$ $(Q = U, D)$ due to the trilinear terms of the up- and down-type squarks become:

$$(\Delta m_{h^0}^2)_U^{max} = \frac{3N_c g_2^2 m_Z^2}{8\pi^2 c_W^2}\left(\frac{m_t^2}{m_Z^2}-\frac{1}{3}\right)^2, \qquad (\Delta m_{h^0}^2)_D^{max} = \frac{N_c g_2^2 m_Z^2}{96\pi^2 c_W^2}. \tag{3.5}$$

This approximation plus the leading log approximation [eq. (3.1)] lies within a few GeV of the full one-loop calculation plotted in fig. 6.

We now turn to the discussion of the gauge-gaugino-Higgs-Higgsino sector. This sector depends only on four free parameters: the mass of the CP-odd scalar m_{A^0} the Higgsino mass parameter μ and the Majorana masses M_k $(k = 1, 2)$ for the $U(1)_Y$ and $SU_L(2)$ gauginos respectively. The parameter freedom is reduced

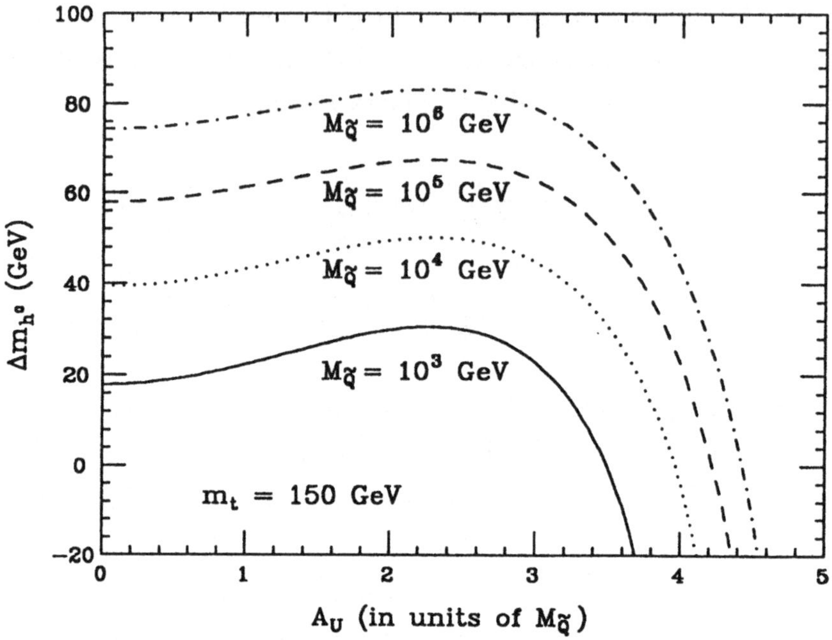

Fig. 6. The one-loop corrections to Δm_{h^0} as a function of the A-parameter. The solid, dotted, dashed and dot-dashed curves denote the full one-loop corrections due to three generations of quarks, leptons and their superpartners for $M_{SUSY} = M_{\tilde{Q}} = M_{\tilde{U}} = M_{\tilde{D}} = 10^3, 10^4, 10^5$ and 10^6 GeV, respectively.

further if we assume that our model is imbedded in a grand unified theory (GUT). Then the gaugino mass parameters are related according to:

$$M_1/M_2 = 5\ \tan^2\theta_W/3\,. \tag{3.6}$$

We would like to display the logarithmic dependence on the scale as well as the behavior of the mass shift for large ratios of the mass parameters. Therefore we parameterize this sector by the CP-odd Higgs mass m_{A^0}, the average mass of the charginos and neutralinos $M_{\tilde{\chi}}^2 \equiv (4\mu^2 + 3M_2^2 + M_1^2)/8$, and the ratio $r_{\tilde{\chi}} \equiv \mu/M_2$. The leading logarithms are:

$$\begin{aligned}(\Delta m_{h^0}^2)_{\beta=\pi/2} = -\frac{g_2^2 m_Z^4}{48\pi^2 m_W^2}\Bigg[&(5 - 10c_W^2 + 32c_W^4)\ln\left(\frac{M_{\tilde{\chi}}^2}{m_Z^2}\right) \\ &- (1 - 2c_W^2 + 2c_W^4)\ln\left(\frac{m_{A^0}^2}{m_Z^2}\right)\Bigg],\end{aligned} \tag{3.7}$$

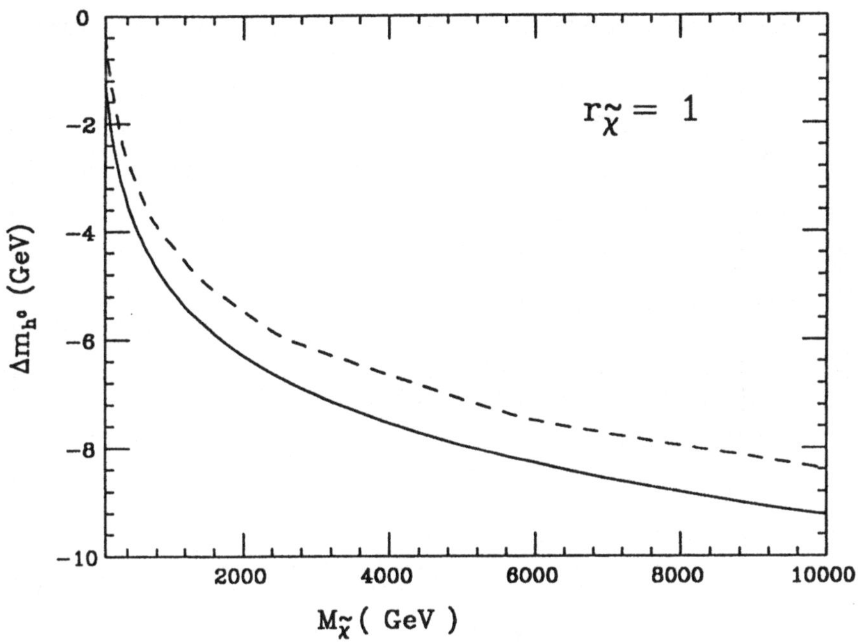

Fig. 7. The one-loop radiative corrections to Δm_{h^0} due to the gauge-gaugino-Higgs-Higgsino sector as a function of $M_{\tilde{\chi}}$ $[M_{\tilde{\chi}}^2 \equiv (4\mu^2 + 3M_2^2 + M_1^2)/8]$ for $r_{\tilde{\chi}} \equiv \mu/M_2 = 1$.

where $c_W \equiv \cos\theta_W$. In fig. 7 we have plotted m_{h^0} as a function of $M_{\tilde{\chi}}$ (solid line). Clearly, the contribution from the gauge-gaugino-Higgs-Higgsino sector is relatively small and negative. The leading logs (dashed line) are again the dominant terms. In the case $r_{\tilde{\chi}} = 1$, this approximation differs from the exact one-loop result only by 1 GeV over the entire range of $M_{\tilde{\chi}}$. Fig. 8 displays the dependence of the mass shift on the ratio $r_{\tilde{\chi}}$. Here we have subtracted the leading logs. We see that the full one-loop radiative correction never differs from the leading log approximation by much more than 1 GeV as long as $M_{\tilde{\chi}} < 1$ TeV. Moreover, note that the contribution of the Higgs sector [second term in eq. (3.7)] is positive but very small and never exceeds a few GeV .

In this section we have seen that the one-loop radiative corrections can become very large. For a top quark mass $m_t > 175$ GeV the corrections become larger than the tree-level result. It is therefore necessary to find a criterion for the validity of our perturbation expansion. By comparing the one-loop result with the largest possible contribution to the 2-loop correction [see also ref. 8] we find:

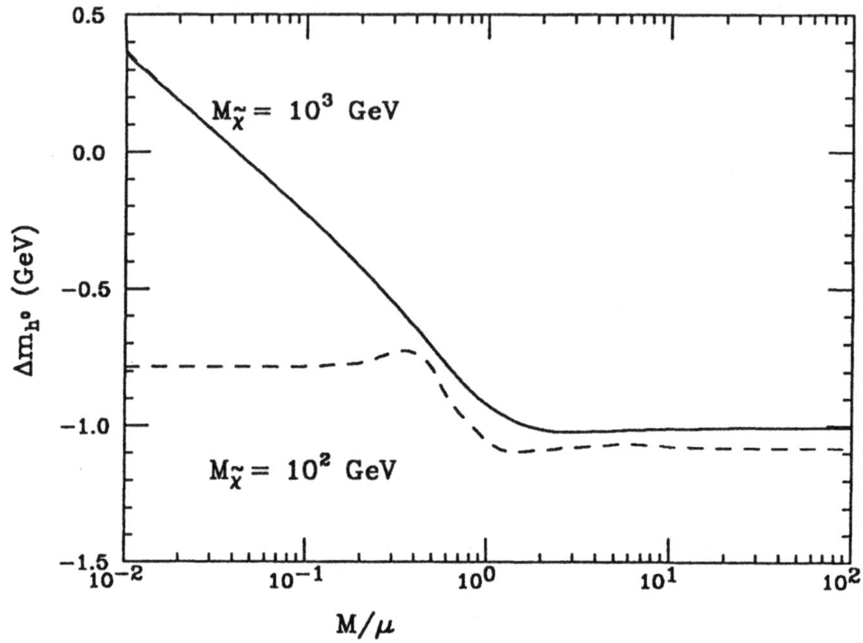

Fig. 8. The difference of the one-loop radiative corrections and the leading log approximation to Δm_{h^0} due to the gauge-gaugino-Higgs-Higgsino sector as a function of the ratio $r_{\tilde{\chi}}$. The solid (dashed) line corresponds to $M_{\tilde{\chi}} = 10^3$ (10^2) GeV.

$$\frac{N_c g_2^2 m_t^2}{16\pi^2 m_W^2} \ln\left[\frac{M_{SUSY}^2}{m_t^2}\right] < 1 \tag{3.8}$$

or $m_t \leq 6 m_W$ (for $M_{\tilde{Q}} \leq 1$ TeV). However, if we were to allow $M_{SUSY} \gg 1$ TeV then the validity of our result would already be suspect for much smaller values of m_t. We therefore shall sum the leading logs to all orders using the technology of the renormalization group equations.

4. Renormalization Group Improvement

We have seen that in MSSM the Higgs self-coupling constant and the gauge coupling constants are related [eq. (2.4)]. However, the parameters of any theory will in general depend on the energy scale ($\sqrt{s}$) at which they are evaluated. This dependence is described by the renormalization group equations (RGE):

$$\frac{\mathrm{d}}{\mathrm{d}t} p_i = \beta_i(p_1, p_2, ..), \text{ where } p_i = g_j^2 (j = 1,2,3), h_f^2, \lambda_k (k = 1, .., 4), \dots . \tag{4.1}$$

Here $t \equiv \ln(s)$, the Yukawa couplings are $h_f \equiv g_2 m_f/(\sqrt{2} m_W)$ and g_j ($j = 3,2,1$) are the gauge couplings of the $SU_c(3) \times SU_L(2) \times U_Y(1)$ gauge groups. Consider first the case of an unbroken supersymmetric theory. The fact that eq. (2.4) is true for arbitrary s implies that the β-functions have to satisfy analogous equations:

$$\begin{aligned} \beta_{\lambda_1} &= \beta_{\lambda_2} = -\tfrac{1}{2}\beta_{\lambda_3} = \tfrac{1}{8}[\beta_{g_2^2} + \beta_{g_1^2}] \\ \beta_{\lambda_4} &= \tfrac{1}{2}\beta_{g_2^2} \end{aligned} \tag{4.2}$$

These β-functions obtain contributions from all the particles of the theory that are present at the scale $\sqrt{s}$. If soft SUSY breaking terms are included, then the particles and their superpartners will no longer be mass-degenerate. As a result the β-functions at a scale below M_{SUSY} will no longer satisfy eq. (4.2) and the gauge coupling constants and the self-coupling constant will evolve differently. In principle, all $SU_c(3) \times SU_L(2) \times U_Y(1)$ multiplets could lie at a different mass scale (above the weak scale the members of these multiplets are mass-degenerate). As $\sqrt{s}$ decreases one would have to modify the β-functions every time a new multiplet disappears. However, we shall only discuss two possible scenarios that are particularly interesting. Probably the most natural scenario is the one where all the soft SUSY mass parameters are of the order of M_{SUSY}. In this case only the SM particles are present at the weak scale. The β-functions are then well-known [9]. This case has already been investigated in ref. 10. We are more interested in the scenario that yields the maximum Higgs mass. This limit is achieved by making all the bosons (fermions) as heavy (light) as possible with the exception of the top-quark, where the quartic increase with the Yukawa-coupling dominates over the logarithmic decrease with m_t. Our low energy theory will then consist of the SM particles plus gauginos ($\widetilde{W}^a, \widetilde{B}$) and Higgsinos ($\widetilde{H}_1, \widetilde{H}_2$). To arrange this, we require that the gaugino masses M_k ($k = 1,2$) and the supersymmetric Higgsino mass parameter μ are all of order the weak scale. The additional interaction terms are:

$$\Delta\mathcal{L} = \frac{i\tilde{g}_2}{\sqrt{2}}\tau^a_{ij}\widetilde{W}^a \widetilde{H}^i_2 H^{j*}_2 + \frac{i\tilde{g}_1}{\sqrt{2}}\delta_{ij}\widetilde{B}\widetilde{H}^i_2 H^{j*}_2 \,, \tag{4.3}$$

with the boundary conditions:

$$\tilde{g}_j(M_{SUSY}) = g_j(M_{SUSY})\,, \qquad (j = 1,2)\,. \tag{4.4}$$

The β-functions in this model are:

$$
\begin{aligned}
48\pi^2\beta_{\widetilde{g}_1^2} &= \Big[(\tfrac{1}{2}N_d + \widetilde{N}_{\widetilde{d}})\widetilde{g}_1^2 + 3N_c h_t^2 + \mathcal{O}(g_1^2 - \widetilde{g}_1^2)\Big]\widetilde{g}_1^2 \\
48\pi^2\beta_{\widetilde{g}_2^2} &= \Big[(\tfrac{1}{2}N_d + \widetilde{N}_{\widetilde{d}} + 4N_{\widetilde{t}} - 22)\widetilde{g}_2^2 + 3N_c h_t^2 + \mathcal{O}(g_2^2 - \widetilde{g}_2^2)\Big]\widetilde{g}_2^2 \\
48\pi^2\beta_{g_1^2} &= (\tfrac{20}{3}N_g + \tfrac{1}{2}N_d + N_{\widetilde{d}})g_1^4 \\
48\pi^2\beta_{g_2^2} &= (4N_g + \tfrac{1}{2}N_d + N_{\widetilde{d}} + 4N_{\widetilde{t}} - 22)g_2^4 \\
48\pi^2\beta_{g_3^2} &= (4N_g - 33)g_3^4 \\
16\pi^2\beta_{h_t^2} &= (\tfrac{9}{2}h_t^2 - 8g_3^2 - \tfrac{9}{4}g_2^2 - \tfrac{17}{12}g_1^2 + \tfrac{3}{2}\widetilde{g}_2^2 + \tfrac{1}{2}\widetilde{g}_1^2)h_t^2 \\
16\pi^2\beta_\lambda &= 12\lambda^2 + \tfrac{3}{16}[2g_2^4 + (g_2^2+g_1^2)^2] - \tfrac{1}{4}(\widetilde{g}_2^2 + \widetilde{g}_1^2)^2 - \widetilde{g}_2^4 - N_c h_t^4 \\
&\quad - \tfrac{1}{2}\lambda(9g_2^2 + 3g_1^2 - 4N_c h_t^2 - 6\widetilde{g}_2^2 - 2\widetilde{g}_1^2)\,.
\end{aligned}
\tag{4.5}
$$

Here the number of generations is $N_g = 3$, the number of Higgs doublets present below the scale $\sqrt{s}$ is $N_d = 1$, the number of Higgsino doublets is $N_{\widetilde{d}} = 2$ and the number of fermionic triplets is $N_{\widetilde{t}} = 1$. In addition the number of complete Higgs-Higgsino doublets is denoted by $\widetilde{N}_{\widetilde{d}} = 1$ [note that $\widetilde{N}_{\widetilde{d}} \leq \min\{N_d, N_{\widetilde{d}}\}$]. Since the low energy theory below M_{SUSY} possesses only one Higgs doublet, there is only one Higgs self-coupling $\lambda = \lambda_2$. Notice that $8\beta_\lambda = \beta_{g_1^2} + \beta_{g_2^2}$ if we impose the supersymmetric boundary conditions and set $h = N_g = 0$. The only remnant of SUSY will be the boundary condition eq. (2.4). The tree-level relations for the masses are:

$$
m_{h^0}^2 = 4\lambda(m_{weak}^2)v^2, \qquad m_Z^2 = \tfrac{1}{2}\Big[g_1^2(m_{weak}^2) + g_2^2(m_{weak}^2)\Big]v^2 \tag{4.6}
$$

and the Higgs mass shift to all orders in the leading logs is:

$$
(\Delta m_{h^0}^2)_{RGE} = m_{h^0}^2 - m_Z^2 = [4\lambda(m_{weak}^2) - \tfrac{1}{2}(g_2^2 + g_1^2)(m_{weak}^2)]v^2\,. \tag{4.7}
$$

Here λ and g_j ($j = 1, 2$) are the solutions of the RGEs with the boundary condition eq. (2.4) at the scale $\sqrt{s} = M_{SUSY}$. Note that M_{SUSY} is the only free parameter in the squark (slepton) sector. The RGEs [eq. (4.5)] can be solved iteratively. To first order in the leading logs the solutions are:

$$\begin{aligned}\lambda(M_{SUSY}) &= \lambda(m^2_{weak}) + \beta_\lambda t,\\ [g_1^2+g_2^2](M_{SUSY}) &= [g_1^2+g_2^2](m^2_{weak}) + (\beta_{g_1^2}+\beta_{g_2^2})t,\\ \text{where } t &\equiv \ln\left[\frac{M^2_{SUSY}}{m^2_{weak}}\right].\end{aligned} \tag{4.8}$$

The shift of the Higgs mass to this order is:

$$(\Delta m^2_{h^0})_{LL} = t\left[4\beta_\lambda - \tfrac{1}{2}(\beta_{g_1^2}+\beta_{g_2^2})\right]v^2 = \frac{g_2^2 m_Z^2}{16\pi^2 c_W^2}t\left[6\frac{m_t^4}{m_Z^4} - 3\frac{m_t^2}{m_Z^2} + 7s_W^4 + \tfrac{13}{3}c_W^4\right] \tag{4.9}$$

Here we have simply recovered the leading logs of eq. (3.1) and of the second term in eq. (3.7). Additionally, there will be a finite contribution to the effective Higgs self-coupling constant and the effective gauge coupling constants from the A-parameter. This can be included by modifying the boundary condition at the scale $\sqrt{s} = M_{SUSY}$:

$$\lambda = \tfrac{1}{8}(g_1^2+g_2^2) + \frac{1}{128\pi^2}\sum_U N_c\left(\frac{a_U A_U^4}{m^4_{\tilde{Q}}} + \frac{b_U A_U^2}{m^2_{\tilde{Q}}}\right), \tag{4.10}$$

where a_U and b_U are defined as in eq (3.4). This result has been deduced by comparison with the one-loop result in eq. (3.3). There is no contribution from the bottom squark to this order since $\mu \ll M_{SUSY}$. This shift of the effective coupling constants has also been obtained explicitly by evaluating the corresponding Feynman diagrams [6]. If we evaluate the Higgs mass shift [eq.(4.7)] with the modified boundary conditions [eq. (4.10)] we find that it agrees with the one-loop result in eq. (3.3) to first order in perturbation theory. However, there are two important differences: first, all the parameters in eq. (4.10) are evaluated at the SUSY scale and the ones in eq. (3.3) are evaluated at the weak scale. Second, the modification of the boundary condition will implicitly change β_λ and thus also the leading log result.

Our goals now are twofold. First, we want to replace the leading log approximations [eq.(3.1) and (3.7)] by the sum of the leading logs to all orders and second,

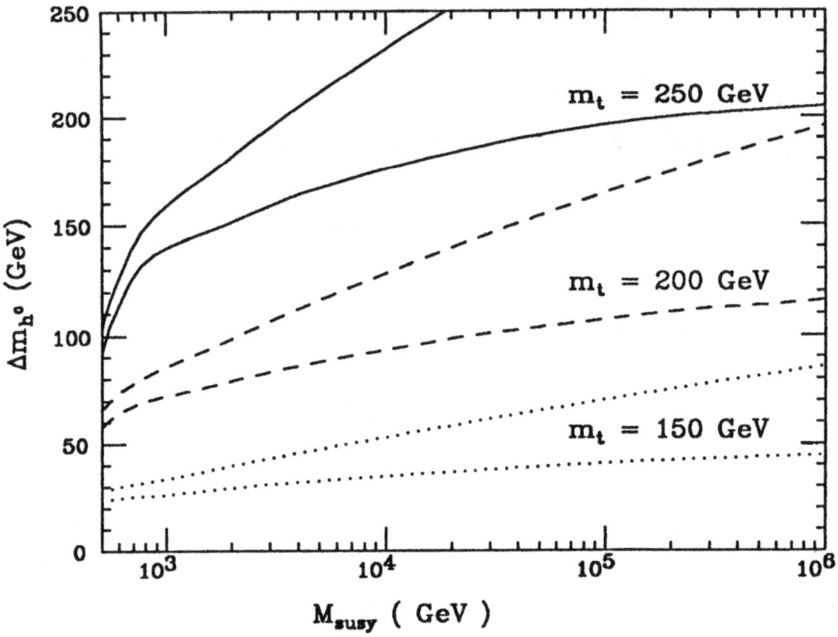

Fig. 9. The radiative corrections to the upper limit of m_{h^0}. The solid, dashed and dash-dotted curves denote the shift of the Higgs mass for $m_t = 150$, 200 and 250 GeV. The A-parameter is $A_U^2 = 6 - 2m_Z^2/m_t^2$. The larger of the two values correspond to the pure one-loop corrections. The smaller values are the one-loop corrections including RGE-improvement.

we want to include the contributions of the soft SUSY-breaking terms (that means the A-parameters) at the SUSY scale where they are generated rather than at the weak scale. That means, that the RGE improvement in eq. (2.2) is obtained by taking the solutions of the RGEs with the boundary condition eq. (4.10) and subtracting the one-loop leading logs $[(\Delta m_{h^0}^2)_{LL}]$ as well as the leading order mass shift due to the A-parameter $[(\Delta m_{h^0}^2)_U]$ given in eq. (3.3):

$$(\Delta m_{h^0}^2)_{RGIMP} = (\Delta m_{h^0}^2)_{RGE} - (\Delta m_{h^0}^2)_{LL} - (\Delta m_{h^0}^2)_U . \tag{4.11}$$

In fig. 9 we have plotted the mass shift $\Delta m_{h^0} \equiv m_{h^0} - m_Z$ as a function of the supersymmetry breaking scale M_{SUSY} for $m_t = 150$, 200 and 250 GeV. Here we have simply taken m_{weak}^2 to be m_t^2 since the largest corrections arise from the top-stop sector, and all SUSY breaking parameters have been set equal to M_{SUSY}. The effect of RGE-improvement is to flatten out the logarithmic increase with the scale.

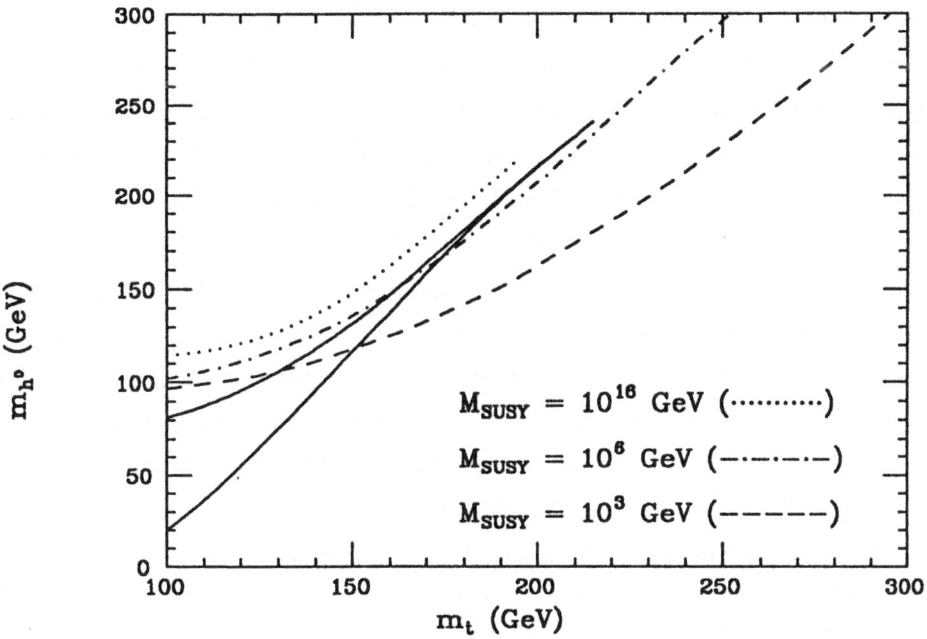

Fig. 10. The upper limit of m_{h^0} as a function of m_t for $M_{\tilde{Q}} = M_{\tilde{U}} = M_{\tilde{D}} = M_{SUSY} = 10^3, 10^6$ and 10^{16} GeV (dashed, dot-dashed and dotted, respectively). Here the chargino and the neutralino masses are of order the weak scale [$(M_2 = \mu = \mathcal{O}(m_Z)$]. The dotted curve is the solution of the RGEs and thus only contains the leading logs to all orders. As the boundary condition we have chosen the largest possible value of λ in the MSSM given in eq. (4.10). This curve ends at 195 GeV because the top-Higgs Yukawa coupling diverges at some energy scale below M_{SUSY}. The solid lines correspond to the solution of the RGEs with only the SM particles present below $M_{SUSY} = 10^{16}$ GeV. In the upper solid curve $\lambda(M_{SUSY})$ is defined as in eq. (4.10). The lower solid curve is the lower limit of the Higgs mass in the SM. It is derived from the requirement that λ remain positive all the way up to the GUT-scale

In fig. 10 we have plotted the one-loop radiatively corrected upper limit of the lightest Higgs mass including RGE-improvement as a function of the top quark mass. The A-parameter is chosen such that it yields the maximum mass shift. The charginos and neutralinos are kept at the weak scale ($\mu = M_2 = m_Z$) and the second Higgs doublet and all the sfermion masses are assumed to be at the SUSY scale [$M_{SUSY} = M_{\tilde{Q}} = M_{\tilde{U}} = M_{\tilde{D}} = m_{A^0} = 10^3$ (10^6) GeV for the dashed (dot-dashed) graph]. The solid and dotted curves are not the full one-loop corrections but are the solutions of the RGEs. In the dotted curve the parameters are chosen as above except that the SUSY scale is moved all the way up to the

GUT scale $M_{SUSY} = M_{GUT} = 10^{16}$ GeV. It is interesting to note that the upper limit of m_{h^0} increases only by about 20 GeV by going from $M_{SUSY} = 10^6$ GeV to $M_{SUSY} = 10^{16}$ GeV. Furthermore, in this scenario we leave the perturbative regime if the top quark mass becomes larger than 195 GeV, as the top Yukawa coupling develops a Landau pole before the GUT-scale is reached. In the two lower curves all the superpartners acquire masses of order M_{SUSY}. The only difference is the boundary condition for λ; in the upper curve λ starts at its maximum value [eq. (4.10)] and in the lower curve λ is chosen to be minimal such that it never goes below zero. This last curve is a lower limit of the Higgs mass in the SM. It is derived from the requirement that minimum of the potential at $v = \sqrt{2}m_W/g_2$ is the absolute minimum, that is λ is positive all the way up to the GUT scale. These two graphs converge for large top quark masses. This can be explained by examining the dominant contributions to β_λ. Here it is important to note that the self-coupling constant λ becomes large at low energies and has to be included. Thus we find:

$$16\pi^2 \frac{\mathrm{d}}{\mathrm{d}t}\lambda = 12\lambda^2 - 3h_t^4 + 6h_t^2\lambda \tag{4.12}$$

We see that λ can only grow for decreasing t as long the right-hand side of eq. (4.12) in negative. Therefore this equation has an infrared fixed point when the right-hand side vanishes. This happens for $h_t^2 = (1+\sqrt{5})\lambda$ or $m_{h^0} = 1.11 m_t$. This crude estimate is in good agreement with the solid and dotted curves in fig. 10.

5. Conclusions

We have calculated the full one-loop radiative correction to the upper limit of m_{h^0} in the minimal supersymmetric model (MSSM). The dominant contributions to these corrections grow logarithmically with the masses of the superpartners. This shift of a particular supersymmetric sector is positive (negative) if the heavy partners are bosons (fermions). Summation of leading logs to all orders in perturbation theory is achieved by employing RGE techniques. Including the RGE-improvement the mass shift actually grows less quickly than $\ln(M_{SUSY}/m_Z^2)$ and eventually converges to a fixed point. This behavior suggests that even with fine-tuning, the mass of the lightest Higgs boson can not exceed the top mass by more than roughly 15%. It is interesting to note that this bound is not far above the lower limit of the Higgs mass in the SM coming from stability constraints of the electroweak vacuum. Thus the overlap of the regions in the m_t–m_{h^0} plane compatible with both MSSM and the SM (assuming that there is no additional new physics between the weak scale and the GUT scale) is rather small. This gives hope for the possibility that with the discovery of the t-quark and h^0 at least one of these two models can be ruled out.

Acknowledgement

I would like to thank Howard Haber for his guidance and contributions to this paper as well as his editorial advice. I would also like to thank Wolfgang Hollik for inviting me to present this talk, and I am grateful to the faculty of the Max-Planck-Institute for their warm hospitality during my stay.

This work was supported in part by the United States Department of Energy.

REFERENCES

[1] See for example, L. Susskind, *Phys. Rep.* **104** (1984) 181.

[2] H.P. Nilles, *Phys. Rep.* **110,** (1984) 1;
H.E. Haber and G.L. Kane, *Phys. Rep.* **117,** (1985) 75;
R. Barbieri, *Riv. Nuovo Cimento* **11** (1988) 1.

[3] For a review and a guide to the literature, see: J.F. Gunion, H.E. Haber, G.L. Kane, and S. Dawson, *The Higgs Hunter's Guide,* (Addison-Wesley, Redwood City, CA, 1990).

[4] H.E. Haber and R. Hempfling, *Phys. Rev. Lett.* **66** (1991) 1815.

[5] Y. Okada, M. Yamaguchi and T. Yanagida, *Prog. Theor. Phys.* **85** (1991) 1;
J. Ellis, G. Ridolfi and F. Zwirner, *Phys. Lett.* **B257** (1991) 83;
J. Ellis, G. Ridolfi and F. Zwirner, *Phys. Lett.* **B262** (1991) 477;
A. Yamada, *Phys. Lett.* **B263** (1991) 233;
P.H. Chankowski, S. Pokorski and J. Rosiek, preprint MPI/91-57 (1991).

[6] Okada, M. Yamaguchi and T. Yanagida, *Phys. Lett.* **B262** (1991) 54.

[7] H. E. Haber and R. Hempfling SCIPP–91/38 (1991).

[8] J.R. Espinoza and M. Quiros *Phys. Lett.* **B267** (1991) 271.

[9] N. Cabibbo, L. Maiani, G. Parisi, R. Petronzio *Nucl. Phys.* **B158** (1979) 295.

[10] R. Barbieri, M. Frigeni, F. Caravaglios, *Phys. Lett.* **B258** (1991) 167.

The Standard Model with Minimal Dynamical Electroweak Symmetry Breaking

C. E. M. Wagner
Max Planck Institut für Physik
Föhringer Ring 6, D-8000 München 40, Germany

Abstract

The interesting possibility that the spontaneous breakdown of the electroweak symmetry relies on the formation of condensates of the quarks of the third generation was recently pointed out by several authors. In these lectures we discuss the recent efforts towards the understanding of the physical implications of such a possibility. The minimal class of models are just equivalent to the Standard Model with specific ultraviolet boundary conditions, the so called compositeness conditions, on the renormalization group trajectories of the physical couplings. Sharp low energy predictions for the top quark and Higgs masses appear within this framework. The possible generalizations of these models are discussed and different proposed dynamical scenarios for the generation of the effective short distance interactions are described.

1 Introduction

Recent experimental data has confirmed to a very high accuracy level the predictions of the Standard Model, a gauge theory based on the symmetry group $SU(3)_C \times SU(2)_L \times U(1)_Y$ [1]. The spectrum is replicated in three families of fermion fields, with each family transforming with equal quantum numbers under the symmetry group. Indeed, although one member of the third family, the so called top quark, has not been observed until now, precise experimental measurements imply not only its presence but also the range of allowed values for its mass [2].

The symmetry group of the Standard Model is spontaneously broken to $SU(3)_C \times U(1)_{em}$ through the so called Higgs mechanism. For the minimal physical realization of this mechanism in the Standard Model, a scalar Higgs field, transforming as a doublet under $SU(2)_L$, is needed. The Higgs field acquires a vacuum expectation value, inducing the breakdown of the electroweak symmetry and giving masses to the gauge bosons through their gauge couplings. Within this framework, the three families of fermions acquire masses through their Yukawa couplings to the physical neutral Higgs field. These couplings are free parameters of the theory and their tree level values are independent of the gauge coupling values. Hence, the Standard Model provides no explanation for the striking hierarchy of masses between the fermions of the first and the third generations,

$$\frac{m_t}{m_e} > 10^5 \tag{1.1}$$

where m_t and m_e are the top quark and electron masses respectively. In fact, although the value of the electron mass is well below the weak scale, $m_e = 511$ KeV, the top quark is probably heavier than the vector bosons mediating the electroweak gauge interactions. As we will see below, the understanding of the heaviness of the top quark is one of the main physical motivations for the dynamical scenarios under discussion.

In spite of its increasing phenomenological success, the Standard Model provides no explanation for the family replication and the particular gauge symmetry structure of the theory, and gives no satisfactory answer to the difficult question of masses. An answer to these questions can only be provided if we assume the Standard Model to be an effective cutoff theory, correctly describing the particle phenomenology only up to a certain high energy scale Λ at which new physics should appear. In addition, a correct phenomenological description requires the inclusion of the gravitational interactions, which become relevant at energy scales of the order of the Planck scale, $M_P \approx 10^{19}$ GeV. Hence, the Planck scale is expected to provide the order of magnitude of the maximum energy scale until which the Standard Model may give a correct phenomenological description of the elementary particles interactions. For a heavy top quark or scalar Higgs, there is another reason to assume the appearence of new physics at some finite scale Λ, which is related to the nature of the continuum limit of the standard model. The continuum limit may be defined as the limit in which the cutoff is removed, that is the ratio of the physical masses to the cutoff goes to zero. In the standard model, the Higgs-Yukawa sector, as well as the $U(1)_Y$ gauge sector, is not asymptotically free. The Higgs Yukawa sector of the theory has an infrared fixed point at vanishing values of the Yukawa couplings and the quartic coupling of the scalar Higgs field. If no other fixed point is present in the theory, the coupling values will be tuned to their fixed point values in the continuum and the theory will be trivial, that is noninteracting. The $U(1)$ gauge - fermion sector has also an infrared fixed point at vanishing renormalized gauge coupling, and the same conclusions as for the Higgs Yukawa sector apply. In other words, if the renormalization group β function does not have an additional zero apart from the trivial one, for any nonvanishing value of the relevant couplings the theory will become strongly interacting at a finite energy scale Λ, signaling the presence of new physics. In these lectures we will call "triviality bound "on a given coupling, for an effective cutoff Λ, to the low energy value of this coupling for which the theory becomes strongly interacting at scales of order Λ.

The existence of new fixed points in a quantum field theory can only be investigaded through nonperturbative methods. Lattice studies of the Higgs Yukawa sector and of the noncompact version of QED have shown complex phase diagrams. In lattice QED, for example, a critical bare gauge coupling have been found, above which the global chiral symmetry of the massless theory is dynamically broken [3]. In the Higgs Yukawa sector other phases than the symmetric (or paramagnetic) and spontaneously broken (or ferromagnetic) phases appear [4],[5]. In spite of these interesting results, there is no hint of a nontrivial ultraviolet fixed point in the Higgs Yukawa theory. In the $U(1)$ gauge - fermion theory case, the analysis is more involved and a definitive answer to this question is still lacking.

An important remark is that, in the presence of asymptotically free couplings, a potentially trivial theory may have a nontrivial continuum limit if a mechanism of reduction of couplings is in effect. In the case of the Higgs Yukawa sector coupled to QCD, for example, a nontrivial continuum limit exists [6]. A reduction of couplings between the top quark Yukawa coupling, the scalar quartic coupling and the QCD gauge coupling occurs if the top mass is approximately 100 GeV. In this scheme, the top quark Yukawa coupling is attracted to a nontrivial fixed point value, which was first found in Ref. [7]. In addition, for moderate values of the top quark mass $m_t < 200$ GeV, the top quark Yukawa coupling becomes strong at energy scales far above the Planck scale, where, as we discussed above, the low energy description is expected to loose its validity. This is similar to what happens with the abelian gauge coupling, which is expected to become strong at very large scales, and hence the so called Landau singularity of the $U(1)$ running coupling constant is usually considered to be of theoretical, rather than phenomenological, interest. If, however, the top quark mass $m_t > 200$ GeV, the top quark Yukawa coupling β function becomes positive at large energy scales, reaching the strong coupling regime at an energy scale of the order or lower than the Planck mass. If, as suggested by lattice and $1/N_C$ studies, no nontrivial ultraviolet fixed point appears at large values of the Yukawa coupling constant, new physics should appear above an effective cutoff scale $\Lambda < M_P$. It is for these relatively heavy top quarks that the minimal dynamical breakdown of the electroweak symmetry takes place.

As we explained above, in the standard model the scale of the gauge bosons and fermions masses is determined by the vacuum expectation value of the Higgs field. Assuming the standard model to give a good phenomenological description up to scales of order Λ, a hierarchy problem arises if we wish the ratio of Λ to the gauge boson masses to be much larger than one [8]. The problem can be rephrased in what is called the fine tuning problem. This is based on the fact that, as will be discussed below, the fermionic radiative corrections to the scalar mass parameters are proportional to the cutoff squared

$$\Delta m^2 \approx -\alpha \Lambda^2 \tag{1.2}$$

where α is a number of order one. Hence, the mass parameter of the scalar field at the scale Λ, m^2, must be fine tuned to a very precise value in order to obtain

$$m^2 + \Delta m^2 = -\mathcal{O}(v^2) \tag{1.3}$$

as required by the standard model, where $v \simeq 175$ GeV, is the vacuum expectation value of the Higgs field. The fine tuning problem can only be solved by pushing the scale at which new physics appears to be of the order of 1 TeV. One of the most interesting ways of giving a solution to this problem is by considering the minimal supersymmetric extension of the standard model [9]. In this case, all quadratic divergences disappear and the fine tuning problem is solved. However, since the quadratic corrections to the scalar mass parameters are proportional to the scale of soft supersymmetry breakdown, the unobserved superpartners masses should be of the order of 1 TeV if a real solution to the fine tuning problem exists within this context.

The effective theory at the effective cutoff scale may include apart from the usual, renormalizable interactions, higher dimensional, usually called irrelevant, interactions. In general, the Lagrangian density may be rewritten as

$$\mathcal{L} = \mathcal{L}^{ren} + \mathcal{L}_I, \tag{1.4}$$

where $\mathcal{L}_I$ includes all irrelevant interactions. As an example, let us consider [10]

$$\mathcal{L}_I = \frac{K}{\Lambda^2}\phi^6 \tag{1.5}$$

in the scalar ϕ^4 theory. Upon renormalization these new irrelevant interactions induce finite corrections to the values of the renormalized couplings, Δm_ϕ^2 and $\Delta\lambda$, but no new physics is induced in the scaling region where cutoff effects may be neglected. However, irrelevant interactions can produce relevant effects, if their inclusion induce a phase transition at some nonvanishing critical value of their associated couplings. In this case, new physics may appear at the critical points, through relevant interactions of composite fields. One well known example of this is the so called Nambu Jona - Lasinio model [11],

$$\mathcal{L}_I = \frac{G}{\Lambda^2}(\bar{\psi}_L\psi_R)(\bar{\psi}_R\psi_L) \tag{1.6}$$

where $\mathcal{L}^{ren}$ includes the usual kinetic term for the massless fermion fields. In this case, a phase transition to a phase in which global chiral symmetry is broken, is induced if the four Fermi coupling G is above a critical value G_c. A complex composite scalar field $H = G\,\bar{\psi}_L\psi_R$, which includes a massless real goldstone boson and a real massive field, appears in the physical spectrum of the theory. In the broken phase the fermion acquires a nonvanishing mass, m_f, through the $<\bar{\psi}\psi>$ condensation. The scaling region, in which all physical masses are much lower than the cutoff scale Λ and the scalar field H becomes a dynamical field, is achieved by tuning G to its critical value. An important feature of this model is that its infrared structure is completely equivalent to the one of the Higgs - Yukawa theory [12]. In addition, in the broken phase, the relevant Yukawa couplings of the fermions with the scalar fields are strong whenever $\Lambda/m_f \leq 10^{15}$. This can be seen as a signature of the composite nature of the scalar field. The Nambu Jona Lasinio model provides the basis for the so called Top Condensate Models, which will be analysed in the next section.

2 Minimal Dynamical Electroweak Symmetry Breaking

As we discussed above, the generation of masses in the standard model relies on the presence of the scalar Higgs sector, which, together with the top quark, has eluded the experimental searches performed so far. Although in the standard model the physical Higgs is thought to be an elementary scalar particle, it is probably providing only an effective description of the more fundamental dynamics responsible for the electroweak symmetry breaking. In the most simple proposed dynamical scenarios, the scalar Higgs is replaced by a set of massless fermions coupled to a new strongly interacting gauge sector. The chiral symmetry breaking is induced by this new strong interactions, called technicolor due to the analogy with QCD, and the Higgs appears as a bound state of the new fermions, which are called techniquarks [13]. Although technicolor theories provide a solution to the fine tuning problem, it is difficult to incorporate a heavy top quark within this scheme. It would be interesting to find a dynamical scheme in which the breakdown of the electroweak symmetry would not require the addition of new fundamental fermions, and in which a heavy top quark could be incorporated in a natural way. The top condensate models are an interesting attempt to provide such a scheme.

The experimetal lower bound on the top quark, $m_t \geq 89$ GeV, has opened the window for a a strongly coupled Higgs - top quark sector. This motivated several authors to inquire into a possible relation between the heaviness of the top quark and the nature of the dynamics leading to the generation of masses in the standard theory. In fact, in analogy to what happens in the Nambu Jona Lasinio model, a strong Yukawa coupling could be the signature of a dynamical mechanism for the electroweak symmetry breaking which relies only on the observed quark and leptons of the standard model, and in which the Higgs field appears as a $t-\bar{t}$ bound state. The basic mechanism for the physical realization of this idea was first proposed by Nambu [14], by making an analogy between the spontaneous breakdown of the electroweak symmetry in the Standard Model and the BCS [15] mechanism in condensed matter theories. Several authors analysed the physical consequenses of such a scenario [16] - [17], and a detailed field theoretical analysis was first done by Bardeen, Hill and Lindner in Ref. [18]. They started with a gauged, $SU(3)_C \times SU(2)_L \times U(1)_Y$ invariant, Nambu Jona Lasinio model,

$$\mathcal{L} = \mathcal{L}_K^{\psi} + \mathcal{L}_{YM} + G\left(\bar{\psi}_L^c t_R^c\right)\left(\bar{t}_R^d \psi_L^d\right) \tag{2.1}$$

where $\mathcal{L}_K^{\psi}$ and $\mathcal{L}_{YM}$ are the kinetic terms for the fermion and Yang Mills fields respectively, $\psi_L^T = (t \quad b)_L$, t and b are the bottom and top quark fermion fields and the indices c and d indicate a sum over color degrees of freedom. In this first simplified formulation only the top quark acquires mass. The masses for the other fermion fields, however, may be generated by introducing the corresponding Yukawa couplings between the fermions and the scalar composite field [18] - [19].

If $G > 0$ the interactions are attractive, and for $G > G_c$, the local chiral symmetry of the theory is broken through a top condensate, $< \bar{t}t > \neq 0$. In the scaling region, a composite scalar doublet,

$$H = G\, \bar{t}_R \psi_L \tag{2.2}$$

appears in the spectrum of the theory. The quantum numbers of these composite fields are exactly equal to the ones of the elementary Higgs field in the standard model, and hence, for $G > G_c$ the $SU(2)_L \times U(1)_Y$ local symmetry is broken to $U(1)_{em}$. Three massless Goldstone bosons, associated to the breakdown of the gauge symmetry are induced, giving masses to the electroweak gauge bosons through the usual Higgs mechanism. In addition, a physical, electrically neutral scalar field appears in the spectrum of the theory. In general, the low energy spectrum is completely equivalent to the Standard Model one, although the reduction in the number of free parameters of the theory increases its predictability. In fact, as we will discuss below, for a given effective cutoff scale Λ, sharp predictions for the scalar Higgs and top masses can be made within this context.

2.1 Large N_C Analysis of the Top Condensate Model

The dynamical properties of the gauged Nambu Jona Lasinio model can only be explored by using nonperturbative methods. A systemathical analysis can be done, for example, by solving the self consistent Schwinger Dyson equations of the theory in the large N_C approximation, where N_C is the number of colors [18]. The critical four Fermi coupling, for example, may be estimated by solving the self consistency equation for the top quark mass. We will first study the model in the so called bubble approximation, that is the large N_C limit, for vanishing $SU(3)_C$ gauge coupling value. The dynamical effects of the inclusion of the $SU(3)_C$ interactions will be discussed below.In this approximation, we obtain

$$m_t = \frac{GN_C}{8\pi^2}\left(\Lambda^2 - m_t^2 \log\left(\frac{\Lambda^2}{m_t^2}\right)\right) m_t \tag{2.3}$$

Hence, for a nontrivial solution of the gap equation, $m_t \neq 0$, the top quark mass is given by

$$m_t^2 \log\left(\frac{\Lambda^2}{m_t^2}\right) = \Lambda^2 - \frac{8\pi^2}{N_C G} \tag{2.4}$$

Observe that, since the left hand side is positive a nontrivial solution only exists if $G > G_c$, with $G_c = 8\pi^2/N_C\Lambda^2$. Since the $\log(\Lambda/m_t)$ is only a slowly varying function of m_t, the natural scale for the fermion mass in the broken phase is just the cutoff scale. A large hierarchy between the cutoff scale and the fermion mass scale requires a very precise fine tuning of the four Fermi coupling to it critical value, which is nothing but the usual fine tuning problem of the standard model.

The relation between this model and the standard Higgs Yukawa model becomes apparent if we rewrite the Lagrangian density in an equivalent form, by the introduction of an auxiliary scalar doublet H

$$\mathcal{L} = \mathcal{L}_K^{\psi} + \mathcal{L}_{YM} + \bar{\psi}_L^b t_R^b H + H^\dagger \bar{t}_R^b \psi_L^b - M_0^2 H^\dagger H, \tag{2.5}$$

where $M_0^2 = 1/G$ and H can be eliminated through a Gaussian integration, or equivalently by its replacement through its equation of motions $H = G\ \bar{t}_R \psi_L$. At this level, the scalar field H is a static field, with no independent dynamics. The physical picture changes, however, once the quatum fluctuations of the fermion fields are taken into account. In the bubble approximation, for example, the scalar fields propagate through fermion "bubbles ". The propagator of the scalar field H, $D^{-1}(p)$, may be obtained by computing the bubble function with external momentum p through the relation[20]

$$D^{-1}(p) = \frac{1}{G} + B(p) \tag{2.6}$$

where $B(p)$ is the bubble function. A nonvanishing kinetic term for the unrenormalized scalar field H appear, together with a correction to its physical mass. The function $B(p)$ is quadratically divergent, but the quadratical divergences are cancelled once the gap equation is taken into account. Observe that the fermion mass is nothing but the vacuum expectation value of the electrically neutral, CP even component of the scalar field $< H^0 >$ and hence once the gap equation is fulfilled the quadratical divergences of the scalar propagator are automatically cancelled.

The propagator of the neutral scalar field H^0 may be explicitly computed, giving

$$D^{-1}(p) = \frac{N_C}{8\pi^2}\left(p^2 - 4m_t^2\right)\log\left(\frac{\Lambda^2}{m_t^2}\right) + \chi(p^2) \tag{2.7}$$

where for $p^2 = \mathcal{O}(m_t^2)$ and $\Lambda \gg m_t$, the function $\chi(p^2)$ is negligible. Hence, the neutral scalar field propagator has a pole at

$$m_{H^0} = 2\, m_t. \tag{2.8}$$

We would like to emphasize that this prediction was obtained in the context of the bubble approximation where important effects, like the QCD corrections, have been neglected. Although these corrections do not change the qualitative physical picture, they have an important incidence on the quantitative relations between physical couplings and masses.

The effective scalar potential for the unrenormalized field H^0

$$V(H^0) = \frac{1}{G}\left(H^0\right)^2 + W(H^0) \tag{2.9}$$

may be computed by making use of the relation[21]

$$\frac{\partial W(< H^0 >)}{\partial < H^0 >} = < \bar{t}t > \tag{2.10}$$

where, from the gap equation

$$< \bar{t}t >= -\frac{N_C}{4\pi^2}\left(\Lambda^2 - m_t^2 \log\left(\frac{\Lambda^2}{m_t^2}\right)\right) m_t. \tag{2.11}$$

Hence, the effective potential reads

$$V(H^0) = \left(\frac{1}{G} - \frac{1}{G_c}\right)\left(H^0\right)^2 + \frac{N_C}{16\pi^2}\left(H^0\right)^4\left\{\log\left(\frac{\Lambda^2}{(H^0)^2}\right) + \frac{1}{2}\right\} \tag{2.12}$$

The minimization of the effective potential, reproduces the gap equation for the vacuum expectation value of the neutral Higgs component $< H^0 >= m_t$. From the effective potential it is clear that for $G < G_c$ the minimum of the potential is at vanishing $< H^0 >$, while for $G > G_c$, nonvanishing values of $< H^0 >$ are preferred. Finally, for $\Lambda \gg m_t$, the effective quartic coupling for the unrenormalized scalar field, $(H^0)^4\lambda_0/2$ is given by

$$\lambda_0 = \frac{N_C}{4\pi^2}\log\left(\frac{\Lambda}{m_t}\right) \tag{2.13}$$

2.2 Electroweak Gauge Bosons and the Higgs Mechanism

Apart from the physical neutral field H^0, three massless Goldstone bosons appear in the spectrum, both in the pseudoscalar channel $A = G\,\bar{t}\gamma_5 t$ as in the charged channels $H^+ = G\,\bar{b}_L t_R$ and $H^- = G\,\bar{t}_R b_L$. Once the weak gauge couplings are restored in the model the electroweak gauge bosons acquire mass by absorbing the Goldstone bosons degrees of freedom. The quantum corrections induced by the fermion fields may be understood as the sum of the interactions of the gauge bosons with the renormalized Goldstone modes plus the single bubble polarization diagram, which restores the transversality of the propagator. The coupling constant of the interaction of the W boson with the renormalized charged Goldstone mode is nothing but the charged Goldstone decay constant, $f_W(p^2)$. Hence, the result is given by [18]

$$\frac{1}{g_2^2} D_{\mu\nu}(p)^{-1} = \left(p_\mu p_\nu/p^2 - g_{\mu\nu}\right)\left[\frac{p^2}{g_2^2(p^2)} - f_W^2(p^2)\right] \tag{2.14}$$

where the value of the gauge boson mass is given by the on -shell condition

$$M_W^2 = g_2^2(M_W^2) f_W^2(M_W^2) \tag{2.15}$$

In the bubble approximation the value of the zero momentum decay constant is given by

$$f_W^2(0) = \frac{1}{4\sqrt{2}G_F} \approx \frac{N_C}{(32\pi^2)} m_t^2 \log\left(\frac{\Lambda^2}{m_t^2}\right). \tag{2.16}$$

A similar expression is obtained for the neutral gauge boson Z^0 with an effective gauge coupling equal to $(g_2^2 + g_1^2)^{1/2}$ and

$$f_Z^2(0) \approx f_W^2(0) \tag{2.17}$$

Hence, the tree level relation between the Z^0 and the $W^\pm$ masses is obtained. In addition, the standard value for the ρ parameter is also reproduced. The low energy electroweak predictions are hence equivalent to those of the Standard Model. However, from Eq. (2.16), a relation between the Fermi scale, the effective cutoff scale and the top quark mass is obtained. Fixing the Fermi scale to its physical value, we obtain a top quark mass of $m_t \approx 165$ GeV for a cutoff scale $\Lambda \approx 10^{15}$ GeV. Let us remark again that neither this result, nor the relation $m_H^0 = 2m_t$, can be considered as predictions of the model, since they were obtained in an approximation in which all gauge corrections, and the dynamical effects of the propagating scalar fields have been neglected. We will perform a detail analysis of these effects in the framework of the effective Lagrangian description.

3 Effective Lagrangian Analysis

For $\Lambda \gg m_t$, the values of the relevant quantities are dominated by large logarithms, and all physical results may be reproduced by doing an effective field theory analysis. We start with the Lagrangian density

$$\mathcal{L}(\Lambda) = \mathcal{L}_K^\psi + \mathcal{L}_{YM} + \bar{\psi}_L^b t_R^b H + H^\dagger \bar{t}_R^b \psi_L^b - M_0^2 H^\dagger H, \tag{3.1}$$

which characterizes the interactions at the large energy scale Λ. The effective theory at the low enrgy scale μ may be obtained by integrating out the short distance fermion effects, which in this context is equivalent to consider the quadratic and large logarithmic corrections induced by the fermion loops

$$\mathcal{L}(\mu) = \mathcal{L}_K^{\psi} + \mathcal{L}_{YM} + \bar{\psi}_L^b t_R^b H + H^\dagger \bar{t}_R^b \psi_L^b$$
$$+Z_H \left|\mathcal{D}_\mu H\right|^2 - \frac{\lambda_0}{2}\left(H^\dagger H\right)^2 - (M_0^2 + \Delta M^2) H^\dagger H, \tag{3.2}$$

where

$$\begin{aligned} Z_H &= \frac{N_C}{(4\pi)^2} \log\left(\frac{\Lambda^2}{\mu^2}\right) \\ \lambda_0 &= 2 Z_H \end{aligned} \tag{3.3}$$

while $\Delta M^2 \approx -1/G_c$. The values of the wave function renormalization constant and of the quartic couplings are normalized so that the effective Lagrangian coincides with Eq.(3.1) at $\mu = \Lambda$. This leads to the following boundary conditions

$$Z_H(\mu \to \Lambda) = 0, \qquad \lambda_0(\mu \to \Lambda) = 0 \tag{3.4}$$

which are called the compositeness conditions. Observe that for $\mu \approx m_t$ we recover the effective potential and the propagator for the physical scalar field that we obtained in the last section.

The Lagrangian can be rewritten in a more conventional way by normalizing the field H so that it has a canonical kinetic term, $H \to Z_H^{1/2} H$. In terms of the renormalized field, the Lagrangian may be rewritten as

$$\begin{aligned} \mathcal{L}(\mu) &= \mathcal{L}_K^{\psi} + \mathcal{L}_{YM} + h_t \bar{\psi}_L^b t_R^b H + h_t H^\dagger \bar{t}_R^b \psi_L^b \\ &+ \left|\mathcal{D}_\mu H\right|^2 - \frac{\lambda}{2}\left(H^\dagger H\right)^2 - m_H^2 H^\dagger H, \end{aligned} \tag{3.5}$$

where the renormalized couplings $h_t = Z_H^{-1/2}$ and $\lambda = Z_H^{-2}\lambda_0$. The compositeness conditions imply the divergence of the renormalized couplings when $\mu \to \Lambda$.

$$\begin{aligned} h_t^2 &= \frac{8\pi^2}{N_C} \frac{1}{\log\left(\frac{\Lambda}{\mu}\right)}, \\ \lambda &= \frac{16\pi^2}{N_C} \frac{1}{\log\left(\frac{\Lambda}{\mu}\right)} \end{aligned} \tag{3.6}$$

The physical Higgs and top quark masses are given by the on shell relations

$$m_t = h_t(m_t)\, v, \qquad m_{H^0}^2 = 2\,\lambda(m_H^0)\, v^2 \tag{3.7}$$

where v is the vacuum expectation value of the renormalized field $v \simeq 175$ GeV. Since in the bubble approximation the relation $\lambda(\mu)/(2h_t^2(\mu)) = 1$ is fulfilled, the relation $m_H^0 = 2m_t$ is recovered.

Since the results of this section only depend on the leading logarithmic corrections, they may be recovered by making use of the renormalization group equations for the renormalized couplings h_t and λ [22], in the bubble approximation,

$$\begin{aligned} 16\pi^2\frac{dh_t}{dt} &= N_C h_t^3 \\ 16\pi^2\frac{d\lambda}{dt} &= -4N_C h_t^4 + 4N_C h_t^2\lambda. \end{aligned} \tag{3.8}$$

where $t = \log(\mu/\Lambda)$. If these equations are solved, with the compositeness boundary conditions for the renormalized couplings, Eq. (3.6) is recovered. Hence, the renormalization group technique provides a simple and powerful tool for the analysis of the physical implication of the theory in the infrared regime.

3.1 QCD Corrections

Although the bubble approximation gives a good qualitative description of the top condensate model, the QCD corrections should be taken into account to achieve correct phenomenological predictions. Actually, QCD effects play a very relevant role in the final quantitative predictions of the model. The inclusion of the strong gauge interactions should be considered as a first step for the analysis of the model, since the electroweak gauge interactions and dynamical Higgs fields effects must be included in order to obtain reliable phenomenological predictions. A large N_C analysis, in the context of ladder QCD was done in Ref. [23]. The top quark and Higgs masses may be computed by solving the corresponding Schwinger Dyson equations for the fermion and scalar propagators in the large N_C limit. As happens in the bubble approximation, in the scaling region the value of all the physically relevant quantities are dominated by large logarithms. Hence, the results of ladder QCD may be recovered by solving the improved renormalization group equations

$$\begin{aligned} 16\pi^2\frac{dh_t}{dt} &= N_C h_t^3 - (N_C^2 - 1)g_3^2 h_t \\ 16\pi^2\frac{dg_3}{dt} &= -\left(11 - \frac{4N_g}{3}\right) g_3^2 \\ 16\pi^2\frac{d\lambda}{dt} &= -4N_C h_t^4 + 4N_C h_t^2\lambda \end{aligned} \tag{3.9}$$

together with the compositeness boundary conditions, where N_g is the number of generations. (Note that, at this level, we already include the effect of the light quarks in the QCD running coupling constant).

For values of $\Lambda = \mathcal{O}(10^{15})$ GeV, the large N_C solutions are dominated by an infrared quasi-fixed point. This is an approximate fixed point of the theory, which would become exact only if the beta function of the QCD coupling were equal to zero [24]. More specifically, the relations implied by this quasi fixed point in the ladder QCD limit are $h_t^2 \approx (N_C - 1/N_C)g_3^2$ and $\lambda \approx g_t^2$. Since $N_C g_3^2$ is a slowly running coupling for energy scales μ of the order or greater than m_t, the running coupling constants are attracted to their infrared quasi fixed point values, whenever they are in their vicinities. This acutally happens for a large range of compositeness scales $\Lambda \simeq 10^{10} - 10^{19}$ GeV. Taking into account this nontrivial infrared structure, we obtain an approximate value for the top quark mass $m_t \approx 300$ GeV and the ratio $m_H/m_t = \sqrt{2}$. This shows, as we anticipated, that the QCD effects can produce significant quantitative corrections to the physical couplings.

It is important to remark that this infrared quasi fixed point value has nothing to do with the nontrivial infrared fixed point value found in Ref. [7]. As we discussed in the introduction, the Pendelton and Ross infrared fixed point, which is reached in the reduction of couplings scheme[6], is an exact solution to the renormalization group equations and occurs for somewhat smaller values of the top quark mass than the infrared quasi fixed point analysed above.

4 Improved Renormalization Group Analysis

The results of the last section can be improved by including the electromagnetic and weak gauge interactions, together with the dynamically generated scalar effects. This can be done by including nonleading order in $1/N$ effects in the self consistency equations for the scalar Higgs and top quark self energies. When the compositeness scale is much larger than the weak scale, the value of the relevant coupling is well determined by computing the leading logarithmic corrections. Hence, the results of this approximation can be reproduced by considering the full one loop renormalization group equations of the standard model [18]:

$$
\begin{aligned}
16\pi^2 \frac{dh_t}{dt} &= \left(\left(N_C + \frac{3}{2}\right) h_t^2 - (N_C^2 - 1) g_3^2 - \frac{9}{4} g_2^2 - \frac{17}{12} g_1^2\right) h_t \\
16\pi^2 \frac{dg_i}{dt} &= \beta_i g_i^3 \\
16\pi^2 \frac{d\lambda}{dt} &= 12\left(\lambda^2 + (h_t^2 - A)\lambda + B - h_t^4\right) \qquad (4.1)
\end{aligned}
$$

where $A = g_1^2/4 + 3g_2^2/4$, $B = g_1^4/4 + g_1^2 g_2^2/8 + 3g_2^4/16$, $\beta_1 = 41/6$, $\beta_2 = -19/6$ and $\beta_3 = -7$.

As in the ladder QCD case, the solutions to the renormalization group equations with the ultraviolet boundary conditions, are strongly focused to the infrared quasi-fixed point present in the theory. The infrared quasi fixed point value for the top quark Yukawa coupling is slightly lower in this case than in the pure ladder QCD case, mainly due to the dynamical Higgs contributions to the renormalization group equation of the top Yukawa coupling. In Fig. 1, we show the solution to the running top Yukawa coupling constant for different ultraviolet boundary conditions. Observe that, for a compositeness scale of the order of $\Lambda = 10^{10} - 10^{19}$ GeV, the top Yukawa coupling is strongly focussed to a small set of infrared values, with corresponding top quark masses of the order of 230 GeV.

The quartic coupling is also attracted to its infrared quasi fixed point value, which as can be seen from Eq.(4.1), gives a relation between the top quark Yukawa coupling to the quartic coupling which translates into a mass ratio $m_{H^0}/m_t \approx 1.1$ The exact numerical values for the top quark and Higgs masses obtained in the different approximations and for different values of the compositeness scale are shown in Table 1 [22].

$\Lambda(GeV)$	10^{19}	10^{15}	10^{11}	10^{7}
m_t (GeV) Bubble Sum	144	165	200	277
m_t (GeV) Planar QCD	245	262	288	349
m_t (GeV) Full RG Eq.	218	229	248	293
m_h (GeV) Full RG Eq.	239	256	285	354

Table 3.1 Predictions for the top quark mass m_t and the Higgs mass m_h in the different approximations described above.

The values for the top quark and Higgs masses obtained by using the full, one loop renormalization group equations are stable under variations of the compositeness scale Λ. It follows from Table 1 that, for $\Lambda \simeq 10^{15}$ GeV, these masses vary less than a 20 % under a variation of the compositeness scale of eight orders of magnitude. In general, for $\Lambda \leq 10^{19}$ GeV,

$$m_t \geq 210 GeV. \tag{4.2}$$

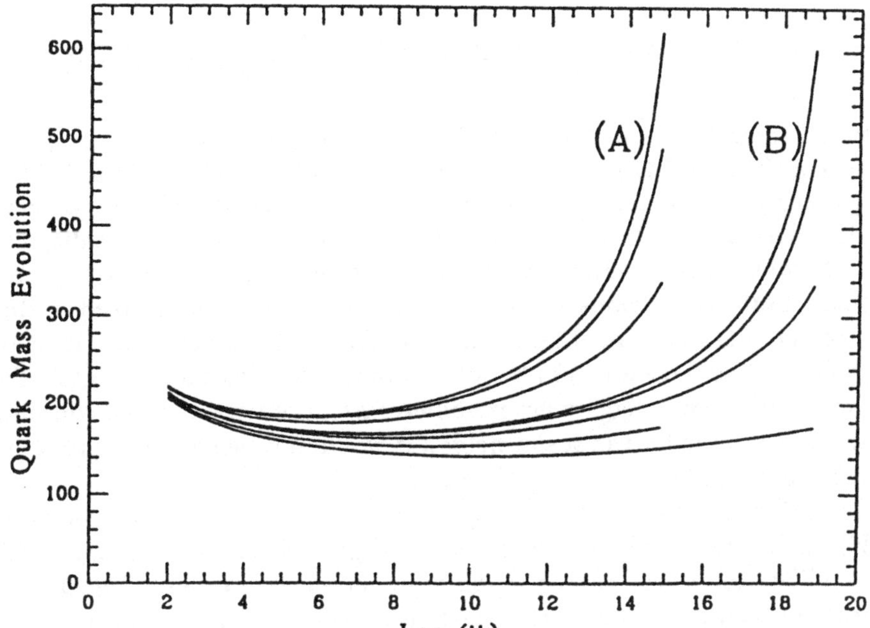

Fig. 1. Improved RG trajectories of the top quark mass ($h_t(\mu)\, v$) as a function of the scale μ, for a compositeness scale (A) $\Lambda = 10^{15}$ GeV and (B) $\Lambda = 10^{19}$ GeV [24].

As we discussed in section 1, for a given effective cutoff scale Λ, the triviality bound on the top quark may be defined as the value of m_t which is obtained assuming that the top quark Yukawa coupling becomes strong at scales of the order of Λ. Since in the dynamical scheme under consideration the remormalized coupling diverge at the compositeness scale, the top quark mass obtained within the top condensate model is consistent with the renormalization group trajectories associated with the triviality bounds on this quantity, for an effective cutoff scale equal to the compositeness scale. As is apparent from Fig. 1 [24], the presence of the infrared quasi fixed point, makes the value of this bound very insensitive to the exact large value of the top quark Yukawa coupling at the effective cutoff scale. The triviality bound on the top quark mass may be interpreted as the maximum allowed value of this quantity in any theory in which no new physics appear up to scales of order Λ, and hence is of great theoretical interest.

5 Sensitivity to the Inclusion of Higher Dimensional Interactions.

The theory described above may be generalized by introducing additional effective interactions at the scale Λ [25]. A general set of higer dimensional interactions can be introduced, so far they are consistent with the symmetries of the theory. The effective Lagrangian at the compositeness scale, for example, may be written as [26] - [27]

$$\mathcal{L} = \mathcal{L}_K^{\psi} + \mathcal{L}_{YM} + G\left(\bar{\psi}_L^c t_R^c + \frac{\chi}{\Lambda^2}\mathcal{D}^{\mu}\bar{\psi}_L^c \mathcal{D}_{\mu} t_R^c\right)\left(1 - f\frac{\mathcal{D}^2}{\Lambda^2}\right)\left(\bar{t}_R^d \psi_L^d + \frac{\chi}{\Lambda^2}\mathcal{D}_{\mu}\bar{t}_R^d \mathcal{D}^{\mu}\psi_L^d\right). \tag{5.1}$$

The additional interactions included in the above Lagrangian, Eq. (5.1), preserve the $SU(3)_C \times SU(2)_L \times U(1)_Y$ symmetry of the theory. The couplings G, f and χ can get arbitrary values so far they are consistent with the generalized gap equation of the model. The model above is expected to be in the same universality class as the gauged Nambu Jona Lasinio model. The additional interactions are expected to induce only finite corrections to the renormalized couplings derived within the framework of the Nambu Jona Lasinio model. Since the theory described by the above Lagrangian, has the same number of free parameters as the ϕ^4 Higgs - Yukawa theory (given by the bare Higgs mass parameter M_0, the bare Yukawa coupling and the bare quartic coupling for the scalar fields), it would not be surprising to find some kind of correspondence between both theories.

The qualitative behaviour of the theory can be studied by doing a large N_C analysis [26], [27]. The coupling χ changes the Higgs - fermion coupling, and hence produce a modification of the gap equation together with a finite modification of the induced wave function renormalization constant and of the induced quartic couplings for the scalar fields. For physical momenta $p^2 << \Lambda^2$, the coupling f induces an explicit kinetic term

$$\mathcal{L}_K^H = G_c^{-1}\frac{f}{\Lambda^2}\mathcal{D}^{\mu}H^{\dagger}\mathcal{D}_{\mu}H \tag{5.2}$$

for the composite field

$$H = G\left(\bar{t}_R^d \psi_L^d + \frac{\chi}{\Lambda^2} \mathcal{D}_\mu \bar{t}_R^d \mathcal{D}^\mu \psi_L^d\right). \tag{5.3}$$

The value of G_c may be computed by solving the self consistent equations for the fermion self energy $\sigma(k) = m_t(1 - \frac{\chi}{\Lambda^2}k^2)$. A nonvanishing solution is obtained if $G > G_c$, with

$$G_c^{-1} = \frac{N_C \Lambda^2}{8\pi^2}\left(1 - \chi + \frac{\chi^2}{3}\right). \tag{5.4}$$

Hence, in the bubble sum approximation the induced wave function renormalization constant reads

$$Z_H(\mu) = G_c^{-1}\frac{f}{\Lambda^2} + Z_H^\chi(\mu) \tag{5.5}$$

while the quartic coupling reads

$$\lambda_0(\mu) = \lambda_0^\chi(\mu) \tag{5.6}$$

where $Z_H^\chi(\mu)$ and $\lambda_0^\chi(\mu)$ are the values of the corresponding induced quantities when $f = 0$. These were first computed by W. Bardeen in Ref. [26], where he obtained

$$Z_H^\chi(\mu) = \frac{N_C}{8\pi^2}\left(\log\left(\frac{\Lambda}{\mu}\right) - \chi + \frac{\chi^2}{8}\right) \tag{5.7}$$

while

$$\lambda_0^\chi = \frac{N_C}{4\pi^2}\left(\log\left(\frac{\Lambda}{\mu}\right) - 2\chi + \frac{3\chi^2}{2} - \frac{2\chi^3}{3} + \frac{\chi^4}{8}\right). \tag{5.8}$$

As we anticipated, the couplings f and χ produce finite corrections to the leading logarithmic results. If we take arbitrary values for the couplings f and χ, we will also get arbitrary values for the renormalized couplings in the infrared. Hence, the generalized Nambu Jona Lasinio model is equivalent to the Higgs Yukawa model, and no real signature of compositeness can be obtained from the infrared structure of the theory [27] - [28].

The results of Ref. [27] imply that once the gauge interactions effects are added, the generalized top condensate model is just equivalent to the Standard Model. The sharp predictions obtained with the ultraviolet compositeness conditions on the renormalization group flow of the renormalized couplings are only consistent with a particular choice of the higher dimensional operators. In fact, it is impossible to determine the magnitude of the coupling associated with the higher dimensional operators without knowing the precise short distance dynamics leading to the effective interactions at the scale Λ. However, when the raio of Λ to the physical mass scale m_t is very large, the finite corrections induced by the higher dimensional interactions are small compared to the leading logarithmic contributions, unless the value of their associated couplings is unnaturally large.

The stability of the low energy predictions is largely enhanced once the strong gauge interactions are included. This is due to the action of the infrared quasi - fixed point in the theory, as has been shown by W. Bardeen in Ref. [26]. The results of his analysis, for $\Lambda \approx 10^{15}$ GeV. are shown in Fig. 2. Observe that, in the range of values under study, both the top quark and the Higgs masses are stable under variations of χ once the gauge and Higgs bound states interaction effects are included. These additional effects have been included by using the bubble approximation to compute the value of the renormalized couplings at scales $\mu \approx \Lambda/5$. These results have been used as new ultraviolet boundary conditions on the renormalization group flow of the relevant couplings, which run according to the the one loop renormalization group equations of the Standard Model. It follows from Fig. 2 that, for moderate values of the couplings associated with the higher dimensional operators, only a small correction to the minimal model predictions is obtained.

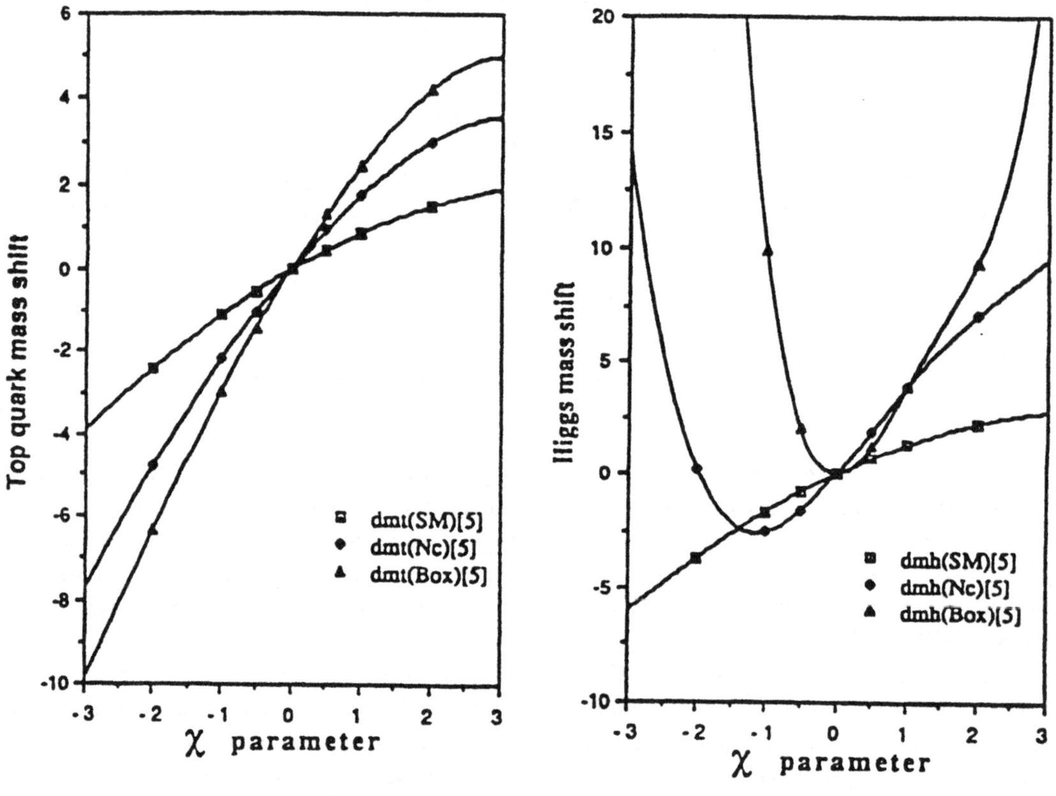

Fig. 2: Top quark and Higgs mass shifts as a function of the numerical parameter χ, for a compositeness scale $\Lambda = 10^{15}$ GeV. The triangles, circles and squares denote the results of the NJL, ladder QCD and the full RG equations respectively [26].

6 Dynamical Scenarios

In the last section we discussed the stability of the top quark and Higgs mass predictions under the inclusion of additional higher dimensional operators. The predictions are stable so far the couplings associated with these additional interactions are of order one. Hence, as we discussed in the last section, the values obtained in the analysis of Ref.[18] give the predictions for a general class of models with moderate values for the higher dimensional interactions couplings and a large compositeness scale Λ. This is a very beautiful property of these models, because it allows us to make low energy predictions without knowing the exact dynamical scenario related to the formation of the top condensate, so far there is a large hierarchy of scales between the weak scale and the compositeness scale. However, it is important to inquire about the nature of the high energy theory leading to the effective Nambu Jona Lasinio lagrangian at the scale Λ.

The simplest dynamical scenario is to assume that the four fermi interactions arise as the residual effect of some heavy gauge boson interactions at the scale Λ [29], [30]. This is based in the fact that a current current interaction like

$$\mathcal{L}_I \simeq -\frac{1}{2m_G^2}\,\bar{\chi}_L^i\gamma^\mu\chi_L^i\;\bar{\chi}_R^j\gamma^\mu\chi_R^j \tag{6.1}$$

where m_G is some heavy boson mass, may be rewritten, after applying a Fierz transformation, like

$$\mathcal{L}_I \simeq \frac{1}{m_G^2}\,\bar{\chi}_L^i\chi_R^j\;\bar{\chi}_R^j\chi_L^i \tag{6.2}$$

The easiest way of recovering the colour factors is hence to consider an SU(3) generalization of the Standard Model gauge group symmetry [30]. Thus, let us consider a theory, which at scales $\mu \gg m_G$ is symmetric under a symmetry group

$$G : SU(3)_1 \times SU(3)_2 \times SU(2)_L \times U(1)_Y \tag{6.3}$$

and we incorporate all the quark and leptons of the Standard Model, with the usual quantum numbers under $SU(2)_L \times U(1)_Y$. In order to break the symmetry group to the standard model one, we introduce a Higgs field Φ transforming as a $(3,\bar{3})$ under $SU(3)_1 \times SU(3)_2$ and which acquires a vacuum expectation value $<\Phi>\simeq$ diag(M,M,M).

We will assume that all leptons are singlets under the $SU(3)$ groups transformations, while the quarks transform as triplets. We will furthermore assume, for simplicity, that all quarks which couple to one of the $SU(3)$ gauge bosons do not couple to the other. After symmetry breaking we will have two vector mass eigenstates: A massless one, G_μ, identified with the gluon and a massive one, B_μ, which we will call, following Ref.[30], "coloron ":

$$\begin{aligned} A_{1,\mu} &= \cos\theta\, G_\mu - \sin\theta\, B_\mu \\ A_{2,\mu} &= \sin\theta\, G_\mu + \cos\theta\, B_\mu \end{aligned} \tag{6.4}$$

In order to preserve the observed QCD current we have to demand that $g_3 = h_1\cos\theta$ and $g_3 = h_2\sin\theta$, where h_i is the $SU(3)_i$ gauge coupling. The effective coupling of the fermions to the massive gauge boson is equal to $g_3\cot\theta$ for those ones coupled to $A_{2,\mu}$ and $g_3\tan\theta$ for those ones coupled to $A_{1,\mu}$. From Eq. (6.1) - (6.2), it is clear that if we want only the top quark to have an effective four Fermi coupling above its critical value, we have to assign the quantum numbers of the quarks in a way that does not respect the family structure of the standard model. One possible assignement is if the left handed top - bottom doublet ψ_L and the right handed top quark transform as triplets under $SU(3)_2$ and as singlets under $SU(3)_1$, while all other quarks and transform as triplets under $SU(3)_1$ and as singlets under $SU(3)_2$. This choice is not anomaly free, but anomalies may be cancelled by adding two additional heavy quarks, which do not affect the low energy dynamics of the theory [30]. In addition we require $\tan\theta << 1$ to obtain the desired hierarchy between effective four Fermi couplings. Ignoring $1/N_C$ corrections, the top - bottom - coloron interactions result in effective four Fermi interactions, just like in the minimal top condensate model, with a four Fermi coupling

$$G \simeq \frac{g_3^2 \cot^2\theta}{M_G^2}, \tag{6.5}$$

where $M_G = \sqrt{h_1^2 + h_2^2}\, M$, is the coloron mass. If these four Fermi interactions were the only relevant ones, the model would just reduce to the minimal top condensate model. The gap equation for the top quark mass would require a fine tuning of the effective coupling so that

$$\frac{N_C}{2\pi}\alpha_3(m_G)\cot^2\theta \approx 1 \tag{6.6}$$

which, for $N_C = 3$, and for large values of m_G, is consistent with our assumption on $\tan\theta$. The interesting question is what are the modification to this picture which are obtained when, instead of replacing the coloron effects by an effective four fermi coupling, we compute the gap equation and the scalar propagator in the presence of the strongly coupled, spontaneously broken, gauge interactions. This was done in Ref [30], with somewhat surprising results. The effects were not only small, but almost negligible. In fact, the modifications are equivalent to those ones which would be obtained through the addition of higher dimensional operators with an associated coefficient $\chi \simeq 1/8$, showing that at least in this class of models, the results of Ref. [18] are recovered.

Althogh the $SU(3)$ generalization to the standard gauge group is the most natural one regarding the color degrees of freedom, a model based on an $U(1)$ extension has been proposed [31]. The idea is to let the quarks and leptons to transform with asymmetric quantum numbers under the group $U_1(1) \times U(1)_2$, just like we did in the $SU(3)$ case, and to have the usual quantum numbers under $SU(3)_c \times SU(2)_L$. The advantage is that the quantum numbers may be selected so that only $< \bar{t}t > \neq 0$ while the anomalies of the theory are cancelled without the need of introducing additional heavy particles in the spectrum. The disadvantage is that, if the $U(1)$ currents transform as color singlets, the effective four fermi couplings differ from the ones of the model of Ref. [18], but they are given by Eq. (6.1), where i and j are color indices. It is easy to see that, in this case, no large N_C expansion for the gap equation is possible, since the quark of a given color receives a mass $m_a \simeq < \bar{\psi}^a \psi^a >$ where no sum over a is understood. One of the aims of Ref.[31] is the solution to the fine tuning problem by lowering the compositeness scale to values of the order of a few TeV. For these values of Λ and with no large N_c expansion at hand, all dynamical calculations have to rely on a truncation of the complete Schwinger Dyson equations, being difficult to estimate the accuracy of the approximation done to evaluate the phenomenological predictions of the model.

A qualitative analysis of the $U(1)$ extended model has been done in Ref [31]. If the mass of the $U(1)$ gauge boson is very large compared to the weak scale, the low energy predictions are expected to be equivalent to the ones obtained in the minimal top condensate model. If, instead, the compositeness scale is pushed not far away from the weak scale, the phenomenology of this model is rather different from the standard model one. In fact, apart from scalar Higgs particles, additional bound states are expected to appear, with masses close to the weak scale. These additional bound states can have important phenomenological consequences, like a modification of the ρ parameter value. If, as suggested in Ref. [31], a negative correction to the ρ parameter appears, the experimental upper bounds on m_t would be weaker than in the standard model case. In addition, since for values of Λ close to the weak scale the leading logarithms are of order one, higher dimensional operators can become relevant in this case, lowering the values for m_t in comparison to those ones obtained in the minimal model for the same values of Λ. In spite of their appeal, reliable computations will be needed to prove the validity of these attractive conjectures. An interesting feature of this model is the necessity of a strongly interacting $U(1)$ theory, which could make connection between the top condensate models and the conjectured nontrivial ultraviolet fixed point of the $U(1)$ theory coupled to fermions [32].

It would be interesting to find an embedding in which quarks and leptons would couple to the additional gauge fields in a way which preserves the usual family replication of quantum numbers [29]. In this case a 3 x 3 matrix $\mathcal{M}$ of up quark condensates, with equal entries in all the elements $\mathcal{M}_{ij}$ will appear, in a flavor democratic way. This matrix can be diagonalized by a rotation over the flavor indices. The diagonalized matrix has only a nonvanishing entry, which could be identified with the top quark condensate, and a massive quark to be identified with the top quark. A model based on this idea is presently under investigation [42].

7 Conclusions for the Minimal Model

In the minimal model, the breakdown of the electroweak symmetry is induced through a mechanism analogous to the BCS mechanism in condensed matter physics. The elementary Higgs field is replaced by effective attractive short range interactions which result in a propagating scalar bound state once the coupling is tuned to its critical value. The infrared structure of the model is equivalent to the standard model one. In fact, the top condensate model is just a particular limiting case of the Standard Model, which is reached for those values of the renormalized couplings which are consistent with the renormalization group trajectories associated with the triviality bounds on the top quark mass. Hence, a strong Yukawa coupling at the compositeness scale Λ appears as a clear signature of compositeness of the Higgs state. The model provides sharp predictions for the top quark and Higgs masses, due to the existence of an infrared quasi - fixed point. For compositeness scales $\Lambda = 10^{10} - 10^{19}$ GeV, the top quark mass $m_t \approx 230$ GeV and $m_H \approx 250$ GeV. The model is attractive not only because of its simplicity, but also because it gives a natural explanation for the relatively large value of the top quark mass.

The predictions of the Minimal Top Condensate Model can be modified by assuming the presence of additional higher dimensional interactions at the compositeness scale Λ. However, when there is a large hierarchy of scales between the effective cutoff and the weak scale, $\Lambda \gg m_t$, the low energy results are dominated by the large logarithmic corrections induced in the minimal gauged Nambu Jona Lasinio model, unless the couplings associated to the higher dimensional interactions are much larger than one. The presence of the infrared quasi fixed point enhance the stability of the low energy predictions under the variation of these additional couplings. Hence, the predictions of Ref. [18] correspond to a general class of models in which the effective theory at the compositeness scale Λ is characterized by attractive four - fermi interactions plus additional higher dimensional interactions with associated adimensional couplings with values of order one. In this sense, the results of Ref. [18] provide the most natural values for the top quark and Higgs masses within this scheme. This is in agreement with the results obtained in the low energy approach of Ref. [33].

Within the minimal top condensate model, the top quark mass value is somewhat larger than the preferred experimental values, coming from the ρ parameter measurements and direct top quark searches, 89 GeV $\leq m_t \leq$ 180 GeV [2]. If the top quark mass is indeed in this range of values, there are good chances to measure it at the Fermilab Tevatron within the next five years. If, instead, the Tevatron search for the top quark proves to be unsuccessful, the range of values predicted by the minimal top condensate models will remain as an interesting open possibility.

We presented different proposed dynamical scenarios for the generation of the effective interactions at the scale Λ. These are based on extensions of the standard model, to a strongly coupled gauge theory, whose symmetry group is spontaneously broken to $SU(3)_C \times SU(2)_L \times U(1)_Y$ at energies far above the weak scale. The four Fermi interactions appear through a Fierz transformation of the current - current interactions associated to the massive gauge boson sector. To generate the hierarchy of masses between the fermions of the first and third generation, an asymmetric assignement of quantum numbers under the additional gauge symmetry group is required. It would be interesting to find an embedding of the enlarged symmetry group in a somewhat larger symmetry group structure, which could give a natural explanation to this seemingly arbitrary assignement. In addition, unless the compositeness scale is pushed to values close to the weak scale, the gauge couplings have to be adjusted very precisely in order to generate the required hierarchy of scales. This is nothing but a refined version of the fine tuning problem.

A possible solution to the phenomenological problems related to the present experimental bounds on the top quark mass may be given by enlarging the model to include a fourth generation. In view of the present experimental bounds on the number of light neutrino families, a mechanism to generate masses for the neutrinos of the fourth generation should be incorporated. This can be done in a natural way, including new effective interactions for the fourth generation of neutrinos, while pushing the compositeness scale to be of the order of 10 TeV [34]. The fourth generation of quarks become heavy, but no effective bound on the top quark mass appears within this model.

There have been several additional extensions of the minimal model which have been studied in the literature, and that we will not discuss in detail. Two Higgs doublet models were analyzed in Refs. [35]-[36]. In Ref [38], a model with horizontal gauge symmetry was analyzed. The top quark mass predictions within this model resemble the ones obtained in Ref.[18]. In Ref.[37], it was shown that, if a somewhat more general set of effective four fermi interactions is included, the first and second generation of fermions can acquire masses in a radiative way, explaining not only the heaviness of the top quark, but also the hierarchy of masses, $m_t \gg m_e$. In addition, the top quark mass value can be considerably lower than the one obtained in the minimal model. The price to pay is the appearence of several scalar bound states, each of them arising through the fine tuning of a four fermi coupling to its critical value. It will be difficult to find a dynamical scenario in which these fine tunings may arise in a natural way.

Perhaps the simplest solution to the naturalness problem, keeping only three generations and a large hierarchy between the compositeness and the weak scales, relies on a supersymmetric extension of the model [39] - [40]. Although in a supersymmetric extension additional elementary scalar particles appear in the spectrum, they are not put by hand but they appear naturally through the requirement of preserving the global supersymmetry of the theory. Within the minimal supersymmetric top condensate model, the chiral Higgs multiplets appear as bound states of the third generation of quarks and their supersymmetric partners, without the necessity of fine tuning. In fact, in a softly broken supersymmetric theory the quadratical divergencies, which lead to the fine tuning problem in the Standard Model, are absent. As in the minimal case, the top quark mass values are strongly focussed to an infrared quasi fixed point, which is reached at somewhat lower values of the top Yukawa coupling than in the minimal case. In addition, the top quark mass is proportional to only one of the two Higgs vacuum expectation values contributing to the Z^0 mass, and hence can acquire significantly lower values than in the minimal case [40]. The SUSY extension of the minimal top condensate model will be the subject of the talk of M. Carena [41].

ACKNOWLEDGEMENTS

I would like to thank R. Bönisch, C. Hill, S. Love, and particularly my collaborators W. Bardeen, M. Carena, T. Clark and K. Sasaki for useful comments and discussions on the work reported here. I would also like to thank the organizers of the Workshop on *Phenomenology of Particle Physics* for inviting me to give this talk.

References

[1] S. L. Glashow, Nucl. Phys. 22 (1961) 579;
A. Salam in *Elementary Particle Theory* ed. N. Svartholm (Almquist and Wiksell, Stocholm 1968);
S. Weinberg, Phys. Rev. Lett. 19 (1967) 1264.

[2] See, for example, P. Langacker and M. X. Luo, Phys. Rev. D44 (1991) 817;
J. Ellis, G. Fogli and F. Lisi, Proceedings of the 1991 Lepton Photon Conference at CERN.

[3] J. B. Kogut, E. Dagotto and A. Kocic, Phys. Rev. Lett. 61 (1988) 2416; Nucl. Phys. B 317 (1989) 253; Nucl. Phys. B331 (1990) 500;
M. Göckeler, R. Horsley, E. Laermann, P. Rakow, G. Schierholz, R. Sommer and U. J. Wiese, Nucl. Phys. B334 (1990) 527.

[4] M. Lüscher and P. Weisz, Nucl. Phys. B290 [FS20] (1987) 25; Nucl. Phys. B295 [FS21] (1988) 65.

[5] A. M. Thornton, Phys. Lett. 214B, 577 (1988);
A. Hasenfratz, W. Liu and T. Neuhaus, Phys. Lett. 236B 229 (1990);
C. Wagner, Ph. D. Thesis, Hamburg University, DESY Preprint 89/083.
L. Lin, I. Montvay and H. Wittig, DESY Preprint 91/011 (1991).

[6] W. Zimmermann, Comm. Math. Phys. 97 (1985) 211;
J. Kubo, K. Sibold and W. Zimmermann, Phys. Lett. B200 (1989) 191;
J. Kubo, Max Planck Inst. preprint MPI - PAE/PTh 8/91 (1991).

[7] B. Pendelton and G. G. Ross, Phys. Lett. 98B, (1981) 291.

[8] See, for example, C. Llewellyn Smith and G. G. Ross, Phys. Lett. 105B (1981).

[9] For a review see: H. Nilles, Phys. Rep. 110C (1984) 1,
H. Haber and G. Kane, Phys. Rep. 117C (1985) 75.

[10] W. Bardeen, Talk presented at the Workshop on CP violation, Aspen, Colorado, July 1991.

[11] Y. Nambu and G. Jona - Lasinio, Phys. Rev. 122 (1961) 345.

[12] T. Eguchi, Phys. Rev. D14 (1976) 2755.

[13] For a review, see T. Appelquist, in *Dynamical Electroweak Symmetry Breaking*, Lectures presented at the VI Jorge Andre Swieca Summer School, Sao Paulo, Brazil, (1991).

[14] Y. Nambu, Enrico Fermi Institute Preprint 89-08(1989).

[15] J. Bardeen, L. Cooper and J. Schrieffer, Phys. Rev. 108 (1957) 1175.

[16] V. Miransky, M. Tanabashi and K. Yamawaki, Phys. Lett. B221(1989)177.

[17] W.J. Marciano Phys. Rev. Lett.62 (1989) 2793; Phys. Rev. D 41 (1990) 219.

[18] W.A. Bardeen, C.T. Hill and M. Lindner, Phys. Rev. D41 (1990) 1647.

[19] J. Fröhlich and L. Lavoura,Phys. Lett. B253 (1991) 218.

[20] W. Bardeen, C. Leung and S. Love, Nucl. PHys. B273 (1986) 649, Nucl. Phys. B 323 (1989) 493.

[21] W. Bardeen, S. Love and V. Miransky, Phys. Rev. D42 (1990) 3514.

[22] C. Hill, Fermilab prep. FERMI-CONF-90/170-T, to appear in the Proceedings of the Workshop on Dynamical Symmetry Breaking, Nagoya, Japan (1990).

[23] S. F. King and S. H. Mannan, Phys. Lett. B241 (1990) 241;
F. Barrios and U. Mahanta, Phys. Rev. D43 (1991) 79.

[24] C.T. Hill Phys. Rev.D 24 (1981) 691.
C. Hill, C. N. Leung and S. Rao, Nucl. Phys. B262(1985) 517.

[25] M. Suzuki, Mod. Phys. Lett. A5, 1205 (1990).

[26] W. Bardeen, Fermilab prep. FERMILAB - CONF - 90/269 - T (1990).

[27] A. Hasenfratz, P. Hasenfratz, K. Jansen, J. Kuti and Y. Shen, Nucl. Phys. B253 (1991) 218.

[28] J. Zinn Justin, Nucl. Phys. B367 (1991) 105.

[29] R. Bönisch, Phys. Lett. B268 (1991) 394; Univ. of Munich prep. LMU 91/03.

[30] C. Hill, Phys. Lett. B266 (1991) 419.

[31] M. Lindner and D. Ross, CERN - TH - 6179 -91, August 1991.

[32] V. A. Miransky, Nuovo Cim. 90A (1985) 149.
S. Raby and G. Giudice, preprint DOE-ER-01545-550 (1991).

[33] V. Zhakarov and E. Paschos, Max Planck Institut preprint, MPI/Ph-91/50.

[34] C. Hill, M. Luty and E. Paschos, Phys. Rev. D43 (1991) 3011.

[35] M. Luty, Phys. Rev. D41 (1990) 1863.

[36] M. Suzuki, Phys. Rev. D41 (1990) 3457.

[37] K. Babu and R. Mohapatra, Phys. Rev. Lett. 66 (1991) 556.

[38] T. K. Kuo, U. Mahanta and G. T. Park, Phys. Lett. B248 (1990) 119.

[39] W. A. Bardeen, T. E. Clark and S. T. Love, Phys. Lett. B 237 (1990) 235.

[40] M.Carena, T. Clark, C. Wagner, W. Bardeen and K. Sasaki, FERMILAB-TH-91-96-T (1991). To appear in Nucl. Phys. B.

[41] M. Carena, these proceedings.

[42] R. Bönisch, Private communication.

The Minimal Supersymmetric Standard Model with Dynamical Electroweak Symmetry Breaking

Marcela Carena

Max Planck Institut für Physik
Werner Heisenberg Institut
Föhringer Ring 6, D-8000 München 40
Fed. Rep. of Germany

Abstract

I analyze the minimal supersymmetric extension of the standard model, in the case in which the electroweak symmetry breakdown relies on the formation of condensates involving the third generation of quarks and their supersymmetric partners. Using Schwinger-Dyson equations as well as renormalization group techniques, I compute the mass of the top quark as a function of the ratio R of the Higgs vacuum expectation values, for different values of the compositeness scale Λ and the soft supersymmetry breaking scale Δ_S. Considering the renormalization group evolution of the scalar quartic couplings, I examine the particle spectra of the Higgs sector in the general case in which two light Higgs doublets appear in the low energy theory. The neutral CP even and charged Higgs masses are given as a function of R and the CP odd Higgs mass, for fixed values of Λ and Δ_S. For $\Lambda = 10^{16}$ GeV and $\Delta_S = 1$ TeV, the characteristic values of the top quark mass are 140 GeV $\leq m_t \leq$ 195 GeV, while the lightest CP even mass must be below 135 GeV. The m_t predictions are only slightly dependent on Δ_S. Moreover, in the dynamical scheme under study, the top quark and Higgs masses lie at the edge of their triviality bounds, setting constraints on the range of mass parameters consistent with possible grand unification scenarios.

1 Introduction

The standard model [1] provides a very good understanding of the strong and electroweak interactions and, so far, it has withstood all the experimental onslaughts. However, there are still a great number of open questions, which need to be answered. In particular, I would like to concentrate on the origin of masses of the fundamental excitations in the model. All these masses arise out of the presumption that the scalar Higgs boson acquires a nonvanishing vacuum expectation value inducing the spontaneous breakdown of the electroweak symmetry. Since the existence of the Higgs particle has not been proven until now, many other alternative explanations for the physical mechanism responsable for the electroweak symmetry breaking have been explored in the literature. Most of the proposed models require the existence of a host of new particles, which have not yet been experimentally observed. In this regard, the top quark condensate models have the beauty of relying only on the spectrum of particles which have already been detected. The basic idea, first suggested by Nambu [2], that the electroweak symmetry breaking could be dynamically generated through attractive fermion self interactions, was then studied by Bardeen, Hill and Lidner [3], and in a somewhat different presentation by Miransky, Tanabashi and Yamawaki [4], as well. Within the top quark condensate models the Higgs sector is replaced by the interactions between the fermions, and the scalar Higgs appears as a composite particle. In fact, based on the increasing experimental lower bound on the top quark mass, it has been realized that the top quark may be sufficiently heavy as to induce the formation of a condensate which catalyses the electroweak symmetry breakdown.

The physical consequences of the minimal dynamical electroweak symmetry breaking mechanism have been discussed in great detail by C. Wagner in the previous talk in this workshop [5]. Thus, I shall review very succinctly the main results and shortcomings which lead to the supersymmetric extension as an interesting alternative. The $SU(3)_c \times SU(2)_L \times U(1)_Y$ gauge invariant Nambu-Jona-Lasinio model [6] considers effective four fermion interactions between the dynamical quarks and leptons of the model. These interactions are thought to describe the residual effect of some unknown high energy dynamics which becomes relevant above the high energy scale Λ, which acts as an effective cutoff for the low energy theory. Due to its heaviness, which implies a relatively strong coupling to the composite Higgs field, the top quark plays an essential role in this dynamical scenario. The mechanism of the dynamical breakdown of the electroweak symmetry can be minimally described by keeping just an effective four fermion interaction, with coupling constant G, involving only the quarks of the third generation [3], [4]. The four fermion coupling constant acquires a critical value, G_C, which separates the region in coupling constant space in which the chiral $SU(2)_L \times U(1)_Y$ symmetry is preserved, from that one in which this symmetry is broken. In the broken phase a top condensate forms, $< \bar{t}t > \neq 0$, inducing mass for the fermions and the electroweak bosons. When the four fermion coupling is close to its critical value, a composite scalar Higgs appears at low energies as a new dynamical degree of freedom. The physical neutral component of the scalar Higgs field is a $t\bar{t}$ bound state, while three composite Goldstone bosons appear as a result of the gauge symmetry breakdown. The massless Goldstone bosons are then absorbed by the electroweak gauge bosons through the usual Higgs mechanism. Sharp predictions for the masses of the physical Higgs boson and the top quark appear in the above dynamical scheme.

The gauged Nambu-Jona-Lasinio model is in the same universality class as the standard model and, in fact, for $\Lambda \gg M_Z$ this model can be viewed as a particular limiting case of the Glashow, Salam and Weinberg theory [3]. In order to get realistic phenomenological predictions for the top quark and Higgs masses, the four fermion coupling must be fine tuned very precisely to its critical value. Considering this as an effective theory generated by an unknown high energy dynamics, it will be difficult to imagine a dynamical scenario which provides naturally such a fine tuning. In addition, for a compositeness scale of the order of $\Lambda \simeq 10^{15}$ GeV, the characteristical value for the top quark mass in the standard model with minimal dynamical electroweak symmetry breaking is $m_t \simeq 230$ GeV [3]. From the present measurements of the ρ parameter [7], it follows that, such value of the top quark mass may be too large to fit within the experimental data.

The minimal supersymmetric extension of the composite Higgs model provides solutions to the above problems [8], [9]. In fact, in the supersymmetric extension the quadratic divergences present in the standard model are cancelled, therefore, no fine tuning of the four fermi coupling constant or, in general, of the Higgs mass parameters is required [10]. Moreover, within the supersymmetric extension, lower values for the top quark mass are predicted [9]. This is due to the fact that there is an infrared quasi fixed point which governs the value of the top quark mass, while determining the low energy value of the top quark Yukawa coupling [11]. In the supersymmetric model, the infrared quasi-fixed point yields a low energy value for the top quark Yukawa coupling which is lower than the standard model one [8]. Furthermore, considering the case in which the two scalar Higgs doublets have nontrivial vacuum expectation values, since both give contributions to the Z^0 mass, their individual vacuum expectation values are lower than in the standard model case. This implies that the value of the top quark mass may be significantly lower than that computed in the framework of the top quark condensate standard model [9]. One shortcoming of the supersymmetric extension is to presume the existence of a host of fundamental particles. However, such new particles are the superpartners of the standard fundamental excitations, thus, they appear naturally through the requirement of preserving the global supersymmetry of the model.

An interesting aspect of the composite Higgs scheme for the dynamical breakdown of the electroweak symmetry is that the low energy values of the top quark Yukawa coupling are consistent with the renormalization group trajectories associated with the triviality bounds on this quantity [3], [9]. Hence, the top quark mass values obtained within this model can be understood as upper bounds for this mass, in any theory in which, apart from the supersymmetric partners, no new physics appears up to the compositeness scale Λ, thus, yielding constraints on the range of mass parameters consistent with possible GUT scenarios [12].

In section 2 I shall introduce the minimal supersymmetric extension of the standard model with dynamical symmetry breaking. I shall use the Schwinger-Dyson equations technique, in section 3, to explore within the bubble sum approximation the critical structure and the low energy dynamical behavior of the model. I shall then devote section 4 to compute, by using renormalization group techniques, the corrections generated after the inclusion of the gauge field interactions. At the same time, I shall analyze the modifications to the top quark mass predictions which are induced by introducing the bottom quark Yukawa coupling effects. The top quark mass will be evaluated as a function of the ratio R of the vacuum expectation values of the two scalar Higgs doublets, for different values of the effective cutoff scale Λ and the soft supersymmetry breaking scale Δ_S. I shall then compute, in section 5, the complete Higgs spectrum as a function of the explicit scalar mass terms and the radiatively corrected scalar quartic couplings, which are determined through their renormalization group flow. The top quark mass will be also computed as a function of the lightest CP even Higgs mass for different values of the compositeness scale, the supersymmetry breaking scale and the tree level value of the neutral CP odd Higgs mass. Finally, in section 6 I shall present the triviality bounds on both the top quark and the lightest CP even Higgs masses as a function of the ratio of the vacuum expectation values R, and for various values of the CP odd mass. It is the aim to achieve some more general conclusions about possible grand unified scenarios and the kinematical accesibility of a new decay mode for the lightest CP even Higgs particle. I reserve section 7 to summarize the main results.

2 Supersymmetric Extension of the Top Quark Condensate Model

The $SU(3)_c \times SU(2)_L \times U(1)_Y$ invariant gauged supersymmetric Nambu-Jona-Lasinio model with explicit soft supersymmetry breaking terms [9], [13] - [15] considers the dynamics associated with the quark chiral superfields condensation. Written in terms of the two composite chiral superfields H_1 and H_2, the action of the model at the scale Λ is given by,

$$\begin{aligned}\Gamma_\Lambda &= \Gamma_{YM} + \int dV \left[\bar{Q}e^{2V_Q}Q + T^C e^{-2V_T}\bar{T}^C + B^C e^{-2V_B}\bar{B}^C\right](1-\Delta^2\theta^2\bar{\theta}^2) \\ &\quad + \int dV \bar{H}_1 e^{2V_{H_1}} H_1 (1 - M_{H_1}^2 \theta^2\bar{\theta}^2) \\ &\quad - \int dS \epsilon_{ij}\left(m_0 H_1^i H_2^j (1 + B_0\theta^2) - g_{T_0} H_2^j Q^i T^C (1 + A_0\theta^2)\right) \\ &\quad - \int d\bar{S} \epsilon_{ij}\left(m_0 \bar{H}_1^i \bar{H}_2^j (1 + B_0\bar{\theta}^2) - g_{T_0} \bar{T}^C \bar{Q}^i \bar{H}_2^j (1 + A_0\bar{\theta}^2)\right), \end{aligned} \tag{2.1}$$

where Q is the $SU(2)_L$ doublet of top and bottom quark chiral superfield multiplets, T^C and B^C are the $SU(2)_L$ singlets charge conjugate top and bottom quark chiral multiplets, respectively, and $dV = d^4x d\theta^2 d\bar{\theta}^2$ and $dS = d^4x d\theta^2$ are the superspace integration measures. Γ_{YM} includes the usual supersymmetric gauge field kinetic term as well as the supersymmetry breaking gaugino mass term, and

$$\begin{aligned} V_Q &= \frac{1}{2}g_3 G^a\lambda^a + \frac{1}{2}g_2 W^i\sigma^i + \frac{1}{6}g_1 Y, \qquad & V_T = \frac{1}{2}g_3 G^a\lambda^a + \frac{2}{3}g_1 Y, \\ V_B &= \frac{1}{2}g_3 G^a\lambda^a - \frac{1}{3}g_1 Y, & V_{H_1} = \frac{1}{2}g_2 W^i\sigma^i - \frac{1}{2}g_1 Y, \end{aligned} \tag{2.2}$$

describe the quark and Higgs multiplets interactions with the $SU(3)_c \times SU(2)_L \times U(1)_Y$ gauge fields. Δ^2 and $M_{H_1}^2$ are explicit soft supersymmetry breaking scalar mass parameters, while A_0 and B_0 provide two soft supersymmetry breaking terms proportional to the scalar trilinear and bilinear terms of the superpotential. In the above, only the Yukawa coupling related to the top quark has been introduced. The most general version, however, should include Yukawa type interactions for all the quarks and leptons of the theory. As a matter of fact, in section 4 I shall extend the analysis including also the Yukawa coupling associated with the bottom quark. Nevertheless, the essential qualitative features of the quark multiplet condensation are accurately described in this first simplified version.

In Eq. (2.1), the chiral superfield H_2 acts as a Lagrange multiplier enforcing the relation

$$H_1(1 + B_0\theta^2) = \frac{g_{T_0}}{m_0} Q T^C (1 + A_0\theta^2). \tag{2.3}$$

Once the auxiliary chiral superfields H_1 and H_2 are integrated out, which results in the replacement of H_1 by its equation of motion, the gauged supersymmetric Nambu-Jona-Lasinio model just in terms of the gauge and quark chiral superfields is recovered. It is instructive to write the above action in components, after integrating out the auxiliary F fields [16] of each chiral superfield,

$$\Gamma = \int d^4x \left[\sum_{\Phi=Q,T^C,B^C} \left(|\mathcal{D}_\mu \Phi|^2 - \Delta^2 |\Phi|^2 + \bar{\psi}_\Phi \not{\mathcal{D}} \psi_\Phi \right) + |\mathcal{D}_\mu H_1|^2 \right.$$
$$- M_{H_1}^2 |H_1|^2 - m_0^2 |H_2|^2 - g_{T_0} \left(\bar{\psi}_L t_R H_2 + h.c. \right) - g_{T_0}^2 \left(|H_2 \tilde{T}^C|^2 + |\epsilon_{kl} H_2^k \tilde{Q}^l|^2 \right)$$
$$\left. - g_{T_0} \delta \epsilon_{kl} \tilde{Q}^l \tilde{T}^C H_2^k - \epsilon_{kl} F_{H_2}^k \left(m_0 H_1^l - g_{T_0} \tilde{Q}^l \tilde{T}^C \right) + ... \right], \qquad (2.4)$$

where only the terms relevant in deriving the Schwinger-Dyson equation for the top quark mass (see section 3 below) have been retained. Φ represents any scalar-quark (squark) field, while ψ_Φ stands for the corresponding quark field. In the equation above, the usual kinetic term for the squarks and quarks fields, as well as a kinetic term for the scalar Higgs H_1 appear. There are also the explicit soft supersymmetry breaking scalar mass terms and a mass term for the scalar Higgs H_2, which comes from the replacement of the auxiliary field F of the chiral superfield H_1 by its equation of motion. The Yukawa interaction of the top quark with the composite scalar field H_2 arises out of the trilinear term in the superpotential, while the two quartic terms couple the scalar Higgs H_2 to the squark fields $\tilde{Q}$ and $\tilde{T}^C$, and appear after integrating out the F components of the corresponding quark chiral superfields. Finally, the effective trilinear coupling $\delta = A_0 - B_0$ comes partly from the original soft supersymmetry breaking term proportional to the scalar trilinear term in the superpotential and partly from the scalar bilinear counterpart. More explicitly, the B_0 contribution has been obtained after using the Euler Lagrange equation of motion for the scalar field H_1, defined by the constraint coming from the F term of the superfield H_2, which acts like a Lagrange multiplier. As it will become clear later, the inclusion of a nonvanishing value of the parameter δ is essential in order to generate nontrivial vacuum expectation values for both neutral scalar Higgs, without inducing an unacceptable light axion in the theory [17]. From the Euler Lagrange equations it follows,

$$H_1 = \frac{g_{T_0}}{m_0} \tilde{Q} \tilde{T}^C, \qquad H_2 = -\frac{g_{T_0}}{m_0^2} \bar{t}_R \psi_L + ... \qquad (2.5)$$

In general, the Higgs superfields appear as bound states of the quark chiral superfields. Observe that integrating out the static scalar field H_2, the usual four Fermi interaction with coupling strength $G = g_{T_0}^2 / m_0^2$ is obtained. The same as in the standard model case, there is a critical value of the four fermion coupling, G_C, above which a dynamical mass for the top quark is generated. As I shall show in the next section, G_C may be determined in a self consistent way by using the Schwinger-Dyson equations for the top quark mass.

3 The Bubble Sum Approximation

In the last years, much effort has been devoted in using the Schwinger-Dyson equations to perform a self consistent summation of diagrams in different approximations in quantum field theory [4], [13], [18], [19]. In particular, in the framework of the top quark condensate model, the Schwinger-Dyson equation for the top quark mass has been computed in the large number of colors (N_C) limit, for vanishing strong gauge coupling constant [3], [8] [9]. This is usually called the bubble sum approximation. In the frame of the minimal supersymmetric standard model under study, the diagrams contributing to the top quark mass self consistent equations are depicted in Fig. 1a and yield,

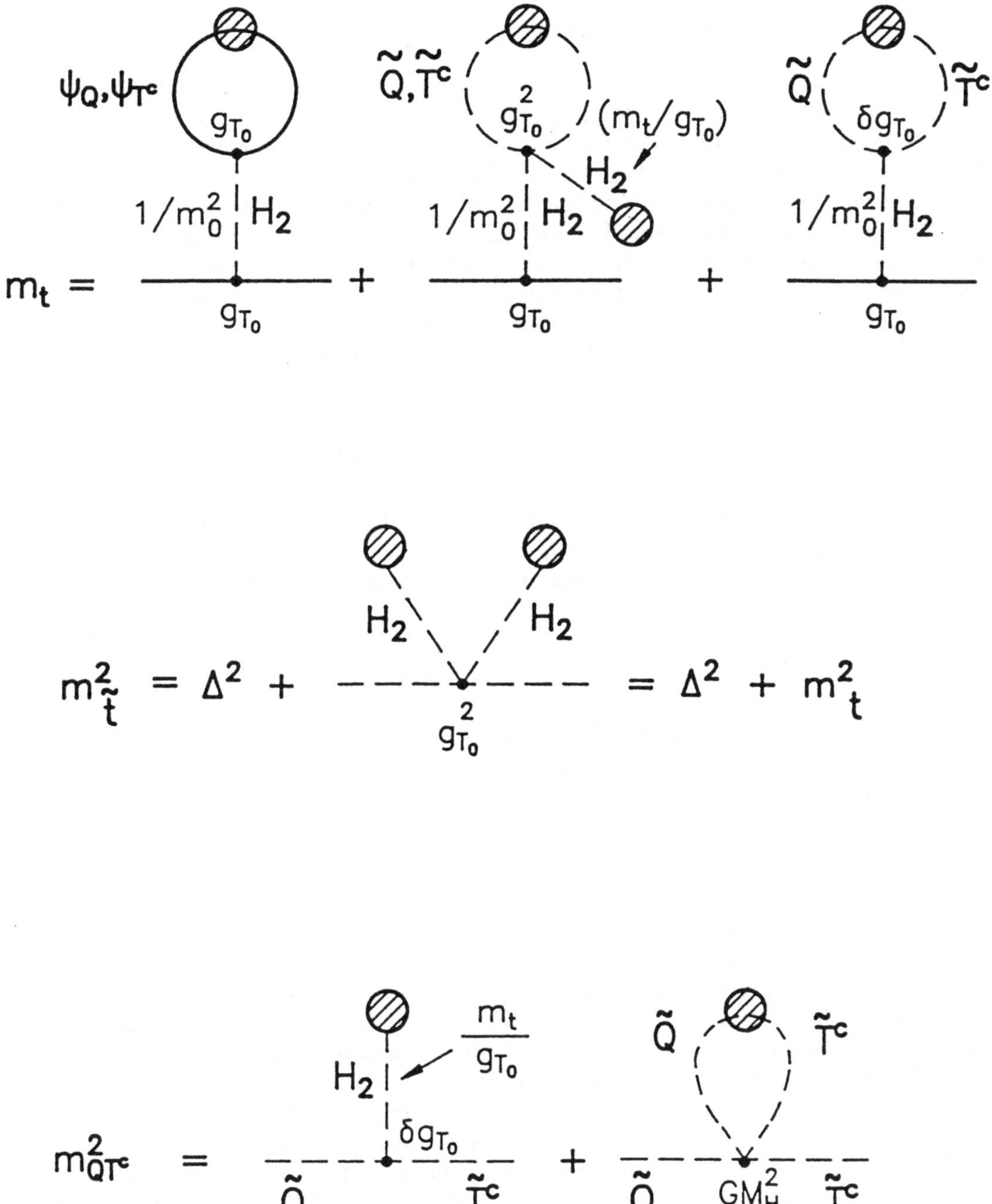

Fig. 1: The Schwinger-Dyson equations in the bubble sum approximation for a) the top quark mass, b) the scalar quark masses and c) the left-right mixing squark mass parameter.

$$m_t = \frac{2m_t G N_C}{16\pi^4}\left[\int d^4p\left(\frac{1}{(p^2+m_{\tilde{t}}^2)} - \frac{p^2+m_{\tilde{t}}^2}{[(p^2+m_{\tilde{t}}^2)^2 - m_{QTC}^4]}\right)\right]$$
$$+\frac{\delta G N_C}{16\pi^4}\int d^4p \frac{m_{QTC}^2}{[(p^2+m_{\tilde{t}}^2)^2 - m_{QTC}^4]}. \qquad (3.1)$$

The first two terms in expression (3.1) are the fermion and squark contributions, while the last term comes from the left-right squark loop induced through the soft supersymmetry breaking term δ. In the above, the usual supersymmetric cancellation of the quadratic divergent terms takes place. From the explicit expression of m_t, it follows that the scalar quark mass, $m_{\tilde{t}}$, and the left-right mixing squark mass parameter, m_{QTC}, need to be computed. The superfield propagator was first derived in Ref. [20] and then generalized by the inclusion of the δ term in Ref. [9]. The squared squark mass is given by the squared of the explicit soft supersymmetry breaking mass contribution plus the squared of the top quark mass, as is shown in Fig. 1b. The value of m_{QTC}^2 is also determined by the corresponding Schwinger-Dyson equation depicted in Fig. 1c. It reads,

$$\delta m_t = m_{QTC}^2\left[1 + \frac{G M_{H_1}^2 N_C}{32\pi^2}\log\left(\frac{\Lambda^4}{(m_{\tilde{t}}^2+\Delta^2)^2 - m_{QTC}^4}\right)\right] \equiv m_{QTC}^2\ \alpha^{-1}, \qquad (3.2)$$

where the logarithmic term is generated by the explicit soft supersymmetry breaking term associated with the scalar field H_1.

For a nontrivial solution of the Schwinger-Dyson equations, the gap equation[9]

$$G^{-1} = \frac{N_C\Delta^2}{16\pi^2}\left[\left(1+\frac{2m_t^2+\delta^2\alpha}{2\Delta^2}\right)\log\left(\frac{\Lambda^4}{(m_{\tilde{t}}^2+\Delta^2)^2 - m_{QTC}^4}\right) - \frac{2m_t^2}{\Delta^2}\log\left(\frac{\Lambda^2}{m_{\tilde{t}}^2}\right)\right]. \qquad (3.3)$$

must be fulfilled. In the limit $\delta \to 0$, $M_{H_1}^2 \to 0$, for example, the critical value for the four Fermi coupling, above which a nonvanishing value for the top quark mass is dynamically generated, is given by

$$G_C^{-1} = \frac{N_C\Delta^2}{8\pi^2}\log\left(\frac{\Lambda^2}{\Delta^2}\right). \qquad (3.4)$$

Furthermore, in the limit in which all the supersymmetry breaking terms vanish, the value of G_C diverges and, therefore, no mass for the top quark can be dynamically generated. In the presence of nonvanishing supersymmetry breaking terms, instead, the quadratic dependence of G_C^{-1} on Λ is replaced by a mild quadratic dependence on the soft supersymmetry breaking scales. The compositeness scale appears only in the logarithm and, hence, the solution to the gap equation gives always a critical value which is of the order of the inverse of the largest soft supersymmetry breaking scale. Thus, as I already said, no fine tuning is necessary in this model.

In the low energy regime, a gauge invariant kinetic term for the chiral superfield H_2 is induced through radiative corrections. In the bubble sum approximation it may be determined by computing the contributions to the H_2 chiral superfield self energy depicted in Fig. 2. The scalar self energy results in a kinetic term and a negative squared

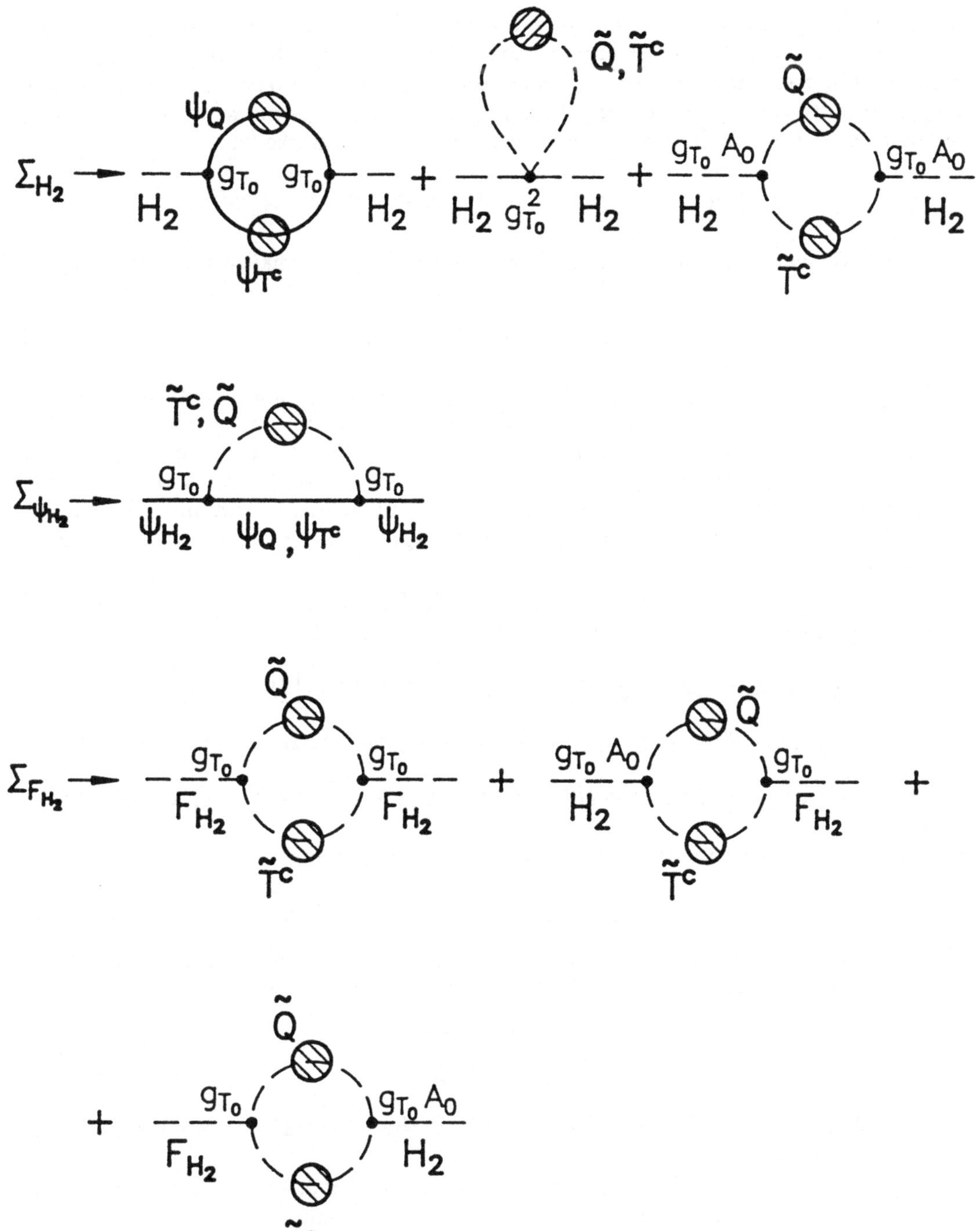

Fig. 2: The Feymann diagrams contributing to the induced kinetic term of the Higgs chiral superfield H_2 in the bubble sum approximation. The chiral superfield self energy is given by $\Sigma_{H_2} + \Sigma_{\psi_{H_2}} + \Sigma_{F_2}$.

mass term proportional to the supersymmetry breaking mass parameter Δ^2, both multiplied by a logarithmically divergent wave function renormalization factor Z_{H_2}. The Higgsino acquires a kinetic term with the same wave function renormalization constant Z_{H_2}, as implied by supersymmetry. Finally, computing the terms which depend on the auxiliary field F_{H_2}, the induced kinetic term can be written like [9]

$$Z_{H_2} \int dV \bar{H}_2 e^{2V_{H_2}} H_2(1 + A_0\theta^2 + A_0\bar{\theta}^2 + (2\Delta^2 + A_0^2)\theta^2\bar{\theta}^2), \tag{3.5}$$

where the H_2 wavefunction renormalization constant, at a normalization scale μ is given by,

$$Z_{H_2} = \frac{g_{T_0}^2 N_C}{16\pi^2} \log\left(\frac{\Lambda^2}{\mu^2}\right) \tag{3.6}$$

and $V_{H_2} = V_{H_1} + g_1 Y$. The wave function renormalization constant is normalized so that at the scale $\mu = \Lambda$ the gauged supersymmetric Nambu-Jona-Lasinio action, Eq. (2.1) is recovered. Hence, $Z_{H_2} \to 0$ when $\mu \to \Lambda$, which is usually called the compositeness condition. Thus, at the compositeness scale Λ, the previous kinetic term, Eq. (3.5), tends to zero and the chiral superfield H_2 has no independent dynamics. For energies lower than the compositeness scale, instead, H_2 appears as an independent dynamical degree of freedom.

Rescaling the chiral superfield H_2 by $H_2(1-A_0\theta^2)/\sqrt{Z_{H_2}}$, so that it has a canonically normalized kinetic term, the low energy effective action is given by

$$\begin{aligned}\Gamma_Z \ = \ & \Gamma_{YM} + \int dV \left[\bar{Q}e^{2V_Q}Q + T^C e^{-2V_T}\bar{T}^C + B^C e^{-2V_B}\bar{B}^C\right](1 - \Delta^2\theta^2\bar{\theta}^2) \\ & + \int dV \bar{H}_1 e^{2V_{H_1}} H_1(1 - M_{H_1}^2\theta^2\bar{\theta}^2) - \int dS \epsilon_{ij}\left(mH_1^i H_2^j(1+\delta\theta^2) - h_t H_2^j Q^i T^C\right) \\ & - \int d\bar{S}\epsilon_{ij}\left(m\bar{H}_1^i \bar{H}_2^j(1+\delta\bar{\theta}^2) - h_t\bar{T}^C\bar{Q}^i\bar{H}_2^j\right) + \int dV \bar{H}_2 e^{2V_{H_2}} H_2(1 + 2\Delta^2\theta^2\bar{\theta}^2).\end{aligned} \tag{3.7}$$

where the renormalized mass and top quark Yukawa coupling,

$$m = m_0/\sqrt{Z_{H_2}}\,, \qquad h_t = g_{T_0}/\sqrt{Z_{H_2}} \tag{3.8}$$

have been defined. Thus, within the bubble sum approximation and for $N_C = 3$, the top quark mass, $m_t = h_t(m_t)\, v_2$ is determined by the expression,

$$m_t^2 \log\left(\frac{\Lambda^2}{m_t^2}\right) = \frac{16}{3}\pi^2 v_2^2 \tag{3.9}$$

Observe that, since m_0 and g_{T_0} have finite values, the renormalized couplings, Eq.(3.8), diverge at the scale Λ.

Once the chiral superfield H_2 is canonically normalized, from the effective action Eq. (3.7) it follows that the effective supersymmetry breaking term proportional to the bilinear term of the superpotential is $B = \delta$, while the trilinear counterpart vanishes, $A = 0$. In addition, the induced explicit supersymmetry breaking mass term for the scalar field H_2, $M_{H_2}^2 = -2\Delta^2$, is negative. Therefore, it may generate a nontrivial minimum for the scalar potential, even when the supersymmetry breaking term B vanishes [10]. However, a nontrivial value of B is crucial to generate a vacuum expectation value for the scalar Higgs H_1, which will then give masses to the bottom quarks and leptons of the theory. The effective values of these supersymmetry breaking terms can be understood by analyzing their renormalization group flow. In the bubble sum approximation, the renormalization group equations for the supersymmetry breaking parameters read,

$$\begin{aligned}\frac{dM_{H_2}^2}{dt} &= -N_C Y_t (m_Q^2 + m_{TC}^2 + M_{H_2}^2 + A^2)\\ \frac{dA}{dt} &= -2N_C Y_t A = \frac{1}{2}\frac{dB}{dt}\end{aligned} \tag{3.10}$$

where $Y_t = (h_t/4\pi)^2$, $t = \log(\Lambda/\mu)$ and $m_Q^2 \equiv \Delta^2 \equiv m_{TC}^2$ are the explicit soft supersymmetry breaking mass parameters for the squark fields that appear in the original action. From the above expressions, it is straightforward to see that the fulfillment of the compositeness condition, $Y_t^{-1}(\Lambda) = 0$, together with the requirement of finiteness of the soft supersymmetric breaking parameters, implies the cancellation at all energy scales of A and of the sum of the squared explicit supersymmetry breaking mass terms for the squarks and the H_2 scalar field. Although the above cancellations at all scales are just a property of the bubble sum approximation, it is important to remark that the same cancellations at the scale Λ are a prediction of the model.

4 The Inclusion of Gauge Interactions: Renormalization Group Analysis

The analysis performed in the above section gives a proper qualitative description of the dynamical behavior of the model. However, in order to incorporate the effects of the gauge interactions, it is necessary to go beyond the bubble sum approximation. Instead of computing gauge field corrections and higher order in $1/N_C$ effects, a better alternative to obtain the low energy physics is to make use of renormalization group techniques, which provide an efficient method of summing an infinite set of diagrams [3], [8], [9], [21]. The renormalization group evolution of the running couplings contains just the leading logarithmic contributions. However, at scales μ much smaller than the effective cutoff of the theory, the only relevant contributions are expected to arise out of such leading logarithmic terms. The full renormalization group equations of the supersymmetric standard model [22], which are needed to compute the value of the top quark mass are,

$$\begin{aligned}\frac{d\alpha_3}{dt} &= 3\frac{\alpha_3^2}{4\pi}\\ \frac{d\alpha_2}{dt} &= -\frac{\alpha_2^2}{4\pi}\\ \frac{d\alpha_1}{dt} &= -11\frac{\alpha_1^2}{4\pi}\\ \frac{dY_t}{dt} &= Y_t\left(\frac{16}{3}\tilde{\alpha}_3 + 3\tilde{\alpha}_2 + \frac{13}{9}\tilde{\alpha}_1 - 6Y_t - Y_b\right)\\ \frac{dY_b}{dt} &= Y_b\left(\frac{16}{3}\tilde{\alpha}_3 + 3\tilde{\alpha}_2 + \frac{7}{9}\tilde{\alpha}_1 - 6Y_b - Y_t\right)\end{aligned} \tag{4.1}$$

with $\alpha_i = 4\pi\tilde{\alpha}_i = g_i^2/4\pi$ and $Y_b = (h_b/4\pi)^2$. The above renormalization group equations provide the proper renormalization group flow for scales μ below Λ and above the soft supersymmetry breaking scale Δ_S.

To compute the value of the top quark mass it is necessary to determine the value of the top quark Yukawa coupling at the energy scale $\mu = m_t$, which can be obtained using the renormalization group evolution, after defining the high energy value of h_t by the appropriate boundary condition of compositeness. Moreover, the boundary conditions for the gauge couplings have to match their low energy experimental values. However, in general, the characteristic soft supersymmetry breaking mass scale is expected to be larger than the electroweak scale. Thus, the low energy effective theory below Δ_S is equivalent to the standard model with one or two Higgs doublets, depending on the values of the mass parameters appearing in the scalar potential. As a result, the renormalization group evolution of the running couplings at scales μ below Δ_S is properly determined by the solutions to the standard model renormalization group equations which are given by [11],

$$\begin{aligned}
\frac{d\alpha_3}{dt} &= 7\frac{\alpha_3^2}{4\pi} \\
\frac{d\alpha_2}{dt} &= \beta_2\frac{\alpha_2^2}{4\pi} \\
\frac{d\alpha_1}{dt} &= -\beta_1\frac{\alpha_1^2}{4\pi} \\
\frac{dY_t}{dt} &= Y_t\left(\frac{24}{3}\tilde{\alpha}_3 + \frac{9}{4}\tilde{\alpha}_2 + \frac{17}{12}\tilde{\alpha}_1 - \frac{9}{2}Y_t - \frac{\alpha_b}{2}Y_b\right) \\
\frac{dY_b}{dt} &= Y_b\left(\frac{24}{3}\tilde{\alpha}_3 + \frac{9}{4}\tilde{\alpha}_2 + \frac{5}{12}\tilde{\alpha}_1 - \frac{9}{2}Y_b - \frac{\alpha_t}{2}Y_t\right),
\end{aligned} \tag{4.2}$$

where $\beta_2 = 3(19/6)$, $\beta_1 = 7(41/6)$ and $\alpha_b = \alpha_t = 1(3)$ if there are two (one) light scalar Higgs doublets in the spectrum.

As far as there is a huge hierarchy of mass scales, $\Lambda \gg M_Z$, the existence of an infrared quasi-fixed point of the top quark Yukawa coupling allows a first rough estimate of its low energy value. In fact, if the supersymmetry breaking scale is of the order of the electroweak scale, the supersymmetric renormalization group equation for the top quark Yukawa coupling yields $h_t(M_Z) \simeq \sqrt{8/9}\, g_3(M_Z)$. Thus, considering $m_t \simeq h_t(M_Z)\, v_2$, the top quark mass expression is given by,

$$m_t \simeq h_t(M_Z) v \frac{R}{\sqrt{1+R^2}} \tag{4.3}$$

where $M_Z^2 = (g_1^2 + g_2^2)v^2/2$ and hence $v = \sqrt{v_1^2 + v_2^2} \simeq 173.5$ GeV, implying an approximate value

$$m_t \simeq 196\ GeV\ \frac{R}{\sqrt{1+R^2}}. \tag{4.4}$$

Therefore, for values of the ratio $R \simeq 1$, it is possible to obtain top quark mass predictions which are significantly lower than those computed in the standard top condensate model.

Considering the running of the Yukawa and gauge couplings determined by their renormalization group equations, it is possible to compute much more precisely the value of the top quark mass [9]. The numerical integration is performed using both set of equations (4.1) and (4.2), while imposing the continuity of the running couplings at the scale $\mu = \Delta_S$. The boundary conditions for the gauge couplings are chosen to be $\alpha_3(M_Z) = 0.115$, $\alpha_2(M_Z) = 0.0336$ and $\alpha_1(M_Z) = 0.0102$, which are consistent with present experimental constraints [7]. For the top quark Yukawa coupling the compositeness condition $Y_t^{-1}(\Lambda) = 0$ is required, while for the bottom quark Yukawa coupling, the boundary condition is determined through the experimental value of the bottom quark mass. In fact, the low energy value of h_b, computed at $\mu = m_b$, is given as a function of R and considering $m_b \simeq 5$ GeV. Once the boundary conditions for the running couplings are fixed, the solutions to the renormalization group equations provide their values at any given energy scale μ. As I shall analyze in the next section, if there is only one light Higgs doublet in the low energy spectrum, it will be composed by a combination of the two scalar Higgs doublets H_1 and H_2 [23] with a mixing angle $\theta_M = \theta$ where $\tan\theta = R$. Then, the top and bottom quark Yukawa couplings appearing in the renormalization group equations (4.2) will be the effective couplings of the light scalar doublet with the top and bottom quarks, h_t^{eff} and h_b^{eff}, respectively. The continuity conditions at the scale Δ_S will read, $h_b^{eff} = h_b \cos\theta$ and $h_t^{eff} = h_t \sin\theta$.

The results I shall present now come out from the study done in collaboration with Bardeen, Clark, Sasaki and Wagner [9], where we have used the renormalization group approach that I detailed above. The top quark mass has been computed as a function of the ratio R for various values of the compositeness and supersymmetry breaking scales.

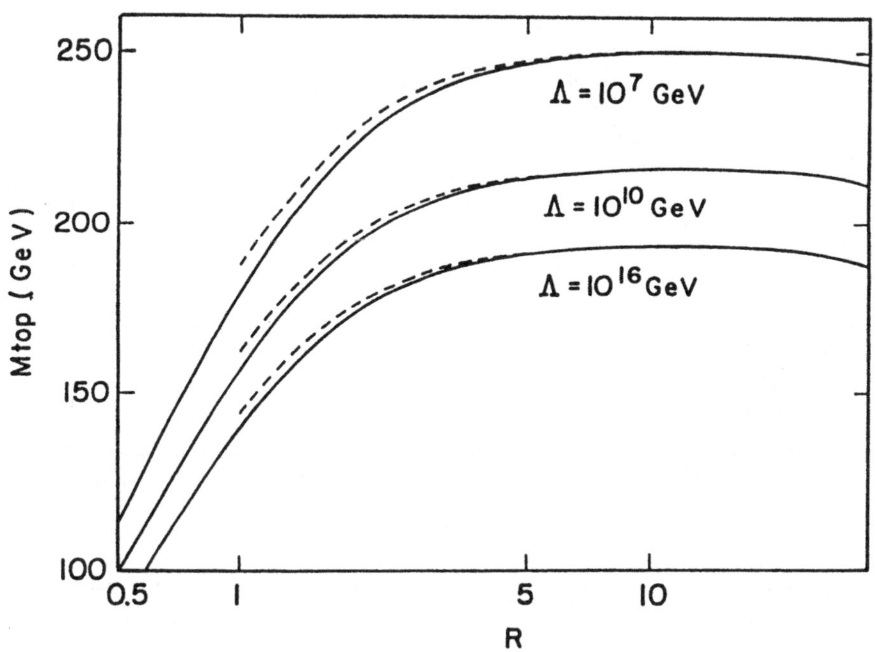

Fig. 3: Top quark mass as a function of the ratio R, for a soft supersymmetry breaking scale Δ_S= 1 TeV and three different values of the compositeness scale Λ, for the case of one light Higgs doublet (dashed line) and two light Higgs doublets (solid line).

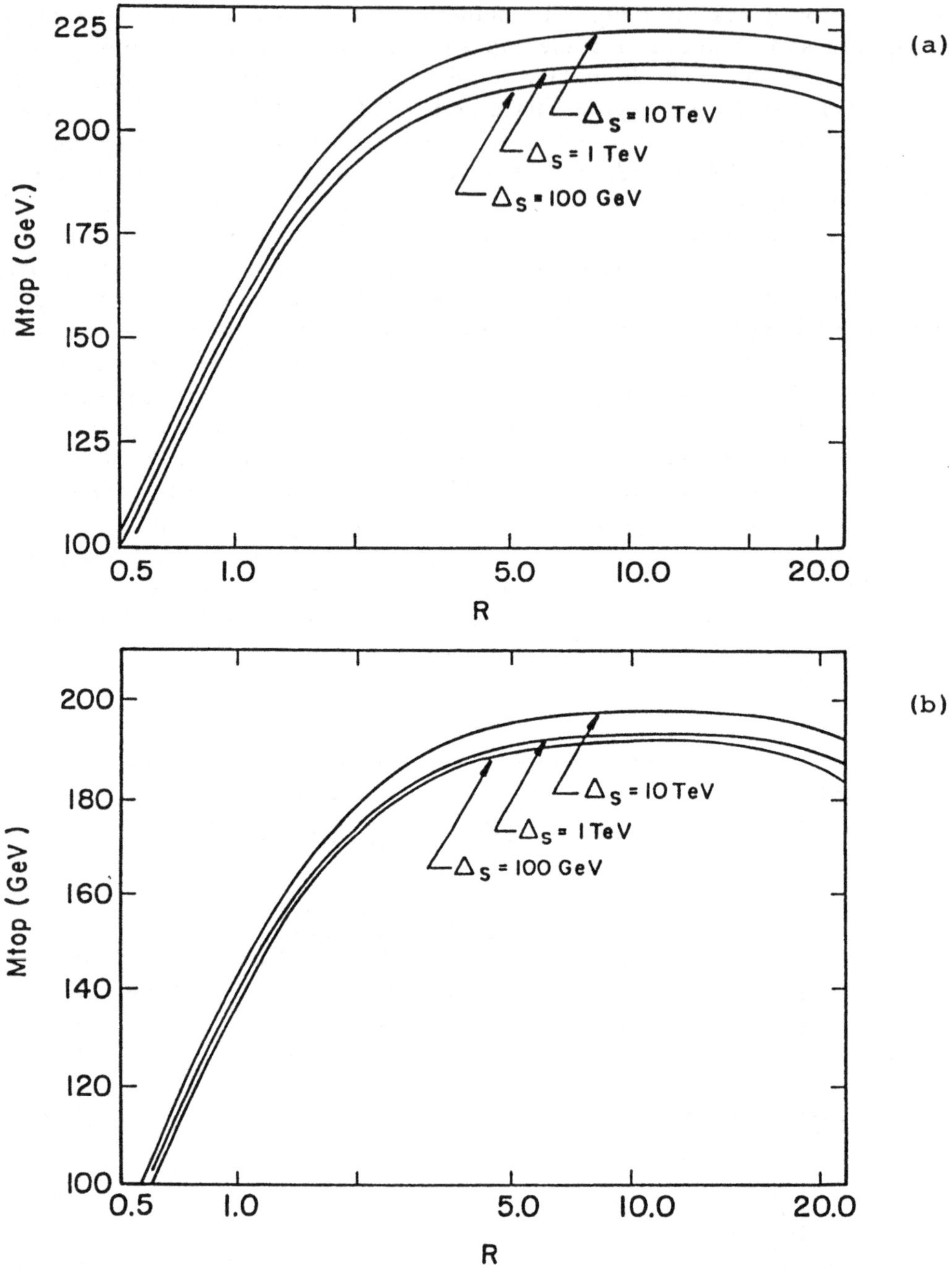

Fig. 4. Top quark mass as a function of the ratio R, for three different values of the soft supersymmetry breaking scale, for the case of two Higgs doublets and a compositeness scale a) $\Lambda = 10^{10}$ GeV and b) $\Lambda = 10^{16}$ GeV.

From Figs. 3 and 4, it is obvious that the top quark mass acquires its lowest value for the smallest value of R. In fact, for the small to moderate range of values of R, it increases monotonically with the ratio. For the large R regime, instead, a slight decrease in the top quark mass value is observed. To understand this behavior it is necessary to analyze the interplay of the two Yukawa couplings. Due to the experimentally fixed value of the bottom quark mass, the bottom quark Yukawa coupling has a dependence on R given by, $h_b(R) = h_b(R=1)\sqrt{(1+R^2)/2}$. Thus, for larger values of R, h_b becomes larger and the infrared quasi fixed point is reached for smaller values of the top quark Yukawa coupling. In addition, the dependence of the vacuum expectation value of H_2 on the ratio, $v_2(R) = v_2(R=1)R\sqrt{2/(1+R^2)}$, is responsible for the increase in m_t observed in the figures for the small and intermediate values of R and shows that v_2 varies only slightly for large values of R. The approximate stability of v_2, together with the decrease in h_t, results in the decrease in the top quark mass for the large R regime. Furthermore, if R becomes too large ($R \geq 36$ for $\Lambda = 10^{16}$ GeV) the bottom quark Yukawa coupling becomes larger than the top quark one, implying an interchange of rolls between the two particles. In our computations, we have set an upper bound on R through the requirement that $h_t > h_b$.

In Fig. 3 we have also plotted the results obtained in the case in which only one light scalar doublet emerges in the low energy theory. For small values of R, the top quark mass predictions for the two light Higgs doublets scenario are slightly lower than those for the one light Higgs doublet case. However, quite in general, the top quark mass values are insensitive to the existence of one or two light Higgs doublets in the spectrum.

In Figs. 4a and 4b, we concentrate on the variation of the top quark mass for different values of the soft supersymmetry breaking scale. Although the values of m_t obtained for Δ_S as large as 10 TeV are slightly larger than those obtained for $\Delta_S =$ 100 GeV or 1 TeV, in general, the low energy predictions for the top quark mass are stable under variations of Δ_S. In all the calculations we have done, the supersymmetric particles were assumed to decouple at the energy scale $\mu = \Delta_S$. The stability of our results for different values of the soft supersymmetry breaking scale is, therefore, an important issue which supports such threshold approximation. From our computations it follows, that for values of the ratio $R \geq 1$ and a soft supersymmetry breaking scale $\Delta_S \simeq 1$ TeV the characteristic values of the top quark mass are 140 GeV $\leq m_t \leq$ 195 GeV for $\Lambda \simeq 10^{16}$ GeV . If, instead, $\Lambda = 10^{10}$ GeV then 160 GeV $\leq m_t \leq$ 220 GeV.

Moreover, in the numerical work done in Ref. [9], we have explored the change in the top quark mass results for different high energy values of the top quark Yukawa coupling. For a compositeness scale $\Lambda \geq 10^{16}$ GeV ($\Lambda \geq 10^{10}$ GeV), a slight variation, of less than 1% (2%) of the top quark mass value is obtained by setting $Y_t(\Lambda) = 0.1$ instead of $Y_t(\Lambda)^{-1} = 0$. This analysis proves that, although the perturbative one loop renormalization group equations are not a reliable tool to determine the evolution of the Yukawa coupling at energy scales μ close to Λ, the action of the infrared quasi fixed point makes the top quark mass predictions quite insensitive to the precise boundary condition on the top quark Yukawa coupling at Λ.

5 Renormalization Group Equations and the Scalar Higgs Spectrum

In deriving low energy supersymmetry from a more fundamental theory such as supergravity or superstrings [22], [24], it has been realized that there are soft supersymmetry breaking parameters which may be introduced in the model without spoiling the cancellation of the quadratic divergences. Such parameters are the gaugino masses, the explicit mass parameters for the scalar components of the chiral superfields and the scalar couplings proportional to the trilinear and bilinear scalar terms in the superpotential. Moreover, there is also the free parameter m appearing in the superpotential. As I have discussed before, the compositeness condition imposes some constraints on the soft supersymmetry breaking parameters at the compositeness scale Λ. In addition to the cancellation of the trilinear coupling $A(\Lambda)$, there is a fixed relation at the compositeness scale among the supersymmetry breaking mass terms for the squarks and the scalar Higgs H_2. Although these conditions increase the predictability of the model, they are not sufficient to determine the low energy values of the mass parameters appearing in the scalar Higgs potential. This may be already seen in the bubble sum approximation, in which the bilinear coupling B, the supersymmetry breaking mass $M_{H_1}^2$ and the mass m remain as free parameters, independent of the characteristic squark mass scale. The inclusion of the gauge interactions does not improve the predictability of the theory. In fact, after they have been introduced, the renormalization group flow of the mass parameters in the scalar potential depends in a complicated way, not only on the soft supersymmetry breaking squarks masses, but on the gaugino masses as well. The low energy values of the mass parameters can only be determined after fixing a particular supersymmetry breaking scheme. However, there is a wide range of possible values for m and the soft supersymmetry breaking parameters at $\mu = \Lambda$ that leads to low energy predictions which are in agreement with experimental and theoretical constraints [24]. Therefore, for the purpose of this study, I shall treat the mass parameters in the scalar potential most generally by considering them as free parameters, with the only requirement given by the fulfillment of the minimization conditions at the weak scale.

The low energy Higgs potential is given by the expression [9],

$$\begin{aligned} V_{eff} &= m_1^2 H_1^\dagger H_1 + m_2^2 H_2^\dagger H_2 - m_3^2\left(H_1^T i\tau_2 H_2 + h.c.\right) \\ &+\frac{\lambda_1}{2}\left(H_1^\dagger H_1\right)^2 + \frac{\lambda_2}{2}\left(H_2^\dagger H_2\right)^2 + \lambda_3\left(H_1^\dagger H_1\right)\left(H_2^\dagger H_2\right) + \lambda_4\left|H_2^\dagger i\tau_2 H_1^*\right|^2, \end{aligned} \qquad (5.1)$$

where $m_3^2 = Bm$ and, as I said before, m_1^2 and m_2^2 take generic values. Actually, I shall then parametrize the solutions in terms of $M^2 = (m_1^2 + m_2^2)/2$ and R instead of m_i, $i = 1, 2, 3$. Quite in general, the parameter m_2^2 decreases with increasing values of h_t. In fact, for particular boundary conditions for the gauginos and scalars masses and for sufficiently large top quark Yukawa couplings, there is a region in parameter space for which not only m_2^2 but also M^2 may become negative. After the inclusion of radiative corrections this will appear as a real possibility, without entering in conflict with the stability requirements for the effective potential (see below). In the following, I shall consider the case $M^2 > 0$, while the case in which M^2 acquires negative values will be included in the next section.

In the expression of the low energy scalar potential, Eq. (5.1), all the relevant quartic couplings compatible with the symmetries of the theory have been considered. As a matter of fact, the inclusion of other independentent quartic terms is protected by either discrete symmetries ($H_1 \to -H_1$; $H_2 \to -H_2$) or a global PQ symmetry. The PQ symmetry is only broken by the mass term m_3^2 in the scalar potential. By considering just the leading logarithmic contributions, no PQ symmetry breaking quartic term $\lambda_5(|H_1^T i\tau_2 H_2|^2 + h.c.)$ will be generated through the mass parameter m_3^2. Hence, λ_5 will not appear in the renormalization group analysis.

At scales μ above Δ_S the quartic couplings λ_i are determined through the D terms [16] of the electroweak gauge superfields,

$$\lambda_1(\mu \geq \Delta_S) = \lambda_2(\mu \geq \Delta_S) = \frac{g_1^2 + g_2^2}{4},$$
$$\lambda_3(\mu \geq \Delta_S) = \frac{g_2^2 - g_1^2}{4}, \qquad \lambda_4(\mu \geq \Delta_S) = -\frac{g_2^2}{2}. \tag{5.2}$$

At low energy scales μ below Δ_S, instead, the quartic couplings may be computed by solving the corresponding renormalization group equations [11],

$$\begin{aligned}
16\pi^2 \frac{d\lambda_1}{dt} &= -6\left[\lambda_1^2 + (4\pi)^2\lambda_1(Y_b - \tilde{\alpha}_1/4 - 3\tilde{\alpha}_2/4) + (4\pi)^4(\tilde{\alpha}_1^2/16 + \tilde{\alpha}_1\tilde{\alpha}_2/8 \right.\\
&\quad \left. + 3\tilde{\alpha}_2^2/16 - Y_b^2)\right] - 2\lambda_3^2 - 2\lambda_3\lambda_4 - \lambda_4^2\\
16\pi^2 \frac{d\lambda_2}{dt} &= -6\left[\lambda_2^2 + (4\pi)^2\lambda_2(Y_t - \tilde{\alpha}_1/4 - 3\tilde{\alpha}_2/4) + (4\pi)^4(\tilde{\alpha}_1^2/16 \right.\\
&\quad \left. + \tilde{\alpha}_1\tilde{\alpha}_2/8 + 3\tilde{\alpha}_2^2/16 - Y_t^2)\right] - 2\lambda_3^2 - 2\lambda_3\lambda_4 - \lambda_4^2\\
-32\pi^2 \frac{d\lambda_3}{dt} &= (\lambda_2 + \lambda_1)(6\lambda_3 + 2\lambda_4) + 4\lambda_3^2 + 2\lambda_4^2 + \lambda_3(4\pi)^2(-3\tilde{\alpha}_1 - 9\tilde{\alpha}_2\\
&\quad +6Y_t + 6Y_b) + (4\pi)^4(9\tilde{\alpha}_2^2/4 + 3\tilde{\alpha}_1^2/4 - 3\tilde{\alpha}_1\tilde{\alpha}_2/2 + 12Y_tY_b)\\
-32\pi^2 \frac{d\lambda_4}{dt} &= \lambda_4(2\lambda_1 + 2\lambda_2 + 8\lambda_3 + 4\lambda_4) + 3\lambda_4(4\pi)^2(-3\tilde{\alpha}_2\\
&\quad -\tilde{\alpha}_1 + 2Y_t + 2Y_b) + 3(4\pi)^4(\tilde{\alpha}_1\tilde{\alpha}_2 + 4Y_tY_b),
\end{aligned} \tag{5.3}$$

with the boundary conditions at $\mu = \Delta_S$ given by Eq. (5.2).

The minimization conditions for the above potential read,

$$\sin(2\theta) = \frac{2m_3^2}{(m_1^2 + m_2^2) + \lambda_1 v_1^2 + \lambda_2 v_2^2 + (\lambda_3 + \lambda_4)v^2}, \tag{5.4}$$

and

$$R^2 = \tan^2\theta = \frac{m_1^2 + \lambda_1 v_1^2}{m_2^2 + \lambda_2 v_2^2} = \frac{m_1^2 + \lambda_2 v^2 + (\lambda_1 - \lambda_2)v_1^2}{m_2^2 + \lambda_2 v^2} \tag{5.5}$$

Concerning the stability conditions, in the absence of radiative corrections, the mass parameter M^2 must be greater than $|m_3|^2$ in order to assure that the potential is stable along the direction $|H_1| = |H_2|$. Once the radiative corrections are taken into account, the stability conditions imply that $\lambda_1 \geq 0$, $\lambda_2 \geq 0$ and $\sqrt{\lambda_1\lambda_2} > -\lambda_3 + |\lambda_4|$ if $\lambda_4 < 0$ or $\sqrt{\lambda_1\lambda_2} > -\lambda_3$ if $\lambda_4 > 0$. These relations are always verified within this model. Hence, the low energy Higgs potential is stable.

After the breakdown of the electroweak symmetry, the neutral components of the scalar fields acquired vacuum expectation values. By solving the eigenvalue equations for the neutral and charged components of the scalar doublets, three neutral scalar eigenstates, h, H and A, and two charged ones, $H^{\pm}$, appear in the low energy spectrum of the theory. The two neutral CP even scalar states have masses given by

$$\begin{aligned} m_{H,h}^2 &= \frac{1}{2}\Big[2M^2 + 3\lambda_2 v_2^2 + 3\lambda_1 v_1^2 + (\lambda_3+\lambda_4)v^2 \\ &\pm \sqrt{(-m_A^2\cos(2\theta) + 2\lambda_1 v_1^2 - 2\lambda_2 v_2^2)^2 + (m_A^2 - 2(\lambda_3+\lambda_4)v^2)^2 \sin^2(2\theta)}\Big] \end{aligned} \quad (5.6)$$

where m_A is the mass of the neutral CP odd scalar state,

$$m_A^2 = 2M^2 + \lambda_2 v_2^2 + \lambda_1 v_1^2 + (\lambda_3+\lambda_4)v^2. \quad (5.7)$$

Moreover, the minimization condition Eq. (5.4) implies $m_A^2 = 2m_3^2/sin(2\theta)$. It is then straigtforward to see that, for nontrivial values of v_1 and v_2, a vanishing value of m_3^2 would imply an unacceptable massless CP odd axion in the theory. Finally, the mass of the charged Higgs eigenstates, m_{ch}, comes out as a function of the CP odd scalar mass as well,

$$m_{ch}^2 = m_A^2 - \lambda_4 v^2. \quad (5.8)$$

There are also a neutral CP odd and two charged massless Golstone modes, which are absorved by the electroweak gauge bosons through the usual Higgs mechanism.

Furthermore, from the diagonalization of the mass matrices of the CP odd and charged Higgs eigenstates, it follows that they have a $H_1 - H_2$ mixing angle $-\theta$. From the above expressions for the scalar masses, Eqs. (5.6), (5.7) and (5.8), it can be readily observed that if M is of the order of the weak scale, two light Higgs doublets appear in the low energy spectrum. If, instead, the mass parameter $M \gg M_Z$, then the masses of the CP odd, charged and heaviest CP even states are approximately given by $2M^2$. In addition, for such large values of M the $H_1 - H_2$ mixing angle for the neutral CP even states is approximately given by θ. Hence, the light CP even Higgs state together with the CP odd and charged Goldstone modes, forms a Higgs doublet ϕ whose expression is given by

$$\phi = H_1\cos\theta + i\tau_2 H_2^* \sin\theta \quad (5.9)$$

An alternative method to analyze the effective low energy model in the limit $M \gg M_Z$, is to assume from the beginning that, at scales below Δ_S, one has the standard model with only one light Higgs doublet [23]. Considering the effective potential for the scalar doublet ϕ,

$$V(\phi) = m_\phi^2 \phi^\dagger\phi + \frac{\lambda}{2}(\phi^\dagger\phi)^2, \quad (5.10)$$

the mass of the physical neutral scalar field ϕ^0 is given by,

$$m_{\phi^0}^2 = 2\lambda v^2 \quad (5.11)$$

The value of the quartic coupling λ at low energies is then computed by solving the corresponding renormalization group equation,

$$\begin{aligned} 16\pi^2\frac{d\lambda}{dt} &= -6\Big[\lambda^2 + (4\pi)^2\lambda(Y_t^{eff} + Y_b^{eff} - \tilde{\alpha}_1/4 - 3\tilde{\alpha}_2/4) + (4\pi)^4(\tilde{\alpha}_1^2/16 \\ &+ \tilde{\alpha}_1\tilde{\alpha}_2/8 + 3\tilde{\alpha}_2^2/16 - (Y_t^{eff})^2 - (Y_b^{eff})^2)\Big], \end{aligned} \quad (5.12)$$

with the boundary condition $\lambda(\Delta_S) = \frac{g_2^2+g_1^2}{4}\cos^2(2\theta)$, where $Y_{b,t}^{eff} = (h_{b,t}^{eff}/4\pi)^2$. The mass of ϕ_0 computed in this way is expected to coincide with the value m_h determined in the general case whenever the mass parameter M is much larger than M_Z.

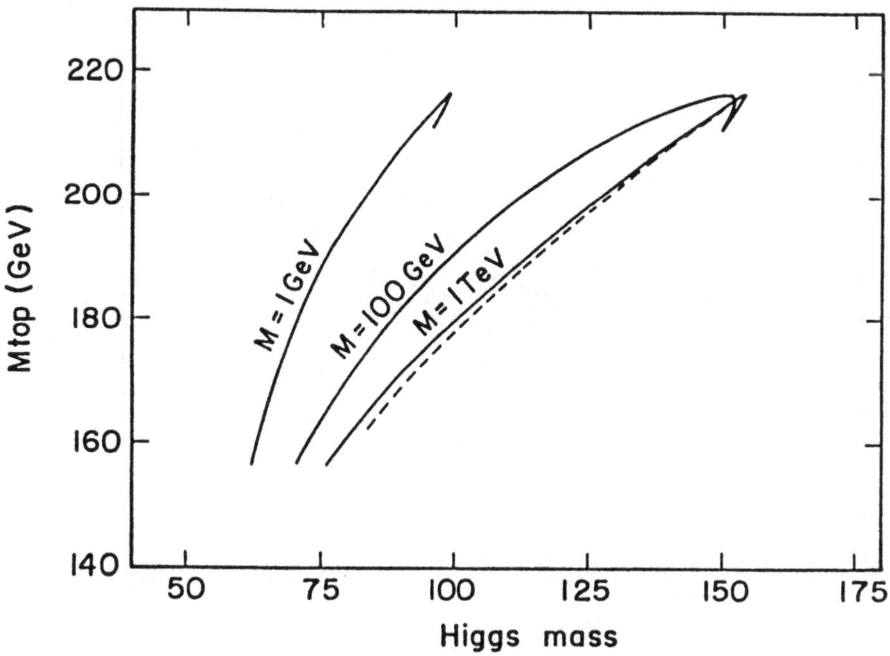

220
200
180
160
140
Mtop (GeV)
M = 1 GeV
M = 100 GeV
M = 1 TeV
50
75
100
125
150
175
Higgs mass

5 (a)

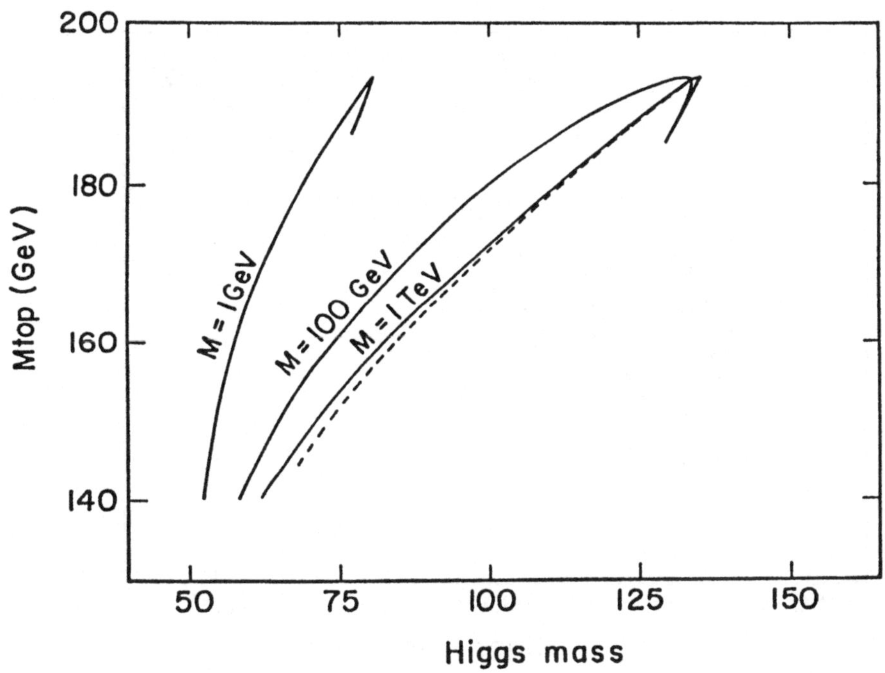

200
180
160
140
Mtop (GeV)
M = 1 GeV
M = 100 GeV
M = 1 TeV
50
75
100
125
150
Higgs mass

5 (b)

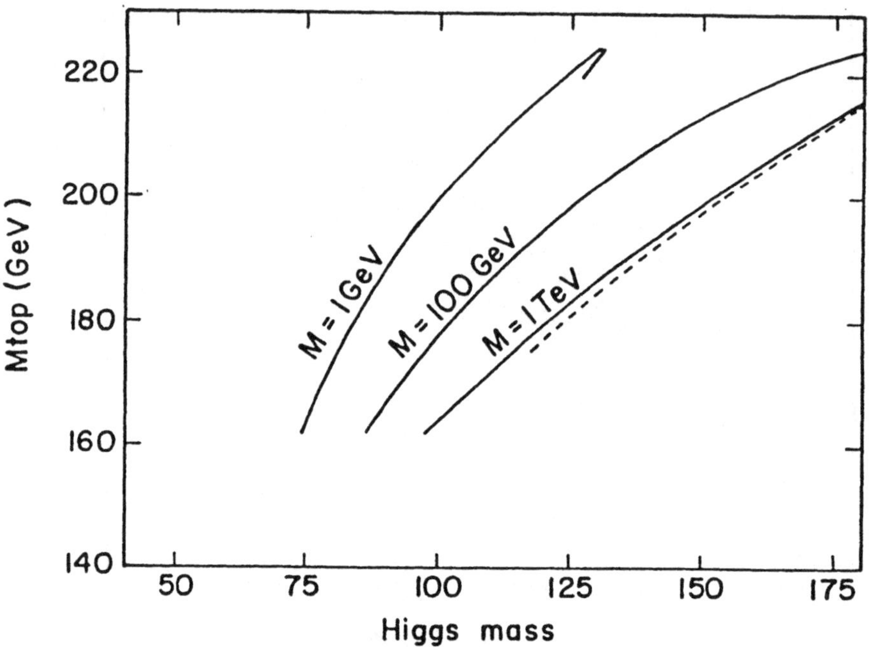

5 (c)

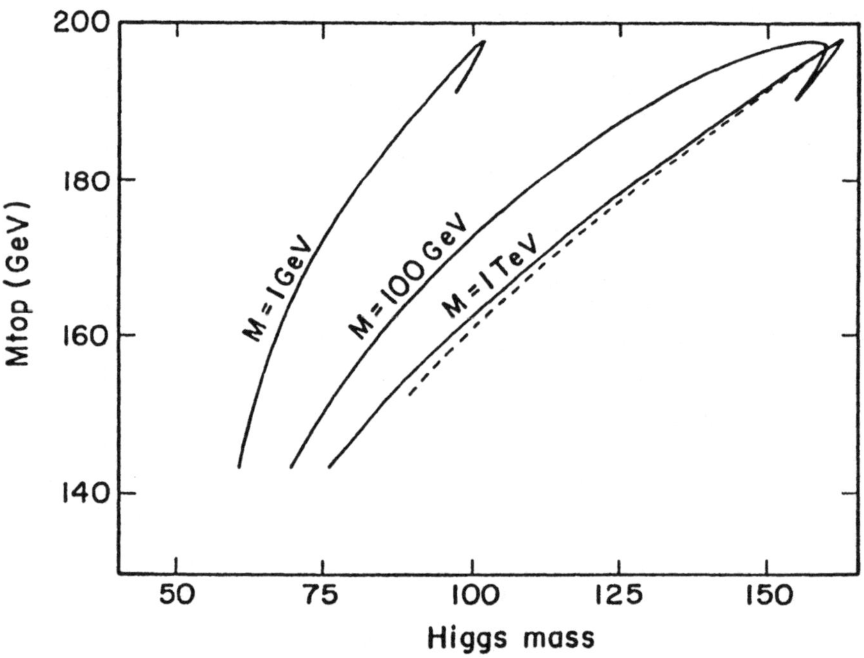

5 (d)

Fig. 5: Top quark mass as a function of the lightest Higgs mass m_h, for three different values of the mass parameter M (solid lines), and the same functional relation for the case of one light Higgs doublet (dashed line), for $\Delta_S = 1$ TeV and a) $\Lambda = 10^{10}$ GeV , b) $\Lambda = 10^{16}$ GeV as well as for $\Delta_S = 10$ TeV and, once more, c) $\Lambda = 10^{10}$ GeV , d) $\Lambda = 10^{16}$ GeV.

The following results have been obtained in Ref. [9], where we computed the scalar masses by using Eqs. (5.6), (5.7) (5.8) and (5.11). The low energy values of the running quartic couplings contributing to those equations have been evaluated at the renormalization scale $\mu^2 = m_i^2$, with m_i being the corresponding scalar mass. For fixed values of the compositeness scale and the soft supersymmetry breaking scale, the top quark mass has only a functional dependence on the ratio R, while the scalar Higgs masses depend on R as well as on the mass parameter M. In Figs. 5a - 5d, the value of m_t is given as a function of the lightest CP even scalar eigenstate mass for three different values of $M > 0$. Moreover, in all cases, that is, for $\Delta_S = 1$ TeV and $\Lambda = 10^{10}(10^{16})$ GeV (Figs. 5 a (b)), and for $\Delta_S = 10$ TeV and the same two values for Λ (Figs. 5 c (d)), we have also computed the above values in the one light Higgs doublet case, which coincide remarkably well with those obtained in the two light Higgs model for $M \gg M_Z$. After the inclusion of radiative corrections, the tree level values of the scalar masses are drastically modified. In fact, from the above figures it follows that for $M > 0$ and $R \geq 1$, the lightest Higgs mass is enhanced to values which are above the present experimental bound, $m_h > 41$ GeV [25].

So far, the value of the ratio of the vacuum expectation values has been assumed to be equal or greater than one. However, analyzing the renormalization group equations of the mass parameters m_i^2 appearing in the scalar potential [22], it is possible to define a lower bound on R. Considering $m_1^2(\Lambda) = m_2^2(\Lambda)$, which is the standard condition in all the supersymmetry breaking schemes analyzed in the literature until now, then, provided $h_t > h_b$ the relation $m_1^2(M_Z) > m_2^2(M_Z)$ is fulfilled. Defining the mass parameter $M_{12}^2 = m_1^2 - m_2^2$ and using the expression of the CP odd scalar mass m_A, the minimization condition, Eq. (5.5), can be written as

$$R^2 = \frac{m_A^2 + (\lambda_1 - \lambda_3 - \lambda_4)v^2 + M_{12}^2}{m_A^2 + (\lambda_2 - \lambda_3 - \lambda_4)v^2 - M_{12}^2}. \tag{5.13}$$

For values of $m_A \gg M_Z$, the above equation implies that $R \geq 1$, which is the lower bound on R in the case in which only one light Higgs doublet appears in the low energy theory. If, instead, one considers the two light Higgs doublet case, the result depends on the exact value of m_A and M_{12}^2. For $M_{12}^2 \simeq 0$, the lower bound on R is obtained for the lowest possible value of m_A compatible with the experimental contraints. This yields a lower bound on m_t of 120 GeV for the range of energy scales, $\Delta \simeq 1 - 10$ TeV and $\Lambda \simeq 10^{10} - 10^{16}$ GeV. In fact, the lower bound on R decreases when Λ decreases and also when Δ_S increases. However, since Y_t increases under these conditions, the lower bound on m_t remains unmodified. Moreover, the lower bound on R increases as a function of M_{12}^2. For $M_{12}^2 = \mathcal{O}(M_Z^2)$ or $M_{12}^2 = \mathcal{O}(\Delta_S^2)$ the $R \geq 1$ restriction is recovered. This is due to the fact that the lowest possible value of R occurs for $m_A \gg M_Z$. Thus, for a soft supersymmetry breaking scale $\Delta_S \simeq 1$ TeV and $\Lambda = 10^{16}$ GeV, the top quark mass is bounded to be $m_t \geq 140$ GeV. Furthermore, if Δ_S is assumed to be of the order of M_Z, the quartic couplings will be approximately given by their supersymmetric expressions and the tree level bounds $R \geq 1$ and $m_h < M_Z|\cos(2\theta)|$ will hold. In such case, for $\Lambda = 10^{16}$ GeV and from the experimental bound on the lightest CP even mass [25], it follows that $R > 1.6$ and hence $m_t > 165$ GeV.

6 Triviality Bounds

For a consistent one loop perturbative treatment of the theory, the renormalized Yukawa coupling should fulfill the condition $h_t^2/4\pi \leq 1$ for all energy scales below the cutoff scale. In fact, an upper bound on the value of the top quark Yukawa coupling may be obtained through the requirement that the Landau singularity associated with this quantity occurs at an energy scale just above the effective cutoff scale. Within the dynamical scheme under analysis, the signature for the realization of the electroweak symmetry breaking is a strong Yukawa coupling at the compositeness scale Λ, which acts as the effective cutoff scale up to which the model is valid as an effective theory. Thus, the values of the renormalized coupling at a given energy scale μ, which are obtained from the renormalization group trayectories associated with the compositeness condition, may be reinterpreted as upper bounds, the so called triviality bounds, on the top quark Yukawa coupling. Furthermore, for a given value of the ratio R and the mass parameter M or, equivalently, the CP odd mass, both m_t and m_h are increasing functions of the top quark Yukawa coupling. Therefore, the values of m_h and m_t that I have presented in the last two sections are, in fact, upper bounds (triviality bounds) on both masses. Most interesting, they define general upper bounds on these quantities in any model which belongs to the same universality class as the one under consideration.

As I have already discussed, the action of the infrared quasi fixed point results in the stability of the low energy predictions under variations of the boundary condition on the top quark Yukawa coupling at the scale Λ. Thus, for large values of Λ the perturbative renormalization group techniques provide a reliable computation of the top quark mass. If $\Lambda \simeq \Delta_S$, instead, the perturbative method gives just a rough estimate of the exact triviality bound on h_t. In spite of that, such an estimate is useful since the existence of a Landau singularity in the one loop running coupling at scales $\mu \simeq \Delta_S$ will indicate the breakdown of the perturbative consistency of the theory.

In the previous section, I showed that the scalar masses have a functional dependence on the mass parameter M. For the CP even eigenstates masses, I shall now rewrite such dependence on M as a function of the CP odd mass m_A. In particular, I am interested in defining the triviality bounds on m_h as a function of the CP odd mass. The expression (5.6) may be written as follows [12],

$$m_{H,h}^2 = \frac{1}{2}\left[m_A^2 + 2\lambda_1 v_1^2 + 2\lambda_2 v_2^2 \pm \sqrt{m_A^4 - 2m_A^2 f_1(\theta,\lambda_i) + f_2(\theta,\lambda_i)}\right]. \qquad (6.1)$$

From the explicit expressions for f_1 and f_2,

$$\begin{aligned} f_1(\theta,\lambda_i) &= \cos(2\theta)(2\lambda_1 v_1^2 - 2\lambda_2 v_2^2) + 2\sin^2(2\theta)(\lambda_3+\lambda_4)v^2 \\ f_2(\theta,\lambda_i) &= (2\lambda_1 v_1^2 - 2\lambda_2 v_2^2)^2 + 4\sin^2(2\theta)\left[(\lambda_3+\lambda_4)v^2\right]^2, \end{aligned} \qquad (6.2)$$

it is straightforward to prove that $f_2 \geq f_1^2$. Thus, for fixed values of Δ_S and R the CP even masses depend only on m_A^2. Moreover, it is ready to show that the minimum value of m_h occurs for $m_A = 0$. This follows from the fact that, $f_2 > f_1^2$ is just the condition to have a minimum at $m_A = 0$ and also that one for which the first derivative of $m_{H,h}$ with respect to m_A is always positive for nonvanishing values of the CP odd state mass. In addition, for the particular case $f_2 = f_1^2$, it can be shown that if $m_A^2 > f_1$, then m_h is independent of m_A, while m_H behaves like in the case $f_2 > f_1^2$. If, instead, $m_A^2 < f_1$, the behavior of m_H and m_h is interchanged.

From the above analysis, a lower bound on the CP even masses values, $m_{H,h}^{min}$, is obtained for a vanishing value of the CP odd mass. However, from the knowledge that $m_{H,h}$ are monotonically increasing functions of m_A, together with the information about the experimental bounds on the CP odd mass (m_A^{exp}), an improved lower bound on $m_{H,h}$ may be determined by replacing $m_A = 0$ by m_A^{exp}. For example, after the inclusion of radiative corrections, for $\Delta_S = 1$ TeV and for values of $R \geq 1$, the experimental lower bounds on the CP odd mass obtained by extrapolating the results of Ref. [25] for values of $m_t = 120, 160, 200$ GeV are $m_A^{exp} \geq 35, 25, 20$ GeV, respectively. Furthermore, from the functional dependence of m_h on m_A, a finite upper bound on the lightest CP even scalar mass is defined in the limit of large values of the CP odd mass, $m_A \gg M_Z$,

$$(m_h^{max})^2 = v^2 \left[2\lambda_1 \cos^4\theta + 2\lambda_2 \sin^4\theta + (\lambda_3 + \lambda_4)\sin^2(2\theta)\right]. \tag{6.3}$$

Therefore, the lightest CP even scalar Higgs mass is bounded by $m_h^{min} < m_h < m_h^{max}$. Considering the running of the quartic couplings compatible with the compositeness condition on the top quark Yukawa coupling, as done before, we obtain [12] the upper and lower bounds on m_h within the top quark condensate model. The improved lower bound depends on the top quark mass through the experimental bound on m_A and can only be considered in a restricted way. It is interesting to notice that, since the upper bound on m_h is defined for $m_A \gg M_Z$, it is effectively equivalent as to consider the model with only one light neutral Higgs particle in the physical spectrum of the theory.

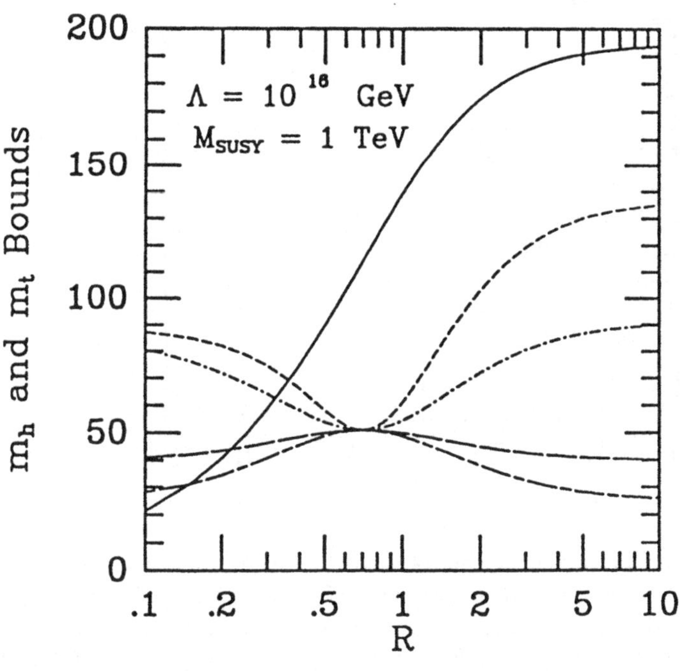

6 (a)

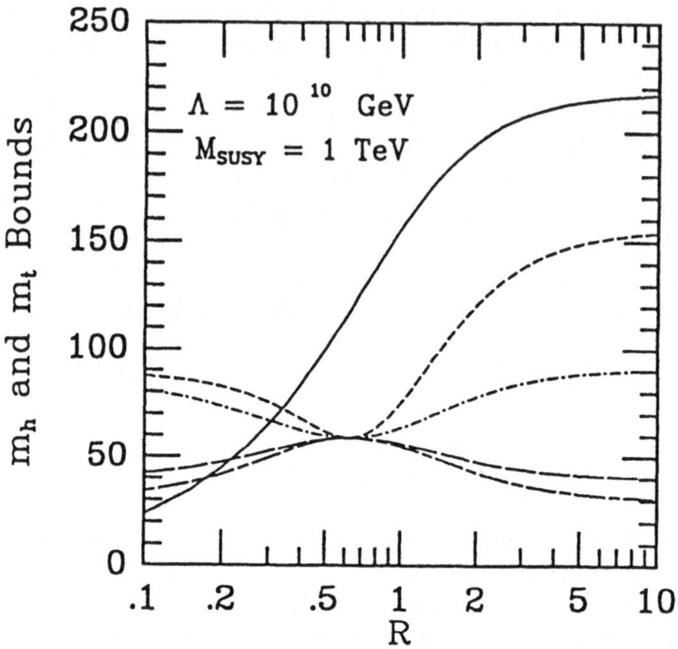

6 (b)

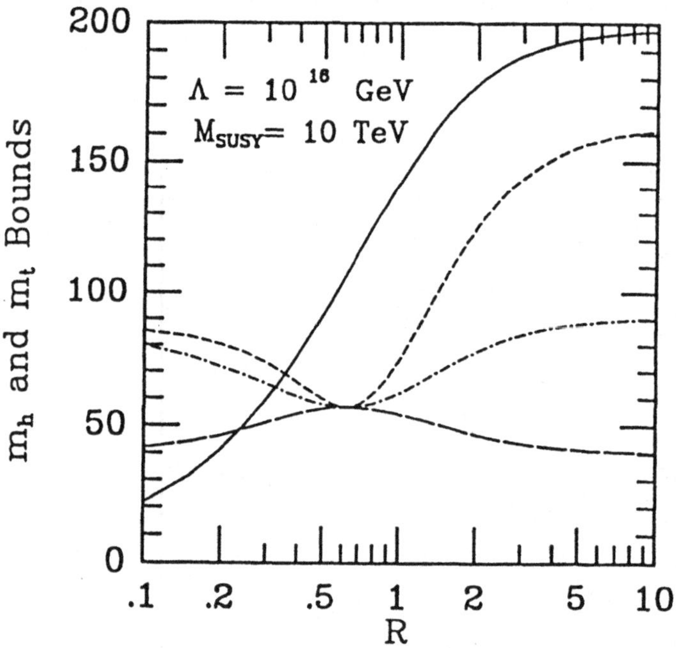

6 (c)

Fig. 6: Triviality bounds on the top quark mass (solid line) and on the lightest CP even Higgs mass as a function of the ratio R, for $m_A = 40$ GeV (long-dashed line), $m_A = M_Z$ (dot-dashed line), $m_A \gg M_Z$ (dashed line) and for an effective cutoff scale and a soft supersymmetry breaking scale a) $\Delta_S = 1$ TeV and $\Lambda = 10^{16}$ GeV, b) $\Delta_S = 1$ TeV and $\Lambda = 10^{10}$ GeV, c) $\Delta_S = 10$ TeV and $\Lambda = 10^{16}$ GeV. The long dash-dashed lines in (a) and (b) are the triviality bounds on m_h for $m_A = 25$ GeV and $m_A = 30$ GeV, respectively.

For fixed values of the top quark mass, the triviality bounds on the top quark Yukawa coupling determine lower bounds on R, R_b, below which the theory will become perturbatively inconsistent at energy scales below the effective cutoff scale. Once more, for $\Lambda = \mathcal{O}(\Delta_S)$ the lower bounds on the ratio of vacuum expectation values, $R_b = 0.23, 0.30, 0.35$ for $m_t = 120, 160, 200$ GeV, respectively, are just a rough estimate. In fact, if the boundary condition on h_t is changed from the compositeness condition to $h_t(\Lambda)^2/4\pi = 1$, the lower bounds on R turn out to be $R_b = 0.28, 0.37, 0.47$, respectively. On the contrary, in the case in which there is a huge hierarchy of energy scales ($\Lambda = 10^{10}$ (10^{16}) GeV), the R_b results are stable under variations of the boundary condition on h_t.

In many recent works [26], it has been remarked that the values of the gauge couplings obtained in the most recent measurements at LEP are consistent with a supersymmetric grand unified scenario with a grand unification scale $M_{GUT} \simeq 10^{16}$ GeV and a soft supersymmetry breaking scale $\Delta_S = 1$ TeV. If this were a signature of the presence of a grand unified theory at the scale M_{GUT}, then m_t and m_h must be below their triviality bounds for an effective cutoff $\Lambda = M_{GUT}$. In Fig. 6a the triviality bounds on m_t and m_h are given as a function of R, for an effective cutoff scale of 10^{16} GeV, for different values of m_A and for a soft supersymmetry breaking scale $\Delta = 1$ TeV [12]. From these results it follows that, if perturbative unification is required at $\Lambda = M_{GUT} = 10^{16}$ GeV, then the value of m_t can not be above 195 GeV, while the lightest Higgs mass must be below 135 GeV. On the other hand, for a top quark mass of $m_t = 120, 160, 200$ GeV, a lower bound on R may be derived, yielding $R_b = 0.75, 1.4, \infty$, respectively.

Before finishing this section let me comment on the possibility of a new decay mode for the lightest CP even Higgs particle. Recent analyses within the minimal supersymmetric standard model [12], [27], indicate that for small values of $m_A \leq 40$ GeV and for small values of R, there is an interesting region of parameters in which $m_h \geq 2m_A$. Therefore, the decay mode $h \to AA$ is kinematically allowed in such region, what could change the strategy of Higgs searches at LEP [28]. The analysis of the triviality bounds on m_h for $\Delta_S = 1$ TeV and $\Lambda = 10^{16}$ (10^{10}) GeV presented in Figs. 6a (b) [12], clearly shows that for $m_A \simeq 40$ GeV, the lightest CP even state mass is bounded to be $m_h \leq 52$ (60) GeV. Hence, for $m_A \simeq 40$ GeV the above phenomenologically interesting region of parameters requires too large top quark Yukawa couplings, for which one loop perturbative consistency is lost at energy scales below the effective cutoff scale. In fact, for each value of Λ there is one value of R for which the relation $f_1^2 = f_2$ is fulfilled, implying that the lightest CP even mass is independent of m_A. For the range of energy scales considered in Figs. 6a and 6b, it occurs that the value of $m_h = m_h^{fix}$, at which the curves for different m_A converge, is a maximum for any value of the CP odd mass below 40 GeV. Thus, the decay mode $h \to AA$ will be allowed only if the condition $m_A < m_h^{fix}/2$ holds. This means that for $\Delta_S = 1$ TeV and the cutoff scale $\Lambda = 10^{10}$ (10^{16}) GeV, the proposed new decay mode for the lightest CP even scalar Higgs into two CP odd states will be possible only for quite small values of m_A, that is, $m_A < 30$ (26) GeV. In principle, since the radiative corrections to the lightest CP even mass increase with Δ_S, this bound on m_A could increase with the supersymmetric threshold, too. However, Fig. 6c, in which $\Lambda = 10^{16}$ GeV and $\Delta_S = 10$ TeV has been considered, shows that the above lower bound on m_A remains stable under variations of the soft supersymmetry breaking scale.

7 Conclusions

I have presented a rather complete analysis on the values of the top quark and Higgs masses compatible with a dynamical breakdown of the electroweak symmetry, in a minimal supersymmetric standard model in which two Higgs doublets have masses below the soft supersymmetry breaking scale Δ_S. For fixed values of the soft supersymmetry breaking scale and the cutoff scale, I have demostrated, in a way independent of the supersymmetry breaking scheme, that the values of the top quark mass depend only on the ratio of the Higgs vacuum expectation values R. For $R \geq 1$, $\Lambda \simeq 10^{16}$ GeV and $\Delta_S = 1$ TeV, the characteristic values of the top quark mass are 140 GeV $\leq m_t \leq$ 195 GeV, a prediction which is only mildly dependent on Δ_S. If $\Lambda = 10^{10}$ GeV, the values of the top quark mass become larger, $160 GeV \leq m_t \leq 220 GeV$. Slightly smaller values of m_t are predicted if the value of R is smaller than one.

The issue of dynamical symmetry breaking is related to the so called triviality bounds on the top quark and Higgs masses. The upper bounds (triviality bounds) on m_t and m_h for any theory in the same universality class as the supersymmetric top quark condensate model are computed as a function of R and for different values of the CP odd state mass m_A. In particular, for an effective cutoff scale of the order of $\Lambda \simeq 10^{16}$ GeV, which appears to be the favored grand unification scale according to recent LEP data, the phenomenologically interesting region of parameters for which $m_A \simeq 40$ GeV and $m_h \geq 2m_A$ is ruled out. The reason is that for this condition to be fulfilled, the top quark Yukawa coupling would become strong at energy scales significantly below M_{GUT}. The possibility of a new decay mode $h \rightarrow AA$, which could induce unexplored decay patterns in the Higgs searches at LEP, appears to be consistent with one loop perturbative unification at a scale of the order of 10^{16} GeV and Δ_S =1 (10) TeV, only if $m_A < 26(29)$ GeV. In the same framework, for perturbative unification to hold, the top quark mass must be lighter than 195 GeV while $m_h < 135$ (160) GeV if $\Delta_S =$ 1 (10) TeV. In addition, if $m_A < M_Z$, the lightest CP even state mass is lower than M_Z. Moreover, due to the presence of the infrared quasi fixed point of the top Yukawa coupling, these results are expected to be stable under higher order loops effects.

Within the supersymmetric extension, the same as in the standard top quark condensate model, the underlying dynamics which generates the effective theory at the compositeness scale remains unspecified. In this talk I have focused on the simplest version of a model with supersymmetric nonrenormalizable interactions which can induce electroweak symmetry breakdown at low energies. It is presumable that a more general model, with additional higher dimensional interactions, which become irrelevant in the low energy regime, may generate similar results [29], [30]. It is also possible that the associated higher dimensional operators will modify the compositeness condition. However, due to the presence of the infrared quasi fixed behavior, the low energy values of the running couplings are expected to remain stable whenever $\Lambda \gg m_t$. In order to test the sensitivity of the top quark mass predictions to the inclusion of higher derivative four Fermi interactions, an explicit computation, following Suzuki parametrization scheme [29], was recently done by Bardeen [31] in the frame of the standard top condensate model. It was shown that, as far as the coefficients associated with the higher dimensional operators are at most of moderate strength, the predictions of the minimal model remain stable. In a recent study, Clark, Love and Ter Veldhuis [32] obtained the same stability property while extending the analysis to the supersymmetric case. Moreover, although no fine tuning is required in the supersymmetric model, if one considers the cutoff to set the scale of the nonrenormalizable interactions, then the four Fermi coupling seems to be huge. However, it is not certain what should determine the scale for the nonrenormalizable dynamics within the supersymmetric scheme. In this respect, there is a work by Ellwanger [33] proposing a model in which the supersymmetric four fermion coupling is derived to be of order $1/\Delta_S^2$. Furthermore, some recent attemps have been done, to define the underlying theory which leads to the standard Nambu-Jona-Lasinio model at the compositeness scale [34]. It remains an interesting possibility to speculate about the physics beyond the effective cutoff scale in the supersymmetric top condensate scenario.

ACKNOWLEDGEMENTS

I am grateful to W. A. Bardeen, T. E. Clark, K. Sasaki and C. E. M. Wagner for an enjoyable and fruitful collaboration which led to most of the results I have presented in this talk. I would also like to thank W. Hollik, R. Rückl and J. Wess, who organized this workshop.

References

[1] S. L. Glashow, Nucl. Phys. 22 (1961) 579;
A. Salam, in Elementary Particle Theory, ed. N. Svartholm, (1968) 367;
S. Weinberg, Phys. Rev. 419 (1967) 1264.

[2] Y. Nambu, Bootstrap Symmetry Breaking in Electroweak Unification, EFI 89-08(1989); BCS Mechanism, Quasi- Supersymmetry, and Fermion Mass Matrix , talk presented at the Kasimirz Conference, EFI 88-39 (1988).

[3] W.A. Bardeen, C.T. Hill and M. Lindner, Phys. Rev. D41 (1990) 1647.

[4] V. Miransky, M. Tanabashi and K. Yamawaki, Mod. Phys. Lett. A4 (1989) 1043, Phys. Lett. B221 (1989) 177.

[5] C. Wagner, The Standard Model with Minimal Dynamical Electroweak Symmetry Breaking, these proceedings.

[6] Y. Nambu and G. Jona Lasinio, Phys. Rev. 122 (1961) 345.

[7] P. Langacker and M. X. Luo, Phys. Rev. D44 (1991) 817;
J. Ellis, G. Fogli and F. Lisi in the Proceedings of the 1991 Lepton Photon Conference.

[8] W. A. Bardeen, T. E. Clark and S. T. Love, Phys. Lett. B 237 (1990) 235.

[9] M. Carena, T. Clark, C. Wagner, W. Bardeen and K. Sasaki, FERMILAB-TH-91-96-T (1991). To appear in Nucl. Phys. B.

[10] E. Cremmer, B. Julia, J. Scherk, S. Ferrara, L. Girardello and P. van Nieuwenhuizen, Phys. Lett. B79 (1978) 231, Nucl. Phys. B147 (1979) 105;
E. Cremmer, S. Ferrara, L. Girardello and A. Van Proeyen, Phys. Lett. B116 (1982) 231, Nucl. Phys. B212 (1983) 413;
S.K. Soni and H.A. Weldon, Phys. Lett. 126B (1983) 215;
L. Alvarez-Gaumé, J. Polchinski and M.B. Wise Nucl. Phys. B221 (1983) 495.

[11] C.T. Hill Phys. Rev. D24 (1981) 691.
C. Hill, C. N. Leung and S. Rao, Nucl. Phys. B262 (1985) 517.

[12] M. Carena, K. Sasaki and C. Wagner, MPI-Ph/91-109.

[13] W.A. Bardeen, C.N. Leung and S.T. Love, Phys. Rev. Lett. 56 (1986) 1230, Nucl. Phys. B273 (1986) 649, Nucl. Phys. B323 (1989) 493.
[14] W. Buchmüller and S.T. Love, Nucl. Phys. B204 (1982) 213.
[15] W. Buchmüller and U. Ellwanger Nucl. Phys. B245 (1984) 237.
[16] J. Wess and J. Bagger, Supersymmetry and Supergravity, Princeton U. Press, Princeton, 1983.
[17] K. Inoue, A. Kakuto and S. Takeshita, Prog. Theor. Phys. 67 (1982) 1889; ibid 68 (1982) 927;
H. P. Nilles, Phys. Rep 110 (1984) 1;
K. Sasaki, Phys. Lett. 199B (1987) 395;
H. Okada and K. Sasaki, Phys. Rev D40 (1989) 3743.
[18] K. Jonhson, M. Baker and R. Willey, Phys. Rev. 136 (1964) B111; ibid 163 (1967) 1699;
S. L. Adler and W. A. Bardeen, Phys. Rev. D4 (1971) 3054;
T. Maskawa and H. Nakajima, Prog. Theor. Phys. 52 (1974) 1326 ; ibid 54 (1975) 860;
R. Fukuda and T. Kugo, Nucl. Phys. B117 (1976) 250;
V. A. Miransky, Nuovo Cim. 90A (1985) 149.
[19] K. I. Kondo, H. Mino and K. Yamawaki, Phys. Rev. D39 (1989) 2430;
T. Appelquist, M. Soldate, T. Takeuchi and L . C. R. Wijewardhana, in Proceedings of the 12th Johns Hopkins Workshop on Current Problems in Particle Theory, ed. by G. Domokos and S. Kovesi-Domokos, World Scientific (1988).
[20] T.E. Clark and S.T. Love, Nucl. Phys. B310 (1988) 371.
[21] C. Hill, Dynamical Symmetry Breaking of the Electroweak Interactions and the Renormalization Group , FERMI-CONF-90/170-T.
[22] J. Bagger, S. Dimopoulos and E. Massó, Phys. Rev. Lett.55 (1985) 920;
L. E. Ibañez, C. Lopez and C. Muñoz, Nucl. Phys. B256 (1985) 299;
B. Gato, J. Leon, J. Perez-Mercader and M. Quiros, Nucl. Phys. B253 (1985) 285;
N. K. Falck, Z. Phys. C30 (1986) 247.
[23] Y. Okada, M. Yamaguchi and T. Yanagida, Prog. Theor. Phys. 85 (1991) 1 ; Phys. Lett. B262 (1991) 54 ;
J. Ellis, G. Ridolfi and F. Zwirner, Phys. Lett. B 257 (1991) 83;
H. Haber and R. Hempfling, Phys. Rev. Lett. 66 (1991) 1815;
R. Barbieri, M. Frigeni and F. Caravaglios, Phys. Lett. B258 (1991) 167.
[24] C. Kounnas, A. Lahanas, D. Nanopoulos, M. Quiros, Nucl. Phys. B236 (1984) 438
J. P. Derendinger, C. A. Savoy, Nucl. Phys. B237 (1984) 307.
[25] D. Decamp et al, Alleph Collab., CERN PPE/91-111, July 1991.
[26] J. Ellis, S. Kelley and D. V. Nanopoulos, Phys. Lett. B260 (1991) 131;
P. Langacker and M. Luo, Phys. Rev. D44 (1991) 817;
U. Amaldi, W. de Boer and H. Füstenau, Phys. Lett. B260 (1991) 240.
[27] J. L. Lopez and D. V. Nanopoulos, Phys. Lett. B266 (1991) 397.
[28] A. Brignole, J. Ellis, G. Ridolfi and F. Zwirner, CERN - TH 6151/91.
[29] M. Suzuki, Mod Phys. Lett. A5 (1990) 1205.
[30] A. Hasenfratz, P. Hasenfratz, K. Jansen, J. Kuti and Y. Shen, The Equivalence of the Top Quark Condensate and The Elementary Higgs Theory, UCSD/PTH-91-06, AZPH-TH/91-12.

[31] W. A. Bardeen, Electroweak Symmetry Breaking: Top Quark Condensates, FERMILAB-PUB-90/269-T.

[32] T. Clark, S. Love and W. ter Veldhuis, Stability of the Top Quark Mass in Supersymmetric Top Condensate Models, PURD-TH-91-06 (1991).

[33] U. Ellwanger, Nucl. Phys. B356 (1991) 46.

[34] C. Hill, Phys. Lett. B266 (1991) 419;
R. Bönisch, Phys. Lett. B268 (1991) 394; München Univ. Preprint LMU 91/03;
M. Lidner and D. Ross, CERN-TH-6179-91.

Lecture Notes in Physics

For information about Vols. 1–374
please contact your bookseller or Springer-Verlag

Vol. 375: C. Bartocci, U. Bruzzo, R. Cianci (Eds.), Differential Geometric Methods in Theoretical Physics. Proceedings, 1990. XIX, 401 pages. 1991.

Vol. 376: D. Berényi, G. Hock (Eds.), High-Energy Ion-Atom Collisions. Proceedings, 1990. IX, 364 pages. 1991.

Vol. 377: W. J. Duschl, S.J. Wagner, M. Camenzind (Eds.), Variability of Active Galaxies. Proceedings, 1990. XII, 312 pages. 1991.

Vol. 378: C. Bendjaballah, 0. Hirota, S. Reynaud (Eds.), Quantum Aspects of Optical Communications. Proceedings 1990. VII, 389 pages. 1991.

Vol. 379: J. D. Hennig, W. Lücke, J. Tolar (Eds.), Differential Geometry, Group Representations, and Quantization. XI, 280 pages. 1991.

Vol. 380: I. Tuominen, D. Moss, G. Rüdiger (Eds.), The Sun and Cool Stars: activity, magnetism, dynamos. Proceedings, 1990. X, 530 pages. 1991.

Vol. 381: J. Casas-Vazquez, D. Jou (Eds.), Rheological Modelling: Thermodynamical and Statistical Approaches. Proceedings, 1990. VII, 378 pages. 1991.

Vol. 382: V.V. Dodonov, V. I. Man'ko (Eds.), Group Theoretical Methods in Physics. Proceedings, 1990. XVII, 601 pages. 1991.

Vol. 384: M. D. Smooke (Ed.), Reduced Kinetic Mechanisms and Asymptotic Approximations for Methane-Air Flames. V, 245 pages. 1991.

Vol. 385: A. Treves, G. C. Perola, L. Stella (Eds.), Iron Line Diagnostics in X-Ray Sources. Proceedings, Como, Italy 1990. IX, 312 pages. 1991.

Vol. 386: G. Pétré, A. Sanfeld (Eds.), Capillarity Today. Proceedings, Belgium 1990. XI, 384 pages. 1991.

Vol. 387: Y. Uchida, R. C. Canfield, T. Watanabe, E. Hiei (Eds.), Flare Physics in Solar Activity Maximum 22. Proceedings, 1990. X, 360 pages. 1991.

Vol. 388: D. Gough, J. Toomre (Eds.), Challenges to Theories of the Structure of Moderate-Mass Stars. Proceedings, 1990. VII, 414 pages. 1991.

Vol. 389: J. C. Miller, R. F. Haglund (Eds.), Laser Ablation-Mechanisms and Applications. Proceedings. IX, 362 pages, 1991.

Vol. 390: J. Heidmann, M. J. Klein (Eds.), Bioastronomy - The Search for Extraterrestrial Life. Proceedings, 1990. XVII, 413 pages. 1991.

Vol. 391: A. Zdziarski, M. Sikora (Eds.), Ralativistic Hadrons in Cosmic Compact Objects. Proceedings, 1990. XII, 182 pages. 1991.

Vol. 392: J.-D. Fournier, P.-L. Sulem (Eds.), Large-Scale Structures in Nonlinear Physics. Proceedings. VIII, 353 pages. 1991.

Vol. 393: M. Remoissenet, M.Peyrard (Eds.), Nonlinear Coherent Structures in Physics and Biology. Proceedings. XII, 398 pages. 1991.

Vol. 394: M. R. J. Hoch, R. H. Lemmer (Eds.), Low Temperature Physics. Proceedings. XXX,XXX pages. 1991.

Vol. 395: H. E. Trease, M. J. Fritts, W. P. Crowley (Eds.), Advances in the Free-Lagrange Method. Proceedings, 1990. XI, 327 pages. 1991.

Vol. 396: H. Mitter, H. Gausterer (Eds.), Recent Aspects of Quantum Fields. Proceedings. XIII, 332 pages. 1991.

Vol. 398: T. M. M. Verheggen (Ed.), Numerical Methods for the Simulation of Multi-Phase and Complex Flow. Proceedings, 1990. VI, 153 pages. 1992.

Vol. 399: Z. Švestka, B. V. Jackson, M. E. Machedo (Eds.), Eruptive Solar Flares. Proceedings, 1991. XIV, 409 pages. 1992.

Vol. 400: M. Dienes, M. Month, S. Turner (Eds.), Frontiers of Particle Beams: Intensity Limitations. Proceedings, 1990. IX, 610 pages. 1992.

Vol. 401: U. Heber, C. S. Jeffery (Eds.), The Atmospheres of Early-Type Stars. Proceedings, 1991. XIX, 450 pages. 1992.

Vol. 402: L. Boi, D. Flament, J.-M. Salanskis (Eds.), 1830-1930: A Century of Geometry. VIII, 304 pages. 1992.

Vol. 403: E. Balslev (Ed.), Schrödinger Operators. Proceedings, 1991. VIII, 264 pages. 1992.

Vol. 404: R. Schmidt, H. O. Lutz, R. Dreizler (Eds.), Nuclear Physics Concepts in the Study of Atomic Cluster Physics. Proceedings, 1991. XVIII, 363 pages. 1992.

Vol. 405: W. Hollik, R. Rückl, J. Wess (Eds.), Phenomenological Aspects of Supersymmetry. VII, 329 pages. 1992.

Vol. 406: R. Kayser, T. Schramm, L. Nieser (Eds.), Gravitational Lenses. Proceedings, 1991. XXII, 399 pages. 1992.

Vol. 407: P. L. Smith, W. L. Wiese (Eds.), Atomic and Molecular Data for Space Astronomy. VII, 158 pages. 1992.

Vol. 408: V. J. Martínez, M. Portilla, D. Sàez (Eds.), New Insights into the Universe. Proceedings, 1991. XI, 298 pages. 1992.

New Series m: Monographs

Vol. m 1: H. Hora, Plasmas at High Temperature and Density. VIII, 442 pages. 1991.

Vol. m 2: P. Busch, P. J. Lahti, P. Mittelstaedt, The Quantum Theory of Measurement. XIII, 165 pages. 1991.

Vol. m 3: A. Heck, J. M. Perdang (Eds.), Applying Fractals in Astronomy. IX, 210 pages. 1991.

Vol. m 4: R. K. Zeytounian, Mécanique des fluides fondamentale. XV, 615 pages, 1991.

Vol. m 5: R. K. Zeytounian, Meteorological Fluid Dynamics. XI, 346 pages. 1991.

Vol. m 6: N. M. J. Woodhouse, Special Relativity. VIII, 86 pages. 1992.

Vol. m 7: G. Morandi, The Role of Topology in Classical and Quantum Physics. XIII, 239 pages. 1992.

Vol. m 8: D. Funaro, Polynomial Approximation of Differential Equations. X, 305 pages. 1992.

Vol. m 9: M. Namiki, Stochastic Quantization. X, 217 pages. 1992.

Vol. m 10: J. Hoppe, Lectures on Integrable Systems. VII, 111 pages. 1992.

Vol. m 11: A. D. Yaghjian, Relativistic Dynamics of a Charged Sphere. XII, 115 pages. 1992.

Vol. m 12: G. Esposito, Quantum Gravity, Quantum Cosmology and Lorentzian Geometries. XVI, 326 pages. 1992.

GPSR Compliance

The European Union's (EU) General Product Safety Regulation (GPSR) is a set of rules that requires consumer products to be safe and our obligations to ensure this.

If you have any concerns about our products, you can contact us on ProductSafety@springernature.com

In case Publisher is established outside the EU, the EU authorized representative is:

Springer Nature Customer Service Center GmbH
Europaplatz 3
69115 Heidelberg, Germany

Batch number: 10640288

Printed by Printforce, the Netherlands